Functional Neuroscience

Springer
New York
Berlin
Heidelberg
Barcelona
Hong Kong
London
Milan
Paris
Singapore
Tokyo

Oswald Steward, Ph.D.
Reeve-Irvine Research Center
University of California, Irvine

Functional Neuroscience

with 306 Illustrations, 89 in color

Springer

Oswald Steward, Ph.D.
Reeve-Irvine Research Center
University of California, Irvine
1105 Gillespie Neuroscience Research Facility
Irvine, California 92697-4292
USA

Library of Congress Cataloging-in-Publication Data
Steward, Oswald.
    Functional neuroscience / Oswald Steward.
        p. cm.
    Includes bibliographical references.
    ISBN 0-387-98543-3 (softcover : alk. paper)
    1. Neurosciences. I. Title.
    [DNLM: 1. Nervous System Physiology. 2. Nervous System—
physiopathology. 3. Nervous System—anatomy & histology. WL 102
S849f 1998]
    RC341.S85 2000
    612.8—dc21
    DNLM/DLC
    for Library of Congress   98-21387
                                                                CIP

Printed on acid-free paper.

Production coordinated by Impressions Book and Journal Services, Inc., and managed by Lesley Poliner; manufacturing supervised by Jacqui Ashri.
Typeset by Impressions Book and Journal Services, Inc., Madison, WI.
Printed and bound by Quebecor/Hawkins, Kingsport, TN.
Printed in the United States of America.

9 8 7 6 5 4 3 2 1

ISBN 0-387-98543-3   Springer-Verlag New York Berlin Heidelberg   SPIN 10680218

To
Kathy, Jessie, and Ossie

Preface

This book is designed to be used as the core neuroscience text for medical students. It presents the key information that is required for a basic understanding of the operation of the nervous system. My goal in writing this text was to integrate and blend two successful styles. One is the style that has dominated the neuroanatomical teaching known as "functional neuroanatomy", in which the focus is on regional anatomy, and how lesions at particular sites disrupt different pathways.The second is the style favored by neurophysiologists known as "systems neuroscience" in which the focus is on the physiological operation of the different functional systems. I use the term "functional neuroscience" to refer to the resulting blend.

The overall organization of the presentation, as reflected by the chapter headings, is traditional. We begin by considering the ground plan of the nervous system (that is, regional anatomy and the organization at different levels of the neuraxis). Then, we consider cellular aspects of neuroscience, fo-

cusing first on the basic cell biology of neurons and glia, and then on signaling processes. In this, I have made certain choices that reflect the way the medical curriculum is organized in most medical schools. For example, there is no chapter that covers resting membrane potential and action potential, or the elemental aspects of synaptic transmission. These topics are usually taught in the physiology courses in most medical curricula. The present book assumes a background in these subjects, and then moves on **to** those aspects of synaptic transmission that are especially important for neurons in the central nervous system (that is, integration of information from multiple inputs, interactions between different kinds of neurotransmitter systems, and integration of excitation and inhibition).

The third section of the book is the heart of what I call functional neuroscience. In this section, each of the key functional systems are considered one by one. We begin with the somatosensory pathways, and then consider the motor pathways and sensory motor integration. The reason for considering somatosensory and then somatomotor is that the two are mapped in parallel throughout the neuraxis, and there important function interactions between sensory and motor components at each level. We then consider the systems that control the output of the somatic motor pathways, the cerebellum and the basal ganglia, and follow with a consideration of the special senses: vision, audition, vestibular system, and the chemical senses (taste and smell). The final section considers the systems that mediate associative and integrative functions, and the "higher" functions mediated by the cerebral cortex.

Some instructors who are used to the "functional neuroanatomy" style may be surprised that there is no separate chapter on the cranial nerves. Instead cranial nerves belonging to a particular functional system are described in the chapter pertaining to that system. The cranial nerves that are involved in somatic sensation are described in the somatosensory section; the cranial nerves that control the muscles of the face are described in the motor system, and the cranial nerves that control autonomic functions are described in the chapter on the autonomic nervous system, and finally there is a separate chapter that considers the control of eye movement.

This book is designed for medical students. I have attempted to keep the presentation as concise as is consistent with accuracy. There is minimal consideration of original data, and the only references are the ones indicated as sources for the figures. The cost of maintaining brevity is that I do not cite the thousands of individuals whose research forms the basis of the material presented herein, or provide a view of how the knowledge was actually derived. Also missing is a consideration of the limitations of our present knowledge. For this reason, instructors will need to provide supplementary material for graduate students or medical students interested in neuroscience research.

HOW TO APPROACH THE SUBJECT OF NEUROSCIENCE

The information that is encompassed by *Functional Neuroscience* is of critical importance to physicians of all types. The importance of the information to a specialist is obvious, however, family physicians must also use this information on virtually a daily basis.

Whenever a patient walks into your office for whatever reason, you will be carrying out a basic neurological exam. This means observing the patient for any signs and symptoms suggestive of disease of the nervous system. When you first see the patient, you check for general demeanor. When you shake their hand, you check their hand strength. You watch their eyes to detect any abnormalities in eye movement. You observe how they hold themselves in order to detect any postural asymmetries. You listen to what they say to detect any difficulties with speech or difficulties with cognition. Part of your routine neurological exam will involve watching your patient move across the room. This will give you an initial indication of their motor status. You will likely do a simple check of reflexes, and you may carry out simple tests of eye movements to track visual stimuli. You may ask to the patient to stand with their eyes closed to check their balance, to touch their nose or ear with their eyes closed, to rapidly pat the back and front of their hands to check for cerebellar function. And then of course, you will listen to what the patients say about any problems they are experiencing.

Everything that you learn in this way is then integrated into an opinion about the patient's functional situation. If you detect no abnormalities, and in the absence of any complaints from the patient, you will conclude that their nervous system is functioning normally. Any complaints, however minor, will cause you to formulate hypotheses about what may be wrong. The important question that you must ask yourselves is, "is there a lesion involving the nervous system, and if so where is it localized?". The answer to these questions will tell you whether you handle treatment yourself or refer the patient to a specialist.

Learning neuroscience is a challenge. One reason is that it is difficult to begin by presenting the "big picture" and then move to the details. Instead, in most cases one can't understand the big picture without knowing the details. Usually it is not until the end of the course that all the pieces start to "come together". This causes anxiety. The most important advice I can give is, be patient. The big picture will eventually become clear.

The other difficulty is in the nature of the information that must be learned. Physicians must understand the pathways that mediate particular functions. One must be able to predict the site of a lesion on the basis of a certain symptom complex, and one must be able to predict the symptoms from a known lesion. This requires integration of a variety of types of information. Students must also learn about the way signaling occurs in the nervous system. This is a different kind of information, more conceptual in nature than the memorization required to know where a particular pathway lies. Some students excel at learning one type of information, others excel at learning the other. It may help to recognize individual differences in order to focus one's efforts for maximal efficiency.

One important aspect of neuroscience is that one can't learn something and then put it aside to move on to something else. At each stage, students must make use of information learned at earlier stages. This becomes apparent when trying to understand the consequences of lesions at particular sites. To understand the consequences of a lesion involving the brain stem, one must know the distribution of the somatosensory pathways, the motor pathways, the organization of the cranial nerve nuclei, the descending autonomic pathways, and the functional role that the cerebellum plays in modulating motor output. The big picture comes from the details. This type of learning is challenging and sometimes frustrating, but IS what you need to know.

An important skill that physicians must develop is the ability to identify structures from histological sections, CT scans, or magnetic resonance images (MRI). This skill requires an understanding of regional neuroanatomy and three-dimensional structure. In this regard, memorization of a particular photograph or drawing is not sufficient, since any section or image may be taken at a slightly different angle or magnification. One must be able to generalize in order to recognize structures in configurations that have not previously been seen. Although sections and diagrams are included in this book, the full development of the ability to identify structures will require practice with a variety images. Students are encouraged to explore image-based self-study programs which are available by subscription, or to use any of a number of human brain atlases.

Acknowledgments

I would like to thank two colleagues have played an especially important role in shaping my philosophy regarding the teaching of neuroscience. The first is Edwin R. Rubel, formerly of the University of Virginia, and now at the University of Washington. Dr. Rubel's lectures on sensory systems were my first exposure to what I now call functional neuroscience. The second key colleague is Dr. Gary Banker who directed the neuroscience course at the University of Virginia, and was a master in the "functional neuroanatomy" style. Both of these individuals as well as other colleagues as the University of Virginia will certainly detect elements of their own teaching style in this book.

The original drawings and diagrams in this book are the work of a single artist, Ms. Ann Dunn. I would like to acknowledge her creativity and her tolerance for the uneven schedules, and what must sometimes have seemed like artistic whims of the author. Thanks also to Paula Falk for help in assembling the figures and to Elaine Lowe, who helped with the proofreading of the manuscript.

Finally, I would like to thank William Day, Senior Medical Editor at Springer-Verlag from the time I began this project through the time that I submitted the manuscript. His confidence in the concept of the book helped to motivate me during those times when progress was slow.

Contents

The Ground Plan of the Nervous System

1

Ground Plan of the Nervous System I

Principles of Central Nervous System Development

This chapter introduces students to the key terminology, grammar, and "syntax" that we use to describe the nervous system. We also describe the principal structures that will be our focus of attention, and the basic layout (ground plan) of the mature central nervous system (CNS). We briefly describe how the nervous system develops. A consideration of CNS morphogenesis is helpful at this point because the basic ground plan is much clearer early in development and becomes obscured as the brain develops its complex, three-dimensional form. Last, we briefly summarize the cellular organization at different levels of the CNS, focusing on the basic organization of neurons within the gray matter, and the key fiber tracts that contain ascending and descending pathways.

THE TERMINOLOGY, GRAMMAR, AND SYNTAX OF NEUROSCIENCE

Descriptors for Relationships

Students must be familiar with the common terminology used to describe interrelationships between parts of the nervous system. Unfortunately, beginning students must deal with some ambiguities in the descriptors that have come into common use.

One set of terms that can be confusing refers to relationships along the body axis. *Dorsal* and *ventral* are terms that have a consistent meaning applied to animals in which the body axis is horizontal—*dorsal* is toward the back, and *ventral* is toward the belly. The terms become confusing with reference to humans, characterized by erect posture; for example, the terms *dorsal* and *ventral,* and *anterior* and *posterior* are used interchangeably when referring to the spinal cord (i.e., dorsal roots and posterior roots are synonymous, ventral median fissure and anterior median fissure are synonymous, etc). When referring to the brain, however, the situation becomes complicated because the neural axis actually takes a 90° turn at the level of the mesencephalon (Figure 1.1). Consequently, in the brain, *dorsal* is toward the top of the head; *ventral* is toward the base.

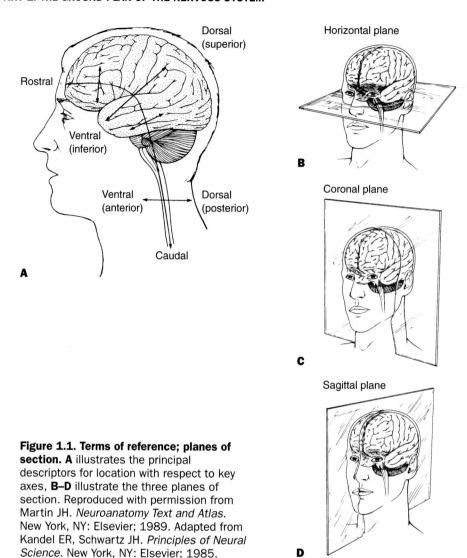

Figure 1.1. Terms of reference; planes of section. A illustrates the principal descriptors for location with respect to key axes, **B–D** illustrate the three planes of section. Reproduced with permission from Martin JH. *Neuroanatomy Text and Atlas.* New York, NY: Elsevier; 1989. Adapted from Kandel ER, Schwartz JH. *Principles of Neural Science.* New York, NY: Elsevier; 1985.

Two other terms that are used to define relationships are *rostral* and *caudal*. In animals with a horizontal body axis, *rostral* is toward the nose; *caudal* is toward the tail. In humans, the 90° turn in the neuraxis also means a 90° turn in the rostrocaudal axis (Figure 1.1).

Standard Planes of Section

We study the internal organization of the nervous system by examining sections that are taken in standard planes. The spinal cord and brainstem are usually viewed in cross section, as if a hot dog were cut into pieces and the pieces were examined end on. The brain may be sectioned in the horizontal, coronal (also called *frontal*), or sagittal plane (Figure 1.1). Students must learn to identify key structures in all three planes of section.

Terminology Pertaining to the Cellular Organization of the Nervous System

The nervous system is made up of two principal types of cells. *Neurons* (nerve cells) are the functional units of the nervous system. *Glia* are support cells. The cell biology of neurons and glia is considered in more detail in Chapters 4 and 5. Only the key details are summarized here.

In their signaling role, neurons perform the following functions:

1. Receive input
2. Integrate information
3. Transmit information over distances
4. Communicate with target cells via the release of neurotransmitters

Different parts of the neuron are specialized for these functions. Neurons receive inputs on their *dendrites*. They integrate information based on the shape of their dendrites and *cell bodies*. They transmit information via *axons*. And they release neurotransmitters at *synaptic terminals* (sometimes referred to as just *synapses* or *terminals*).

When we use the phrase "neurons are found in . . ." we are referring to *neuronal cell bodies*. Sometimes we just use the term *cells* to refer to neuronal cell bodies. When we say, "Parkinson's disease involves a loss of cells in the substantia nigra," we mean that the disease causes the death of neurons whose cell bodies lie in the substantia nigra.

In the peripheral nervous system, collections of nerve cell bodies are called *ganglia;* collections of axons are termed *nerves*.

In the CNS, areas that contain nerve cell bodies are called *gray matter;* areas that contain axons are termed *white matter.* Areas in which there are synaptic connections between axons and dendrites are termed *neuropil.*

For teaching purposes, histological sections of the brain are stained in one of two ways: (1) Weigert stain highlights white matter tracts; gray matter and neuropil are unstained. (2) Nissl stains highlight gray matter; white matter and neuropil are unstained. Students must be able to identify which stain is being used so as to be able to identify key structures in stained sections.

In gray matter of the CNS, there are two basic types of organization: nuclear and laminar. Cortical structures are laminated (e.g., cerebral cortex, olfactory cortex, and cerebellar cortex). Other regions have a nuclear organization (e.g., spinal cord, medulla, midbrain, diencephalon, and basal ganglia).

Individual nuclei contain collections of neurons of a similar type. Layers may be made up of nerve cell bodies or neuropil that contain particular types of connections.

Axons that interconnect one region with another tend to travel together through the CNS. Collections of axons with similar origins and terminations are termed *tracts* or *fiber tracts* (the term *fibers* is used synonymously with *axons*). The term *lemniscus* also refers to collections of axons but is less common. Collections of axons of different types are sometimes called a *fasciculus* (because the axons fasciculate). In the spinal cord, the terms *funiculus* and *column* refer to tracts or collections of tracts.

The terms *afferent* and *efferent* have both a specific and a general meaning. When the terms are used alone, a specific meaning is implied: *afferents* refer to sensory fibers that enter the CNS; *efferents* refer to motor fibers that project from the CNS. When used with modifiers (e.g., *cortical afferents and efferents*), the terms mean *inputs* and *outputs*. That is, afferents are inputs; efferents are outputs.

MORPHOLOGICAL SUBDIVISIONS OF THE NERVOUS SYSTEM

The two principal subdivisions of the nervous system are the CNS, which includes the spinal cord, brain, and retina, and the peripheral nervous system, which includes sensory and autonomic ganglia and peripheral nerves.

The CNS is divided into six subdivisions along the rostrocaudal neuraxis (Figure 1.2), described here in ascending order (caudal to rostral). It is important to recall that these are gross regional subdivisions that are useful for describing location in the neuraxis but that each subdivision includes components with quite different functions.

1. The *spinal cord* (a rarely used term for the spinal cord is *myelon*) extends from the *foramen magnum* (the hole in the skull

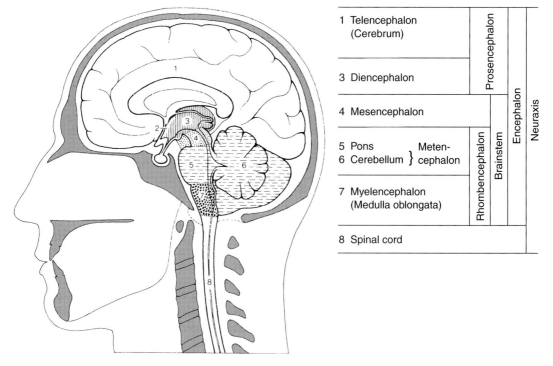

Figure 1.2. Morphological subdivisions of the CNS. 1, The telencephalon (in white), **2,** boundary between telencephalon and diencephalon, **3,** diencephalon, **4,** mesencephalon, **5,** tegmentum of the pons, **6,** cerebellum, **7,** medulla oblongata, **8,** spinal cord. From Nieuwenhuys R, Voogd J, van Huijzen C. *The Human Central Nervous System.* Berlin: Springer-Verlag; 1988.

through which the spinal cord extends) to the terminus of the cord. The spinal cord is divided into sacral, lumbar, thoracic, and cervical segments (again, in caudal to rostral order).

2. The *myelencephalon* corresponds to the *medulla.* It extends from the junction with the spinal cord to the caudal border of the pons.

3. The *metencephalon* includes the *pons* and *cerebellum* and the portion of the brainstem lying between the pons and cerebellum.

4. The *mesencephalon* corresponds to the *midbrain.* It begins at the rostral border of the pons and continues to the junction of the diencephalon.

5. The *diencephalon* includes the *thalamus* and *hypothalamus.*

6. Finally, the *telencephalon* includes the cerebral hemispheres, the *basal ganglia,* and the *olfactory bulbs.* The basal ganglia are located deep within the substance of the telencephalon and so are not visible from the exterior.

There are other collection terms. The *brainstem* includes the medulla, pons, cere-bellum, and midbrain. The *cerebrum* includes the diencephalon and telencephalon. Other collection terms refer to regional subdivisions that occur during early development (see later). Beginning students need to become familiar with all of these terms because subsequent chapters assume a familiarity with this vocabulary.

Different Subdivisions Develop From a Common Ground Plan During Nervous System Morphogenesis

The ground plan of the nervous system can be understood more easily by considering how the different regions differentiate during early development. In this way, students can first appreciate the basic ground plan and then see how the complex organization of the mature human nervous system develops from that basic ground plan.

The nervous system first appears during early embryogenesis (the third embryonic week) as a thickening of the ectoderm on the dorsal surface of the embryo, which is termed the *neural plate;* this specialized region extends throughout the anterior–posterior axis of the developing embryo. As development proceeds, the lateral portions of the neural plate begin to elevate and infold, eventually fusing dorsally to form a *neural tube* (see Figure 1.3). Fusion begins at the region of the medulla and proceeds in both the rostral and caudal directions. As fusion occurs, the ectoderm lying lateral to the neural plate fuses and the neural tube separates from the ectoderm and becomes internalized. The process through which the neural tube forms is termed *neurulation.*

A special group of cells at the boundary between the lateral surface ectoderm and neural ectoderm separate and assume a position dorsal and lateral to the forming neural tube. These are the cells of the *neural crest.* As development proceeds, the cells of the neural crest migrate away from their initial location into the periphery where they give rise to the neurons and glia of the peripheral nervous system and the adrenal medulla.

The Emergence of Subdivisions: The Three-Vesicle and Five-Vesicle Brain

During the fourth week of gestation, three enlargements (vesicles) appear in the rostral neuraxis; these are termed the rhombencephalon, the mesencephalon, and the prosencephalon (Figure 1.4). The brain at this stage is termed the *three-vesicle brain.* The rhombencephalon, or hindbrain, will become the metencephalon and myelencephalon; the mesencephalon will become the mesencephalon of the adult; the prosencephalon, or forebrain, will become the diencephalon and telencephalon.

By the fifth week of gestation, five vesicles can be seen in the brain, which correspond to the five principal divisions of the adult CNS. The brain at this stage is termed the *five-vesicle brain* and exhibits the basic ground plan of the mature nervous system.

The Flexures of the Neural Tube

An important feature of brain development is the appearance of three *flexures,* or bends (Figure 1.4). At the junction between the spinal cord and rhombencephalon, the neural tube bends anteriorly, forming the cervical flexure. At the junction between the metencephalon and myelencephalon the neuraxis bends back upon itself, forming the pontine flexure. The neuraxis then bends anteriorly again at the level of the mesencephalon, forming the cephalic flexure. It is because of the cephalic flexure that the longitudinal axis of the forebrain (diencephalon and telencephalon) is at a right angle to the axis of the hindbrain and spinal cord, and therefore, the terms *dorsal/ventral* and *rostral/caudal* have points of reference in the forebrain that are different from those in the hindbrain and spinal cord.

Morphogenesis of the Different Subdivisions of the Nervous System

Spinal Cord

Histogenesis

When the neural plate first closes to form the neural tube, the tube is composed of a single layer of elongated epithelial cells (the *neuroepithelium,* see Figure 1.5). These cells are the germ layer for the nervous system. The neuroepithelial cells extend processes from the *central canal* to the basal lamina at the outer margin of the neural tube. When the neuroepithelial cells divide, they withdraw their processes and round up, so that they line the lumen of the central canal. After mitosis, the daughter cells reextend their processes to the basal lamina at the outer margin of the tube.

Because the dividing cells collect next to the lumen of the central canal, and the elongated epithelial cells extend processes distally to the basal lamina, the neuroepithelium has a stratified appearance. But because all of the cells are of the same type, the neuroepithelium is said to be a *pseudostratified epithelium* in contrast to the ectoderm, which is a *stratified epithelium* in the true sense.

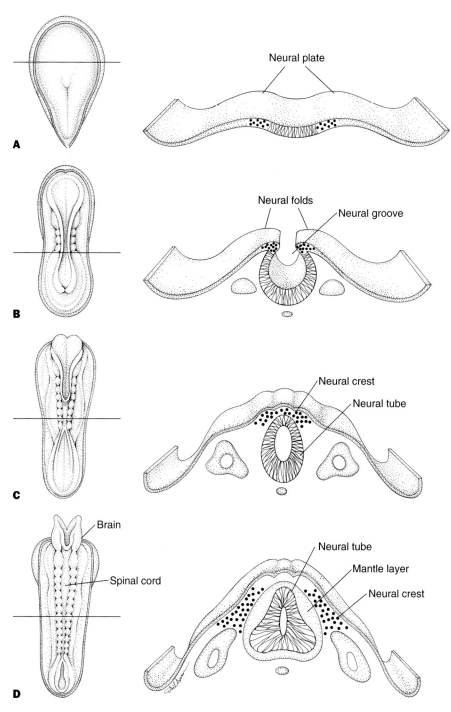

Figure 1.3. Neural plate, neural tube, and neural crest. Left, dorsal views of the developing embryo: **A,** at 18 to 19 days, **B,** 20 to 21 days, **C,** 22 to 23 days, and **D,** 24 to 25 days of gestation. Right, appearance of the neural plate, neural tube, and neural crest in cross section (taken at approximately the level indicated by the lines [left]). After Heimer L. *The Human Brain and Spinal Cord*. New York, NY: Springer-Verlag; 1994. Those drawings in turn were adapted primarily from Cowan WM. The development of the brain. Sci Am 1979;241:3.

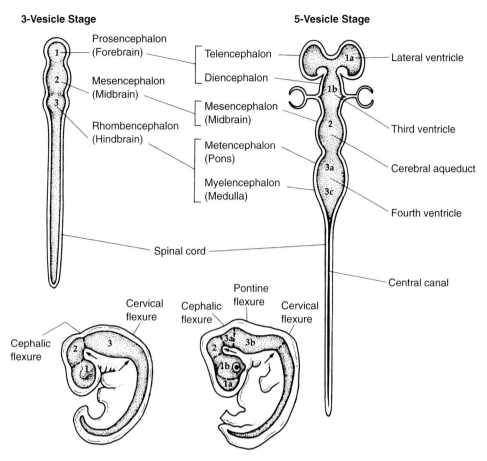

3-Vesicle Stage

5-Vesicle Stage

Figure 1.4. The three-vesicle and five-vesicle brain and flexures. During the fourth week of gestation, three enlargements (vesicles) develop: the rhombencephalon, mesencephalon, and prosencephalon. This is the three-vesicle brain. The rhombencephalon, or hindbrain, becomes the metencephalon and myelencephalon; the mesencephalon becomes the mesencephalon of the adult; the prosencephalon, or forebrain, will become the diencephalon and telencephalon. By the fifth week of gestation, five vesicles can be seen, which correspond to the five principal divisions of the adult CNS. The three principal flexures are indicated. Reproduced with permission from Martin JH. *Neuroanatomy Text and Atlas*. New York, NY: Elsevier; 1989 and Kandel ER, Schwartz JH, Jessel TM. *Principles of Neural Science*, 3rd ed. New York, NY: Elsevier; 1991.

Eventually, the dividing cells undergo a terminal mitosis, after which one or both daughter cells leave the proliferative cycle. The young neurons then accumulate at the periphery of the neuroepithelial layer to form a *mantle layer* or *intermediate zone* (Figure 1.3D). As more and more cells accumulate in the intermediate zone and begin to differentiate, the neural tube begins to thicken. At this stage, three distinct circumferential zones are evident: an internal *neu-roepithelial layer* that is still one cell thick, an *intermediate zone* containing neurons that are beginning to extend processes, and a *marginal zone* adjacent to the basal lamina in which the long fiber tracts of the cord will develop.

Important features of the developing spinal cord and, indeed, the developing nervous system in general are the *radial glial cells*, which are specialized astrocytes (see Chapter 5). As the neural tube thickens,

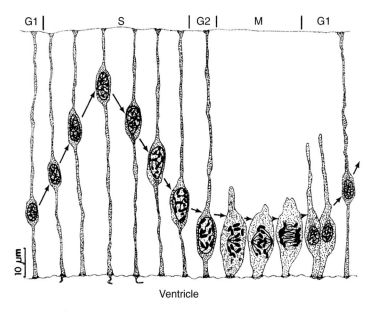

Ventricle

Figure 1.5. Histogenesis in the neural tube. The neural plate and neural tube are made up of a single layer of elongated epithelial cells (the neuroepithelium). These cells are the germ layer for the nervous system. Neuroepithelial cells extend processes from the *central canal* (in the spinal cord) or ventricle (in general) to the basal lamina at the outer margin of the neural tube (S and G2). During division (M) they withdraw their processes and round up so that they line the lumen of the central canal. After mitosis, the daughter cells reextend their processes to the basal lamina at the outer margin of the tube (G1). Reproduced with permission from Jacobson M. *Developmental Neurobiology*, 3rd ed. New York, NY: Plenum Press; 1991.

some of the neuroepithelial cells differentiate into radial glia that extend processes all the way from the central canal to the basal lamina. Young neurons and glia migrate along the processes of radial glia to their final destination in the developing intermediate zone.

Morphogenesis

As the wall of the neural tube begins to thicken, the dorsal and ventral segments of the wall thicken disproportionately so that a longitudinal fissure appears along the lateral wall of the neural tube; this is the *sulcus limitans*. The portion of the neural tube that is dorsal to the sulcus limitans is termed the *alar plate;* the ventral portion is termed the *basal plate* (Figure 1.6). As the cord develops, the neurons in the alar plate differentiate into groups of cells that will receive somatosensory input from the periphery (somatic afferents) that grow into the cord via the dorsal roots; neurons in the basal plate

differentiate into motor neurons (also called *motoneurons*) that populate the ventral horn. Motoneurons give rise to axons that exit the cord via the ventral roots.

As the spinal cord continues to thicken, the portions of the cord containing sensory and motor neurons serving the limbs thicken disproportionately. These particularly thick areas are termed the *cervical* and *lumbar enlargements* (Figure 1.7) and actually contain a larger number of neurons than segments of the cord serving the trunk. The different segments of the spinal cord start out with the same numbers of neurons, and differences in cell number come about because of a greater degree of cell death in the portions of the cord that do not supply the limbs.

The final morphological appearance of the cord depends on two factors: (1) the location and differentiation of the resident neurons, and (2) development of the long tracts in the marginal zone. At maturity, the gray matter of the cord is differentiated into a dorsal and

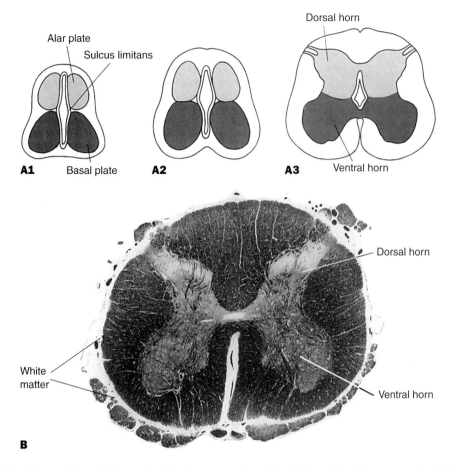

Figure 1.6. Alar and basal plates give rise to neurons that populate dorsal and ventral horns, respectively. A, Transverse sections through the spinal cord at three stages of development. The dorsal and ventral segments of the neural tube thicken disproportionately so that a longitudinal fissure appears (the sulcus limitans). The portion of the neural tube that is dorsal to the sulcus limitans is termed the *alar plate;* the ventral portion is termed the *basal plate.* Neurons in the alar plate differentiate into groups of cells that will receive sensory input (cells that reside in the dorsal horn of the spinal cord); neurons in the basal plate differentiate into motor nuclei (ventral horn). **B,** Cross section through the mature spinal cord illustrating the distribution of gray matter and white matter. Reproduced with permission from Martin JH. *Neuroanatomy Text and Atlas.* New York, NY: Elsevier; 1989.

ventral horn (Figure 1.6) and is surrounded by the white matter. The white matter contains the ascending and descending fiber tracts.

One final aspect of cord morphogenesis should be mentioned. As the body increases in length, the spinal canal grows, but the spinal cord does not increase proportionately in length. For this reason, in the adult nervous system the end of the spinal cord (termed the *conus*) is at the level of the last lumbar segment (Figure 1.7). Segmental nerve roots extend into the caudal spinal canal at the *cauda equinus.* The existence of the space around the cauda equinus permits the physician to withdraw spinal fluid (a *spinal tap*) without risking damage to the spinal cord itself.

Developmental Malformations of the Spinal Cord

Developmental malformations of the spinal cord arise because of defects in neural

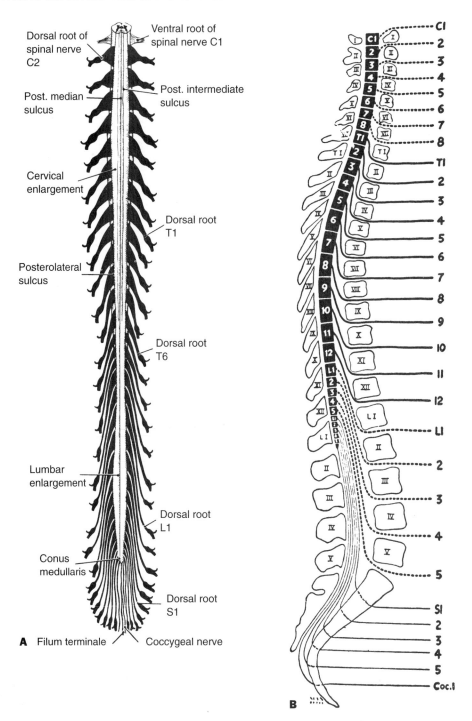

Figure 1.7. Cervical and lumbar enlargements: relationship between spinal segments and vertebral column. A, Posterior view of the mature spinal cord. The portions of the cord containing sensory and motor neurons serving the limbs are relatively larger than other segmental levels; these are termed *cervical* and *lumbar* *enlargements*. **B,** Schematic illustration of the position of spinal cord segments with respect to the vertebrae. The conus of the spinal cord is located at about L1. Reproduced with permission from Carpenter, MB, Sutin, J; *Human Neuroanatomy*. Baltimore, Md: Williams & Wilkins; 1976.

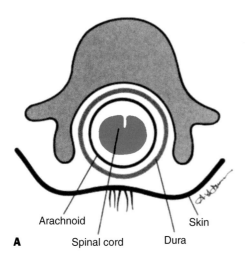

Arachnoid   Skin

A   Spinal cord   Dura

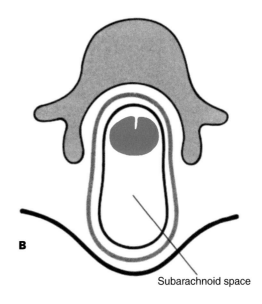

B

Subarachnoid space

Figure 1.8. Defects in closure. A, Spina bifida is the condition in which the vertebral column fails to close. The spinal cord itself may be normal in this condition. **B,** If the meninges protrude so that there is an enlarged subarachnoid space, the condition is termed meningocele. **C,** Myelomeningocele refers to the condition in which portions of the spinal cord itself protrude. After Heimer L. *The Human Brain and Spinal Cord*. New York, NY: Springer-Verlag; 1994. Drawings in Heimer were in turn adapted from Escourolle R, Poirier J. *Manual of Basic Neuropathology*. Rubinstein LJ, trans. Philadelphia, Pa: WB Saunders; 1978.

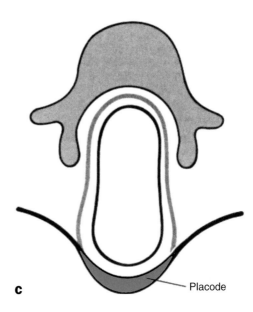

C   Placode

tube closure (Figure 1.8). The general term for the failure of neural tube closure is *rachis-chisis*. If the tube fails to close throughout the brain, the brain fails to develop *(anencephaly)*. If the tube fails to close in the region of the spinal cord, the result is termed *myeloschisis*. *Spina bifida* is the term that applies when the vertebral column fails to close (Figure 1.8A). In this situation, the meninges and some-times the spinal cord may protrude, forming a bulge (Figure 1.8B). If only the meninges protrude from the open spinal canal, the condition is termed *meningocele* (Figure 1.8B). When the protrusion also involves elements of the nervous system, the condition is termed *myelomeningocele* (Figure 1.8C). If there is failure of closure of the vertebral col-umn but no protrusion, the condition is termed *spina bifida occulta* (Figure 1.8A). Spina bifida may be accompanied by myeloschisis.

Myelencephalon (Medulla Oblongata)

The myelencephalon is continuous with the spinal cord, and the histogenesis of the myelencephalon follows a course similar to that of the spinal cord. Young neurons migrate out from the proliferative matrix in the neuroepithelial layer to differentiate in an intermediate zone (Figure 1.9). The principal difference is the tremendous expansion of the central canal leading to the fourth ventricle. This expansion flattens out the alar and basal plates so that they assume an orientation not unlike that present in the neural plate before closure (i.e., the alar plate lies almost directly lateral to the basal plate). Otherwise, the fundamental ground plan is similar, so cells involved in motor functions (cranial nerve nuclei) derive from the basal plate, and cells involved in sensory functions derive from the alar plate.

In addition to the neurons with primarily sensory and motor functions, other cell groups form in the medulla. Included among these are the cells of the *inferior olivary nucleus,* which arise from the alar plate and migrate to a ventral position, and a large, central gray region termed the *reticular formation.*

Metencephalon (Cerebellum and Pons)

The cerebellum and pons develop almost as an appendage to the basic ground plan; histogenesis in the core of the metencephalon initially follows a course similar to that in more caudal regions (Figure 1.10). This

Figure 1.9. Development of the medulla. A, Transverse sections through the medulla at three stages of development. The central canal opens up to create the fourth ventricle, which flattens out the neural plate. As a result, sensory neurons that receive somatic and visceral afferents are located laterally; motoneurons that supply the body and viscera (somatic motor and visceral motor) are located medially. **B,** Transverse section through the medulla in the mature CNS. Reproduced with permission from Martin JH. *Neuroanatomy Text and Atlas.* New York, NY: Elsevier; 1989.

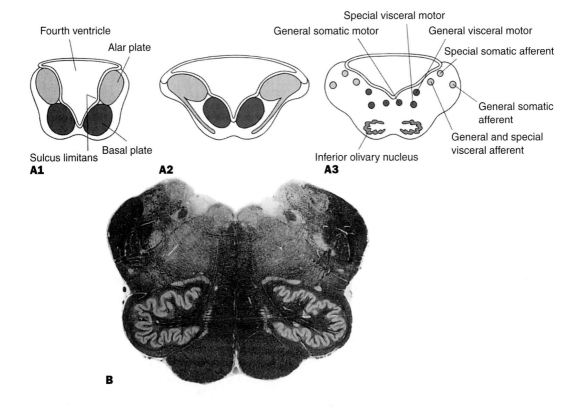

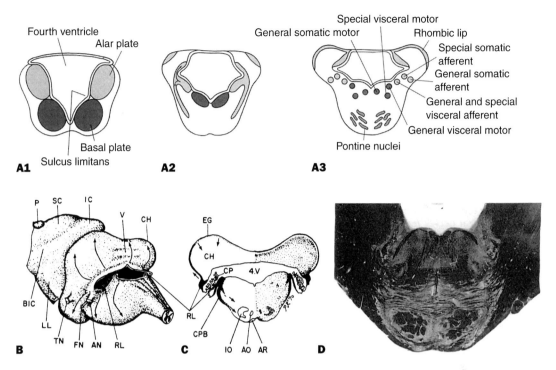

Figure 1.10. Development of the metencephalon. Histogenesis in the core of the metencephalon initially follows a course similar to that in more caudal regions; compare drawings of cross sections at three stages of development in **A** with Figure 1.9. In addition, however, the alar plates on each side expand dorsally and then medially over the fourth ventricle to form the rhombic lips; as development proceeds, the rhombic lips join at the midline to form the cerebellar plate. The cerebellum then expands upward from the cerebellar plate as the resident neurons differentiate. Overall appearance of the metencephalon is shown as viewed from the side **(B)** and in cross section **(C)**. Long arrows in **B** indicate the direction of migration of cells from the rhombic lips up into the cerebellum and down into the core of the metencephalon, where the cells give rise to several nuclei. V indicates vermis of cerebellum; CH, cerebellar hemisphere; RL, Rhombic lip; EG, external granule layer; 4V, fourth ventricle. Other abbreviations are not relevant here. **D**, A section through the metencephalon showing the mature tegmentum and the pons. Embedded within the pons are important collections of neuronal cell bodies (pontine nuclei). **A1–A3** and **D** are from Martin JH. *Neuroanatomy Text and Atlas*. New York, NY: Elsevier; 1989. Drawings in lower panel are from Sidman RL, Rakic P. Neuronal migration with special reference to developing human brain; a review. Brain Res 1973;62:1-35.

core region develops into the *tegmentum* of the pons. An important additional feature, however, is the development of a specialized region in the dorsal and lateral part of the metencephalon termed the *rhombic lips*. The rhombic lips are formed as alar plates on each side and expand dorsally and then medially over the fourth ventricle; as development proceeds, the rhombic lips join at the midline to form the *cerebellar plate*. The cerebellum then expands upward from the cerebellar plate as the resident neurons differentiate.

An important aspect of cerebellar histogenesis is the existence of a secondary proliferative zone during the late stages of cerebellar development. The precursors of one of the important resident neurons of the cerebellum (the granule cells) migrate from the rhombic lip, around the edges of the developing cerebellum, to collect in a thin layer on the surface of the cerebellum. This layer is termed the external granule layer. This zone is a secondary proliferative matrix that continues to produce cells until very late in development (about 10 years of age in humans).

An important clinical note pertains to the external granule layer; it is the site of origin of the most common childhood tumor of neural origin—the *medulloblastoma.* The cells making up this tumor are thought to be transformed cells of the secondary germinal matrix of the external granule layer.

The pons is formed by collections of axons growing to and from the cerebellum. Embedded within the pons are important collections of neuronal cell bodies *(pontine nuclei)* that are an important component of cerebellar circuitry.

Mesencephalon

The mesencephalon retains the same basic ground plan as the spinal cord and myelencephalon. The dorsal alar plate differentiates into the *superior* and *inferior colliculi,* which form the roof *(tectum)* overlying the central canal (Figure 1.11). The superior and inferior colliculi form important relay stations in the visual and auditory pathways, respectively. The narrow central canal extending throughout the mesencephalon connects the fourth ventricle with the third ventricle of the diencephalon (see later). The mature central canal of the mesencephalon is termed the *aqueduct of Sylvius,* or the *cerebral aqueduct.*

The basal plate differentiates into the *tegmentum* (the area of the mesencephalon below the cerebral aqueduct). The neurons in the tegmentum collect into important nuclei that include the red nucleus and substantia nigra, as well as a centrally located reticular formation. Important ascending and descending fiber tracts grow through the marginal zone, especially at the ventral and lateral margin where the *cerebral peduncle* develops—the massive fiber bundle that carries descending projections from the cerebral cortex.

Figure 1.11. Development of the mesencephalon. The basal plate develops into a tegmentum much like that of the metencephalon. The alar plate differentiates into the superior and inferior colliculi, which form the roof (tectum) overlying the central canal. The narrow central canal extending throughout the mesencephalon connects the fourth ventricle with the third ventricle of the diencephalon. The mature central canal of the mesencephalon is termed the *cerebral aqueduct* . **B** illustrates a transverse section through the mesencephalon in the mature CNS. Reproduced with permission from Martin JH. *Neuroanatomy Text and Atlas.* New York, NY: Elsevier; 1989.

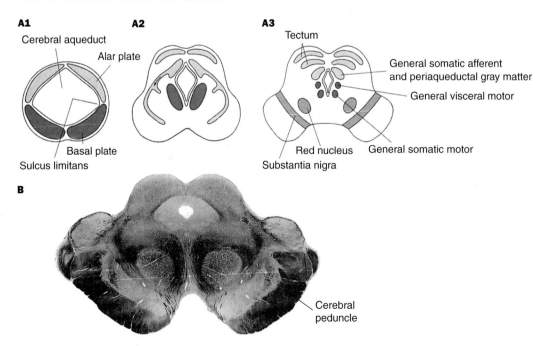

Diencephalon

In terms of histogenesis, most of the diencephalon develops similarly to more caudal regions, but the development of the three-dimensional architecture of the region is a bit more complex. In the first place, the central canal expands into a tall, thin cavity surrounded by walls formed by dorsally expanded basal and alar plates. Hence, it is most convenient to view the region as it would appear in midsagittal section (that is, a section that passes through the expanded central canal) (see Figure 1.12).

The dorsal walls of the diencephalon (alar plate derivative) develop into the *thalamus* (Figure 1.12). The two halves of the thalamus remain separated so that in the mature nervous system, there is a tall, thin central cavity (the *third ventricle*) that begins at the caudal border of the diencephalon and extends to the boundary with the telencephalon. The ventral portion of the diencephalon (basal plate derivative) develops into the *hypothalamus,* which comes to form the floor and lower walls of the third ventricle, and the *subthalamus,* which develops below the thalamus in more lateral locations.

The boundary between the thalamus and hypothalamus is formed by the *hypothalamic sulcus,* which can be thought of as a continuation of the sulcus limitans. As is the case in more caudal regions, the sulcus limitans separates a ventral region that can be considered motor (in this case, the hypothalamus, which is involved in autonomic regulation), and a dorsal region that is sensory (the thalamus) contains nuclei that relay information from somatosensory, visual, and auditory systems to the cortex.

A very important derivative of the diencephalon is the *optic vesicle,* which gives rise to the neural structures of the eye (Figure 1.13). The optic vesicle extends as an outpouching from the rostral and lateral portion of the diencephalon. It extends rostrally until it contacts the overlying surface ectoderm, remaining in contact with the diencephalon via the *optic stalk.* Each optic vesicle then invaginates to form a double-walled optic cup composed of neuroectoderm and surface ec-

toderm. Initially, the two layers are separated by the *intraretinal space;* then the two layers fuse as development proceeds, bringing the neuroectoderm and surface ectoderm into apposition. The neuroectoderm differentiates into the *neural retina,* whereas the former surface ectoderm differentiates into the *pigmented epithelium.* Nerve fibers from the retina then grow through the optic stalk back into the brain to form connections with visual structures in the thalamus and tectum.

The apposition between the neural retina and the pigment epithelium is an area of weakness throughout life. Concussive impacts to the eye, as well as other conditions, can lead to separation of the two layers *(retinal detachment).* There is important metabolic interdependence between cells of the neural retina and the pigment epithelium; thus, if the retina is separated from the pigment epithelium for any length of time, the retina degenerates, leading to blindness. For this reason, retinal detachment calls for immediate measures to surgically reappose the two layers.

Telencephalon

Because of the tremendous growth of the telencephalon during development, the basic ground plan becomes so distorted as to be barely recognizable. Nevertheless, the same basic histogenic and cytogenic processes occur as in other regions.

The earliest-forming cells of the telencephalic vesicle are born in the neuroepithelial layer that lies adjacent to the central lumen of the vesicle. Cells migrate from the neuroepithelium out into the walls of the vesicle. The dorsal portions of the vesicle differentiate into the cerebral cortex (Figures 1.12 and 1.14).

The ventral regions of the telencephalic vesicle give rise to important components of the basal ganglia, which include the *caudate* and *putamen nuclei* (Figure 1.14). The other important component of the basal ganglia (the *globus pallidus*) arises from the diencephalon. The components of the *limbic system,* which include the amygdala and the hippocampus, develop from the ventrome-

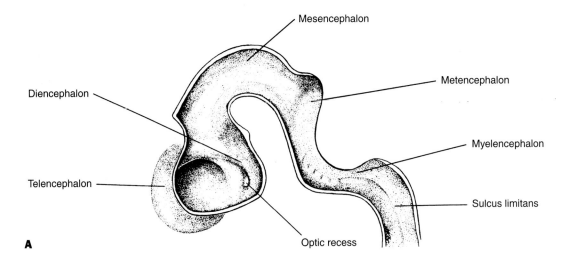

A

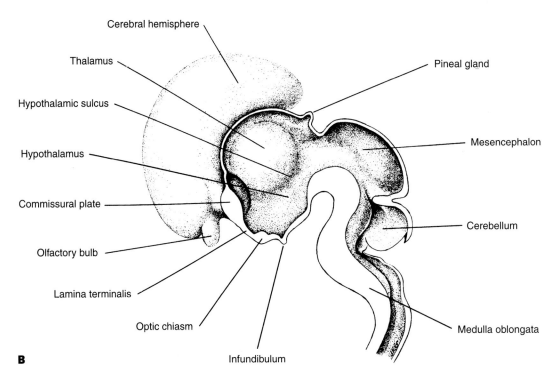

B

Figure 1.12. Development of the diencephalon. A view of the medial surface of the right half of the brain as seen in midsagittal section. **A,** 11-mm human embryo; **B,** 43-mm human embryo. The dorsal part of the diencephalon (alar plate derivative) develops into the thalamus. The ventral part develops into the hypothalamus. The boundary between the thalamus and hypothalamus is formed by the hypothalamic sulcus, which can be thought of as a continuation of the sulcus limitans. From Heimer L. *The Human Brain and Spinal Cord.* New York, NY: Springer-Verlag; 1994.

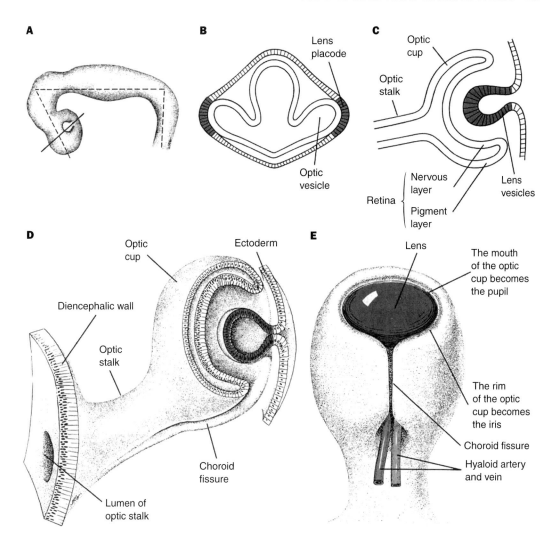

Figure 1.13. Development of the eye. The eye develops from the optic vesicle, which extends as an outpouching from the rostral and lateral portion of the diencephalon. **A** illustrates a drawing of the brain at about three weeks gestation. The solid line indicates the plane of a cross section through the developing diencephalon. **B** illustrates the appearance of the diencephalon at the level indicated. **C** illustrates the way the surface epithelium infolds so as to give rise to the lens vesicle. **D** and **E** are three-dimensional drawings of the developing optic cup. From Heimer L. *The Human Brain and Spinal Cord.* New York, NY: Springer-Verlag; 1994. **C** and **D** are modified from Mann JC. *The Development of the Human Eye.* London: Cambridge University Press; 1950.

dial portion of the telencephalic vesicle, which invaginates into the central cavity of the telencephalic ventricle (Figure 1.14).

As the telencephalic vesicle continues to develop, the neuroepithelial layer differentiates into a ventricular epithelium; cellular proliferation continues in a germinal matrix adjacent to the ventricular epithelium. Young neurons migrate from the germinal matrix out into an intermediate layer (now termed the *cortical plate*) where the cells differentiate and begin to form connections.

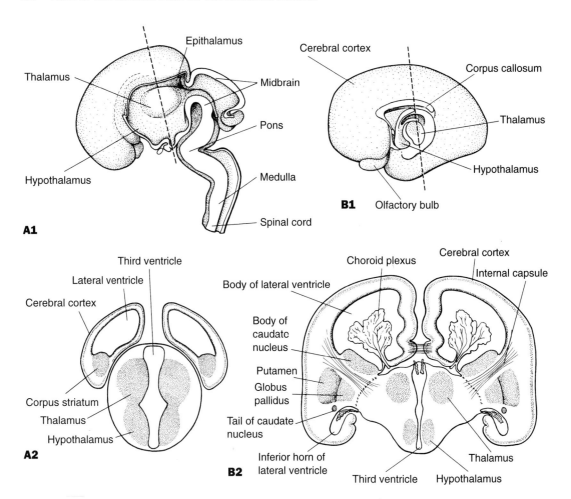

Figure 1.14. Development of the telencephalon. The upper drawings **(A1, B1)** illustrate medial views of the brain at an early stage of development, when the cerebral cortex is beginning to form, and a later stage when the basic plan of the mature brain can be discerned. The bottom panels **(A2, B2)** show transverse sections. Reproduced with permission from Martin JH. *Neuroanatomy Text and Atlas*. New York, NY: Elsevier; 1989. Adapted from Patten BM. *Human Embryology*. New York, NY: McGraw-Hill; 1968.

Development of the Cortical Mantle

As more and more cells accumulate in the cortical mantle, the cortex expands tremendously in thickness and area (Figure 1.14). The expansion is so great that accommodating the cortex within the cranium requires that the cortex become extensively folded (imagine putting a deflated automobile inner tube into the shell of a basketball). The result of this folding is the highly characteristic pattern of folds *(gyri)* separated by fissures *(sulci).*

As the cortex expands, the characteristic *frontal, parietal, occipital, temporal,* and *insular*

lobes appear (Figure 1.15). The frontal, parietal, and temporal lobes expand disproportionately to the insular lobe and expand over the insula. The portions of the external lobes overlying the insula are termed *opercula* (coverings).

Development of the Major Fiber Systems

Telencephalic development continues with the development of the major fiber systems. Axons leaving the cortical mantle first collect in a white matter layer adjacent to the ven-

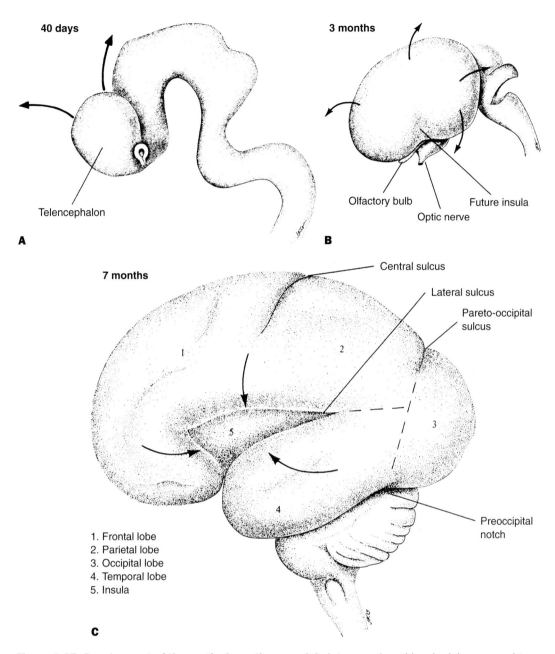

40 days

Telencephalon

A

3 months

Olfactory bulb

Optic nerve

Future insula

B

7 months

Central sulcus

Lateral sulcus

Pareto-occipital sulcus

Preoccipital notch

1
2
3
4
5

1. Frontal lobe
2. Parietal lobe
3. Occipital lobe
4. Temporal lobe
5. Insula

C

Figure 1.15. Development of the cortical mantle.
The drawings illustrate the appearance of the brain at different gestational ages: **A,** 40 days; **B,** 3 months; **C,** 7 months gestation. As the cortex expands, the characteristic frontal, parietal, occipital, temporal, and insular lobes expand to cover the insula. The portions of the external lobes overlying the insula are termed *opercula* (coverings). From Heimer L. *The Human Brain and Spinal Cord.* New York, NY: Springer-Verlag; 1994.

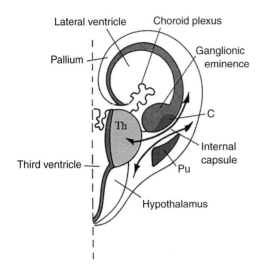

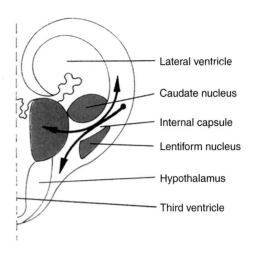

Figure 1.16. Formation of the principal white matter tracts of the telencephalon. Cross sections through the diencephalon and telencephalon at the stage of development illustrated in Figure 1.14B. Axons growing to and from the cortex penetrate through the basal ganglia, separating the caudate nucleus (C) from the putamen (Pu) and form the major fiber system called the *internal capsule*. From Heimer L. *The Human Brain and Spinal Cord.* New York, NY: Springer-Verlag; 1994.

tricle. A large contingent of cortical axons grows to connect the two halves of the cerebral cortex, forming the *corpus callosum.*

Descending projections grow from the cortical white matter toward the diencephalon, collecting in the *internal capsule* (Figure 1.16). Fibers from the frontal lobe collect in the anterior limb of the internal capsule, which grows through the structures composing the basal ganglia. In this way, the fibers of the internal capsule come to separate the caudate and putamen nuclei, which are actually part of the same nucleus (see Chapter 16). Fibers from posterior cortical regions collect in the posterior limb of the internal capsule, growing between the thalamus and the *globus pallidus.* The descending axons then collect in the cerebral peduncle. Simultaneous with the formation of descending projections, connections develop between the thalamus and cerebral cortex, again, growing through the internal capsule.

Development of the Ventricular System

As the cortical mantle expands in area and becomes folded, the central canal also ex-pands beneath the cortical mantle, forming a ventricular system (Figure 1.17). In the mature brain, the ventricular system is thus a system of fluid-filled cavities.

The final three-dimensional form of the ventricles is determined by the characteristic expansion and folding of telencephalic structures. The lateral expansions of the telencephalic vesicle lead to the formation of two prominent *lateral ventricles.* The *anterior horn* of the lateral ventricle extends into the frontal lobe, the *central portion* underlies the parietal lobe, the *posterior horn* extends into the occipital lobe, and the *temporal horn* extends into the temporal lobe. The lateral ventricles communicate with the third ventricle via the interventricular *foramen of Monro.* The third ventricle in turn communicates with the fourth ventricle via the *cerebral aqueduct.*

Summing Up

With this brief overview of CNS organization in hand, we can now begin to fill in the details. We begin by considering the cellular organization at different levels of the nervous system.

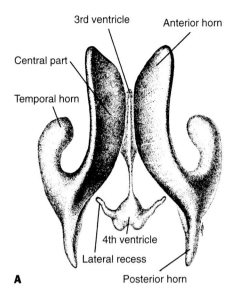

3rd ventricle

Anterior horn

Central part

Temporal horn

4th ventricle

Lateral recess

A

Posterior horn

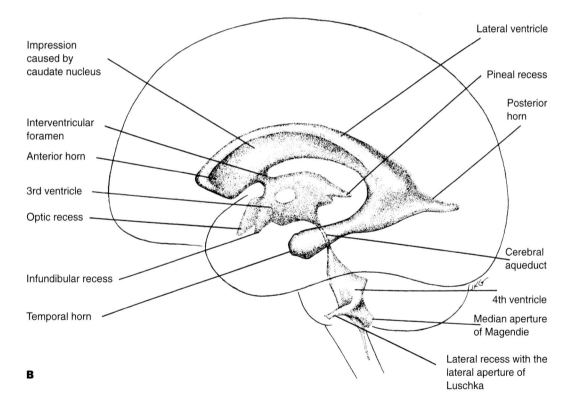

Lateral ventricle

Impression
caused by
caudate nucleus

Pineal recess

Posterior
horn

Interventricular
foramen

Anterior horn

3rd ventricle

Optic recess

Cerebral
aqueduct

Infundibular recess

4th ventricle

Median aperture
of Magendie

Temporal horn

B

Lateral recess with the
lateral aperture of
Luschka

Figure 1.17. Development of the ventricular system. As the cortical mantle expands in area and becomes folded, the central canal also expands beneath the cortical mantle, forming a ventricular system, which is made up of fluid-filled cavities with a complex, three-dimensional form. **A** illustrates the ventricles as they would appear from a dorsal view; **B** illustrates a lateral view. From Heimer L. *The Human Brain and Spinal Cord*. New York, NY: Springer-Verlag; 1994.

2

Ground Plan of the Nervous System II

Levels of the Neuraxis

In this chapter we outline the basic organization of the CNS level by level, beginning with the spinal cord. Each section begins with a description of the gross morphology of the region in the adult nervous system, then defines the cellular composition of the principal subdivisions of the gray matter and identifies the principal fiber tracts that pass through the level.

SPINAL CORD

Surface Features of the Spinal Cord

The surface of the cord is indented by a prominent *ventral* (or *anterior*) *medial fissure,* which runs the length of the cord (Figure 2.1). The ventral medial fissure permits a ready identification of ventral versus dorsal surfaces.

Nerve roots emerge at regular intervals from the dorsolateral and ventrolateral surface of the cord. The *dorsal roots* carry sensory fibers into the cord; the *ventral roots* carry motor axons that project to muscles. The dorsal and ventral nerve roots merge outside the spinal canal into the respective peripheral nerves.

The spinal cord is not of uniform diameter throughout its length; areas that are relatively enlarged indicate regions of specialized function. There are two enlargements, the *cervical enlargement* (C3 through T1) and the *lumbar enlargement* (L1 through S2), which correspond to the segments of the cord that serve the upper and lower limbs, respectively. The cord decreases progressively in diameter from the lumbar enlargement to the *conus* (the terminus of the sacral cord).

Less prominent surface features include the shallow groove running along the dorsal midline of the spinal cord, which marks the position of the *dorsal median septum* (Figures 2.1 and 2.2), and a shallow groove that marks the boundary between the *cuneate* and *gracile fasciculi* in the cervical cord (see later).

The Spinal Cord Has a Segmental Organization

The spinal cord is organized in a segmental fashion so that different segments are concerned with the function of restricted segments of the body. Each segment is linked to

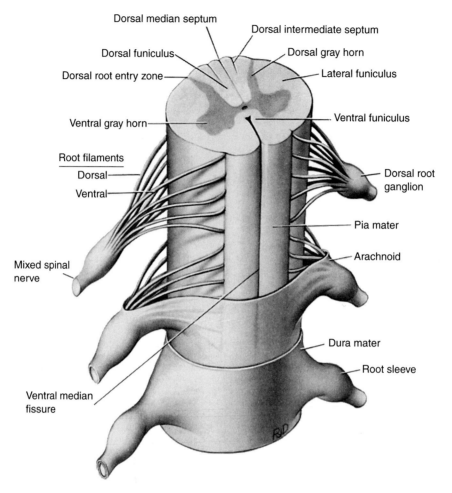

Figure 2.1. Surface features of the spinal cord. The drawing illustrates a three-segment-long region of the spinal cord. The relationship between the meninges and the spinal roots is indicated. Reproduced with permission from Carpenter MB. *Core Text of Neuroanatomy.* Baltimore, Md: Williams & Wilkins; 1983.

a particular region of the body by the sensory and motor nerve roots that emerge from the segment (Figure 2.1). There are 8 cervical segments associated with nerve roots C1 through C8, 12 thoracic segments associated with nerve roots T1 through T12, 5 lumbar segments associated with nerve roots L1 through L5, and 5 sacral segments associated with nerve roots S1 through S5 (Figure 2.1).

As noted in Chapter 1, the disproportionate growth of the body in comparison with the cord results in a displacement of the cord; as a result, the segments come to be displaced with respect to the point of exit of the nerve roots (see Figure 1.7). The collection of nerve roots that extend into the ventralmost portion of the vertebral column beyond the terminus of the cord is termed the *cauda equina* because it resembles a horse's tail.

It is important to realize that there are no actual boundaries that define the segments. Instead, the nerve cell bodies and white matter tracts are present as columns that extend throughout the cord. The divisions and segments provide us with a terminology to define and discuss the location of lesions affecting the cord.

The segmental organization of the cord is in distinction to higher levels of the nervous system, which are concerned with integrat-

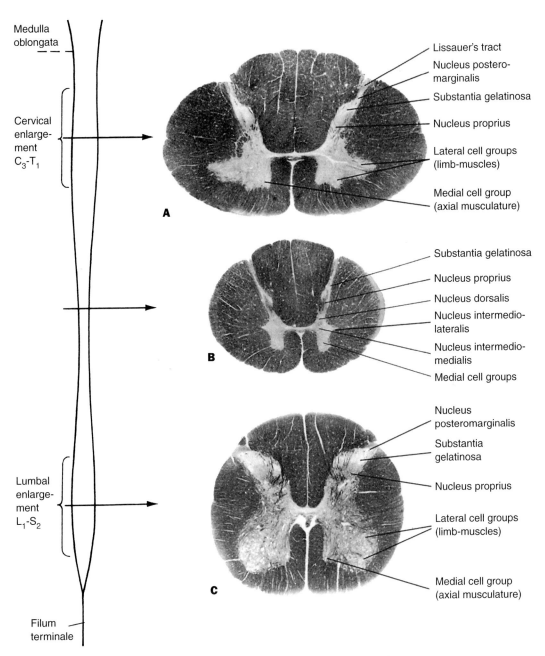

Figure 2.2. Configuration of gray and white matter regions of the spinal cord. A, B, and **C** illustrate cross sections through the cervical **(A),** thoracic **(B),** and lumbar **(C)** regions. From Heimer L. *The Human Brain and Spinal Cord.* New York, NY: Springer-Verlag; 1994. The Weigert-stained sections are from Gluhbegovic N, Williams TH. *The Human Brain and Spinal Cord.* Philadelphia, Pa: JB Lippincott; 1980.

ing information that arises from different segmental segments and coordinating output. The higher levels are thus involved in *suprasegmental* control. The different classes of sensory inputs and motor outputs are distinguished on the basis of the type of tissue that is innervated (somatic or visceral).

General somatic efferents (GSE, motor) innervate structures embryologically derived from the *somites*. Somatic motoneurons project directly to striated muscles. *General somatic afferents* (GSA, sensory) provide sensory input from the same regions supplied by the somatic efferents.

General visceral efferents (GVE) are the motor outflow of the sympathetic and parasympathetic divisions of the autonomic nervous system. The visceral motoneurons of the sympathetic division are found in the intermediolateral gray column in the thoracic and lumbar portions of the spinal cord. The visceral motoneurons of the parasympathetic division are found in the sacral portion of the spinal cord. *General visceral afferents* (GVA) travel from the viscera to the spinal cord via the same peripheral nerves that carry visceromotor fibers.

Internal Organization

The basic internal organization of the spinal cord is similar within each segment, although there are differences that allow one to distinguish between different levels. The gray matter of the cord is located centrally in a roughly butterfly-shaped mass (Figure 2.3). White matter tracts surround the gray matter.

Identifying Levels

The configuration of the gray and white matter differs at different levels: (1) the ventral horns are especially prominent in the cervical and lumbar enlargements; (2) the width of the white matter decreases from the rostral to the caudal; (3) the dorsal column is divided into gracile and cuneate fasciculi rostral to T6; (4) there is a distinct *in-*

termediolateral column from T1 to L3 that contains preganglionic cell bodies of the sympathetic nervous system, and a related column from S2 to S4 that contains preganglionic cell bodies of the parasympathetic nervous system.

The Gray Matter

The gray matter can be divided into three regions: a thin *dorsal horn,* or *posterior horn,* which extends toward the entry zone of dorsal root axons; a larger *ventral horn,* which extends ventrolaterally. Between the two is an *intermediate zone,* sometimes called the *intermediate horn.* The dorsal horn is the sensory recipient zone; the ventral horn contains motoneurons that project to striated muscles. These "horns" are in fact continuous columns of gray matter that extend throughout the length of the spinal cord. These columns are further subdivided into nuclear groupings that are also longitudinal columns that extend for variable distances.

The *marginal zone* (or *nucleus posteromarginalis*), *substantia gelatinosa, nucleus proprius,* and *nucleus dorsalis* (or *Clark's column*) are zones in which sensory fibers terminate. The neurons in these different zones give rise to different components of the central somatosensory pathways (see Chapter 10).

The *intermediolateral nucleus* is present between T1 and L2 and contains the cells of origin of the preganglionic projections of the sympathetic division of the autonomic nervous system (see Chapter 27). The *motor nuclei* contain motoneurons that project to muscles (see Chapter 14).

Neurons of the gray matter that are neither sensory nor motor are *interneurons;* these are the most numerous neuronal type in the gray matter of the cord. Interneurons have relatively short axonal projections. Interneurons that project only within the same segment are *intrasegmental interneurons;* interneurons that project to neighboring segments are *intersegmental interneurons,* or more commonly, *propriospinal neurons;* interneurons that project to the contralateral side of the cord are *commissural interneurons.*

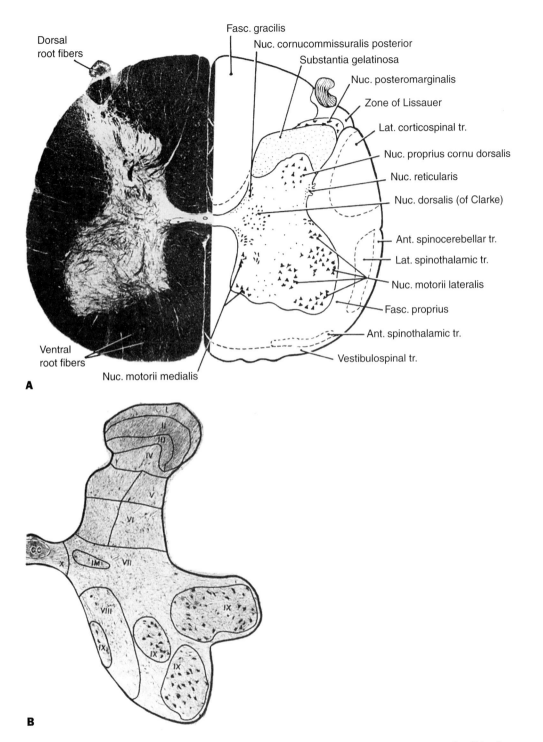

Figure 2.3. Cellular organization of the spinal cord. A illustrates the cellular organization of the lumbar spinal cord. On the left is the segment as seen in Weigert stain. On the right is a schematic illustration of important cell groups and fiber tracts. **B** is a photomicrograph of a Nissl-stained section through the gray matter of the lumbar spinal cord, which reveals neuronal cell bodies. Ant indicates anterior; Fasc, fasciculus; Lat, lateral; Nuc, nucleus; tr = tract. Roman numerals indicate Rexed's laminae; Reproduced with permission from Carpenter MB. *Human Neuroanatomy.* Baltimore, Md: Williams & Wilkins; 1976.

White Matter

There are three principal subdivisions of the spinal cord white matter: (1) the *dorsal columns,* (2) the *lateral columns,* and (3) the *ventral columns.* Alternate terms for the columns are dorsal, lateral, and ventral *funiculi* (singular form, *funiculus*), respectively.

The dorsal columns lie between the dorsal root entry zone and the midline; the columns are actually fiber tracts that carry one component of the ascending sensory projection from the spinal cord to the brain. One column is visible in the thoracic and lumbar cord, whereas two distinct columns can be distinguished in the cervical cord. The lateral-most pair of columns carry ascending fibers from the cervical portions of the cord; the medial-most columns carry input from lower regions. These fiber tracts are also called cuneate and gracile *fasciculi* (singular form, *fasciculus*). The cuneate fasciculus is lateral, the gracile is medial. Axons in the dorsal columns terminate in the *cuneate* and *gracile nuclei* in the medulla (see Chapter 10).

The lateral columns lie between the dorsal root entry zone and the ventral root exit zone. The dorsal part of the lateral funiculus contains descending fibers of the *rubrospinal, reticulospinal,* and *lateral corticospinal tracts* that originate in the *red nucleus, reticular formation* of the brainstem and mesencephalon, and *cerebral cortex,* respectively. Each of these plays a role in motor function (see Chapter 15). Also in the dorsolateral funiculus are ascending fibers of the *spinocerebellar tract.* The ventral part of the lateral funiculus contains descending reticulospinal fibers and ascending fibers of the *spinothalamic, spinoreticular, spinotectal,* and *ventral spinocerebellar systems,* all of which carry ascending sensory information from different types of sensory receptors. There are no distinct boundaries between the different fiber systems, although each system is distributed in a characteristic position in the lateral column.

The ventral columns lie in the zone between the ventral root exit zone and the ventral midline of the cord. The ventral columns contain descending fibers of the *ventral corticospinal tract* and the *vestibulospinal tract,* both of which are important for motor function (see Chapter 15).

As with the gray matter, the appearance of the white matter is different at different segmental levels. For example, the white matter tracts increase in size at higher levels of the cord. The reason is that fiber tracts at any given segment contain the ascending projections from and descending projections to all levels below that point.

Each of the fiber systems and tracts described above are components of functional systems. The origin, termination, and functional significance of each fiber system will be discussed in subsequent chapters.

Lesions of the Spinal Cord

Lesions of gray matter cause sensory or motor deficits at particular segmental levels (segmental deficits). Lesions of white matter interrupt communication between the brain and the body, causing sensory deficits and/or paralysis below the level of the lesion.

THE BRAINSTEM AND CEREBELLUM

The brainstem includes three rostrocaudal levels: the *medulla, pons,* and *midbrain* (myelencephalon, metencephalon, and mesencephalon). The brainstem can also be divided into three longitudinal zones: a *basilar* portion (most ventral), the *tectum* in the midbrain, and the *tegmentum.* The basilar portion includes the *medullary pyramid* in the medulla oblongata, the pons, and the *cerebral peduncle* in the midbrain, all of which are important white matter tracts. The tectum forms the roof of the midbrain and includes the *superior* and *inferior colliculi* (singular form, *colliculus*), two gray matter areas that are important way stations for visual and auditory systems, respectively. The tegmentum is everything between the tectum and the basilar portion and includes all cranial nerve nuclei, the gray matter regions of the central core (the reticular formation), and the ascending and descending fiber tracts that pass through the gray matter.

Surface Features and Gross Morphology of the Brainstem

Surface landmarks mark the location of important structures within the substance of the brainstem. The landmarks are ridges, bulges, or bumps for which there are a number of colorful terms: pyramid, olive, tuberculum, trigone, colliculus, and body (see later). As a general rule of thumb, if a name does not include a term designating a nucleus or fiber tract, it is probably a surface landmark. Students should be able to distinguish between the terms that refer to surface landmarks and the terms that refer to underlying structures.

The Ventral Surface

The transition between the spinal cord and medulla (the *spinomedullary transition*) occurs at about the point of origin of the first cervical nerve root. The point of transition is marked on the basal surface of the brain by the *pyramidal decussation*—the point at which descending fibers of the corticospinal tract cross the midline and join the lateral corticospinal tract of the spinal cord (Figure 2.4 and Chapter 13). Above the level of the pyramidal decussation, the *pyramids* form the basal surface of the medulla, disappearing into the pons at the *pontomedullary junction.* The pyramids contain the descending fibers of the corticospinal tract.

Just dorsal/posterior to the pyramids on the ventrolateral surface of the medulla is the *olive.* The olive is an oval protrusion just above the exit zone of the glossopharyngeal

Figure 2.4. Ventral aspect of the brainstem. Reproduced with permission from Carpenter MB, Sutin J. *Human Neuroanatomy.* Baltimore, Md: Williams & Wilkins; 1983.

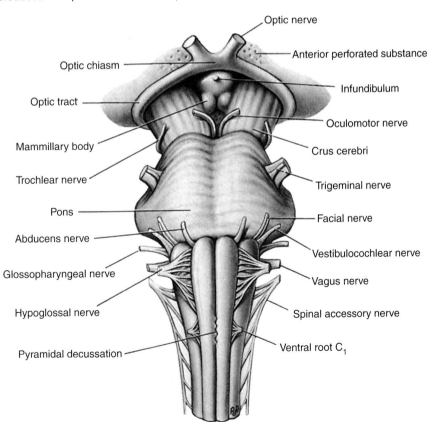

nerve and marks the location of the *olivary nuclei,* which provide an important projection to the cerebellum.

The most prominent landmark of the ventral surface of the brainstem is the pons (l. bridge). This enormous band of fibers wraps around the base of the brainstem and then extends dorsally on each side as the *middle cerebellar peduncle*—the largest of the three cerebellar peduncles that carry projections to and from the cerebellum (see later).

Rostral to the pons, the basal surface of the brainstem is occupied by the *cerebral peduncles,* which are gigantic fiber tracts that contain descending projections from the cortex. Between the peduncles is the *interpeduncular fossa,* a cavity that extends up to a ceiling formed by the base of the midbrain tegmentum. When the vasculature has been removed, this ceiling is marked by numerous perforations that are holes through which small arteries enter the brainstem. For this reason, the area is called the *posterior perforated space.* The junction between the brainstem and diencephalon occurs at the posterior boundary of the *mammillary bodies,* two small spherical mounds on the basal surface of the brain between the cerebral peduncles. The mammillary bodies mark the location of the *mammillary nuclei* of the hypothalamus (see Chapter 28).

Cranial Nerves

The exit zones for most of the cranial nerves can be seen on the ventral and lateral surfaces of the brainstem. Students should familiarize themselves with the location of the cranial nerves and be able to identify each nerve in drawings or intact brains. Beginning at the caudal end of the brainstem, the roots of the *hypoglossal nerve* (XII) leave the brainstem between the pyramid and the olive and then collect into the hypoglossal nerve. The *abducens nerve* (VI) emerges between the pyramid and the olive, just behind the pons.

The roots of the *spinal accessory nerve* (XI), *vagus nerve* (X), and *glossopharyngeal nerve* (IX) emerge from just above the olive, the latter near the posterior margin of the pons. The *facial nerve* (VII) and *vestibulocochlear nerve* (VIII) emerge just behind the posterior margin of the pons, between the glossopharyngeal nerve and the abducens nerve.

The *trigeminal nerve* (V) emerges directly from the pons midway along the ventrolateral surface. Finally, the *oculomotor nerve* (III) emerges from the base of the midbrain tegmentum, between the cerebral peduncles just rostral to the pons.

The Dorsal Surface

The surface features and gross morphology of the dorsal surface of the brainstem can be seen only when the cerebellum is removed (Figure 2.5). Again, moving from caudal to rostral, the spinomedullary transition occurs caudal to the *obex*—the V-shaped posterior boundary of the fourth ventricle. The dorsal columns (the cuneate and gracile fasciculi) can be followed into the transition area where they end in the *cuneate and gracile tuberculi* (singular, *tuberculum*), which are bulges on the dorsal surface of the brainstem, just behind the fourth ventricle. These bulges mark the location of the *cuneate* and *gracile nuclei.* Just ventral to the cuneate tuberculum is the olive.

Rostral to the obex, the brainstem opens into the fourth ventricle. The floor of the fourth ventricle is shaped like a rhombus and is thus called the *rhomboid fossa.* In the floor of the fourth ventricle, just rostral to the obex, is the *area postrema.* The area postrema represents one of the important *circumventricular organs* that lack a blood–brain barrier. This is an important chemoreceptive region that triggers vomiting in response to circulating emetic substances.

Also in the floor of the fourth ventricle are several small bumps that represent the locations of cranial nerve nuclei; these include the *vagal* and *hypoglossal "trigones"* and the *facial colliculus.* Also evident are striations termed the *stria medullaris* of the fourth ventricle (there is also a stria medullaris of the

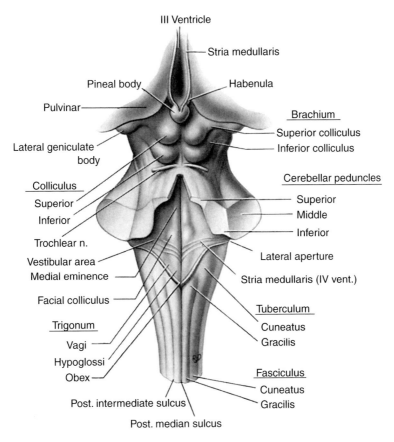

Figure 2.5. Dorsal aspect of the brainstem. Reproduced with permission from Carpenter MB, Sutin J. *Human Neuroanatomy.* Baltimore, Md: Williams & Wilkins; 1983.

third ventricle). The stria medullaris of the fourth ventricle is a collection of small bundles of myelinated axons that project from the *arcuate nucleus* in the brainstem to the cerebellum.

The walls and ceiling of the fourth ventricle are formed by the inferior, middle, and superior cerebellar peduncles, which represent the three major tracts carrying fibers to and from the cerebellum. The *inferior cerebellar peduncle* (also called the *restiform body*) enters the posterior–medial portion of the cerebellum. The *middle cerebellar peduncle* (also called the *brachium pontis*) extends dorsally from the pons, around the lateral edge of the brainstem, entering the cerebellum on its lateral ventral margin. The *superior cerebellar peduncle* (also called the *brachium conjuntivum*) emerges from the anterior medial cerebel-

lum and travels rostrally to enter the mesencephalon, caudal to the inferior colliculus.

Just rostral to the point at which the superior cerebellar peduncle enters the mesencephalon, the *trochlear nerves* (IV) emerge from the dorsal surface of the brainstem. Just rostral to the emergence of the trochlear nerve lie the paired inferior and superior colliculi (also called the *corpora quadrigemina*). The rostral boundary of the brainstem is at the rostral end of the superior colliculus.

The three-dimensional relationships between the different subdivisions of the brainstem are especially well revealed by midsagittal views (Figure 2.6). In these views it is possible to see the continuity of the gray matter of the tegmentum through the medulla and pons, and the continuity of the cerebral aqueduct and fourth ventricle.

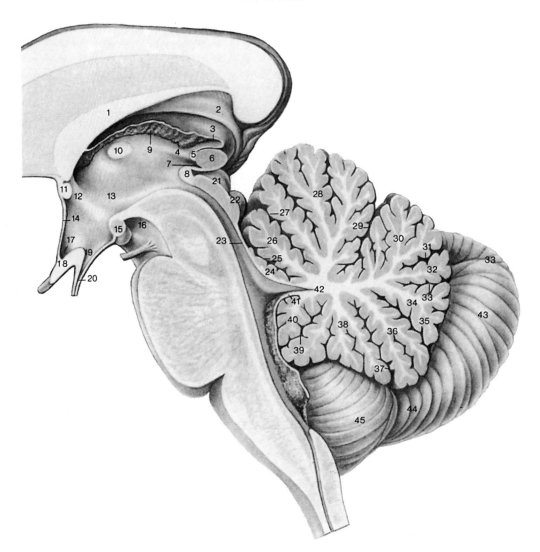

Figure 2.6. Midsagittal view of the brainstem and diencephalon. 1 and 2 indicate the fornix; 4, habenula; 5, habenular commissure; 6, pineal gland; 7, pineal recess; 8, posterior commissure; 9, choroid plexus of the third ventricle; 10, massa intermedia of the thalamus; 11, anterior commissure; 12, column of the fornix as it enters the hypothalamus; 13, hypothalamic sulcus; 14, lamina terminalis; 15, mamillary body; 16, interpeduncular fossa; 17, optic recess; 18, optic chiasm; 19, infundibular recess; 20, infundibulum (stalk of pituitary); 21, superior colliculus; 22, inferior colliculus; 23, superior medullary velum overlying the cerebral aqueduct, 24-45, structures in the cerebellum. From Nieuwenhuys R, Voogd J, van Huijzen C. *The Human Central Nervous System*. Berlin: Springer-Verlag; 1988.

Internal Organization of the Brainstem

The brainstem has both segmental and suprasegmental components. The brainstem contains somatosensory relay nuclei and motoneurons that serve muscles in the head and face and also has suprasegmental functions.

Organization of Components Related to Segmental Functions

The regions devoted to segmental functions have sensory and motor components that differ in two ways from the sensory and motor components of the spinal cord.

1. Motoneurons and sensory recipient neurons in the brainstem are collected in discrete nuclei rather than in continuous columns of cells. However, nuclei with similar functional properties are found in roughly the same cross-sectional position in the brainstem. In this sense, brainstem nuclei are said to be localized in discontinuous columns of cells that extend along the longitudinal axis of the brainstem.

2. In addition to the general somatic afferents and efferents (GSA and GSE), and general visceral afferents and efferents (GVA and GVE) that are present in the spinal cord, there are three additional "special" components. *Special visceral efferents* (SVE, motor) innervate striated muscles that derive embryologically from *the branchial arches (gills);* these are termed *branchiomeric tissues. Special visceral afferents* (SVA) are the afferent limb of the "special senses" that are considered visceral (taste and olfaction). *Special somatic afferents* (SSA) are the afferent limb of the special senses that are considered somatic (auditory, vestibular, and visual). This information is useful in that it provides a framework for understanding the characteristic topographic location of the different nuclei in the brainstem.

Rules of Topography

In the spinal cord the GSA nuclei lie dorsally, whereas GSE nuclei lie ventrally. The cells of origin of visceral efferents lie in an intermediate location.

In the brainstem, the topographic organization is similar. In the portion of the brainstem beneath the fourth ventricle, the GSA nuclei lie dorsolaterally, whereas the GSE nuclei lie ventromedially. There is an additional topographic organization of the branchiomeric and visceral components (Figure 2.7). At any given level, SSA and SVA recipient zones tend to lie medial to the GSA nuclei. GVE and SVE nuclei lie medial to the GSE nuclei. In this way, the different types of sensory and motor components are found in separate, discontinuous columns of cells that extend throughout the longitudinal extent of the brainstem (see Figures 1.9 and 1.10).

One prominent exception to the topographic rule is the *facial nucleus*—a special visceral efferent nucleus that is found in an unexpected location. This exception is ex-

Figure 2.7. Topographic organization of the sensory and motor columns of cranial nerve nuclei in the medulla. Roman numerals refer to cranial nerves. Reproduced with permission from Brodal A. *Neurological Anatomy*, 3rd ed. New York, NY: Oxford University Press; 1981.

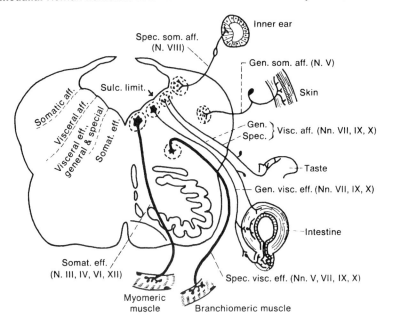

plained by the fact that neurons of the facial nucleus migrate from dorsal regions to a more ventral and lateral location during early development. The arching course of the facial nerve axons over the *abducens nucleus* traces the approximate route of this migration.

Table 2.1 summarizes the organization of afferent and efferent nuclei associated with different cranial nerves, and the nuclei that contribute to them. The exact location of the nuclei within the brainstem and their sensory and motor functions are considered in more detail in later chapters that consider the nuclei in the context of the sensory or motor system that they serve.

Organization of Components Related to Suprasegmental Functions

The suprasegmental functions of the brainstem are of several types. There are (1) structures that play a role in regulating segmental motor output (the reticular formation and the red nucleus), (2) structures that are involved in processing and transmitting sensory information to higher centers (the reticular formation and the inferior and superior colliculi), (3) nuclei that are related to the cerebellum (the olivary nuclei and red nucleus), and (4) nuclei that are components of the basal ganglia (the *substantia nigra*).

With this overview in mind, the organization of the brainstem can best be summarized

Table 2.1. Functional components of the cranial nerves

Number/cranial nerve	Component	Nucleus of origin or destination
I Olfactory nerve	SVE	Olfactory cortical structures
II Optic nerve	SSE	Lateral geniculate nucleus, superior colliculus
III Oculomotor nerve	GSE	Oculomotor nucleus
	GVE	Edinger-Westphal nucleus
IV Trochlear nerve	GSE	Trochlear nucleus
V Trigeminal nerve	SVE	Motor nucleus of V
	GSA	Main sensory and spinal nuclei of V
VI Abducens nerve	GSE	Abducens nucleus
VII Facial nerve	SVE	Facial nucleus
	GVE	Superior salivary nucleus
	GVA	Solitary nucleus
	SVA	Solitary nucleus
	GSA	Spinal nucleus of V
IX Glossopharyngeal nerve	SVE	Nucleus ambiguus
	GVE	Inferior salivary nucleus
	SVA	Solitary nucleus
	GSA	Spinal nucleus of V
X Vagus nerve	SVE	Nucleus ambiguus
	GVE	Motor nucleus of X
	GVA	Solitary nucleus
	SVA	Solitary nucleus
	GSA	Spinal nucleus of V
XI Accessory nerve	SVE	Nucleus ambiguus and accessory nucleus
XII Hypoglossal nerve	GSE	Hypoglossal nucleus

by considering a series of sections through the different rostrocaudal levels (Figures 2.8 through 2.11). You will learn the organization of the different functional systems in subsequent chapters. Then you can return to the illustrations of this chapter to review the spatial relationships between different nuclei and tracts. By the end of your course of study, you should be able to identify, in histological sections or drawings, each structure, nucleus, and tract indicated by italics.

The Spinomedullary Junction

At the *spinomedullary junction,* features of the spinal cord can still be seen together with features of the medulla (Figure 2.8). The cuneate and gracile fasciculi (the dorsal columns) are still visible; however, one can also see portions of the cuneate and gracile nuclei where the fibers of the cuneate and gracile fasciculi terminate. The lateral columns are also present; but the most prominent landmark of the region is the pyramidal decussation in which the fibers in the pyramids cross the midline in a broad band and join the lateral corticospinal tract in the lateral columns. The caudal portions of the *medullary pyramids* themselves are also visible.

The gray matter at the level of the spinomedullary junction is crisscrossed by the fibers of the pyramidal decussation. The lateral portions and ventral-most portions of the ventral horn can still be discerned, but the central portions are largely obscured by the crossing fibers. The dorsal horn is greatly augmented due to the presence of the gray matter of the *spinal trigeminal nucleus.*

The Medulla at the Level of the Sensory Decussation

Further rostrally, the principal features of the medulla emerge. At this level, the ventral portion of the medulla is occupied by the pyramids. The lateral regions are occupied by descending fiber tracts that enter the cord and by the spinal tract of the trigeminal nucleus (cranial nerve V). A characteristic landmark of this portion of the medulla is the *de-cussation of the medial lemniscus,* the point at which the axons from the cuneate and gracile nuclei cross the midline to ascend in the *medial lemniscus.* These fibers follow a broad, curving course from the cuneate and gracile nuclei, crossing the midline to enter the medial lemniscus on the opposite side of the brain. These fibers are termed the *internal arcuate fibers.*

The principal gray matter regions include the cuneate and gracile nuclei in the dorsomedial portion of the medulla, the spinal trigeminal nucleus that occupies a position that is reminiscent of the dorsal horn of the spinal cord, and a prominent central gray zone between the gray masses of the cuneate nucleus. The more ventrolateral gray zones contain substantial numbers of myelinated axons and thus appear as a mixed gray/white matter region termed the *medullary reticular formation,* so named because the mixture of nuclei and fibers makes the region appear reticulated.

The Medulla at the Level of the Olive

Further rostrally, the dorsal surface of the medulla opens into the fourth ventricle. At this level, the most prominent white matter regions include the *inferior cerebellar peduncle* on the dorsolateral margin of the medulla, the pyramids on the basal surface, and the prominent medial lemniscus, which now forms a thick column of white matter that occupies the central core of the ventral medulla.

Figure 2.9 shows the most prominent gray matter regions, including the *inferior olivary nucleus* in the ventrolateral medulla (the nucleus that underlies the olive) and the gray matter of the reticular formation in the dorsolateral portions of the medulla. Lateral to the reticular formation is the spinal trigeminal nucleus.

The dorsal portion of the medulla (the region forming the floor of the fourth ventricle) is occupied by several cranial nerve nuclei: the *hypoglossal nucleus,* near the midline; the *dorsal motor nucleus* of the vagus nerve,

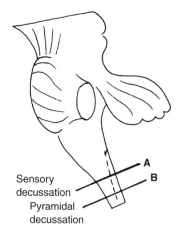

Sensory
decussation
Pyramidal
decussation

Figure 2.8. The spinomedullary junction.
Decussation of the medial lemniscus; pyramidal
decussation. **A,** Cross section at the level of the
sensory decussation; **B,** Cross section at the
level of the motor decussation. From Heimer L.
The Human Brain and Spinal Cord. New York, NY:
Springer-Verlag; 1994. The photomicrographs are
from Gluhbegovic N, Williams TH. *The Human
Brain: A Photographic Guide.* Philadelphia, Pa: JB
Lippincott; 1980.

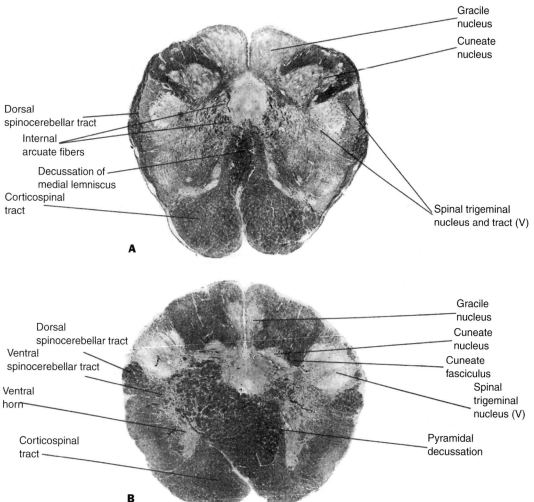

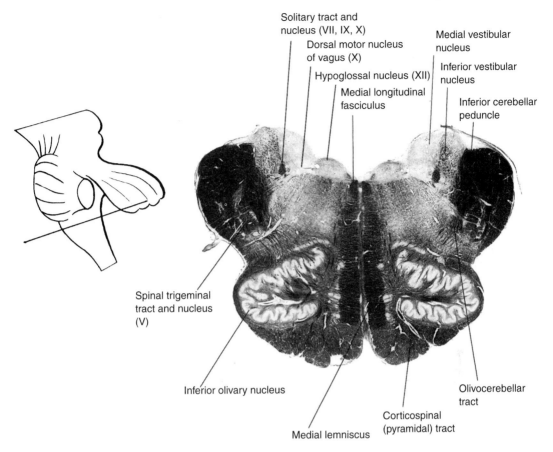

Figure 2.9. The medulla at the level of the olive.
From Heimer L. *The Human Brain and Spinal Cord.* New York, NY: Springer-Verlag; 1994. The photomicrograph is from DeArmond S, Fresco M, Dewey M. *Structure of the Human Brain: A Photographic Atlas.* New York, NY: Oxford University Press; 1976.

just lateral to the hypoglossal nucleus; and the *vestibular nuclei,* beneath the dorsolateral wall of the fourth ventricle. The *solitary nucleus and tract* lie just ventral to the vestibular nuclei.

The Pons and Middle Cerebellar Peduncle

Figure 2.10 illustrates a cross section through the brainstem at the level of the middle cerebellar peduncle. The cellular organization of the brainstem at the level of the pons can be understood easily if one keeps in mind that the cerebellum and its associated pathways form around the basic core structure that is a continuation of the

medulla. That core structure is termed the *tegmentum of the pons.* Sitting upon the tegmentum is the massive structure of the cerebellum. The fiber tracts passing between cerebellum and the brainstem are termed the *cerebellar peduncles.* The pons itself is a massive collection of fibers beneath the tegmentum that continue dorsally as the middle cerebellar peduncle. Hence, it is helpful to consider the cellular organization of the tegmentum, cerebellum, and pons separately.

The gray matter in the *pontine tegmentum* is occupied by the reticular formation; the dorsal region that forms the floor of the fourth ventricle is occupied by cranial nerve nuclei (the vestibular nuclei laterally and the abducens nucleus medially). The facial nucleus lies in lateral tegmentum. Fibers of the facial nerve emerge from the dorsal margin of the nucleus and arch dorsomedially over and around the abducens nucleus and then travel ventrolaterally to exit the brainstem. The bump in the floor of the fourth ventricle formed by the abducens nucleus and facial nerve is called the *facial colliculus.*

The white matter of the pons is made up of transverse and longitudinal fiber systems. The transverse fibers in the pons continue into the middle cerebellar peduncle, forming the *pontocerebellar tracts.* Most of the longitudinal fibers are from the corticospinal tract. One component of the longitudinal fibers leaves the descending tracts to terminate in scattered collections of gray matter within the pons, forming the *pontine nuclei.* These are the fibers of the *corticopontine tract.*

The Cerebellum

The cerebellum is virtually a separate "little brain." It is made up of a laminated cortex and nuclei in the core called *deep cerebellar*

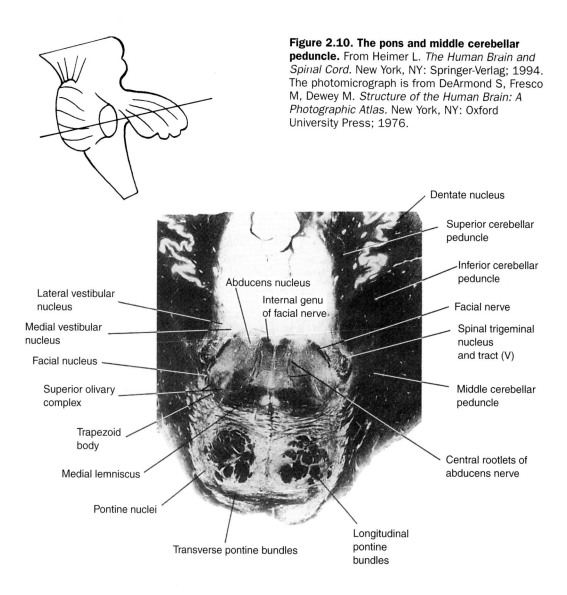

Figure 2.10. The pons and middle cerebellar peduncle. From Heimer L. *The Human Brain and Spinal Cord.* New York, NY: Springer-Verlag; 1994. The photomicrograph is from DeArmond S, Fresco M, Dewey M. *Structure of the Human Brain: A Photographic Atlas.* New York, NY: Oxford University Press; 1976.

nuclei. The cerebellum plays a key role in regulating motor output. Axons conveying information to and from the cerebellum are located in three pairs of cerebellar peduncles: the inferior, middle, and superior cerebellar peduncles.

The cerebellum fills the space overlying the brainstem ventral to the *tentorium* of the dura mater. The space is termed the *posterior fossa;* the larger space rostral to the tentorium, which contains the diencephalon and telencephalon, is the *anterior fossa.* The cerebellum is attached to the brainstem by the three sets of cerebellar peduncles. The fourth ventricle lies between the ventral cerebellum and the brainstem (Figure 2.6).

Internal Organization of the Cerebellum

The cerebellum is made up of a cortex and a collection of deep cerebellar nuclei, which we consider in more detail in Chapter 17. The elongated miniature gyri running horizontally across the cerebellar cortex are termed *folia.* Each folium is made up of a fold of gray matter surrounding a central core of white matter (Figure 2.6).

The Midbrain

The midbrain is made up of a tegmentum and a tectum (roof) represented by the inferior and superior colliculi. The central canal marks the boundary between the tectum and tegmentum (Figure 2.11).

Cellular Regions

The cellular regions of the tegmentum include cranial nerve nuclei and groups of neurons collected in a *mesencephalic reticular formation.* One of the landmarks of the region is a large nucleus termed the *red nucleus* that gives rise to axons that project into the spinal cord, which are important for descending motor control. Another important cell group in the midbrain is the substantia nigra, which lies just ventral to the red nucleus. Neurons in the substantia nigra project to the basal ganglia. Surrounding the central canal is a collection of neurons termed the *periaqueductal gray matter.*

The tectum is made up of two paired structures termed the *inferior colliculus* and *superior colliculus.* The inferior colliculus is a relay nucleus for auditory information. The superior colliculus receives input from the eye and plays a key role in coordinating eye movements for visual tracking

Cranial nerve nuclei of the midbrain include the *trochlear nucleus* in the caudal midbrain and the *oculomotor complex* in the rostral midbrain. The oculomotor complex, which includes the *oculomotor nucleus* and the *Edinger–Westphal nucleus,* lies near the midline, just above the medial longitudinal fasciculus. The root of cranial nerve III, which carries the axons of both nuclei, can be followed through the substance of the midbrain to its exit point between the cerebral peduncles (the interpeduncular fossa).

Fiber Tracts

The principal white matter tract within the midbrain is the massive *basis pedunculi* (base of the peduncle), also called the *crus cerebri,* which contains descending fibers of the corticospinal tract. The tegmentum and basal portion of the midbrain are together termed the *cerebral peduncle.* Although, used properly, the term *cerebral peduncle* includes the tegmentum, the term is often used casually to refer to the white matter of the basis pedunculi.

The other important white matter tracts are in the tegmentum and include the medial lemniscus, located ventrolaterally, and the lateral lemniscus, which carries ascending fibers from auditory relay nuclei to the inferior colliculus. A small but important tract is evident near the midline: the *medial longitudinal fasciculus;* this tract carries ascending and descending connections between cranial nerve nuclei controlling eye movement (see Chapter 21).

THE DIENCEPHALON

The diencephalon has two major parts (the hypothalamus and thalamus) and a minor

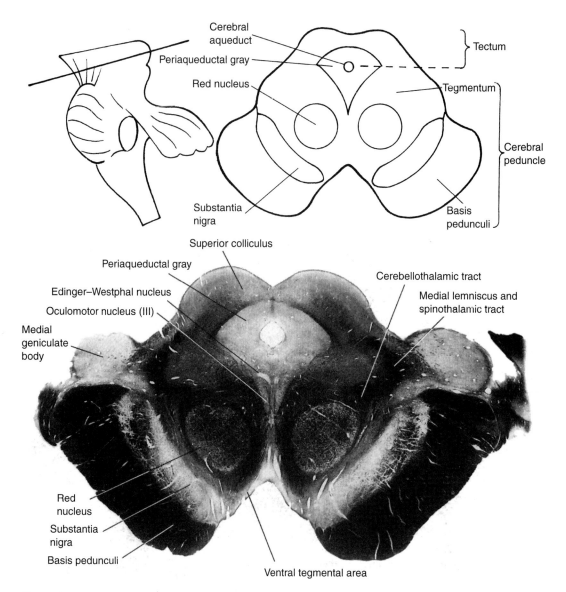

Figure 2.11. The midbrain at the level of the superior colliculus. From Heimer L. *The Human Brain and Spinal Cord.* New York, NY: Springer-Verlag; 1994. The photomicrograph is from Nauta WH, Feirtag M. *Fundamental Neuroanatomy.* New York, NY: WH Freeman; 1996.

part (the epithalamus), which is made up of the pineal gland and a structure termed the *habenula.* The basic structural organization of the diencephalon can be seen in the midsagittal section illustrated in Figure 2.6.

The caudal boundary of the diencephalon is at the caudal margins of the mammillary nuclei. On the dorsal surface, the boundary is between the superior colliculus, which is

part of the mesencephalon, and the *pineal gland,* which is part of the diencephalon (see Figure 2.6).

The rostromedial boundary of the diencephalon can be indicated by a line drawn between the interventricular foramen and the optic chiasm. More laterally, the diencephalon merges with the *ventral forebrain,* which is part of the telencephalon.

The dorsal boundary of the diencephalon is the third ventricle as it extends over the ovoid mass of the thalamus.

The lateral boundary of the diencephalon is formed by the fibers of the *internal capsule,* which is a thick, fan-shaped band of white matter that contains axons passing to and from the cortex.

At the level of the diencephalon, the cerebral aqueduct opens up into the third ventricle. The floor of the ventricle is formed by the hypothalamus. The walls of the ventricle are formed by the thalamus.

The Thalamus

The thalamus is a collection of nuclei, all of which project to the cerebral cortex. The thalamus is divided into medial, anterior, and lateral subdivisions by a thin lamina of white matter termed the *internal medullary lamina.* Small collections of neurons are embedded in the laminae (the *intralaminar nuclei*). Encapsulating the thalamus is a thin band of white matter termed the *external medullary lamina.* Divisions and nuclei are indicated in Figure 2.12.

Different thalamic nuclei receive input from different sensory systems. The ventral tier of nuclei (ventral posterior medial, ventral posterior lateral) receive somatosensory input. The lateral geniculate nucleus receives visual input. The medial geniculate nucleus receives auditory input. These are termed *specific sensory relay nuclei.* Other thalamic nuclei receive a mixture of inputs.

Each of the different nuclei within the thalamus projects to a different part of the cerebral cortex. In turn, the cortex sends a

Figure 2.12. Organization of the thalamus. Three-dimensional view of the thalamus. Major nuclei are indicated. The inset indicates the location of the thalamus within the brain. Reproduced with permission from Martin JH. *Neuroanatomy Text and Atlas.* New York, NY: Elsevier; 1989.

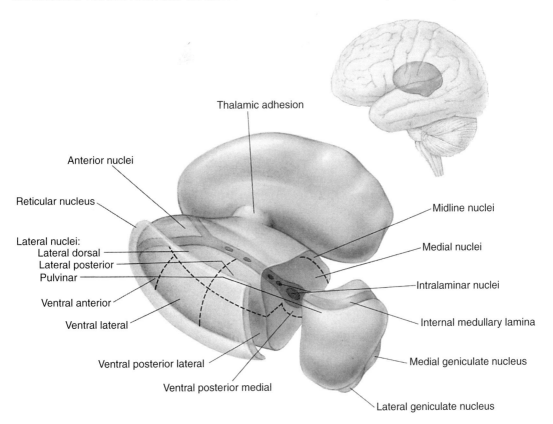

reciprocal projection back to the respective thalamic nucleus. The interconnections between the thalamus and the cortex travel through the *internal capsule;* all thalamocortical and corticothalamic connections are ipsilateral.

The Hypothalamus

The hypothalamus is made up of a complex matrix of nuclei and fibers (Chapter 28). Extending from the base of the hypothalamus is the stalk of the pituitary (the *infundibular stalk*). The stalk carries important axonal projections from nuclei in the hypothalamus to the posterior pituitary. The stalk also contains the vasculature that constitutes the *hypothalamo–hypophyseal portal system*. This important vascular system carries the hypothalamic releasing factors that are released into the portal system by specialized nerve endings and carried to the anterior pituitary, where they stimulate the release of their target hormones (see Chapter 28).

The Epithalamus

The epithalamus includes the pineal gland and the habenula. The pineal gland is a midline structure that is composed of glial cells and specialized parenchymal cells *(pinealocytes)*. It receives its innervation from the *superior cervical ganglion.*

Associated with the habenula are three small fiber tracts, all of which are visible landmarks in midsagittal views of the diencephalon (see Figure 2.6). The *stria medullaris thalami* is a small white matter tract that extends rostrally from the habenula, forming a ridge along the dorsomedial thalamus. The stria medullaris thalami carries fibers from the *septal nuclei* (which lie at the base of the septum pellucidum) to the *habenular nuclei.* The *habenular commissure* is a small bundle of axons that crosses the midline in the rostral portion of the base of the pineal gland, carrying crossing fibers of the stria medullaris to the contralateral habenular nucleus. The *pos-*

terior commissure is a small bundle of axons that crosses the midline in the caudal base of the pineal gland, carrying projections between the two superior colliculi and also fibers interconnecting the two sides of the pretectum.

THE TELENCEPHALON

The telencephalon is made up of the cerebral cortex and the basal ganglia, which include the *caudate nucleus, putamen nucleus,* and *globus pallidus,* and also includes components of the limbic system (the *hippocampus, amygdala,* and *septum*).

The Cerebral Cortex

The largest subdivision of the cerebral cortex is the *neocortex,* also called the *isocortex.* There are five lobes of the neocortex: the *frontal, parietal, occipital, temporal,* and *insular,* the last of which is covered over by the operculi of the other lobes (see Chapter 1 and Figure 2.13). Students should familiarize themselves with the principal sulci and gyri indicated in Figure 2.14.

Cellular Organization of the Cortex

The cortex is made up of layers that contain different populations of nerve cell bodies and their axons and dendrites. The neocortex, which occupies most of the cortex in humans, has six layers. The term *allocortex* refers to cortical areas that have fewer than six layers. Allocortex includes *archicortex* (hippocampus) and *paleocortex,* which includes olfactory cortex and cortical regions in the temporal lobe that project to the hippocampus.

The layers of the neocortex are designated according to the principal cellular constituents (Figure 2.15):

- Layer I, the *molecular layer,* contains dendrites of neurons whose cell bodies lie in deeper

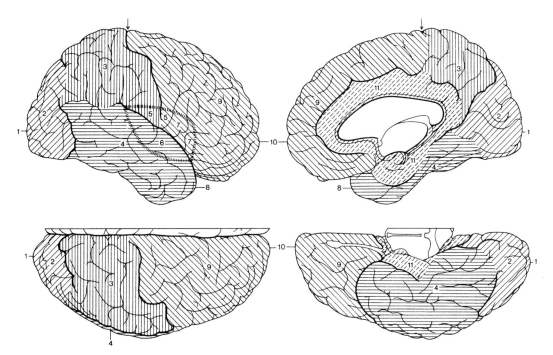

Figure 2.13. The five lobes of the cerebral hemisphere. The frontal lobe lies rostral to the *central sulcus* (9 and 10). The central sulcus extends along the lateral surface of the cortex from the midline to the *lateral fissure* (also called the *Sylvian fissure*), which is the boundary between the frontal lobe and the temporal lobe. The parietal lobe lies caudal to the central sulcus (3). The occipital lobe (2) is separated from the parietal lobe by the *parieto-occipital fissure.* The temporal lobe lies below the lateral fissure (4 and 6). The insular lobe (or *insula)* lies deep within the lateral fissure. The insula cannot be seen unless the portions of the frontal, parietal, and temporal lobes that overlie the insula are removed (the opercula, or coverings). The dashed line indicates the approximate outline of the *circular sulcus,* which separates the insula from the surrounding portions of the cortex. 11 indicates the *cingulate gyrus* (above the corpus callosum) and the hippocampus in the medial part of the temporal lobe. From Nieuwenhuys R, Voogd J, van Huijzen C. *The Human Central Nervous System.* Berlin: Springer-Verlag; 1988.

layers and axons that form synaptic connections with the dendrites.

- Layer II, the *external granular layer,* contains the cell bodies of small neurons whose dendrites radiate in a stellate pattern. The cells are termed *granule neurons* or *stellate neurons.*
- Layer III, the *external pyramidal layer,* contains the cell bodies of small pyramidal neurons, so named because their cell bodies have a pyramidal shape. The dendrites of layer III pyramidal neurons extend to layer I.
- Layer IV, the *internal granular layer,* contains the cell bodies of small neurons whose dendrites radiate in a stellate pattern.
- Layer V, the *internal pyramidal layer,* contains the cell bodies of large pyramidal neurons. The dendrites of layer V pyramidal neurons extend to layer I.

- Layer VI, the *multiform layer,* contains neurons with a variety of forms.

Cytoarchitectonic Divisions of the Cortex: Brodmann's Areas

Although all parts of the neocortex have six layers, certain layers are more prominent in some cortical regions. These differences form the basis for dividing the cortex into different *cytoarchitectonic regions (Brodmann's areas)* (Figure 2.16).

The different cytoarchitectonic areas of the cortex correspond to functional areas. The areas that are most often referred to by number include the following:

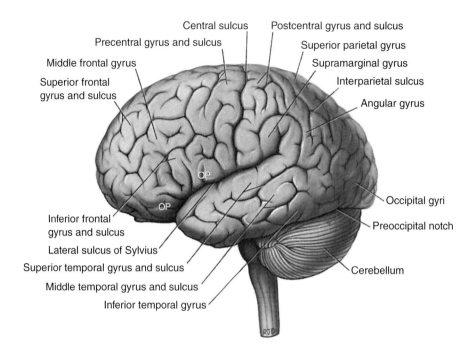

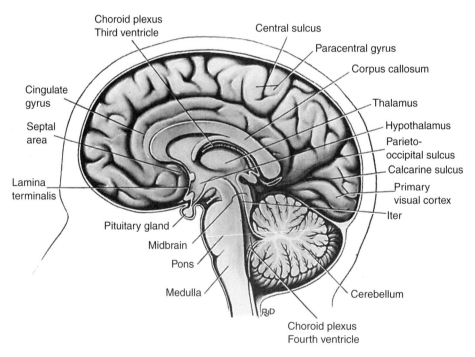

Figure 2.14. Principal sulci and gyri. Reproduced with permission from Noback CR, Demarest RJ. *The Human Nervous System.* New York, NY: McGraw-Hill; 1975.

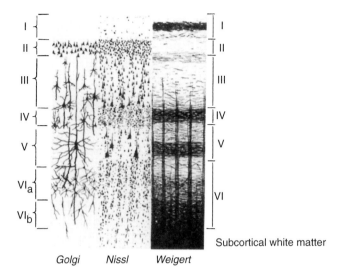

I
II
III
IV
V
VI_a
VI_b

I
II
III
IV
V
VI

Subcortical white matter

Golgi   Nissl   Weigert

Figure 2.15. Cellular organization of the neocortex. The cellular organization of the cerebral cortex as it would be seen using three different histological stains: the Golgi stain, which stains neurons in their entirety; a Nissl stain, which stains neuronal cell bodies; and the Weigert stain, which stains axons. Reproduced with permission from Carpenter MB. *Human Neuroanatomy.* Baltimore, Md: Williams & Wilkins; 1976; after Brodmann K. *Vergleichende Lokalisationlehre der Grosshirnrinde in ihren Prinzipien Dargestellt auf Grund des Zellenbaues.* Leipzig: JA Barth; 1909.

- Areas 1,2, and 3: primary somatosensory cortex
- Area 4: motor cortex
- Areas 5 and 6: premotor and supplementary motor cortex
- Area 17: primary visual cortex
- Areas 18 and 19: visual association cortex
- Area 41: auditory cortex

There are two fiber pathways, termed *commissures,* that connect the two halves of the cortex. The most prominent of the white matter tracts interconnecting the two halves of the cortex is the *corpus callosum.* Midline views (see Figure 2.6) reveal the extent of this thick fiber system. The *splenium* is the bulb-shaped posterior portion of the corpus callosum, and the *body* extends from the splenium to the point that the corpus callosum makes a sharp bend. The bend itself is termed the *genu,* and the thin ventral continuation of the corpus callosum is termed the *rostrum.* A thin lamina called the *lamina terminalis* extends from the rostrum toward the base of the brain, ending near the optic chiasm.

The other fiber tract connecting the two hemispheres is the *anterior commissure,* which crosses the midline between the rostrum of the corpus callosum and the descending column of the fornix.

Another prominent landmark of the medial wall of the telencephalon is a thin lamina termed the *septum pellucidum,* which extends between the corpus callosum and the fornix. This thin sheet of tissue separates the two lateral ventricles. At the base of the septum pellucidum lie the *septal nuclei,* which are part of the limbic system.

Cortical Inputs and Outputs

The cortex receives inputs from thalamic nuclei and certain brainstem nuclear groups. However, many of the inputs are from other cortical regions. The outputs from the cortex derive primarily from pyramidal neurons. These are divided into three types:

1. *Projection fibers* that leave the cortex entirely and project to noncortical areas
2. *Ipsilateral association fibers* that project to other cortical areas on the same side of the brain
3. *Commissural fibers* that project to the opposite side via the corpus callosum

Projection fibers include the following types:

1. Corticothalamic
2. Corticospinal and corticobulbar (motor)
3. Cortical projections to various brainstem nuclei (mostly motor control)
4. Cortical projections to the basal ganglia

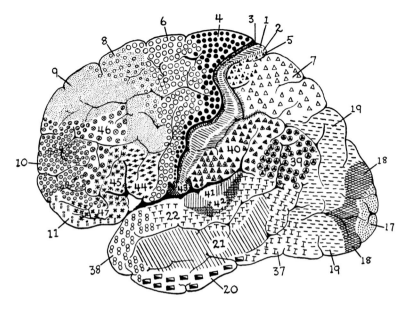

Figure 2.16.
Brodmann's areas. The cytoarchitectonic map of the human cortex constructed by Brodmann. Reproduced with permission from Carpenter, M.B. *Human Neuroanatomy.* Baltimore, Md: Williams & Wilkins; 1976; after Brodmann K. *Vergleichende Lokalisationlehre der Grosshirnrinde in ihren Prinzipien Dargestellt auf Grund des Zellenbaues.* Leipzig: JA Barth; 1909.

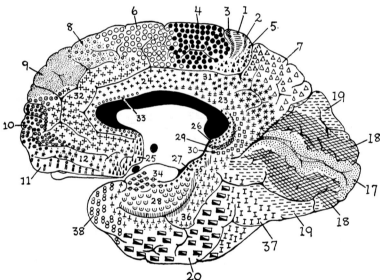

The cortical efferents listed in (1) through (3) project via the *internal capsule.* Cortical projections to the basal ganglia travel via the *external capsule.*

The Telencephalic Components of the Basal Ganglia

Deep in the cortex are the telencephalic components of the basal ganglia, which include the caudate nucleus, the putamen, and the globus pallidus (Figure 2.17). These nuclear structures lie between the cortical white matter and the diencephalon. Also included in the basal ganglia are the subthalamus of the diencephalon and the substantia nigra of the midbrain. These latter are considered part of the basal ganglia because these nuclei, together with the caudate nucleus, putamen, and globus pallidus, represent a system of in-

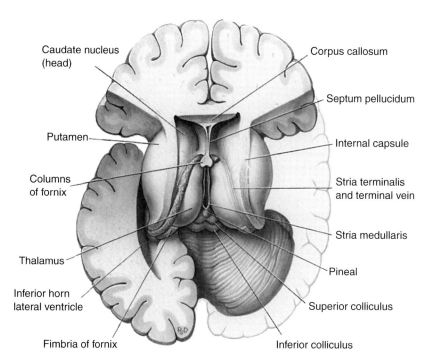

Figure 2.17. Location of the basal ganglia. A brain dissection that reveals the location of the thalamus, internal capsule, basal ganglia, and ventricular system. Reproduced with permission from Carpenter MB. *Human Neuroanatomy.* Baltimore, Md: Williams & Wilkins; 1976.

Caudate nucleus (head)

Corpus callosum

Putamen

Septum pellucidum

Columns of fornix

Internal capsule

Thalamus

Stria terminalis and terminal vein

Inferior horn lateral ventricle

Stria medullaris

Fimbria of fornix

Pineal

Superior colliculus

Inferior colliculus

terconnected cell groups that play a key role in the control of movement.

The cellular organization of the basal ganglia will be considered in more detail in Chapter 16.

The Telencephalic Components of the Limbic System

The telencephalic components of the limbic system include the hippocampus, the amygdala, and the septal nuclei.

The hippocampus is found in the core of the temporal lobe. It sends a prominent fiber projection to the mammillary nuclei via the fornix. It receives a massive projection from the cortex that originates in the *parahippocampal gyrus.*

The amygdala is a nuclear group embedded within the rostral tip of the temporal lobe. The amygdala gives rise to the fiber tract called the *stria terminalis,* which curves around the thalamus, following the boundary between the caudate nucleus and the thalamus.

The *septal nuclei* are a small collection of nuclei located within the base of the *septum pellucidum*. They receive ascending input via the medial forebrain bundle and give rise to ascending projections to the habenula (which travel via the stria medullaris thalami) and the hippocampus (which travel via the fornix).

3

Central Nervous System Vasculature, Blood–Brain Barrier, and Cerebrospinal Fluid

The proper operation of the nervous system depends on two fluid systems: the vasculature and the cerebrospinal fluid (CSF). The importance of the vasculature is self-evident. Like any other tissue, the cells of the brain depend on a continuous supply of oxygenated blood. Interruption of the blood supply to any part of the nervous system *(ischemia)* leads to rapid neuronal death. This is true for nerves and fiber tracts as well as for areas containing neuronal cell bodies. Insufficiency of blood supply can lead to transient cellular dysfunction. It is important for the clinician to know the distribution of the major vessels, and the signs and symptoms that result from an interruption of blood flow in the different vessels.

This chapter outlines the distribution of the vessels. A thorough understanding of the symptoms that result from disruption of blood flow requires an understanding of the function of the regions that are supplied, which is provided in subsequent chapters.

The other important fluid system of the brain is the (CSF) system. Disruption of CSF flow or absorp-

tion leads to a damming up of the CSF in the ventricles, which in turn leads to increases in intracranial pressure. It is important for the physician to be able to recognize the signs of such blockade so that treatment can be initiated to minimize damage to neural tissue.

ARTERIAL SUPPLY OF THE BRAIN

The blood supply to the brain derives from the paired internal carotid and vertebral arteries (Figure 3.1); these in turn give rise to major branches that supply different parts of the brain. The area supplied by a given branch artery is termed that artery's *arterial territory* and represents the area that is compromised if blood flow in that artery is interrupted.

It is useful to divide our consideration on the basis of the systems that originate from the two major supplies. Hence, the *internal*

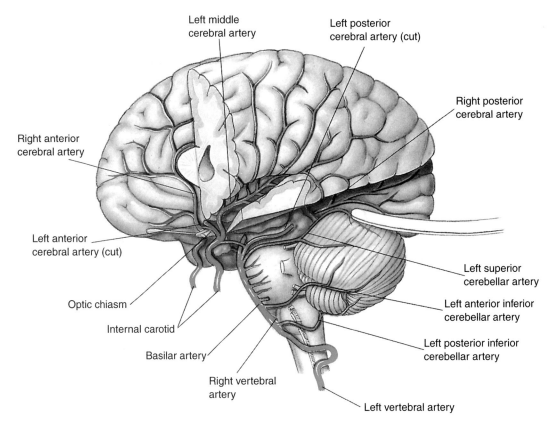

Figure 3.1. The arterial supply to the brain. The blood supply to the brain derives from the two major arterial systems; the internal carotid system is the set of arteries that derives from the internal carotid artery; the vertebrobasilar system is the set of arteries that derives from the vertebral arteries and from the basilar artery, into which the two vertebral arteries converge. Part of the left hemisphere has been removed in order to show the distribution of the vessels that lie deep within the hemisphere. From Heimer L. *The Human Brain and Spinal Cord*. New York, NY: Springer-Verlag; 1995.

carotid system is the set of arteries that derive from the internal carotid artery; the *vertebrobasilar system* is the set of arteries that derive from the *vertebral arteries* and from the *basilar artery* into which the two vertebral arteries converge. We also briefly consider how the two systems communicate with each other at the *circle of Willis,* via the posterior communicating arteries.

There is a recurring organizational theme in which the arterial supply of a particular level can be divided between a *medial supply* that derives from *paramedian vessels,* which penetrate at or near the midline, and a *lateral supply* that originates from *circumferential vessels.* This organization is especially prominent in the spinal cord and brainstem, but the basic pattern can also be seen in the arterial supply to forebrain regions.

The Vertebrobasilar System

The vertebral arteries arise from the subclavian artery and ascend along the spinal column to enter the cranium. They travel along the ventrolateral surface of the medulla and then fuse at the *pontomedullary junction* into the medially located basilar artery that lies at the base of the pons. Just before the vertebral arteries on the two sides fuse, each vertebral artery gives rise to two important branches: (1) the *anterior spinal artery* and (2) the *posterior inferior cerebellar (PICA) artery* (Figure 3.2). Physicians use the acronym PICA as a word (pronounced "Píka").

The anterior spinal arteries originate from the medial aspect of the vertebral arteries just before the vertebral arteries fuse into the basilar artery. The two anterior spinal arteries descend caudally and medially along the base of the medulla and then fuse into a centrally located *ventral spinal artery,* which continues along the ventromedial aspect of the spinal cord.

The medial versus lateral supply systems of the medulla are illustrated in Figure 3.3. The medial wedge of the medulla is supplied by branches from the anterior spinal arteries. These are termed *paramedian branches.* The next most lateral segment is supplied by pen-etrating branches from the vertebral arteries (Figure 3.3). The lateral portion of the medulla is supplied by a system of circumferential vessels that arise from the vertebral arteries.

The PICA originates as the last major branch of the vertebral arteries before they fuse into the basilar artery. Less commonly, the PICA originates from the basilar artery. The PICA on each side courses circumferentially around the medulla to supply the lateral aspect of the inferior cerebellum (see Figure 3.3). Branches also supply the *choroid plexus* of the fourth ventricle.

The basilar artery begins at the junction of the two vertebral arteries and then extends along the base of the pons until it divides into the posterior cerebral arteries (Figure 3.2). Along its course, the basilar

Figure 3.2. The distributions of the vertebrobasilar and internal carotid supply systems. The vertebral arteries arise from the subclavian artery, ascend along the spinal column to enter the cranium. They travel along the ventrolateral surface of the medulla and then fuse at the pontomedullary junction into the medially located basilar artery that lies at the base of the pons. The major branches from the system are described in the text. Students should be familiar with the location and distribution of all the named branches. Reproduced with permission from Martin JH. *Neuroanatomy Text and Atlas.* New York, NY: Elsevier; 1989.

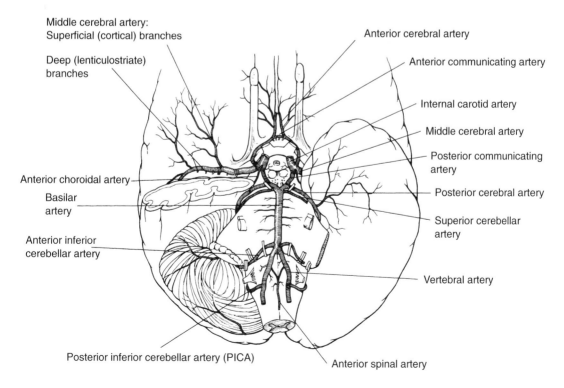

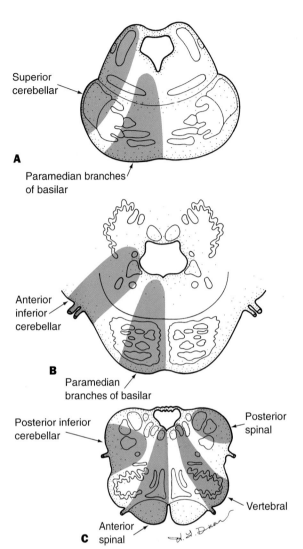

Superior
cerebellar

A

Paramedian branches
of basilar

Anterior
inferior
cerebellar

B
Paramedian
branches of basilar

Posterior inferior
cerebellar

Posterior
spinal

Vertebral

Anterior
spinal

C

Figure 3.3. Arterial territories in the medulla and pons. The medial versus lateral supply systems of the medulla are illustrated in **C**. The medial wedge of the medulla is supplied by branches from the anterior spinal arteries. The next most lateral segment is supplied by the vertebral arteries. The lateral portion of the medulla is supplied by a system of circumferential vessels, the largest of which are the posterior inferior cerebellar arteries. The same basic organization exists in the pons (**A** and **B**).

artery gives rise to (1) the *anterior inferior cerebellar artery (AICA);* (2) the *internal auditory artery,* (3) the *superior cerebellar artery,* and (4) the *posterior cerebral artery.* These are all paired arteries that supply more or less symmetrical regions on the two sides of the brain (lateral supply system). Along its course, the basilar artery also gives rise to small *pontine branches* that supply the medial pons (medial supply system).

The AICA arises near the caudal end of the basilar artery as the first major branch after the junction of the vertebral arteries. It ascends dorsally along the posterior margin of the middle cerebral peduncle to supply the inferior portion of the cerebellum around the point of entry of the peduncle.

The internal auditory artery arises either from the basilar artery or as a branch of the AICA. It supplies the inner ear via the internal auditory canal.

Pontine arteries arise as small branches from the basilar artery and penetrate into the overlying pons.

The superior cerebellar arteries arise from the rostral end of the basilar artery near the rostral end of the pons. These ascend along the rostral margin of the pons to supply the superior surface of the cerebellar hemisphere. A useful landmark in this area is the passage of the third cranial nerve between the superior cerebellar arteries and the posterior cerebral arteries.

The posterior cerebral arteries arise from the rostral terminus of the basilar artery. These arteries follow the posterior basal surface of the cerebral cortex and supply most of the occipital pole (Figure 3.4).

Thalamoperforating branches arise from the rostral end of the basilar artery and the initial segment of the posterior cerebral artery. These small arteries penetrate the brain through the *posterior perforated space* (which forms the roof of the *interpeduncular fossa*) to supply the mesencephalon, hypothalamus, and thalamus. These perforating arteries thus represent the medial supply system to the mesencephalon and diencephalon.

The *posterior communicating arteries* extend rostrally from the posterior cerebral arteries

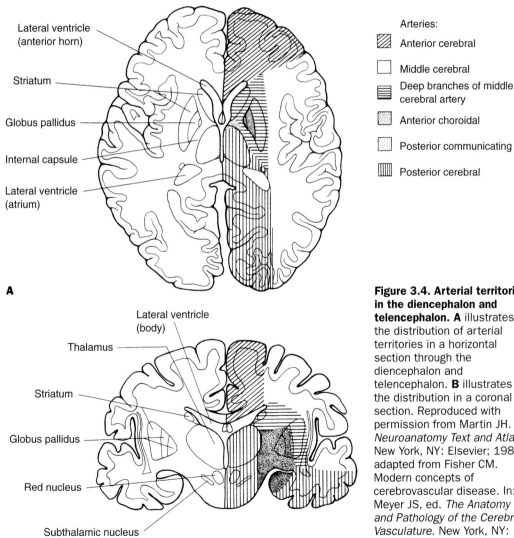

A

B

Figure 3.4. Arterial territories in the diencephalon and telencephalon. **A** illustrates the distribution of arterial territories in a horizontal section through the diencephalon and telencephalon. **B** illustrates the distribution in a coronal section. Reproduced with permission from Martin JH. *Neuroanatomy Text and Atlas.* New York, NY: Elsevier; 1989; adapted from Fisher CM. Modern concepts of cerebrovascular disease. In: Meyer JS, ed. *The Anatomy and Pathology of the Cerebral Vasculature.* New York, NY: Spectral Publications; 1975.

to anastomose with the internal carotid arteries, forming the posterior component of the circle of Willis (Figure 3.2).

The Internal Carotid System

Each internal carotid artery enters the cranial vault at the carotid foramen, turns sharply rostrally and crosses the cavernous sinus, then turns dorsally to penetrate the dura. The opthalmic artery branches from the carotid artery as it crosses the cavernous sinus and extends rostrally along the optic nerve to enter the orbit. After penetrating the dura, the main artery breaks up into a number of branches. The principal branches are the *anterior cerebral artery* and *middle cerebral artery* (MCA). Smaller branches include the posterior communicating artery, and the *anterior choroidal artery* (Figure 3.2).

The anterior cerebral arteries originate from the internal carotid artery, extend rostrally and medially toward one another (Figure 3.2), and then arch dorsally along the medial wall of the hemisphere, following the corpus callosum. Where the arteries from the two sides come together at the midline, there is a small communicating artery between them (the anterior communicating artery).

The proximal segment of the anterior cerebral artery gives rise to perforating branches that enter the brain at the anterior perforated space to supply the striatum *(striate arteries)* and anterior hypothalamus. The initial segment also gives rise to the *recurrent artery of Heubner,* which enters the brain at the anterior perforated space to supply the rostral limb of the internal capsule and nearby portions of the striatum. The rest of the striatum is supplied by the MCA (see later).

The main branch of the anterior cerebral artery follows the curvature of the corpus callosum along the medial wall of the cerebral cortex (Figure 3.5). Just rostral to the genu of the corpus callosum, the artery branches to give rise to the *pericallosal artery,* which follows the corpus callosum, and the *callosomarginal artery,* which travels along the dorsal boundary of the cingulate cortex. These two arteries are important landmarks in cerebral angiography.

The anterior cerebral artery supplies most of the medial wall of the cerebral cortex rostral to the parieto-occipital fissure and the anterior part of the corpus callosum (see Figure 3.4). The most obvious symptom of a stroke involving the anterior cerebral artery is paralysis of the contralateral leg and foot, owing to the disruption of blood flow to the portion of the motor cortex that supplies the lower limb.

The middle cerebral artery (MCA) is the largest of the branches of the internal carotid artery. The MCA extends laterally along the base of the hemisphere in the lateral fissure and then breaks up into a number of branches that supply the entire lateral part of the hemisphere (Figure 3.6). The arterial territory of the MCA is illustrated in Figure 3.4.

Because the MCA is the largest branch of the internal carotid artery, *cerebral emboli* (detached blood clots) conveyed by the internal carotid often enter and lodge within the MCA, blocking blood flow *(occlusive stroke)*. Occlusion of the stem of the MCA results in contralateral hemiplegia and hemianesthesia (because of damage to the primary somatosensory and motor cortices), and sometimes homonymous hemianopsia (because of damage to the optic radiations in the internal capsule and deep white matter). MCA occlusion on the left side may also lead to global aphasia (inability to speak or understand language) because of damage in the language areas of the hemisphere.

The proximal portion of the MCA gives rise to numerous small branches that penetrate the brain dorsally, especially at the anterior perforated space. These penetrating branches are called the *striate arteries,* or *lenticulostriate arteries;* they supply the striatum, globus pallidus, and anterior limb of internal capsule.

The striate arteries are a common site of cerebral hemorrhages. Symptoms depend on

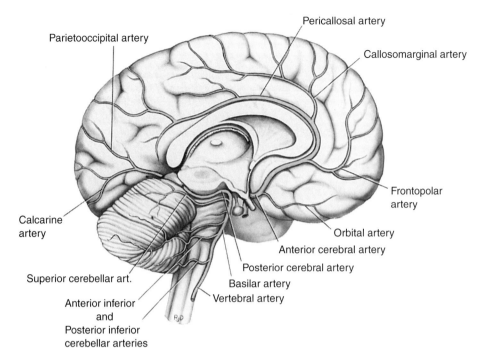

Figure 3.5. The arterial supply to the medial surface of the brain. The anterior cerebral artery follows the curvature of the corpus callosum along the medial wall of the cerebral cortex. Just rostral to the genu of the corpus callosum, the artery branches to give rise to the pericallosal artery, which follows the corpus callosum, and the callosomarginal artery, which travels along the dorsal boundary of the cingulate cortex. These two arteries are important landmarks in cerebral angiography. Reproduced with permission from Carpenter, M.B.; *Human Neuroanatomy.* Baltimore, Md: Williams & Wilkins; 1976.

which terminal branches are affected, but the most prominent symptoms are the result of damage to the internal capsule, which affects ascending sensory and descending motor pathways.

Anastomoses

There are *anastomoses* between the terminal branches of the major cerebral arteries. The extent of these varies. When anastomoses are prominent, they can protect the *shared border zones* or *watershed zones* from stroke.

Penetrating arteries usually lack anastomoses and are termed *end-arteries.* The deep regions they supply are termed *end-zones.* Occlusion of the end-arteries may cause small infarcts *(lacunae),* which result in selective symptoms.

Venous Drainage

From the capillary beds in the brain, veins collect and drain into a series of interconnected sinuses. Students should be familiar with the location of each of the sinuses as well as the *great vein of Galen* (Figure 3.7).

The *superior sagittal sinus* is located at the dorsal margin of the *falx cerebri* (the dural sheet that lies at the dorsal midline between the cerebral hemispheres). The *inferior sagittal sinus* is located at the ventral margin of the *falx cerebri.* The *straight sinus* extends caudally along the *tentorium cerebelli* (the dural

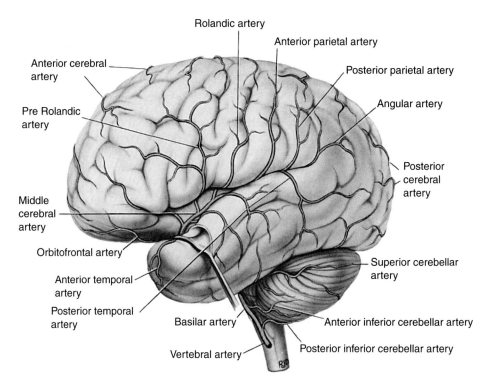

Figure 3.6. The arterial supply to the lateral surface of the brain. The MCA extends laterally along the base of the hemisphere in the lateral fissure and then breaks up into a number of branches that supply the entire lateral part of the hemisphere. Reproduced with permission from Carpenter, M.B.; *Human Neuroanatomy.* Baltimore, Md: Williams & Wilkins; 1976.

sheet between the cerebrum and cerebellum) to connect the inferior sagittal sinus with the *sinus confluens* at the occipital pole of the brain. The *transverse sinus* follows the posterior border of the tentorium cerebelli and then drains into the *sigmoid sinus*. The sigmoid sinus continues through the skull to drain into the *internal jugular vein*. The *cavernous sinus* lies at the base of the brain in the sella turcica, surrounding the pituitary gland. The *superior* and *inferior petrosal sinuses* extend from the cavernous sinus to the sigmoid sinus. The great vein of Galen receives venous blood from deep structures within the cerebrum and drains into the straight sinus.

CIRCULATION OF THE SPINAL CORD

The arterial supply to the spinal cord originates from two sources: (1) the vertebral arteries, which give rise to the anterior (ventral) spinal and posterior (dorsal) spinal system and (2) the radicular arteries (Figure 3.8). The vertebral supply is especially important for the cervical cord. The radicular supply provides circulation to each segmental level.

The Anterior (Ventral) Spinal and Posterior (Dorsal) Spinal Systems

As the anterior spinal arteries descend along the spinal cord, they give rise to two types of branches: (1) *medial branches,* which penetrate into the spinal cord along the *ventromedial fissure,* and (2) *circumferential branches,* which encircle the lateral aspect of the cord and give rise to small branches that penetrate into the white matter to supply the ventral and lateral portion of cord. In this way, the organization of the vascular supply from the

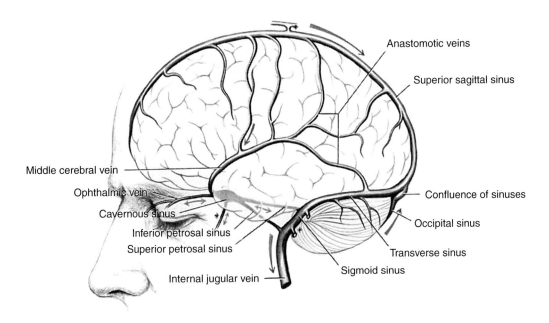

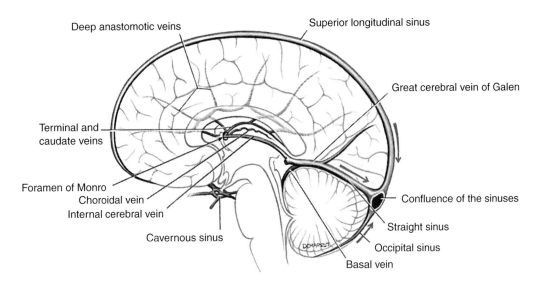

Figure 3.7. Organization of veins and sinuses. Reproduced with permission from Nobak CR, Demarest RJ. *The Human Nervous System: Basic* *Principles of Neurobiology*. New York, NY: McGraw-Hill; 1975.

anterior spinal artery again recapitulates the theme already described of medial versus lateral supplies.

The arterial supply of the dorsal portion of the spinal cord derives from the dorsal spinal artery. This artery branches off from the pos-

terior inferior cerebellar artery. The dorsal spinal arteries from the two sides wrap around the spinomedullary junction, travel together caudally along the dorsal surface of the spinal cord, and then fuse at low cervical levels. The fused artery then continues to

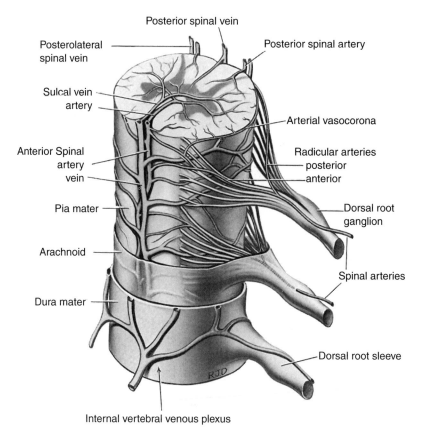

Figure 3.8. Arterial supply and venous drainage of the spinal cord. Reproduced with permission from Carpenter, M.B.; *Human Neuroanatomy.* Baltimore, Md: Williams & Wilkins; 1976.

midthoracic levels. Penetrating branches from the dorsal spinal artery supply the dorsal aspect of the spinal cord.

The Radicular Arteries

Radicular arteries are the major source of blood to the spinal cord. Radicular arteries follow the peripheral nerves into the vertebral canal, divide into dorsal and ventral branches to follow the dorsal and ventral roots, and then anastomose with the dorsal or ventral spinal arteries.

STROKE

Stroke, or *cerebrovascular accident,* refers to neurological symptoms and signs resulting from diseases involving blood vessels. Strokes are of two types: (1) *occlusive* (when blood flow is blocked) and (2) *hemorrhagic* (when there is bleeding from a vessel). The symptoms from hemorrhagic stroke are due both to the loss of blood supply and to the *space-occupying lesion* formed by the extravasated blood.

Ischemia refers to an insufficiency of blood supply, usually due to hypoperfusion. Ischemia deprives tissue of oxygen and other nutrients (especially glucose) and prevents the removal of toxins. *Anoxia* refers to an oxygen deprivation only.

Strokes are often preceded by *transient ischemic attacks* (TIAs), which signal temporary hypoperfusion (usually as a result of temporary occlusion). Occlusion often results from sclerotic plaques. Plaques in some locations can be surgically removed, and thromboses can be treated medically. Thus, it is important to identify warning signals (such as TIAs) before damage becomes permanent.

Cerebral embolisms (detached blood clots) most often enter the MCA. Symptoms resulting from occlusion depend on which terminal branches are affected. Occlusion of the stem results in sensory and motor deficits affecting the contralateral side of the body (hemianesthesia and hemiplegia), visual deficits affecting the contralateral visual field of both eyes (homonymous hemianopsia), and language disorders (global aphasia) if the lesion is on the left side.

ARTERIOVENOUS MALFORMATIONS

Occasionally, the vasculature may develop abnormally so that arteries and veins have direct shunts between them made up of a tangle of vessels (*arteriovenous malformations* [abbreviated as AV malformations]). These vessels are often abnormally weak and subject to rupture. Arteriovenous malformations can be treated surgically with varying degrees of success.

Areas of weakness in a vessel wall can also balloon out under pressure, forming *berry aneurysms* (so called because of their shape). These are also subject to rupture. The most common site of berry aneurysms is the anterior portion of the circle of Willis (90% of the berry aneurysms are found there). These can be treated surgically by clipping the neck of the aneurysm.

CEREBROVASCULAR PHYSIOLOGY

Blood flow through cerebral vessels is regulated by extracellular signals. High levels of neuronal activity increase metabolic demand and lead to increases in blood flow as a result of changes in arteriole diameter. When P_{CO_2} increases, brain arterioles dilate. Conversely, increases in P_{CO_2} cause constriction of the arterioles. The signals that mediate these effects are thought to include local extracellu-

lar concentrations of potassium and adenosine, and changes in extracellular pH. The changes in metabolism that occur as a result of activity form the basis for positron emission tomography (PET) scans.

Hemorrhage often leads to *reactive vasospasm* of cerebral surface vessels, through mechanisms that are not entirely understood. The vasospasm leads to decreased blood flow and ischemic brain damage.

The Blood–Brain Barrier

The blood–brain barrier refers to the fact that there is limited diffusion between the blood and the interstitial fluid of the brain. Small lipophilic molecules can cross the blood–brain barrier directly. Other molecules that enter the brain do so as a result of specific transport systems (carrier-mediated transport). This is in contrast to the permeability characteristics of the vasculature in the remainder of the body, where a barrier system does not exist.

The blood–brain barrier is formed by capillary endothelial cells (Figure 3.9). In the CNS there are tight junctions between these cells. This is in contrast to the situation in the remainder of the body, where the vascular endothelial cells are either fenestrated or have low-resistance junctions. Also, endothelial cells of the brain exhibit little endocytosis and transcellular transport of materials from the vessel to the parenchyma.

The nature of the blood–brain barrier determines what drugs enter the brain, and drug therapies must take this into account. For example, one treatment for Parkinson's disease seeks to increase the concentrations of the neurotransmitter dopamine because it is the dopamine systems that are affected by the disease. Dopamine itself will not cross the blood–brain barrier, but its immediate precursor, L-dopa, will. Thus, therapy to increase dopamine concentrations involves administration of L-dopa.

Some areas of the brain do not have a blood–brain barrier. The barrier is absent in the pituitary, presumably because it would

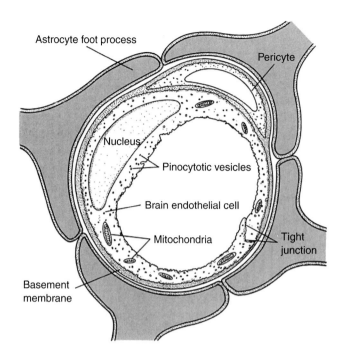

Astrocyte foot process

Pericyte

Nucleus

Pinocytotic vesicles

Brain endothelial cell

Mitochondria

Tight junction

Basement membrane

Figure 3.9. The blood–brain barrier.
The blood–brain barrier is formed by capillary endothelial cells. In the brain, there are tight junctions between these cells, in contrast with the endothelial cells surrounding peripheral vasculature, which are either fenestrated or have low-resistance junctions. From Kandel ER, Schwartz JH, Jessell TM. *Principles of Neural Science*, 3rd ed. New York, NY: Elsevier; 1991; adapted from Fishman RA. *Cerebrospinal Fluid in Diseases of the Nervous System*. Philadelphia, Pa: WB Saunders; 1980.

otherwise block the passage of neurosecretory products into the circulation. It is also absent around several circumventricular structures that are chemoreceptive, including the vascular organ of the lamina terminalis, the subfornical organ, the pineal gland, and the area postrema (Figure 3.10). Regions without a barrier are isolated from the rest of the brain by *tanycytes* (specialized astrocytes), which are coupled to one another by tight junctions.

The Cerebrospinal Fluid System

The ventricles and the subarachnoid space surrounding the surfaces of the brain and spinal cord are filled with CSF. CSF is produced in the choroid plexus in the ventricles, flows throughout the ventricular system, and then into the subarachnoid space at the *foramen of Magendie,* which is located medially at the caudal end of the fourth ventricle (Figure 3.10), and the foramina of Luschka, which open into the subarachnoid space from the lateral recess of the fourth ventricle (a useful mnemonic: Magendie is medial, Luschka is lateral).

CSF is reabsorbed at specialized structures termed *arachnoid granulations* that extend from the subarachnoid space into the dural sinuses. Arachnoid granulations are especially prominent in the superior sagittal sinus.

The *ionic composition* of the CSF is similar to that of blood serum. However, there are important differences. Most important, CSF normally contains protein at a concentration of less than 50 mg/mL (in contrast to a protein concentration in serum of 7000 mg/mL). Higher than normal protein concentrations in CSF are indicative of pathology, as are white or red blood cells.

Hydrocephalus

CSF flow is vectorial—from the ventricles where the CSF is produced, especially the lateral ventricles, into the subarachnoid space, and then into the dural sinuses, where the fluid is reabsorbed. When this flow is blocked, the result is a damming up of CSF leading to an increase in intracranial pressure and an expansion of the ventricles—a condition termed *hydrocephalus*. The increased

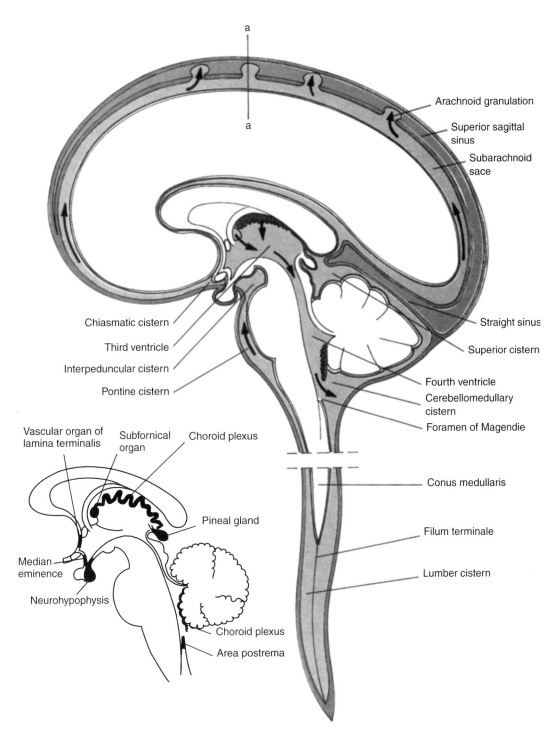

Figure 3.10. The cerebrospinal fluid system. The flow of CSF within the ventricles and subarachnoid space. The inset shows the locations of various circumferential organs that lack a blood–brain barrier. From Heimer L. *The Human Brain and Spinal Cord*. New York, NY: Springer-Verlag; 1995.

The inset is modified from Weindl A. Neuroendocrine aspects of circumventricular organs. In: Ganong WF, Martini L, eds. *Frontiers in Neuroendocrinology*. New York, NY: Oxford University Press; 1973.

pressure and ventricular expansion can damage brain tissue.

Two types of hydrocephalus are recognized: *noncommunicating* (or *obstructive*) and *communicating* (Figure 3.11). Noncommunicating hydrocephalus results from a blockage of the normal flow of CSF through the ventricular system. The most common site of blockage is the cerebral aqueduct, but there can also be blockage at the foramina of the fourth ventricle. The result is increased intraventricular pressure. Any condition in which there is a disruption of CSF reabsorption without a blockade of CSF flow through the ventricles is termed communicating hydrocephalus. Both forms of hydrocephalus cause increases in intracranial pressure, which is treated by implanting a *ventricular shunt*.

Hydrocephalus often occurs as a result of developmental anomalies or can occur as a result of the growth of a tumor in later life. When hydrocephalus occurs early in development, the cranium enlarges dramatically (Figure 3.11). When hydrocephalus occurs in later life, the cranium is fixed and so the in-

Figure 3.11. Hydrocephalus results from a blockade of CSF flow or reabsorption. A, Noncommunicating hydrocephalus results from a blockage of the normal flow of CSF through the ventricular system. The most common site of blockage is the cerebral aqueduct. **B,** Communicating hydrocephalus is any condition in which there is a disruption of CSF reabsorption without a blockade of CSF flow through the ventricles. Both forms of hydrocephalus cause increases in intracranial pressure. From Waxman SG, deGroot J. *Correlative Neuroanatomy*. Norwalk, Conn: Appleton & Lange; 1995.

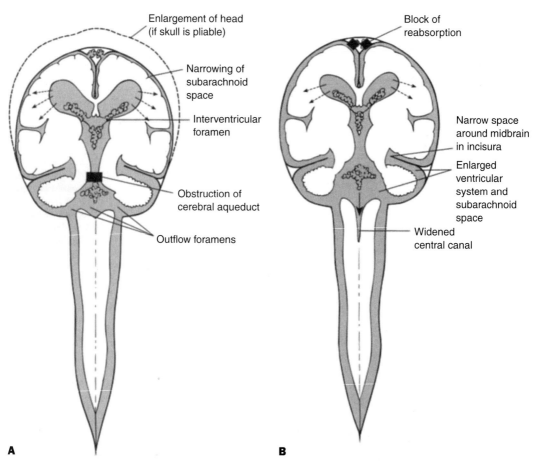

Enlargement of head
(if skull is pliable)

Narrowing of
subarachnoid
space

Interventricular
foramen

Obstruction of
cerebral aqueduct

Outflow foramens

Block of
reabsorption

Narrow space
around midbrain
in incisura

Enlarged
ventricular
system and
subarachnoid
space

Widened
central canal

A   **B**

creased pressure results in expansion of the ventricles with compression of brain tissue.

One poorly understood condition that is most often seen in aged individuals is the condition of *normal pressure hydrocephalus*. In this condition, there is an enlargement of the ventricles and a thinning of the cortical mantle without an accompanying increase in intracranial pressure. Although intracranial pressure is not abnormally high, these patients still sometimes benefit from a ventricular shunt.

It is important for the physician to always recall the hydrodynamic properties of the CSF and the brain. The brain and spinal cord and the CSF are enclosed in an inelastic container (the skull and vertebral column) with fixed internal walls (the dura and the skull). If pressure is elevated, sudden release of pressure at any site can lead to a shifting of the soft contents within the inelastic container. It is for this reason that a lumbar puncture can produce catastrophic results in a patient with elevated intracranial pressure associated with a focal mass in the cranial vault. Specifically, the cerebrum may shift down past the falx cerebelli, resulting in a herniation of the uncal region of the temporal lobe, which compresses the midbrain and cerebral peduncles (an *uncal herniation*), or the tonsils of the cerebellum may shift down through the foramen magnum, compressing the medulla and spinal cord (a *tonsillar herniation*).

2

Cellular Basis of Neuronal Signaling

4

Cell Biology of Neurons

Certainly the most important functions of the brain are the receipt, transmission, and processing of information. For this reason, much of our knowledge about neurons has to do with their signaling capabilities. Nevertheless, neurons are cells, and some of their most important functional properties arise from their cellular characteristics and from cell– cell interactions that are not directly related to signaling activities. For example, synapses play a key role in transmitting information from one cell to the next. However, they also have other functions in regulating the structure, functional properties, and very existence of the cells that they contact. These trophic or regulatory functions are of considerable importance for understanding the consequences of injury or disease of the nervous system, because injury or disease can result in a cascade of effects caused by altered intercellular regulation.

In this chapter we consider neurons as cells, focusing on those aspects of cellular existence that are unique in neurons, or that figure importantly in neuronal function. Special attention is given to those structural features that are directly related to physiological functions. We begin by summarizing the basic structural features of neurons that underlie neuronal function. We then consider the types of synaptic connections that neurons form with one another. Finally, we consider the neuronal cytoskeleton and the intracellular transport mechanisms that deliver the key proteins involved in neuronal function.

NEURONAL POLARITY

As discussed in Chapter 1, neurons *receive* input, *integrate* information, *transmit* information over distances via action potentials, and *communicate* with target cells via release of neurotransmitters.

The key point with respect to neuronal cell biology is that different parts of the neuron are specialized for its different functions. Figure 4.1 illustrates the principal parts of a typical neuron in the cerebral cortex. The key parts of the neuron are the (1) cell body, (2) dendrites, (3) axon, and (4) synaptic terminals of the axon, which contact other neurons. Mapping function to structure, we find that:

1. Dendrites and neuronal cell bodies are the primary receiving elements of the neuron and the site of integration.

2. The axon is specialized to transmit information over long distances.

3. Presynaptic terminals are specialized for communication with target cells through the release of neurotransmitters.

The statement that dendrites receive information and axons transmit information to other neurons is the *principle of dynamic polarization,* first codified by the most famous neuroanatomist of the nineteenth century, Santiago Ramón y Cajal. Modern neuroscientists now recognize that this inherent

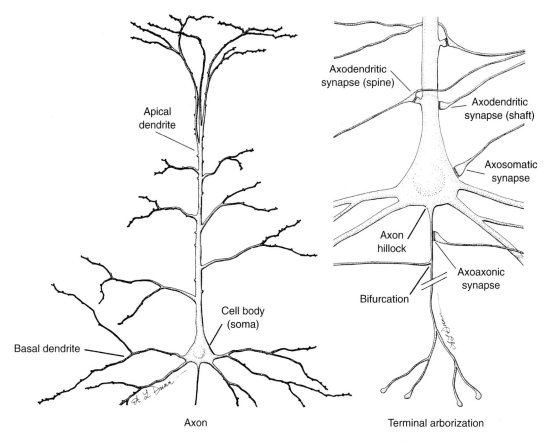

Figure 4.1. Diagram of the structural components and types of synaptic interactions on a typical CNS neuron (a pyramidal neuron in the cerebral cortex).

"functional polarity" of neurons is the result of the specialized structure and molecular architecture of each of the different structural "compartments."

NEURONAL CELL STRUCTURE AND ITS RELATION TO FUNCTION

Different types of neurons are categorized according to their form, especially the form and number of dendrites and axons (which are together called *neurites*). The simplest classification is based on the total number of neurites.

- Neurons that give rise to a single neurite are termed *unipolar;*
- Neurons that give rise to two neurites are *bipolar;*

- Neurons that give rise to multiple neurites are *multipolar.*

Unipolar neurons include dorsal root ganglion cells, which have a single process that *bifurcates* (gives rise to two branches). One of the two processes projects into the spinal cord; the other is the peripherally projecting sensory nerve. In fact, however, dorsal root ganglion cells are *pseudounipolar.* During development the cell is bipolar, and as the cell matures the two processes fuse.

Most neurons in the CNS are multipolar and exhibit a complex form that is highly characteristic for the cell type. The recognition of cell-type–specific differences in form was an important early discovery that helped neuroanatomists define the cellular organization of the nervous system (see Box).

It is remarkable that the recognition of the cellular organization of the nervous system, and essentially all the early information about interneuronal connections in the CNS, depended on the discovery of a technique that is still not understood—the *Golgi technique*. The Golgi technique was discovered accidentally in the mid 1800s by the Italian scientist Camilio Golgi and was used extensively by him and others in the field, including his rival Santiago Ramón y Cajal in Spain. The power of the Golgi stain comes from the fact that it stains individual neurons in their entirety but impregnates only a small percentage of the total number of neurons. As a result, the form of individual neurons can be recognized even when the neurons are embedded in a tissue environment heavily populated with neurons (Figures 4.2 and 4.3). Golgi and Cajal shared the Nobel Prize for Physiology and Medicine in 1911 for their pioneering discoveries about the cellular organization of the nervous system, made possible in large part by the Golgi stain. Interestingly, only one of the two (Cajal) actually believed that Schwann's *cell theory* actually applied to the nervous system, and that the elements being stained were individual cells. It is a quirk of history that Golgi was one of the strongest proponents of the rival *reticular theory*, which held that the nervous system was actually a continuous reticulum, not a collection of individual cells, and that neuronal cell bodies were nodes in this reticulum. This illustrates an important fact: scientists can make immensely important contributions to our knowledge even though their interpretations turn out to be incorrect.

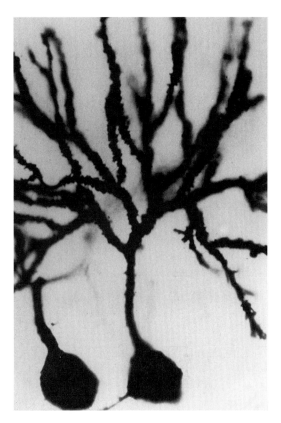

Figure 4.2. A cerebellar Purkinje cell stained by the Golgi method. The axon of the cell can be seen emerging from the lower left portion of the cell body. The dendrites extend into the molecular layer where incoming axons form synapses on the small knobs termed *spines*. Courtesy of E. W. Rubel, University of Washington.

Differences between neuron types that were first recognized on the basis of characteristic differences in either dendritic form or axonal arborization patterns are often correlated with important differences in connectivity or function (for example, the neurotransmitter used by particular cells).

As will be seen in later sections, the differences between neuron types that were first recognized on the basis of characteristic differences in either dendritic form or axonal arborization patterns have been confirmed and elucidated by the results of modern physiological and molecular research techniques. Usually, differences in form are correlated with important differences in connectivity or function (for example, which neurotransmitter is used by particular cells). Thus, as is so often the case, form implies function.

THE CYTOLOGY OF THE NERVE CELL

Almost all neurons have a similar underlying cellular organization, and this organization plays a key role in neuronal function.

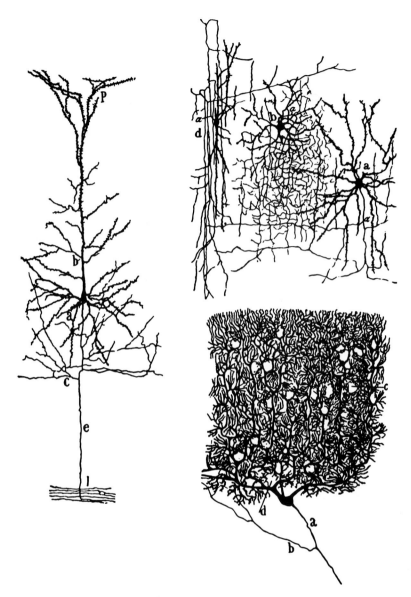

Figure 4.3. Different types of neurons can be recognized on the basis of their characteristic form. These are drawings of different types of cells stained by the Golgi method. On the left is a pyramidal cell of the cerebral cortex. Above right are several examples of short-axon (Golgi type II) cells in the cerebral cortex. Below right is a cerebellar Purkinje cell with its remarkable fan-shaped dendritic tree. From Ramón y Cajal S. *Histologie du Systeme Nerveux de l'Homme et des Vertebres* (1911). Cajal Institute. Madrid, Spain: Consejo Superior de Investigaciones Cientificas; 1955.

The Cell Body, or Soma

The *cell body (soma; pl. somata)* is the trophic center of the neuron. It contains the nucleus and the principal protein synthetic machinery of the neuron. Axons have essentially no ability to synthesize protein because they do not contain ribosomes or significant amounts of RNA throughout most of their length. Thus, axons depend entirely on proteins produced in the cell body, which are delivered into the axon by important transport processes. Dendrites do contain both mRNA and ribosomes, and this protein synthetic machinery plays an important role in dendritic function; nevertheless, most of the proteins required by dendrites are synthesized within the cell body.

Because of the large concentrations of ribosomes, neuronal cell bodies stain heavily with basic dyes such as thionin and cresyl violet, which are routine stains used for neuropathological studies. In many of the larger neurons, thionin and cresyl violet stains reveal clumps of heavily stained material, termed *Nissl bodies*, after F. Nissl who initially described them (Figure 4.4). Nissl bodies represent the stacks of rough endoplasmic reticulum visible at the electron microscopic level. The basic dyes that reveal the Nissl bodies are termed *Nissl stains.*

The distribution of Nissl substance is an indication of the protein synthetic activity of the neuron. When conditions lead to decreases in protein synthesis, there are alterations in Nissl staining. For example, when

Figure 4.4. Neurons in the brainstem reticular formation stained with cresyl violet (a Nissl stain). Clumps within the large neuron are Nissl bodies.

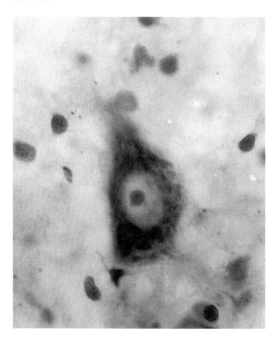

neurons are *axotomized*, they often exhibit a response termed *chromatolysis*, which is characterized by a dispersal of Nissl substance. During the time that the Nissl substance is dispersed, protein synthesis is decreased.

Neuronal cell bodies also contain a prominent *Golgi apparatus*, like other secretory cells. As in other cell types, an important function of the Golgi apparatus of neurons is terminal glycosylation of proteins synthesized in the rough endoplasmic reticulum. Neuronal cell bodies also contain the organelles found in other cell types, such as mitochondria, lysozomes, and vesicular structures, including coated vesicles.

The Nucleus

A substantial proportion of the genome is continually transcribed in neurons. As a result of the high level of transcriptional activity, neurons are *euchromatic;* that is, the nuclear chromatin is dispersed. In contrast, glial cells are *heterochromatic;* their chromatin is usually distributed nonhomogeneously, with clumps of chromatin on the internal face of their nuclear membrane. This characteristic provides one useful way to distinguish neurons from glial cells microscopically.

Axons

Axons are specialized for transmission of information over considerable distances. Axons can reach immense length. In fact, single axons of some vertebrates can reach meters in length.

Two general types of cells are recognized based on their axonal projections. Neurons that project out of the region containing the cell body are termed *projection neurons,* or *Golgi type I.* Neurons that project only locally in the region containing the parent cell body are termed *interneurons,* or *Golgi type II.*

Neurons usually have only one axon. The axon may give rise to many branches (Figure 4.1), but all branches originate from a single stem process. Branching occurs through the formation of *collaterals;* these daughter branches are of approximately the same diameter as the parent.

Some neurons give rise to collateral branches that project back to the region of the cell body (*recurrent collaterals*). These recurrent collaterals often terminate on interneurons (usually inhibitory interneurons), providing a substrate for *recurrent inhibition*. Occasionally, recurrent collaterals may terminate on the same cell that gives rise to the axon, forming *autapses*.

The *initial segment* of the axon begins at the neuronal cell body and ends where the *myelin sheath* begins. There is often a small bulge in the cell body where the axon emerges, which is termed the *axon hillock*. The initial segment is characterized by a submembranous, electron-dense undercoating. A similar undercoating is present under *nodes of Ranvier*. Both sites have high concentrations of sodium channels, and the submembranous undercoating may be associated with these channels. The initial segments of many neurons of the CNS are the site of synaptic connections, most or all of which are thought to be inhibitory.

The principal portion of the axon between the initial segment and the terminal arborizations, termed the *axis cylinder*, has fairly uniform structural characteristics throughout its length.

Axons exhibit complex arborization patterns upon reaching their target area. These are termed *terminal arborizations* (Figure 4.1). Many of these arborizations are highly characteristic for given cell types. For example, the "basket" plexus formed by inhibitory interneurons around the cell bodies of other neurons is characteristic (Figure 4.5).

Dendrites

Dendrites are the principal receiving element of the neuron. The dendritic membrane is a mosaic of synaptic specializations for the receipt of synaptic contacts from other neurons. Because most dendrites conduct electrotonically rather than via action potentials, dendritic form directly determines the integrative properties of the dendrite.

Dendritic form is important for cell-type classification. For this reason, a number of terms have evolved to define different types of dendrites and different portions of the dendritic structure. Sites on dendrites are

Figure 4.5. The axonal plexus of a basket cell in the cerebellum. Neurons elaborate characteristic basket endings around the cell bodies of Purkinje cells (indicated by stippling). From Ramón y Cajal S.

Histologie du Systeme Nerveux de l'Homme et des Vertebres (1911). Cajal Institute. Madrid, Spain: Consejo Superior de Investigaciones Cientificas; 1955.

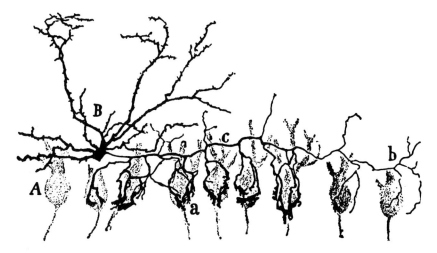

termed *proximal* or *distal,* depending on their distance from the cell body. In some specialized cells, different types of dendrites are recognized. For example, cortical pyramidal cells give rise to a single large *apical dendrite,* which emerges from the apex of the pyramid-shaped cell body, and a number of *basal dendrites* (Figure 4.1).

Dendrites contain all of the organelles found in axons and also contain ribosomes. The presence of polyribosomes in dendrites indicates that local protein synthesis plays an important role in dendritic function. The discovery that dendritic polyribosomes are selectively localized beneath synaptic sites suggests that the proteins that are locally synthesized may be important components of the postsynaptic region.

Most dendrites do not possess high concentrations of Na^+ channels in their membranes and thus do not propagate action potentials. Instead, dendrites behave as *cable conductors.* The cable properties of the dendrite are determined by the form of the dendrite. Thus, dendritic form (size and length) defines the nature of electrical integration carried out by the particular cell type.

Table 4.1 summarizes some of the key features of axons and dendrites as described in the preceding paragraphs.

Dendritic Spines

Many CNS neurons receive most of their synaptic inputs on *dendritic spines*. Spines are mushroom-shaped extensions from the dendritic shaft, each of which is the site of a synaptic junction. Spines can be seen at the light microscopic level when neurons are stained appropriately (see, for example, Figure 4.3), but their true nature is apparent only at the electron microscopic level, where it can be seen that presynaptic terminals synapse directly on the heads of spines (Figure 4.6).

Spines have fascinated neuroscientists since their discovery, but their function remains enigmatic. What is known is that the spine is a sensitive indicator of the integrity and activity of the associated presynaptic terminal. Deafferentation leads to spine loss, as do conditions that cause persistent decreases in activity. Some types of mental retardation (*Down syndrome* for example) are characterized by profound spine loss in the cerebral cortex, and abnormalities in the spines that survive. The abnormalities in spines are thought to indicate disruptions of synaptic connections.

SYNAPSES

Sites of Synaptic Connections

Synapses terminate in a number of locations on cells (Figure 4.1). *Axodendritic synapses* are synaptic contacts between an axon and a dendrite. They occur on spines and on dendritic shafts. *Axosomatic synapses* are contacts

Table 4.1. Features of axons and dendrites

	Electrical properties	Channels/receptors	Origin	Form
Axons	Propogate action potentials	Voltage-gated Na^+ and K^+ channels; voltage-gated Ca^{2+} channels in terminals	One per cell	Uniform diameter
Dendrites	Electrotonic conduction	Ligand-gated (ionotropic) and metabotropic receptors	Multiple	Tapering

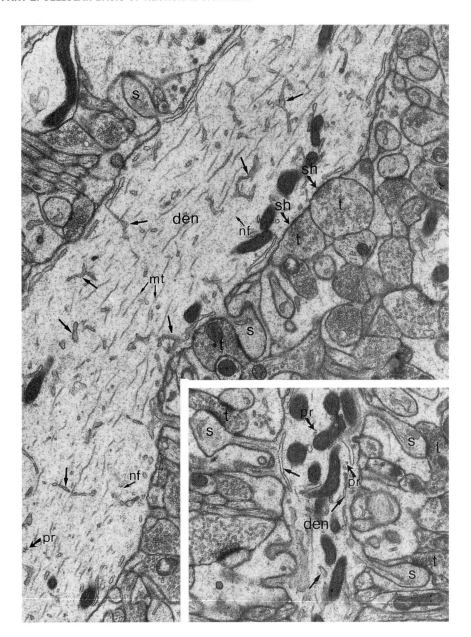

Figure 4.6. Electron micrographs illustrating dendrites of CNS neurons (cerebellar Purkinje cells). A large proximal dendrite fills most of the field. Inset shows a thin distal dendritic segment with several spines (s). den indicates dendrite; s, spine; sh, shaft synapse; t, presynaptic terminals; mt, microtubules; pr, polyribosomes. From Steward O. *Principles of Cellular, Molecular, and Developmental Neuroscience.* New York, NY: Springer-Verlag; 1989.

between an axon and a cell body. These are contacts between one axon and another. Such contacts occur on axon initial segments and on synaptic terminals. *Dendrodendritic synapses* are contacts between two dendrites. Such synapses are rare in most brain regions, but when they occur, there are often reciprocal connections (that is, there are synapses that communicate in each direction between the two dendrites).

Ultrastructure of Synapses

The development of the electron microscope in the early 1950s made it possible for the first time to actually see the ultrastructure of synapses. In general, presynaptic terminals contain either vesicles (if the synapse is chemical), or membrane specializations appropriate for electrical synapses (the latter being rare in vertebrates). At chemical synapses, the vesicles are accumulated at a *presynaptic specialization*, which is the release site (Figures 4.6 and 4.7).

On the postsynaptic side, each synaptic site is marked by a *postsynaptic membrane specialization*, which appears as a darkly stained, amorphous mass in routine electron micrographs (Figures 4.6 and 4.7). The postsynaptic membrane specialization represents the active zone of the synapse and contains the neurotransmitter receptors, ionophores (channels specific to particular ions), recognition and adhesion molecules, and a variety of other molecules involved in synaptic function.

The form of presynaptic terminals varies widely (Figure 4.7). Presynaptic terminals at the terminus of an axon are called *synaptic terminals*, or *terminal boutons*. Synapses that are located along the course of an axon are called *en passant* (in passing, see Figure 4.7B). Some presynaptic terminals are quite small (less than 1 µm in diameter) and contain only a few vesicles. Others are large, highly differentiated structures. Examples of the latter include cerebellar mossy fibers and *calyces of Held* in the anteroventral cochlear nucleus (Figure 4.7C).

Two general types of synapses are recognized (Figure 4.8):

1. *Asymmetric synapses* have a dense postsynaptic membrane specialization that is termed the *postsynaptic density* and usually have round vesicles. Asymmetric synapses are found on dendritic spines and shafts and on cell bodies.

2. *Symmetric synapses* have thin postsynaptic membrane specializations and flattened or pleomorphic (irregularly shaped) vesicles. Symmetric synapses are found on dendritic shafts, cell bodies, and axon initial segments.

Asymmetric synapses are thought to be excitatory, whereas symmetric synapses are thought to be inhibi in electron microsco tem (E.G. Gray) clas. symmetric synapses as spectively (type I being The type I/type II and as ric terminology is used i. the modern literature. Ac   ꜩes of asymmetric and symmetric   ꜩapses can be seen in Figure 4.7.

Terminal size correlates with synaptic potency. Large terminals are very powerful, and may be "command" inputs, in the sense that firing of the presynaptic input always results in postsynaptic firing. Small terminals usually do not produce sufficient depolarization to discharge neurons unless there is a summation of input (simultaneous release by a number of converging terminals).

Uncommon Synaptic Interactions

Axoaxonic synapses are in fact a *synaptic triad* because the presynaptic terminal that is postsynaptic to the first terminal also has a postsynaptic target. These are also termed *serial synapses*. Serial synapses are very important in the sites of termination of retinal ganglion cell axons in the lateral geniculate nucleus.

Dendrites can also be presynaptic to other dendrites. For example, in the olfactory bulb, granule cells have no axons. Instead, they form reciprocal dendrodendritic *synapses* with the dendrites of mitral cells. Each dendrite possesses presynaptic release sites with accumulations of vesicles that are apposed to postsynaptic membrane specializations of other dendrites. Presynaptic and postsynaptic specializations exist side by side, forming reciprocal synapses that form the basis for *local circuit interactions* between neurons without axons.

MOLECULAR ANATOMY OF NEURONS

The different functional domains of the neuron arise because of differences in molecular composition. Dendrites can function

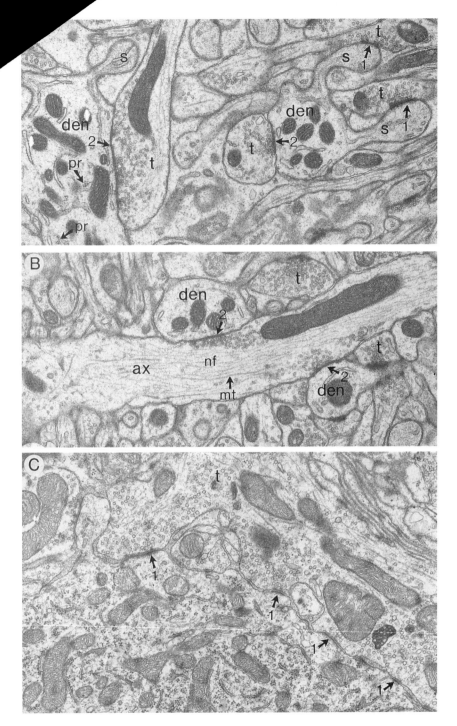

Figure 4.7. Electron micrographs of different types of synapses. A, Asymmetric and symmetric synapses in the cerebellar cortex. Asymmetric synapses (1) are characterized by a dense postsynaptic membrane specialization; symmetric synapses (2) have very thin postsynaptic membrane specializations. pr indicates polyribosomes. **B,** Axon forming en passant synapses with dendrites in the cerebellum. nf indicates neurofilaments; mt, microtubules; ax, axon. **C,** Large synaptic ending (calyx of Held) in the avian cochlear nucleus. This large presynaptic terminal forms numerous asymmetric synaptic junctions on the postsynaptic cell. From Steward O. *Principles of Cellular, Molecular, and Developmental Neuroscience.* New York, NY: Springer-Verlag; 1989.

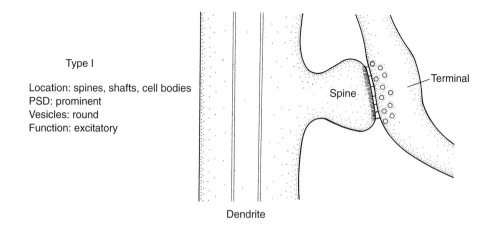

Type I

Location: spines, shafts, cell bodies
PSD: prominent
Vesicles: round
Function: excitatory

Dendrite

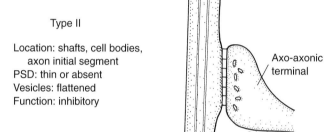

Type II

Location: shafts, cell bodies,
         axon initial segment
PSD: thin or absent
Vesicles: flattened
Function: inhibitory

Figure 4.8. Characteristics of asymmetric (type I) and symmetric (type II) synapses.

as the receiving elements of the neuron because of the presence of *neurotransmitter receptors* and other molecules at the site of synaptic contact.

Because dendrites usually possess relatively few *voltage-gated Na⁺ channels,* they do not propagate action potentials (see Chapter 8). Dendrites do, however, possess *voltage-gated Ca²⁺* channels and at least some Na⁺ channels, and recent studies have revealed that dendrites produce both sodium and calcium "sparks" consisting of Na⁺ or Ca²⁺ spikes that are spatially restricted and do not propagate. These "sparks" are thought to allow for an amplification of synaptic currents.

In contrast, the plasma membrane of the axon is specialized for the long-distance propagation of *action potentials* because it

contains voltage-gated sodium channels. The plasma membrane of synaptic terminals contains the sodium channels that are present in the axis cylinder, and also voltage-gated calcium channels at release sites. The latter play a key role in neurotransmitter release.

THE NEURONAL CYTOSKELETON

The cytoskeleton of neurons has a morphology that is similar to that of other cells. The three visible elements of the cytoskeleton are *microtubules;* 10-nm filaments, which in neurons are termed *neurofilaments;* and *microfilaments.* However, the molecular composition of 10-nm filaments and microtubules is different in neurons than in other cell types (Table 4.2).

Table 4.2. Cytoskeletal structures

Type	Size	Composition	Location
Microfilaments	4–5 nm	Actin (43 kd)	Growth cones and spines
Neurofilaments	10 nm	70 kd (core) 140 kd (sidearms) 220 kd (sidearms)	Axons and proximal dendrites
Microtubules	23 nm	Tubulin (52 and 56 kd) MAPS (sidearms)	Axons and dendrites

Microfilaments (about 7 nm in diameter) are actin filaments and are particularly prominent in the highly motile portions of neurons (growth cones of axons and dendrites), and in dendritic spines. Spines may also be at least somewhat motile, in that spines are thought to change their shape as a result of synaptic activity. A filamentous lattice made up of actin filaments is the principal cytoplasmic feature of the head and neck of dendritic spines.

Neurons contain intermediate filaments (10 nm) of a specialized type termed *neurofilaments,* which are made up of a unique set of three proteins of about 70 kd, 140 kd, and 220 kd. The actual molecular weight of the protein components varies across species. The 70-kd subunit forms the core and is similar but not identical to the protein constituents of intermediate filaments in other cell types. The neurofilaments differ in morphology from intermediate filaments of other cell types in that they possess side arms that are made up of the two higher molecular weight proteins. These side arms are thought to participate in the interactions between neurofilaments and other elements of the cytoskeleton, particularly microtubules.

Neurofilaments are thought to confer structural rigidity. They are most numerous in proximal dendrites and in the axis cylinder, which are the most stable portions of the respective neurites. Neurofilaments are not found in the growing tips of axons or in dendritic spines, which are thought to be dynamic structures. Based on this distribution and the fact that neurofilaments are bio-chemically very stable, it has been proposed that neurofilaments confer structural rigidity to the elements in which they reside.

Neurofilaments are stained by the *reduced silver methods* such as the Bodian or Bielschosky stains (both of which are named after the neuroanatomists who developed them). These stains are used routinely for neuropathological diagnosis. These "silver stains" reveal *neurofibrils* in axons and proximal dendrites that are actually bundles of neurofilaments (Figure 4.9).

Neurofilaments are important for neuropathological diagnosis. Several degenerative diseases result in the appearance of *neurofibrillary tangles,* or *Alzheimer's bodies,* along with *neuritic plaques.* The neurofibrillary tangles can be stained using the reduced silver stains or antibodies that are available commercially (Figure 4.10A, B). The neurofibrillary tangles are accumulations of modified neurofilaments organized in a very unusual and highly characteristic way. The filaments within the tangles are organized in pairs, and the filament bundles assume a helical organization in an almost crystalline structure (Figure 4.10C). These are termed *paired helical filaments* and are considered diagnostic of neurodegenerative disease.

At this time, *Alzheimer's disease* can be definitively diagnosed only by neuropathological evaluation. The disease is manifested by the presence of neurofibrillary tangles and neuritic plaques throughout the hippocampus and cerebral cortex. Neurofibrillary tan-

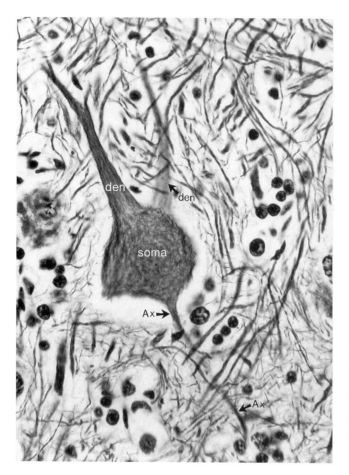

Figure 4.9. Neurofilaments in a neuron of the spinal cord (Bodian stain). The axon emerges from the lower right hand portion of the cell (Ax). Axons of other neurons are visible in the field. Proximal dendrites (den) are also stained. From Steward O. *Principles of Cellular, Molecular, and Developmental Neuroscience.* New York, NY: Springer-Verlag; 1989.

gles are also observed in pathological material from patients with Down syndrome, *postencephalitic parkinsonism,* and *amyotrophic lateral sclerosis.* Neuritic plaques are especially characteristic of Alzheimer's disease and are thought to be the accumulated remains of neuronal processes that have degenerated. One of the key components of the plaques is *amyloid protein,* and one of the current hypotheses about the nature of Alzheimer's disease focuses on defective amyloid metabolism.

Microtubules are composed of tubulin; both alpha- and beta-tubulin are present, together with microtubule-associated proteins that are unique to neurons (see later).

Microtubules are found in both axons and dendrites and mediate the "rapid phase" of intracellular transport. Recent experiments using high magnification cinematography have revealed that particles are translocated along microtubules in axons at very high rates. Plant-derived alkaloids such as colchicine (used for the treatment of gout) and the vinca alkaloids vincristine and vinblastine (used for cancer chemotherapy) bind to tubulin and disrupt microtubules, interfering with rapid transport.

Neuronal microtubules have unique accessory proteins called *microtubule-associated proteins* (MAPs). The complement of MAPs is related to the location of the microtubules. Microtubules in dendrites have a high molecular weight MAP (MAP2) that is essentially absent from axons. Axons have lower molecular weight MAPs in the 50- to 60-kd range, termed TAU *proteins,* and immunocytochemical studies using antibodies against

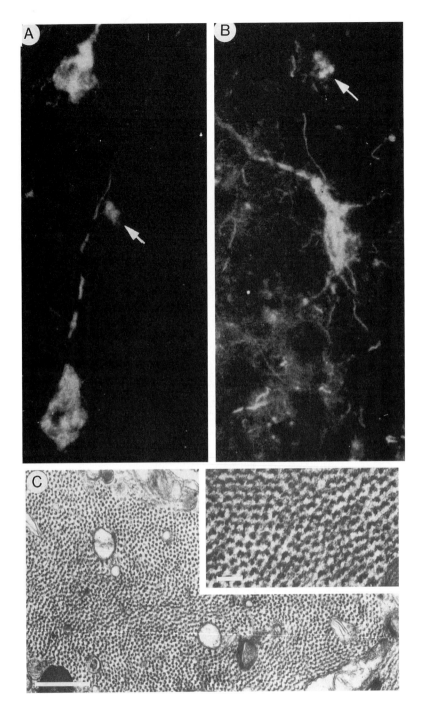

Figure 4.10. Neurofilament pathology in Alzheimer's disease. A and **B** show neurofibrillary tangles in the pyramidal cell layer of the hippocampus, as revealed by immunocytochemistry using antibodies to neurofilament proteins. **C,** An electron micrograph of paired helical filaments within a tangle in a degenerating neuron from an Alzheimer's disease patient. Inset shows a high-magnification view of an ordered array of paired helical filaments. **A** and **B** are from Dahl D, Selkow DJ, Pero RT, et al. Immunostaining of neurofibrillary tangles in Alzheimer's senile dementia with a neurofilament antiserum. *The Journal of Neuroscience.* 1982;2:113–119. Copyright Society for Neuroscience. **C** is from Metuzals J, Montpetit V, Clapin DF, et al. Arrays of paired helical filaments in Alzheimer's disease. In: Bailey GW, ed. *40th Annual Proceedings of the Electron Microscopy Society of America.* Baton Rouge, La: Claitor's Publishing Division; 1982:348–349.

the TAU proteins reveal predominantly axonal staining. The accessory proteins associated with microtubules may play a role in the differential routing of material for transport into dendrites versus axons.

AXOPLASMIC TRANSPORT AND THE DISTRIBUTION OF MATERIALS WITHIN NEURONS

Dendrites have only limited capabilities for local synthesis of proteins, and axons have essentially none; thus, axons and dendrites depend on the transport of proteins produced in the neuronal cell body.

Transport mechanisms are distinguished in two ways: by the direction of transport and by the characteristic rate of transport. In terms of direction, transport is classified as *orthograde*, or *anterograde* (away from the cell body), or *retrograde* (toward the cell body). In terms of speed, there are *slow transport* and *fast transport* processes (see Table 4.3).

Most of what is known about transport in neurons pertains to axonal transport. Transport processes have been studied by taking advantage of the fact that neuronal proteins are synthesized primarily in cell bodies. Hence, when radioactive amino acids are injected into regions containing neuronal cell bodies, the amino acids are taken up and in-

corporated into protein. It is then possible to follow the migration of the labeled proteins. An example of an experiment of this type is illustrated in Figure 4.11. Using this approach, the rates and characteristics of the transport processes were defined.

There is a form of rapid transport that moves material down the axon at a rate of up to 400 mm/day (Figure 4.11). The material transported via rapid transport is associated with small membrane vesicles, and the material is delivered preferentially to synaptic terminals. Biochemical studies indicate that many of the proteins that travel in the rapid phase of transport are membrane glycoproteins. Only a relatively small amount of the total protein delivered into the axon is conveyed by rapid transport.

The bulk of the material transported into the axon moves by the so-called slow component of axoplasmic transport at a rate of about 1 to 2 mm/day. Structural proteins of the axon and the cytoskeleton move via slow transport, and different proteins of the cytoskeleton move at slightly different rates. Actin and neurofilament proteins move at one rate (termed *slow component B*), whereas tubulin moves at a slightly different rate (*slow component A*).

Retrograde transport moves material from the ends of axons and dendrites toward the cell body. Again, there are both fast and slow

Table 4.3. Neuronal transport systems

Transport system	Rate	Material conveyed	Molecular motor	Function
Slow	1–2 mm/d	Cytoskeletal components	Unknown	Replenishment of the axonal cytoskeleton and cytoplasm
Rapid orthograde	Up to 400 mm/d	Membrane-bound organelles	Kinesin	Delivery of material to synaptic terminals
Rapid retrograde	60–100 mm/d	Membrane-bound organelles	Dynein	Delivery of material that has been taken up by the terminal (trophic factors)

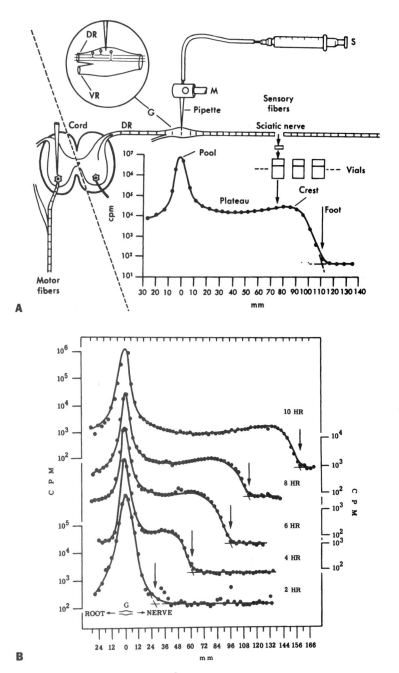

Figure 4.11. Rapid axonal transport. The characteristics of axonal transport were evaluated by injecting radiolabeled amino acids into regions containing the cell bodies of particular neurons (e.g., the dorsal root ganglion or the ventral horn of the spinal cord). **A** illustrates the basic experimental setup. The precursor is incorporated into protein within the neuronal cell bodies and transported down the axon. The rate of transport can be evaluated by harvesting the nerves at different times after the injection, sectioning the nerves into segments, and determining the amount of radioactivity in each segment. Labeled proteins move down the axon as a wave. **B** illustrates that the wave front is at different positions along the nerve at different times. A plateau of radioactivity is apparent in the nerve proximal to the wave front. The scales on the left-hand side refer to the 2-hour and 10-hour samples, and partial scales are illustrated on the right for the 4-, 6-, and 8-hour samples. **A** is from Ochs S. Axoplasmic transport in peripheral nerve and hypothalamo-neurohypophyseal systems. In: Porter JC, ed. *Hypothalamic Peptide Hormones and Pituitary Regulation.* New York, NY: Plenum Press; 1977:13–40. **B** is from Ochs S. Fast transport of materials in mammalian nerve fibers. *Science* 1972;176:252–254. Copyright 1972 by the AAAS.

components; the average rate of fast retrograde transport seems to be different from that of fast anterograde transport. The characteristics of retrograde transport were initially defined by injecting the protein tracer horseradish peroxidase into the terminal zones of particular neurons. The protein is taken up by endocytosis at nerve terminals and transported back to the cell body. Many types of protein can be taken up and transported in a retrograde direction, although certain proteins may be taken up and transported more efficiently than others.

Rapid orthograde and retrograde transport are mediated by different molecular motors that move organelles unidirectionally along microtubules. Different protein motors are responsible for moving materials along microtubules in orthograde and retrograde directions. The best understood of these cel-

lular motors is *kinesin*—a molecule that associates with microtubules and has ATPase activity. Kinesin is a molecule with a pair of globular heads that bind microtubules, and a tail region that binds to the organelle that is to be transported (Figure 4.12). Kinesin translocates materials toward the "plus" ends of microtubules. Virtually all microtubules in axons are oriented with their plus ends distal, and thus kinesin is thought to be one of the key motors for orthograde transport. A different molecule named *dynein* is thought to mediate retrograde transport along microtubules.

Retrograde transport provides a mechanism for the cell body to sample the environment around its synaptic terminals. This transport is thought to play an important role in trophic regulation of neurons by their targets. In many systems, target cells synthe-

Figure 4.12. Rapid orthograde and retrograde transport are mediated by different molecular motors that move organelles unidirectionally along microtubules. Membrane proteins and proteins for release that are destined for axonal transport are synthesized in the cell body and packaged into transport vesicles within the Golgi apparatus. These vesicles associate with motor proteins such as kinesin (blow-up in the lower portion of the figure) that mediate binding of the vesicle to microtubules. The motor proteins then transport the vesicle in an orthograde direction along the microtubule, toward the synaptic terminal. Proteins that are destined for retrograde transport are taken up by synaptic terminals and packaged into vesicles. These vesicles become associated with a different motor protein (dynein) that transports vesicles in the retrograde direction along the microtubule. The model for kinesin (lower right) is from Hirokawa N, Pfister KK, Yorifuji H, Wagner MC, Brady S, Bloom GS. Submolecular domains of bovine brain kinesin identified by electron microscopy and monoclonal antibody decoration. *Cell.* 1989;56:867–878.

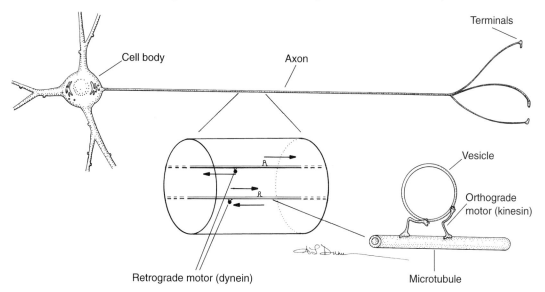

size and release protein factors that are taken up by the presynaptic element and transported in a retrograde direction. The best characterized *target-derived trophic factor* is nerve growth factor. In the sympathetic nervous system, where nerve growth factor was discovered, a continuous supply of the factor is essential for the survival of the sympathetic neurons.

How Do We Know What We Know? The Methods of Functional Neuroscience

One of the central dogmas of neuroscience is that function is mediated by interconnected structures that form a neural system. It follows then that the foundation of our knowledge about how the brain operates lies in our understanding of connectivity—the wiring diagram of the brain. The title of a chapter by Floyd Bloom says it succinctly: "The Gains in Brain Are Mainly in the Stain" (see Bloom FE. The gains in brain are mainly in the stain. In: Worden FG, Swazey JP, Adelman G, eds. *The Neurosciences Paths of Discovery.* Cambridge, Mass: MIT Press; 1975:211–227). By that, he meant that many of the dramatic advances in neuroscience have come from the development of neuroanatomical methods— various stains that have helped us define not only the patterns of connectivity, but also the basic characteristics of the neurons that make up a system (transmitter type, etc.). Here we provide a brief summary of the principal neuroanatomical techniques that have been used to determine patterns of connectivity.

The Golgi Technique

The Golgi technique has already been mentioned and illustrated (Figures 4.2 and 4.3). It is truly remarkable how important a role this technique has played in defining the cellular organization of the nervous system. The technique is still used today, although it has to some extent been supplanted by techniques that involve intracellular injections of tracer substances. The limitation of the Golgi technique is that it stains neurons randomly; hence it is difficult to use the technique for systematic assessments of the connections of particular neuron groups.

Selective Silver Staining

An important technical advance was made in the early 1950s by Walle Nauta and colleagues. The Golgi technique and the Bodian stain for neurofilaments are based on the fact that neurons have an affinity for silver. This affinity is substantially increased when neurons are in the process of degenerating. Nauta took advantage of this property of degenerating neurons to develop selective silver staining techniques that revealed degenerating axons. The experimental approach was to produce lesions in particular structures and then trace the axons that degenerated as a result of the lesion. Hence the technique was based on the important principle of neuronal cell biology that axons depend on the cell body for their survival. This technique, which came to be known as the *Nauta technique,* was used extensively throughout the 1950s.

An important technical improvement on the Nauta technique was made in 1963 by L. Heimer and R. Fink, working in Nauta's lab. This new method, which came to be known as the *Fink-Heimer technique,* was more selective than what came to be known as the *original Nauta technique* and had the important advantage that degenerating synapses were especially well revealed (Figure 4.13). The Fink-Heimer technique was used extensively throughout the 1960s and the first half of the 1970s.

Transport-Based Tract-Tracing Techniques

The next important advance was the development of techniques that were based on anterograde and orthograde transport. For example, to define the orthograde projections from one area to another, radioactive

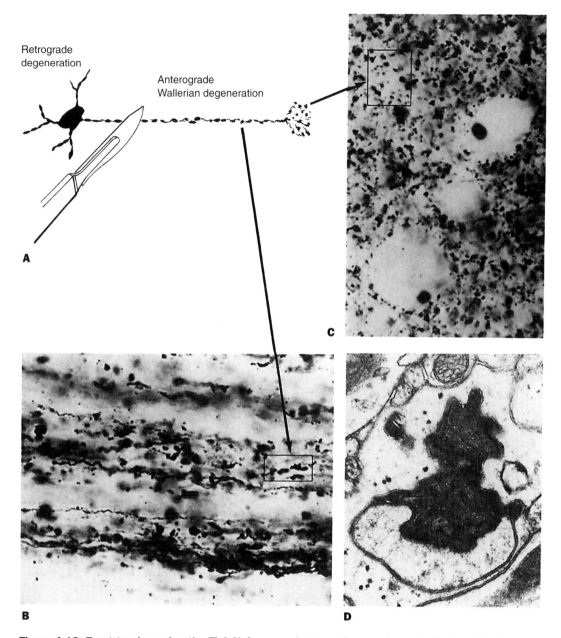

Figure 4.13. Tract tracing using the Fink-Heimer technique. The basic experimental strategy is illustrated in **A**. In this experiment, a selective lesion was made in the piriform cortex of the rat, causing anterograde degeneration of the axon (also called *Wallerian degeneration*). **B** illustrates degenerating axons, and **C** illustrates degenerating synaptic terminals in the amygdala. Silver impregnation is often complemented by electron microscopic evaluation in order to determine the types of synapses that are involved **(D)**. From Heimer L. *The Human Brain and Spinal Cord.* New York, NY: Springer-Verlag; 1994. **B–D** are from de Olmos JS, Ebbesson SOE, Heimer L. Silver methods for the impregnation of degenerating axoplasm. In: Heimer L, Robards M, eds. *Neuroanatomical Tract-Tracing Methods.* New York, NY: Plenum Press; 1981:117–170.

amino acids are injected into an area containing neuronal cell bodies. The resident neurons will take up these protein precursors, synthesize proteins, and transport the labeled proteins into the axons and terminals (Figure 4.14). The distribution of the labeled axons and terminals can then be defined using autoradiography. Figure 4.14 illustrates

an example from the work of S. B. Edwards—one of the individuals responsible for the initial development of autoradiographic tract-tracing techniques. In this study, Edwards and his colleague C. K. Henkel used the technique to define the projection from the superior colliculus to the oculomotor nucleus—a pathway that plays a key role in the

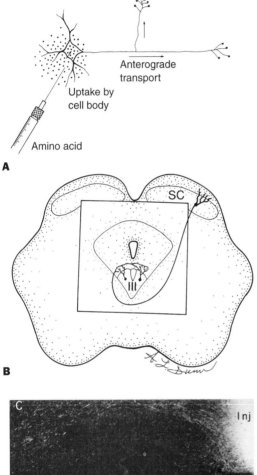

A

B

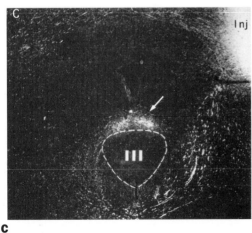

C

Figure 4.14. Tract-tracing techniques based on orthograde axonal transport. A, Anterograde tracing involves the injection of small quantities of tracer (in this case radioactive amino acid) into an area containing neuronal cell bodies. The labeled amino acid is incorporated into proteins that are transported to axons and synaptic terminals. **B** and **C,** To define the projection from the superior colliculus to the oculomotor nucleus (schematized in **B**), radioactive amino acids were injected into the superior colliculus on the right side (**C,** Inj.). After several days, labeled proteins are transported throughout the axons of neurons in the injection site. Histological sections were prepared, coated with a photographic emulsion, and exposed for several weeks. Slides with sections and emulsion are developed, and the sections on the slides are stained. **C** illustrates a dark-field photograph of the area indicated by the box in **B**. In dark-field illumination, silver grains (which indicate the presence of radioactive label) appear as bright spots overlying labeled axons (on the right-hand side) and synaptic terminals (just dorsal to the oculomotor nucleus, III). **A** is from Blackstad TW, Heimer L, Mugnaini E. Experimental neuroanatomy: general approaches and laboratory procedures. In: Heimer L, Robards M, eds. *Neuroanatomical Tract-Tracing Methods.* New York, NY: Plenum Press; 1981:1–53. **B** and **C** are from Edwards SB, Henkel CK. Superior colliculus connections with the extraocular motor nuclei in the cat. *J Comp Neurol.* 1978;179:451–468. Reprinted by permission of Wiley-Liss, Inc., a subsidiary of John Wiley & Sons, Inc.

control of eye movement (see Chapter 21). Recent adaptations of the *anterograde tract-tracing technique* involve the use of other tracers, for example, plant lectins (phaseolus leukoagglutinin) that are taken up and transported by neurons.

A similar approach has been used to define the cells of origin of particular projections by taking advantage of retrograde transport. Indeed, much of our information

about the cell types that give rise to particular pathways has been derived in this way. Early studies used horseradish peroxidase (HRP). The experimental approach involves making small injections of HRP into a region containing synaptic terminals; the molecule is taken up by terminals and transported in a retrograde direction (Figure 4.15). The cells of origin of the terminals can then be identified using histochemical techniques that re-

Figure 4.15. Identifying the cells of origin of pathways using retrograde transport techniques. A, HRP tracing involves the injection of small quantities of HRP into an area containing synaptic terminals. The molecule is taken up and transported back to the cell body. **B,** The cells of origin of the corticospinal tract of the mouse were identified by injecting HRP into the thoracic level of the spinal cord. The protein molecule was

taken up by the axons that had been deliberately damaged during the injection and the molecule was transported back to the cell bodies in the cortex. The presence of the HRP was then revealed using a histochemical technique based on the HRP enzyme activity. **A** is from Heimer L. *The Human Brain and Spinal Cord.* New York, NY: Springer-Verlag; 1994.

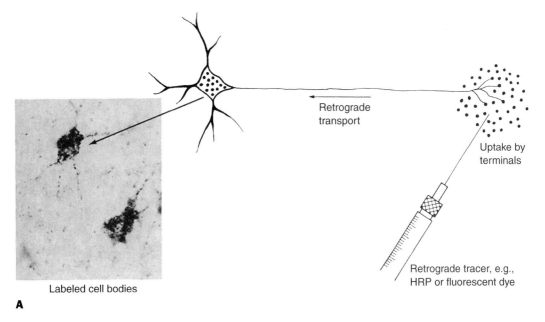

Labeled cell bodies

A

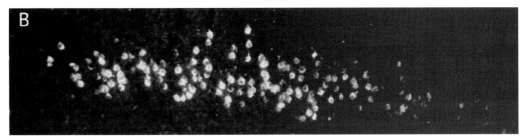

B

veal the presence of the HRP molecule. For example, Figure 4.15B illustrates the use of this technique to define the cells of origin of the corticospinal tract. More recent approaches involve the use of fluorescent tracers, which are also taken up and transported in a retrograde fashion.

Another important category of neuroanatomical tract-tracing techniques is based on stains that reveal particular neurotransmitters or neurotransmitter enzymes. These techniques allow one to determine not only the patterns of connectivity of particular cell types, but also the neurotransmitter used by the pathway. These will be considered further in Chapter 6.

PATHOLOGIES THAT AFFECT TRANSPORT PROCESSES

Exposure to any one of a number of compounds can lead to peripheral neuropathies as a result of the disruption of axonal transport. For example, *n*-hexane, acrylamide, *p*-bromophenylacetylurea, zinc pyridinethione, and β,β iminodipropionitrile have neurotoxic effects that seem to be related to transport blockade. Many neurotoxins that affect transport were identified as a result of the appearance of peripheral neuropathies in individuals who had been exposed to the substances. In addition, chronic treatment with disulfiram (Antabuse) can cause peripheral neuropathy that is correlated with a disruption of transport. Vitamin E deficiency and diabetes are also associated with peripheral neuropathies that are thought to involve disrupted transport.

A blockade of axonal transport is also in evidence in a number of retinopathies and optic neuropathies. Since rapid transport depends on oxidative metabolism, occlusion of arteries supplying the nerve fiber layer of the eye leads to transport blockade. Increased intraocular pressure also leads to transport blockade as a result of compression of the nerve fiber bundles at the lamina cribosa. Papilledema also results in axonal swelling in the anterior part of the optic nerve head. The axonal changes induced by elevated intraocular pressure or papilledema can be reversed if the causitive factors are eliminated quickly, but if the transport blockade persists, degeneration of the distal axons can occur, leading to permanent visual impairment.

Retrograde transport may be one of the mechanisms for the entry of viruses into the CNS. Rabies virus spreads along peripheral nerves to the CNS from the inoculation site. In fact, the earliest demonstration of retrograde transport was in studies that sought to define how rabies virus entered the nervous system. Herpes simplex virus is transported to neurons in the trigeminal nerve and other sites in the CNS after infection of the lips. The virus remains latent in trigeminal ganglia between recurrences of the herpetic lesions, and when activated, travels along the peripheral branches of axons to the lips or conjunctiva.

In the same way that transport mechanisms provide a way for viruses to enter the nervous system, they may eventually provide a means to treat the infections. For example, application of antiviral agents to the terminal fields of affected axons may lead to uptake, retrograde transport, and delivery of the agents to the affected cells in a highly specific fashion. Also, gene therapeutic approaches may someday take advantage of the selective uptake and transport of genetically-modified viruses.

5

Cell Biology of Glia

Despite the preeminent role that neurons play in brain function, glial cells are actually much more numerous than neurons. Glia outnumber neurons by about 10 to 1 and make up about half the volume of the nervous system. Glial cells were first recognized as a distinct cell type in the mid 1800s. The name *neuroglia*, coined by Rudolph Virchow, means *nerve glue*, because these cells were thought to play a "supportive" role in brain function, filling the space that was not occupied by neurons.

Today, it is recognized that neurons and glia are highly interdependent and influence each other's development, differentiation, and physiological function. In this chapter, we consider some of the roles glia play in neuronal function.

GLIAL CELLS OF THE CNS

There are three types of CNS glia that can be distinguished by size and embryonic origin: (1) *astrocytes* (also called *astroglia*), (2) *oligodendrocytes* (also called *oligodendroglia*), and (3) *microglia*. Astrocytes and oligodendrocytes originate from the neuroepithelium. Microglia originate from the mesoderm.

Astrocytes

Astrocytes are present in both gray and white matter and are highly differentiated cells. Their cell bodies are about the size of the cell bodies of small neurons, and like neurons, astrocytes have extensive processes (Figure 5.1).

In gray matter, thin *vellate* (veil-like) astrocytic processes surround neuronal cell bodies, dendrites, and synapses (Figures 5.1 and 5.2). Processes of astrocytes also contact capillaries, where they form flattened structures termed *glial end-feet,* which surround the capillary endothelium. In the white matter, the processes of astrocytes intertwine between bundles of axons and also form glial end-feet on capillaries.

The astrocytes in gray matter differ in appearance from those in white matter. The astrocytes in white matter are termed *fibrous astrocytes* because they have large numbers of *glial filaments* (Figure 5.3). Astrocytes in gray matter are termed *protoplasmic astrocytes* because they have fewer filaments. There may also be molecular differences between the astrocytes in the two locations.

The glial end-feet on blood vessels are bounded by *basal lamina;* thus the glial end-foot represents the boundary between CNS tissue and non-CNS tissue (Figure 5.3). In this way, glial end-feet form a perivascular *glial-limiting membrane,* or *glia limitans,* similar to that found at the surfaces of the brain (see later). A similar glial-limiting membrane is present next to the meningeal sheath in

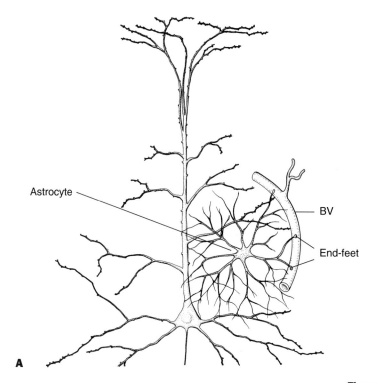

A

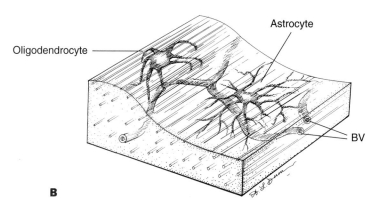

B

Figure 5.1. Relationship between astrocytes and neuronal elements. A, Protoplasmic astrocytes in gray matter. Note that the processes of the astrocyte contact neurons (see also Figure 5.2) and form end-feet on blood vessels (BV). **B,** Fibrous astrocyte in white matter. An intrafascicular oligodendrocyte is also illustrated (for additional details on intrafascicular oligodendrocytes, see Figure 5.8).

the optic nerve, which is a CNS tract (Figure 5.3B).

The anatomical relationship between astrocytic end-feet and blood vessels led to speculations that the glial processes might represent the substrate of the blood–brain barrier. However studies in the late 1960s by M. Brightman and T. Reese convincingly showed that large protein molecules such as horseradish peroxidase readily diffused past

the glial end-feet, and that diffusion was limited by *tight junctions* between capillary endothelial cells. Thus, the blood–brain barrier is formed by the vascular endothelial cells, with the glial end-feet subserving some other function (see later).

Astrocytes also form the glial-limiting membrane (glia limitans) at the surface of the brain (Figure 5.4). This continuous sheet of glial processes and cell bodies arises in part

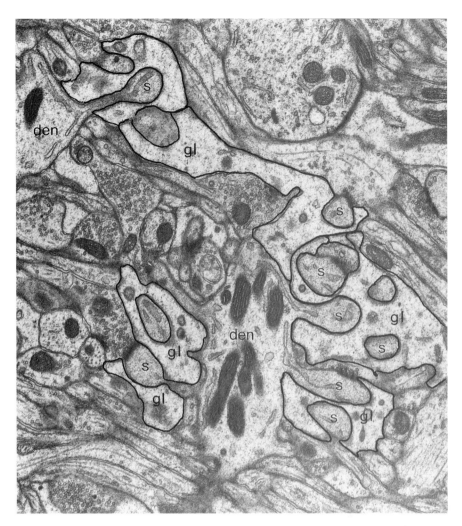

Figure 5.2. Astrocytic processes surround neuronal cell bodies, dendrites, and synapses. This electron micrograph illustrates the relationship between the processes of protoplasmic astrocytes and the elements of the neuropil in the cerebellum. The membranes of the astrocytes are outlined. den indicates dendrite; s, spine; gl, glial process. From Steward O. *Principles of Cellular, Molecular, and Developmental Neuroscience.* New York, NY: Springer-Verlag; 1989.

from astrocytes at the surface, and in part from deeply situated astrocytes that send their processes toward the surface. The external surface of the glia limitans is bounded by the basal lamina, which is situated between the glia limitans and the *pia mater.*

Astrocytes can be identified by specific molecular markers. The intermediate filaments in astrocytes are composed of a protein that is unique to astrocytes: *glial fibrillary*

acidic protein (GFAP). This permits astrocytes to be unambiguously identified using immunocytochemistry (Figure 5.3).

Specialized Forms of Astrocytes

There are specialized types of glial cells that are unique to particular brain regions. All are derivatives of the astrocyte lineage.

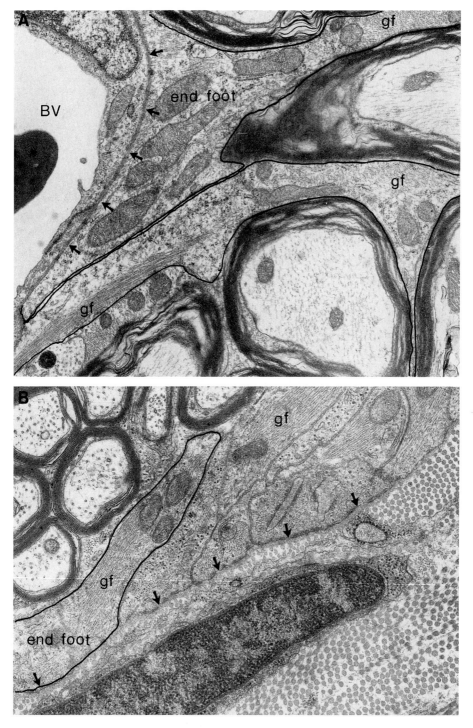

Figure 5.3. Astrocytic processes form end-feet on capillaries and on the basement membrane of the optic tract. A, Relationship between glial end-feet and blood vessel (BV); gf, glial filaments. **B,** The end-foot of an astrocyte forms a glial-limiting membrane (arrows), which abuts the meningeal sheath in the optic nerve—a CNS tract. The membranes of the astrocytes are outlined. From Steward O. *Principles of Cellular, Molecular, and Developmental Neuroscience.* New York, NY: Springer-Verlag; 1989.

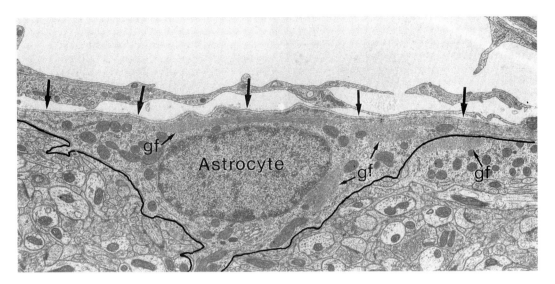

Figure 5.4. Astrocytes form the glial-limiting membrane, or glia limitans, at the surface of the brain. The membrane of the astrocyte is outlined. gf, glial filaments. The gf on the right indicates the filaments within the next astrocyte process that abuts against the first. From Steward O. *Principles of Cellular, Molecular, and Developmental Neuroscience.* New York, NY: Springer-Verlag; 1989.

Bergman glia, also called *Golgi epithelial cells,* are a special type of astrocyte that are located in the cerebellar cortex; the cell bodies of Bergman glia are positioned just below the Purkinje cell layer, and they send a long process (called a *Bergman fiber*) radially through the molecular layer to the pial surface (Figure 5.5). During early development, migrating granule cells follow the processes of Bergman glia from the external granule layer to the internal granule layer.

Müller cells of the retina (also known as *retinal gliocytes*) are also types of astrocytes. The cell bodies of Müller cells are located in the internal nuclear layer near the cell bodies of the ganglion cells. Müller cells send out elaborate processes to the retinal layers. These processes fan out to form the *external limiting lamina* of the retina. The processes of Müller cells also encapsulate photoreceptor processes, neural elements, and extend to form the *internal limiting membrane* on the vitreal aspect of the retina.

Pituicytes of the neurohypophysis are also specialized astrocytes. Their processes end mostly on endothelial cells of the vascular sinuses.

Ependymal cells are found lining the ventricles as an epithelial layer one cell thick. The cells contact one another through gap junctions, and their ventricular faces have numerous microvilli and cilia. The motile cilia are thought to facilitate the movement of cerebrospinal fluid. During early development, the cells often have a basal process that extends into the substance of the nervous system. In the spinal cord and other locations with a central canal, these processes are radially oriented. Such cells are termed *tanycytes, ependymal astrocytes,* or *ependymoglial cells.*

Functions of Astrocytes

Astrocytes are thought to play an important role in the following general areas:

1. Regulation of the ionic composition of the extracellular fluid by taking up ions and neurotransmitter substances that have been released by neurons

2. Provision of trophic support for neurons (synthesizing and releasing neurotrophic factors)

3. Contributions to cellular compartmentation in densely packed neuropil zones by forming

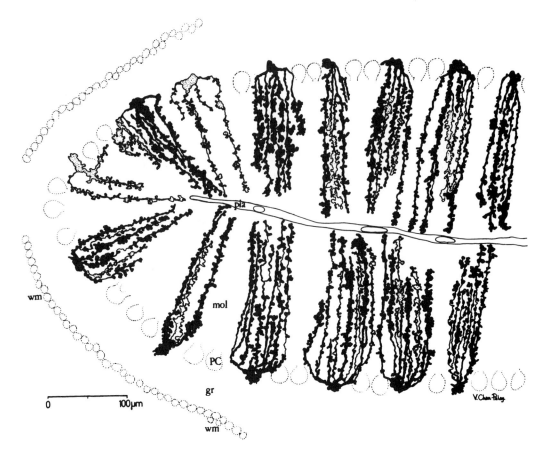

Figure 5.5. Bergman glia: Composite drawing of 14 Bergman glia in the cerebellar cortex. mol indicates molecular layer; PC, Purkinje cell layer; gr, granule layer; wm, white matter. From Palay SL, Chan-Palay V. *Cerebellar Cortex Cytology and Organization.* Berlin, Germany: Springer-Verlag; 1974.

diffusion boundaries between adjacent neuronal processes

4. Mediation of the exchange of materials between capillaries and neurons

5. Response to injury by forming a *glial scar* in areas in which neurons have been killed, and by helping to remove degeneration debris, particularly degenerating synaptic terminals

6. Formation of a physical substrate that guides migrating neurons during early development. Astrocytes may also provide a substrate for long-distance growth of axons.

Astrocytes Buffer the Extracellular Environment of the Brain

The proper function of neurons depends on the ionic composition of the extracellular fluid. High concentrations of potassium in the extracellular medium depolarize neurons and can induce abnormal activity. Hence, there must be a mechanism for the removal of the potassium that is released as a consequence of neuronal activity. Indeed, this mechanism must be a highly effective one because potassium efflux associated with intense neuronal activity can transiently increase extracellular potassium to approximately 10 to 20 times its normal level. Astrocytes are thought to serve in this capacity.

Astrocytes regulate extracellular ion concentrations through a process called *passive spatial buffering*. Astrocytes are highly permeable to potassium. Hence, when potassium is released by neurons, the potassium diffuses into astrocytes at focal areas of high extracel-

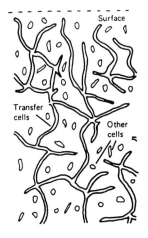

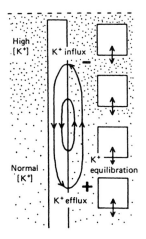

Figure 5.6. Passive spatial buffering.
Astrocytes regulate extracellular ion concentrations through a process called *passive spatial buffering*. Potassium that is released by neurons diffuses into astrocytes and then on into adjacent astrocytes, resulting in a rapid dispersal of the potassium over a wide area. Arrows indicate the pattern of current and ion fluxes that would occur in response to focal increases in extracellular potassium concentration such as might arise following local neuronal activity. From Gardner-Medwin AR. Analysis of potassium dynamics in mammalian brain tissue. *Journal of Physiology* 1983;335:393–426.

lular concentration and then on into adjacent astrocytes (Figure 5.6). The result is a rapid dispersal of the potassium over a wide area.

Astrocytes Take Up Neurotransmitters From the Extracellular Space That Have Been Released by Neurons

Neurotransmitters that have been released by neurons are taken up from the extracellular space by astrocytes. This uptake plays a role in removing neurotransmitters from the synaptic cleft so as to limit the duration of the synaptic potential. The relative contributions of glial cells versus neuronal reuptake mechanisms in physiological settings may vary according to the situation and with different types of neurotransmitter.

The processes of astrocytes also encapsulate neuronal elements, limiting diffusion of neuroactive molecules between neighboring neuronal elements. Astrocyte processes are thought to prevent the diffusion of various substances (ions and neurotransmitters) from one synaptic site to another, thus preventing *cross-talk* between adjacent synapses. This may be especially important in densely packed neuropil zones, where synaptic terminals lie adjacent to one another (see Figure 5.2).

Astrocytes Form a Potential Substrate for Communication Between Neurons and Capillaries

Astrocytic processes surround neurons (see earlier discussion) and also surround capillaries that are present in the brain parenchyma (Figure 5.7). This relationship has led to the speculation that astrocytes may mediate communication between neurons and vascular elements. One possibility is that astrocytes aid in distributing nutritive substances to neurons. Astrocytes may also convey material from the neurons to the capillaries. These materials could be by-products of cellular metabolism or could represent a means that would allow neurons to communicate their metabolic needs to the vasculature. An important aspect of cerebral blood flow is that it is closely coupled to neuronal metabolism, which is in turn closely coupled to neuronal activity. Astrocytes would be an ideal substrate for such communication, and there is currently considerable research concerning this possibility.

Astrocytes Respond to CNS Injury

After damage to the nervous system, astrocytes exhibit a *reactive response* involving an upregulation of expression of a number of genes, followed by a substantial hypertrophy. This reactive response can be documented by immunostaining brain sections with antibodies against GFAP (Figure 5.7).

Reactive astrocytes play a key role in recreating the glial-limiting membrane that serves as the boundary between CNS tissue and non-CNS tissue. Reactive astrocytes also play a role in removing dying synaptic terminals from postsynaptic cells and in synthesizing

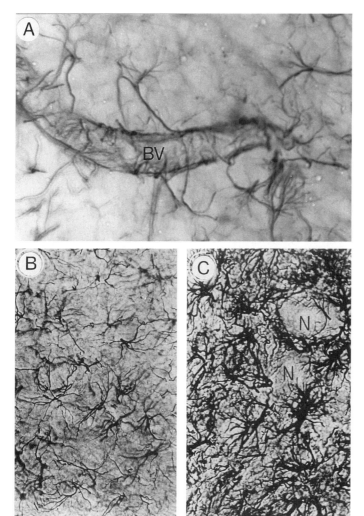

Figure 5.7. Relationship between astrocytes and blood vessels; reactive response of astrocytes following injury. A, Photomicrograph of a section through the brain that has been immunostained for GFAP. Astrocytes are heavily stained. Astrocytic processes can be seen to extend to cover the blood vessel passing through the field. **B,** Astrocytes in the uninjured rat spinal cord, as revealed by GFAP immunostaining. **C,** Reactive astrocytes near an injury site in the spinal cord. From Reier PJ. Gliosis following CNS injury: the anatomy of astrocytic scars and their influences on axonal elongation. N indicates neuronal cell body. In: Federoff S, Vernadakis A, eds. *Astrocytes: Cell Biology and Pathology of Astrocytes.* New York, NY: Academic Press; 1986.

various neurotrophic factors that induce neuronal growth following injury.

Astrocytes Play a Role in Guiding Neuronal Migration During Early Development

Early in development, astrocytes are distributed in a radial fashion in the developing neural tube; their processes extend from the central canal to the pial surface. At this stage, the glia are termed *radial glia.* Young neurons migrate along radial glial processes from the proliferative zone into the area where they will differentiate. After migration is complete, these glial cells retract their long radial processes and differentiate into astrocytes.

Astrocytes may also play a role in guiding growing axons. During the formation of long tracts, axons appear to grow along the surfaces of the radial glial processes. In this way, astrocytes form a matrix around which subsequent neuronal organization takes place.

Astrocytes Possess Receptors for Neurotransmitters, Neuromodulators, and Hormones

The role of receptors in modulating glial function is not known; it is thought that these receptors provide a means by which neurons can regulate astroglial metabolism or gene expression.

Oligodendrocytes

Oligodendrocytes are also present in gray matter and white matter. The oligodendrocytes in white matter are called *intrafascicular oligodendrocytes;* these cells are the myelin forming cells of the CNS (Figure 5.8). The oligodendrocytes in gray matter are termed *perineuronal oligodendrocytes* because they are found in close contact with neuronal cell bodies as *satellite cells.*

The *myelin sheath* is a specialization of the plasma membrane of the oligodendrocyte that wraps itself around the axon. Individual oligodendrocytes extend a number of processes, each of which contacts an individual axon (Figure 5.8). Each process then elaborates a myelin sheath around the axon that is contacted (Figure 5.9).

Myelin is formed in segments that extend for a characteristic distance along individual axons (usually a few hundred micrometers). The myelin sheaths from individual oligodendrocytes are then organized in tandem along the axon. The breaks between the myelin sheaths from individual myelin-forming cells are termed *nodes of Ranvier.* The region between the nodes is termed the *internode.*

The myelin sheath plays a key role in regulating the speed of action potential propagation because the sheath acts as insulation around the axon. Ionic currents associated with the nerve impulse cannot flow across the myelin, and therefore flow across the nerve membrane at the nodes of Ranvier. It is this property that results in *saltatory conduction,* in which the action potential jumps

Figure 5.8. Relationship between oligodendrocytes and neuronal elements. A, Intrafascicular oligodendrocytes are the myelin-forming cells in CNS white matter. A single oligodendrocyte provides the myelin sheath for several axons. **B** illustrates a relatively early stage in the process of myelin formation to emphasize the fact that the myelin consists of several wrappings of the plasma membrane of the oligodendrocyte. As the myelin matures, the membrane wraps become more closely apposed (see Figure 5.9). **C,** Perineuronal oligodendrocytes are located in gray matter and are so named because they are found in close contact with neuronal cell bodies as satellite cells.

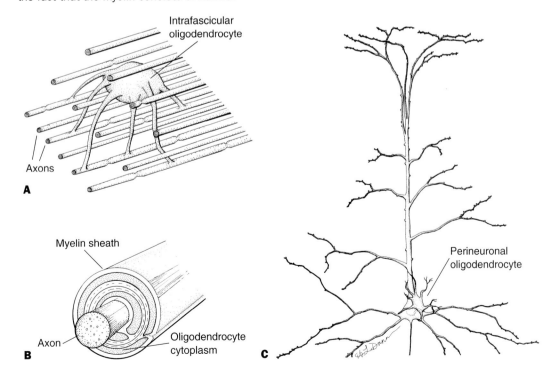

Intrafascicular oligodendrocyte

Axons

A

Myelin sheath

Axon

Oligodendrocyte cytoplasm

B

Perineuronal oligodendrocyte

C

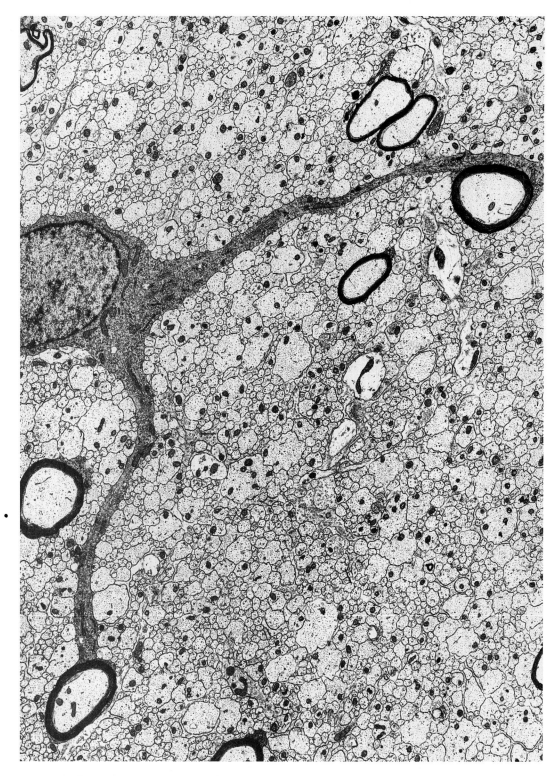

Figure 5.9. Electron micrograph illustrating a process from an oligodendrocyte contacting an axon. Individual oligodendrocytes extend a number of cytoplasmic processes, each of which contacts an individual axon. Each process then elaborates a myelin sheath around the axon that is contacted. **A1** and **A2** indicate two axons; **Ol,** oligodendrocyte cell body. Reproduced with permission from Waxman SG. *Correlative Neuroanatomy.* Norwalk, Conn: Appleton and Lange; 1996.

from one node to the next, thus dramatically increasing the net speed of conduction.

Perineuronal Oligodendrocytes Are Satellite Cells to Neurons

Perineuronal oligodendrocytes have a close physical relationship with neuronal cell bodies and dendrites (Figure 5.10). This fact suggests that the oligodendrocytes interact with the neurons in some way. Indeed, the term *satellite cell* seems particularly appropriate for the perineuronal oligodendrocyte. The nature and functional significance of the interaction is not known.

Glial Cells of the Peripheral Nervous System

In the peripheral nervous system (PNS), myelin is formed by *Schwann cells*; hence, Schwann cells represent a fourth category of glia. In contrast to oligodendrocytes, Schwann cells form myelin around only one axon, and the cell body of the Schwann cell lies adjacent to the axon that it myelinates. In *unmyelinated nerves,* many individual axons may be enveloped by the cytoplasm of a single Schwann cell (Figure 5.11). Schwann cells also encapsulate the terminal arborizations of motor nerves. In this location, they are also called *teloglia.*

The Regulation of Myelin Formation

The extent of myelination varies considerably in different neuronal systems. In the PNS, motor and sensory nerves are heavily myelinated; sympathetic nerves are unmyelinated. There are also differences in the extent of myelination of different pathways in the CNS. Because demyelinating disease is fairly common and severely debilitating, it is important to understand how the extent of myelination is regulated. An important question is whether it is the myelinating cell that determines the extent of myelination, or the axon.

Answers to this question have been provided by A. Aguayo and his colleagues. The approach involved the use of heterologous transplants. A segment of the sciatic nerve, which is usually heavily myelinated, was removed and replaced with a segment from a sympathetic nerve, which is lightly myelinated. In the graft, the axons degenerate because they have been separated from their cell bodies; the Schwann cells survive, however, and can remyelinate axons that regenerate into the graft. When sciatic nerve axons regenerated into a graft containing Schwann cells from sympathetic nerves, a heavy myelin sheath was formed. Thus, Schwann cells from a sympathetic nerve form a heavy myelin sheath when they contact axons of the appropriate type. This indicates that the extent of myelin formation is regulated by the axon.

The technique of heterologous transplantation has provided an important tool for investigating the mechanisms of human genetic disorders that result in abnormalities in myelin formation. For example, by transplanting human nerve segments into immunosuppressed mice, it was found that a human genetic disorder *metachromatic leukodystrophy* was due to a defect in Schwann cells.

Schwann cells promote axonal regeneration. Santiago Ramón y Cajal, in his classical studies of degeneration and regeneration in the nervous system, observed that axonal regeneration in the CNS is quite limited, whereas in the PNS, axon growth readily occurs. A major difference between nerves in the PNS and CNS is the presence of Schwann cells in the PNS, hence Cajal proposed that Schwann cells stimulated nerve regeneration.

Modern studies have fully confirmed this hypothesis. For example, using the sorts of grafting techniques already described, Albert Aguayo and his colleagues found that when CNS nerves that contain only oligodendrocytes (the optic nerve) are grafted into the sciatic nerve, regeneration of the axons of the sciatic nerve is severely disrupted. Grafts that contain Schwann cells support regeneration, however. Similarly, when peripheral nerves containing Schwann cells, such as the sciatic

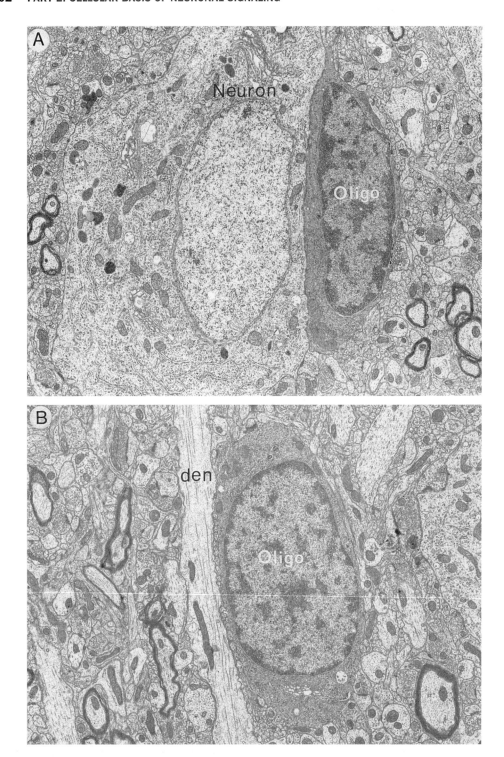

Figure 5.10. Perineuronal oligodendrocytes have a close physical relationship with neuronal cell bodies and dendrites. A illustrates an oligodendrocyte (Oligo) that is closely apposed to a neuronal cell body. **B** illustrates a similar relationship between an oligodendrocyte and a dendrite (den). From Steward O. *Principles of Cellular, Molecular, and Developmental Neuroscience.* New York, NY: Springer-Verlag; 1989.

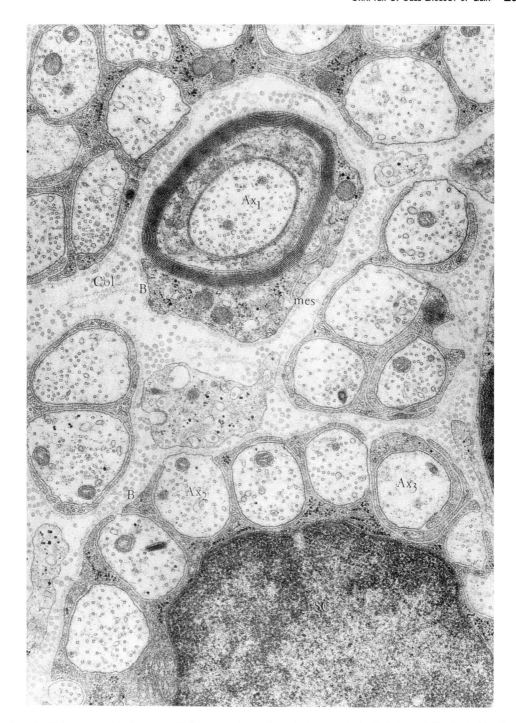

Figure 5.11. Relationship between Schwann cells and axons in peripheral nerves. This electron micrograph illustrates a cross section of the sciatic nerve of an adult rat. Ax_1 is a myelinated axon sectioned transversely through the paranodal region. The other axons in the field are unmyelinated but are surrounded by the cytoplasm from Schwann cells. This is well illustrated by the Schwann cell in the lower portion of the figure (SC), which encloses several axons. Reproduced with permission from Peters A, Palay SL, Webster HDeF. *The Fine Structure of the Nervous System: The Neurons and Supporting Cells.* Philadelphia, Pa: WB Saunders; 1976.

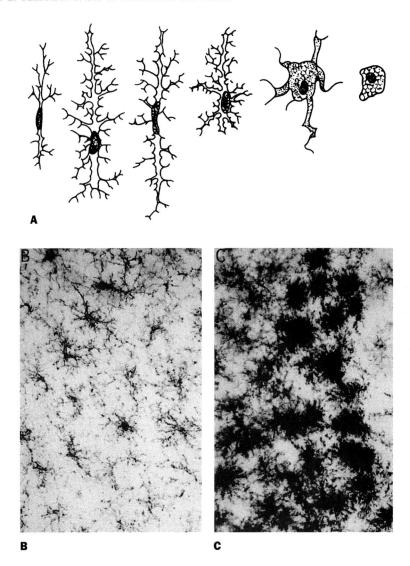

Figure 5.12. Microglial activation in response to CNS injury. A illustrates the progressive transformation of resting microglia into activated cells and then into microglia-derived macrophages. **B** illustrates the appearance of microglial cells in the hippocampus of a normal mouse; the cells are immunostained with an antibody against the Mac1 antigen. **C** illustrates activated microglia in a hippocampus in which there is ongoing axonal and terminal degeneration as a result of a lesion that destroyed a major input pathway. **A** is from Kruetzberg GW. Microglia: a sensor for pathological events in the CNS. *Trends Neurosci.* 1996;19:312–318. **B** and **C** are from Schauwecker PE, Steward O. Genetic influences on cellular reactions to brain injury: activation of microglia in denervated neuropil in mice carrying a mutation (Wld^s) that causes delayed Wallerian degeneration. *J Comp Neurol.* 1996;380:82–94. Reprinted by permission of Wiley-Liss, Inc., a subsidiary of John Wiley & Sons, Inc.

Table 5.1. Properties of Glial Cells

Glial cell type	Location	Characteristic cytology	Function
Protoplasmic astrocytes	Gray matter	Stellate form; long, thin processes that insinuate between neuronal cell bodies, dendrites, and synapses; processes form glia limitans at brain surface and glial end-feet on blood vessels	Spatial buffering of ions (esp. K^+); uptake of neurotransmitters; encapsulate neuronal processes to prevent diffusion of transmitters from one synapse to another
Fibrous astrocytes	White matter	Stellate form; long, thin processes that intertwine between axons	Spatial buffering of ions?
Intrafascicular oligodendrocytes	White matter	Extend long processes that form myelin sheath around axons of CNS neurons	Myelination
Perineuronal oligodendrocytes	Gray matter	Small, process-bearing cells that lie immediately adjacent to neuronal cell bodies and dendrites	Unknown
Microglia	Gray matter, white matter	Resemble perineuronal oligodendrocytes in form; in response to injury, processes withdraw and cells become phagocytic	Normal brain, unknown; injured brain, phagocytosis
Schwann cells	PNS	Cell bodies lie along axons; cells give rise to extensive plasma membrane sheets that enwrap axons to form the myelin sheath	Myelination

nerve, are implanted into the CNS, a vigorous regenerative response can be observed. It may eventually be possible to take advantage of the growth-promoting activity of grafts to promote the regeneration of CNS axons.

Microglia

Microglia are diffusely distributed throughout gray and white matter, where they appear as small cells with short, thin processes. Microglia express many of the same proteins as *macrophages,* including the *complement receptor,* and they are thought to be ontogenetically related to cells of the mononuclear phagocyte lineage. This identification is strongly supported by the fact that microglia play a key role in the response to nervous system injury.

Microglia are activated by virtually any pathological insult to the nervous system. Indeed, their responses are typically the first cellular responses that are seen after injury. The early response to injury involves a strong induction of expression of many of the genes that the cells normally express, including the genes that produce the proteins that are characteristic of macrophages. The subsequent response appears to be graded, depending on the severity of the injury. Minor injuries that cause some degeneration of axons or synaptic terminals induce a partial response. In this state, microglial gene expression is upregulated and the cells hypertrophy (Figure 5.12). When there is frank degeneration of neuronal cell bodies, microglia are transformed into cells that are indistinguishable from phagocytic macrophages. In this state, microglia are considered to be *microglia-derived brain macrophages.*

A three-stage activation scheme has been proposed, and Figure 5.12 illustrates the morphological changes that occur during the transformation. Microglia in the normal brain are considered to be *resting,* or *quiescent.* In the first stage of activation, in which there is a substantial hypertrophy of microglial processes, the cells are called *activated microglia.* In the end stage, when cells have rounded up so that they are indistinguishable from phagocytic macrophages, the cells are called *microglia-derived brain macrophages.*

Table 5.1 summarizes the characteristics of the different types of glia that have been discussed in the preceding.

CLINICAL DISORDERS INVOLVING GLIA

Most brain tumors are of glial or connective tissue origin, and tumors derived from the astrocyte lineage (astrocytomas) are the most common.

The development of immunochemical markers for differentiated astrocytes has provided pathologists with a way to make more sophisticated diagnoses of CNS tumors than previously possible. As already noted, differentiated astrocytes contain large numbers of glial filaments composed of GFAP. Cells that are less differentiated contain less GFAP. Thus, the amount of GFAP correlates with neoplastic potential. For this reason, immunocytochemical stains for GFAP have come into common use in neuropathology as a way of grading the neoplastic potential of tumors.

Oligodendrocytes are affected by a number of disease processes. Demyelinating diseases such as multiple sclerosis are the best example. Certain genetic disorders also lead to abnormal myelination.

6

Presynaptic Mechanisms Underlying Neurotransmitter Synthesis, Storage, Release, and Inactivation

Neurons communicate with each other and with other cells by using one of two forms of *neurotransmission:* electrical or chemical. In the CNS of higher vertebrates, chemical transmission predominates. An understanding of chemical neurotransmission is important for physicians because most drugs that affect the nervous system do so by modifying chemical neurotransmission. You will learn more about the mechanisms of drug action in the pharmacology component of your curriculum. Here, we consider the basic cell biological mechanisms that underlie neurotransmitter synthesis, storage, release, and inactivation.

We also briefly consider how we know which neurotransmitters are used by particular populations of neurons. Most of our knowledge is based on the development of *histochemical* and *immunocytochemical* staining techniques for particular neurotransmitter systems. These techniques have been very important for defining the distribution of neurotransmitter systems in the human CNS because many of the techniques can be used on pathological material. These techniques have provided important information about the types of neurons that are lost as a result of the different neurodegenerative diseases (e.g., Huntington's disease, Parkinson's disease, and Alzheimer's disease).

NEURONS COMMUNICATE BY RELEASING CHEMICAL NEUROTRANSMITTERS FROM SPECIALIZED SYNAPTIC TERMINALS

Neurons lie at one end of a continuum of secretory cells that communicate with other cells through the release of chemical substances (Figure 6.1). Neurons release their neurotransmitters onto a particular postsynaptic site on another cell (the site of contact between the presynaptic terminal and the postsynaptic neuron). At the other end of the continuum are endocrine cells that release hormones into the general circulation. Intermediate between these extremes are the *neurohormonal systems* that give rise to axons that terminate on blood vessels (for example, the neurons in the *supraoptic* and *paraventricular* nuclei of the hypothalamus).

Axons from these neurons project to the posterior lobe of the pituitary (the neurohypophysis), where they synapse on blood vessels so as to release the neuropeptide hormones into the general circulation (this is discussed further in Chapter 27).

There are also a few types of neurons whose synaptic terminals end blindly near other neurons rather than forming traditional synapses. Such terminals release their transmitters into a local extracellular environment rather than upon a specific postsynaptic site. This form of release can be thought of as being intermediate in character to neurohormonal release and the more common targeted release onto particular postsynaptic sites.

There are agreed-upon criteria for considering a substance as a neurotransmitter:

Figure 6.1. The continuum of stimulus-coupled release mechanisms. Hormones are released into the circulation (BV, blood vessel) by endocrine cells. Neurohormones are released into the circulation at specialized endings on blood vessels. Neurotransmitters are usually released upon a specialized postsynaptic membrane specialization, although in a few cases they may be released into the extracellular space. From Steward O. *Principles of Cellular, Molecular, and Developmental Neuroscience.* New York, NY: Springer-Verlag; 1988.

Hormones   Neurohormones   Neurotransmitters

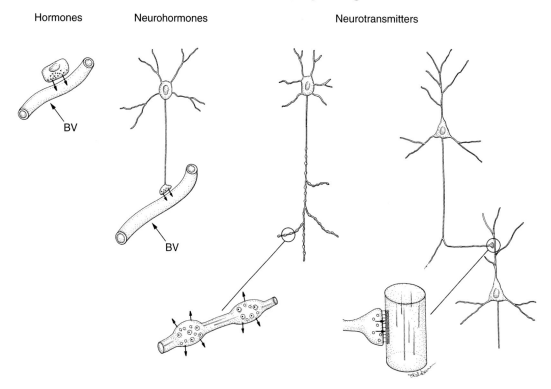

1. *Localization.* The substance should be localized in nerve endings.
2. *Release.* The substance should be released when the terminals are activated.
3. *Physiological identity.* When applied to the postsynaptic site, the substance should produce the same response as the natural neurotransmitter that is released from the terminal.
4. *Pharmacological identity.* Pharmacological agents should have identical effects on the action of the putative neurotransmitter when it is applied artificially, and on the synaptic response elicited by the natural neurotransmitter.
5. Other supportive evidence that a given substance is a neurotransmitter includes the presence of *high-affinity uptake* mechanisms for the substance in synaptic terminals, and the existence of receptors in association with particular pathways.

A note on terminology: When referring to neurons according to the transmitters that they use, the suffix *ergic* is used together with the name of the neurotransmitter. Thus, neurons that use the excitatory amino acid glutamate are termed *glumatergic;* neurons that use peptides are termed *peptidergic;* those that use acetylcholine are termed *cholinergic;* those that use catecholamines are *catecholaminergic;* and so forth.

CELL BIOLOGICAL MECHANISMS THAT UNDERLIE NEUROTRANSMITTER REGULATION

For any neurotransmitter, the neuron must (1) *synthesize* or *accumulate* the neurotransmitter substance; (2) *deliver* the neurotransmitter to the release site (the presynaptic terminals); (3) *store* the neurotransmitter substance in a form that is ready to be released (in vesicles); (4) *release* the neurotransmitter in response to the appropriate stimulus (an action potential invading the presynaptic terminal); and (5) *inactivate* the neurotransmitter or *remove* it from the synaptic cleft. An understanding of these processes is important for the physician because each step may be affected by disease processes, drugs, or toxins and is a potential site for therapeutic intervention.

As we will see, there are important differences in the way these tasks are accomplished for the different neurotransmitters used by CNS neurons. Because of these differences, it is convenient to divide neurotransmitters into three categories: (1) neurotransmitters that are synthesized by specific enzymes, (2) *amino acid neurotransmitters,* and (3) *peptide neurotransmitters* (see Table 6.1). In what follows, we first consider how the different neurotransmitters are dealt with and then consider the topic of neurotransmitter release.

Neurotransmitters That Are Synthesized by Specific Enzymes

Acetylcholine

Acetylcholine (ACh) was identified as a neurotransmitter in the 1920s, when it was found that ACh mimicked the action of the natural neurotransmitter of parasympathetic fibers in the heart (criterion 4 in the list in the last section). It is now recognized that ACh is used at the neuromuscular junction, at ganglionic synapses in the sympa-

Table 6.1. Classes of neurotransmitter

1. Neurotransmitters that are synthesized by specific enzymes

   Acetylcholine
   The catecholamines (epinephrine, norepinephrine, and dopamine)
   Serotonin
   GABA

2. Amino acid neurotransmitters

   The excitatory amino acids (glutamate and aspartate)
   Glycine

3. The neuropeptides

   Pro-opiomelanocortin-derived
   Pro-enkaphalin-derived
   Pro-dynorphin-derived

thetic and parasympathetic nervous system, and in a number of important CNS pathways. ACh pathways are particularly prominent in the caudate nucleus, basal forebrain, and brainstem.

Acetylcholine is synthesized from acetyl-CoA and choline by the enzyme *choline acetyltransferase* (Figure 6.2). Synthesis depends on the uptake of choline, whereas acetyl-CoA is produced by the neuron. Choline acetyltransferase (ChAT) is a cytoplasmic protein that is present within the synaptic terminal; hence, ACh is synthesized within the cytoplasm and then is taken up and concentrated in synaptic vesicles (Figure 6.2). It is estimated that the

concentration of ACh within the vesicles is 880 mM.

ACh is degraded by the enzyme *acetylcholinesterase* (AChE), which is present within the synaptic cleft. In the CNS, AChE is synthesized by the cholinergic neurons themselves, not by the cells that receive cholinergic input. This is in contrast to the situation in the peripheral nervous system, where AChE is produced by *cholinoceptive* muscle cells.

Mapping Cholinergic Pathways

ChAT and AChE are useful molecular markers of central cholinergic neurons. The distribution of CNS neurons that use acetylcholine as a neurotransmitter was first de-

Figure 6.2. The cholinergic synapse. The synthetic pathway for ACh. Numbers indicate potential sites of drug action: 1, uptake of precursor; 2, synthesis of ACh by choline acetyltransferase; 3, uptake into vesicle; 4, release; 5, interaction of neurotransmitter with receptors in the postsynaptic membrane; and 6, degradation by acetylcholinesterase (AChE).

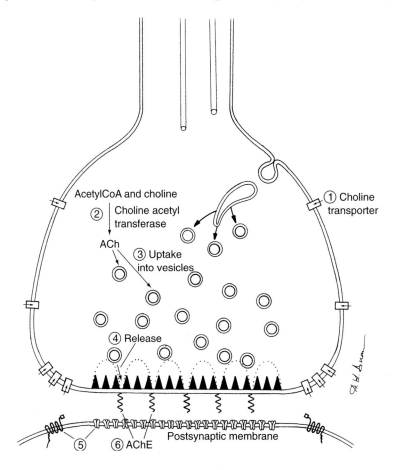

fined using a simple histochemical method for AChE. This stain highlights cholinergic cell bodies, axons, and terminal fields. More recently, immunocytochemical techniques have been developed that take advantage of antibodies against choline acetyltransferase.

Catecholamines

Catecholamines include *epinephrine* (adrenaline), *norepinephrine* (noradrenaline), and *dopamine* (Figure 6.3). Epinephrine is released by the adrenal gland and may be used as a neurotransmitter by a few neurons in the CNS. Norepinephrine is the neurotransmitter in the postganglionic fibers of the sympathetic nervous system, and for certain nuclei in the brainstem. Dopamine is the neurotransmitter used by the *nigro-striatal pathway* from the substantia nigra to the caudate-putamen complex and is also used by neurons in several nuclei in the brainstem.

Each catecholamine is synthesized by a particular rate-limiting enzyme from a catecholamine precursor (Figure 6.3). The rate-limiting enzyme for dopamine is dopa decarboxylase, the rate-limiting enzyme for norepinephrine is tyrosine hydroxylase, and so forth. The synthetic enzymes for the catecholamines are localized in *dense core vesicles,* which are present within presynaptic terminals (Figure 6.4). Other proteins are also present, along with high concentrations of ATP. The contents of the vesicles, including the synthetic enzymes, may be coreleased with the transmitter.

The action of catecholamines is terminated by reuptake into the terminal (see later). This reuptake is mediated by high-affinity *transporters.* Glial cells may also play a role in removing the transmitter, and glia possess enzymes that degrade neurotransmitter substances. The transmitter that is taken up by synaptic terminals can be reutilized; that which is taken up by glial cells is presumably degraded.

Mapping Catecholaminergic Pathways

Catecholamine pathways in the CNS were first mapped using a *histofluorescent* technique developed by B. Falck and N. A. Hillarp. The technique is based on the fact that the neurotransmitters themselves can be complexed with other molecules so as to fluoresce. Because catecholamine neurotransmitters are present throughout the cell bodies, axons, and even dendrites of neurons that use the substances, the histofluorescent technique reveals the distribution of catecholaminergic systems in great detail.

The original histofluorescent techniques have been largely supplanted by immunocytochemical techniques that use antibodies against the synthetic enzymes. All catecholaminergic neurons contain tyrosine hydroxylase, which therefore is a useful marker for catecholamine neurons in general (Figure 6.5). Dopamine β-hydroxylase is a useful selective marker for noradrenergic neurons, but distinguishing neurons that use dopamine is a bit more complicated because dopa decarboxylase is present in both dopaminergic and noradrenergic neurons.

Figure 6.3. Metabolic pathway for the synthesis of catecholamines.

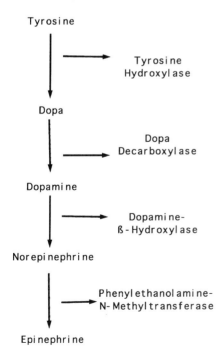

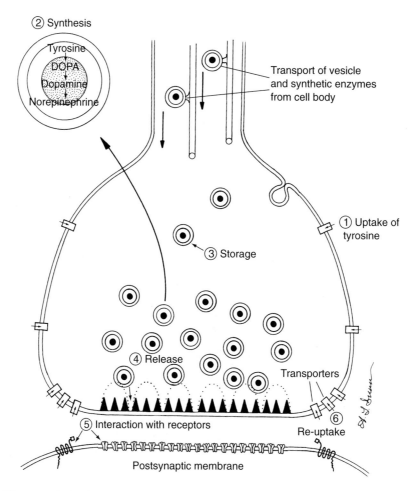

Figure 6.4. The catecholaminergic synapse. The synthetic pathway for catecholamines. Numbers indicate potential sites of drug action: 1, synthetic enzymes (see Figure 6.3); 2, storage in vesicles; 3, release; 4, interaction of neurotransmitter with receptors in the postsynaptic membrane; and 5, reuptake.

Serotonin

Serotonin or 5-hydroxytryptamine (5-HT) is the neurotransmitter used by neurons of the *raphe nuclei* of the brainstem. It is synthesized from the amino acid tryptophan by the enzyme *tryptophan hydroxylase*. The cell biological mechanisms that underlie serotonin synthesis, storage, release, and inactivation are, in general, similar to those of the chatecholamines. In particular, tryptophan hydroxylase is located in presynaptic terminals; the synthetic enzyme and the neurotransmitter are stored in amine granules. Also, like the catecholamines, serotonin's action is terminated by reuptake.

Mapping Serotonergic Pathways

Serotonin pathways were initially mapped using histofluorescent techniques. Today, however, the pathways are typically traced with immunocytochemical techniques.

γ-Aminobutyric Acid

γ-Aminobutyric acid (GABA) is the principal neurotransmitter used by inhibitory interneurons in most brain regions. GABAergic interneurons project locally and form basket-type endings around the perikarya of other neurons. GABA-containing synapses also terminate upon axon initial segments

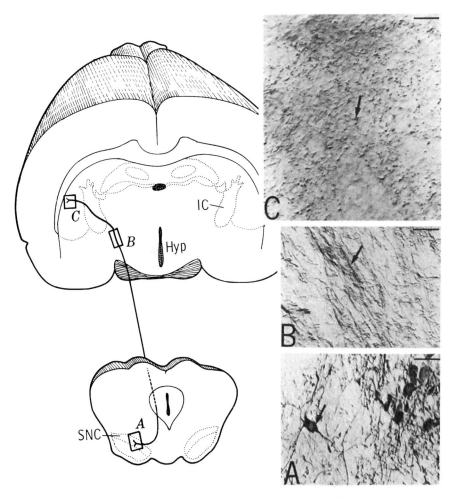

Figure 6.5. An example of the use of immunocytochemistry to map central catecholaminergic pathways. This figure illustrates the localization of tyrosine hydroxylase in dopaminergic neurons of the nigrostriatal system as revealed using immunocytochemistry. Neurons in the substantia nigra pars compacta (SNC) project to the midbrain via a tract that travels through the lateral hypothalamus. **A** illustrates labeled neuronal cell bodies in the SNC; **B** illustrates labeled axons passing through the hypothalamus; **C** illustrates the labeled synaptic terminals in the striatum. Reproduced with permission from Pickel VM. Immunocytochemical methods. In: Heimer L, Robards M. *Neuroanatomical Tract-Tracing Methods.* New York, NY: Plenum Press; 1981:483–509.

and upon dendritic shafts. Cerebellar Purkinje cells are also GABAergic as are many neurons in the basal ganglia. GABAergic synapses form type II (symmetric) synapses, and, using appropriate fixation conditions, the presynaptic terminals contain flattened vesicles (see Chapter 4).

GABA is synthesized from the amino acid glutamate by the enzyme *glutamic acid decarboxylase.* The enzyme is localized in synaptic terminals, and the neurotransmitter itself is accumulated into vesicles that appear flattened under appropriate fixation conditions. GABA's action is terminated primarily by reuptake via specific transporters located in the presynaptic membrane (Figure 6.6).

Mapping GABAergic Pathways

There are no simple histochemical or histofluorescent techniques for detecting GABA

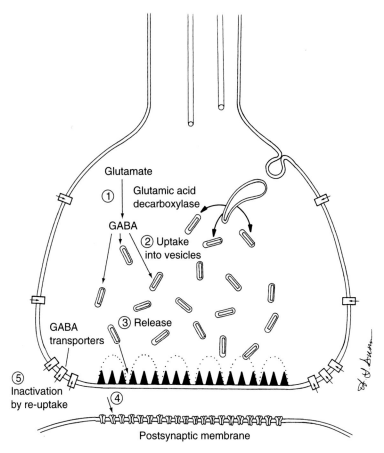

Figure 6.6. The GABAergic synapse. The synthetic pathway for GABA. Numbers indicate potential sites for drug action: 1, synthesis of GABA by glutamic acid decarboxylase (GAD); 2, uptake into vesicles; 3, release; 4, interaction with receptors; and 5, reuptake by GABA transporters.

or its enzymes. However the development of specific antibodies for glutamic acid decarboxylase and GABA itself has made it possible to define the distribution of GABAergic neurons and their synaptic terminals using immunocytochemistry.

The Amino Acid Neurotransmitters

Glutamate

Glutamate is thought to be the most common excitatory neurotransmitter in the CNS. Aspartate may also play a neurotransmitter role alone or in conjunction with glutamate (i.e., as cotransmitters). Hence, both are clas-

sified as excitatory amino acid (EAA) neurotransmitters.

The properties of other neurotransmitter substances that have provided means for experimental investigation are not available with the EAAs. The EAAs that serve a neurotransmitter function are derived from metabolic pathways that are present in all cells (the Krebs cycle), not from unique synthetic enzymes. Thus, it has not been feasible to develop antibodies to unique synthetic enzymes for immunocytochemical studies. Moreover, the EAAs themselves are not uniquely localized in neurons that use them as transmitters; a crucial problem has been distinguishing neurotransmitter pools from metabolic pools.

Despite the lack of appropriate methods for localizing EAA neurotransmitters, the evidence that the EAAs do serve as neurotransmitters is very strong: (1) EAAs are released as a result of physiological activation of CNS pathways. (2) The excitatory amino acids meet the physiological and pharmacological identity criteria, that is, when applied iontophoretically, the amino acids have physiological and pharmacological properties that are similar to the natural neurotransmitter. (3) Specific EAA receptors are found in association with the pathways (see Chapter 7).

Rather than being synthesized locally, the EAA neurotransmitters are taken up into the synaptic terminals and then into synaptic vesicles by high-affinity transporters (Figure 6.7). It is thought that the EAAs are then stored in the small, clear vesicles present in excitatory synaptic terminals; however, a small number of investigators still question this dogma. The same transporters that concentrate the EAAs in the first place are also thought to be responsible for removing the EAAs from the synaptic cleft after release.

Mapping Glutamatergic Pathways

Glutamatergic neurons were initially identified by taking advantage of the fact that these neurons possess selective high-affinity uptake systems. When exposed to radiolabeled glutamate or aspartate, these neurons accumulate the labeled amino acid; the distribution of the neurons can then be mapped

Figure 6.7. The glutamatergic synapse. Glutamate is taken up and concentrated in vesicles rather than being synthesized. Numbers indicate potential sites for drug action: 1, uptake into terminal; 2, uptake into vesicles; 3, release; 4, interaction with receptors; 5, reuptake.

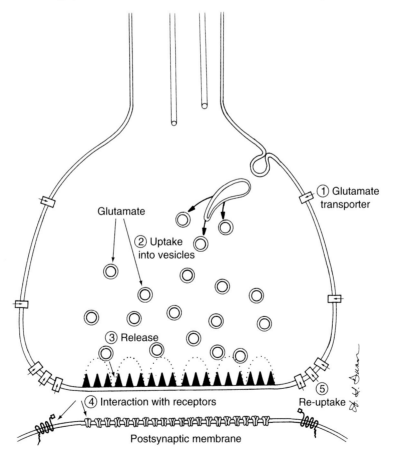

using autoradiography. Much of what we know about the distribution of these pathways is actually based on the differential distribution of EAA receptors (see Chapter 7).

Glycine

Glycine is thought to operate as an inhibitory neurotransmitter for some neurons in the spinal cord, medulla, and pons. As with other EAA neurotransmitters, no specific synthetic or degradative enzymes have been identified for the glycine that is used as a neurotransmitter. Hence, the cell biological mechanisms that underlie the accumulation of glycine in vesicles and the removal of the neurotransmitter after release are thought to be similar to the mechanisms that exist in glutamatergic synapses.

Mapping Glycinergic Pathways

Glycinergic neurons were initially identified by taking advantage of the fact that these neurons possess selective high-affinity uptake systems. When exposed to radiolabeled glycine, these neurons accumulate the labeled glycine, and the distribution of the neurons can be mapped using autoradiography. More recently, antibodies have been produced that have been useful for immunocytochemistry. Using these techniques, it has been shown that glycine and GABA are used by different populations of inhibitory neurons in the CNS.

The Neuropeptides

The *neuropeptides* are small peptide molecules containing from 3 to about 100 amino acid residues. Among the neuropeptides are the *neurohormones,* including the hypothalamic releasing factors; indeed, oxytocin and vasopressin were among the first of the neuropeptides to be recognized and characterized.

A particularly interesting subgroup of neuropeptides are the *endorphins,* which are endogenous, opiate-like substances. These substances have in common an *opioid core* composed of the amino acid sequence Tyr-

Gly-Gly-Phe-Met or Tyr-Gly-Gly-Phe-Leu. Neurons that use endorphins as their neurotransmitter play a key role in regulating transmission along pain pathways (see Chapter 12). Opiates act at the synapses that use endorphins.

Synthesis of the peptide neurotransmitters requires the translation of a specific mRNA, and posttranslational processing that may require several additional enzymatic reactions (Figure 6.8). Like other proteins, neuropeptide precursors are synthesized in the neuronal cell body. The initial neuropeptide gene product is a *preproprotein.* The "pre" sequence is the recognition sequence that directs the preproprotein through the rough endoplasmic reticulum membrane. This "pre" sequence is cleaved cotranslationally prior to the completion of synthesis, so that the *proprotein* comes to reside in the cisterns of the rough endoplasmic reticulum. The proprotein is then translocated from the cisternae into the Golgi apparatus, where it is packaged into *secretory granules.*

The biologically active neuropeptides are derived from different proproteins by sequential proteolysis, sometimes followed by other posttranslational modifications such as acetylation. There are three principal proproteins in mammals that are the precursors for several neuropeptides (see Figure 6.9). Thus, *pro-opiomelanocortin (POMC)* contains the sequences for the opioid peptide β-*endorphin,* as well as the sequences for *adrenocorticotropin (ACTH),* β-*lipotropin,* and α-, β-, *and* γ-*melanocyte stimulating hormone. Proenkephalin* contains the sequences for *met-enkephalin, leu-enkephalin, met-enkephalin-arg-phe,* and *met-enkephalin-arg-gly-leu,* and *prodynorphin* contains the sequences for α/β-*neo-endorphin, dynorphin A,* and *dynorphin B.*

Posttranslational processing may occur in the cell body or within the axon. For example *pro-oxytocin* and *provasopressin* are produced by proteolysis of the proproteins within secretory vesicles. Some of the processing occurs as the vesicles are transported down the axons. By regulating how the proproteins are processed, neurons can regulate not only the amount, but also the

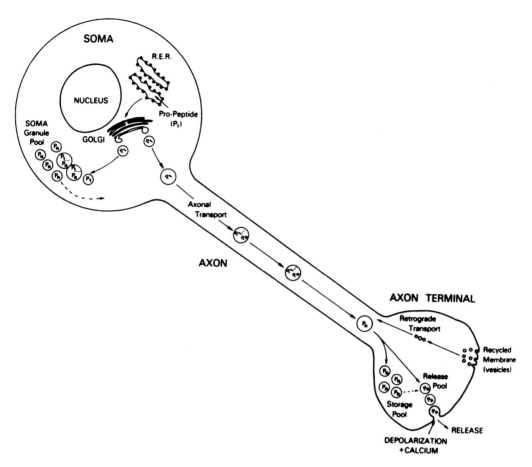

Figure 6.8. Synthesis, processing, and release of neuropeptides. Synthesis of the propeptide (P1) occurs on the rough endoplasmic reticulum in the neuron soma. The propeptide is packaged into secretory granules in the Golgi apparatus. Posttranslational processing occurs in the secretory granules. This processing can occur either in the soma or during axonal transport. From Gainer H, Sarne Y, Brownstein MJ. Biosynthesis and axonal transport of rat neurohypophyseal proteins and peptides. *J Cell Biol.* 1977;73:366–381.

blend of neuropeptides available in the secretory compartment (the secretory granules).

The mechanisms for inactivating the neuropeptides remain to be defined. Neuropeptide inactivation may involve enzymatic hydrolysis in the synaptic cleft, but the direct evidence for this hypothesis is weak. Several enzymes have been identified that are capable of hydrolyzing neuropeptides, but it has not been established that these enzymes are actually present in the synaptic cleft. Furthermore, because even the simple tripeptides are neuroactive, hydrolysis would have to be thorough to effectively terminate the action of the transmitter. In some cases, hydrolysis could simply change the biological activity of the neuropeptide rather than inactivating it.

Mapping Peptidergic Pathways

The key approach that led to the elucidation of the many neuropeptide systems in the CNS was the development of specific antibodies for each peptide that could be used for immunocytochemistry. Using these techniques, it was found that the hypothalamus

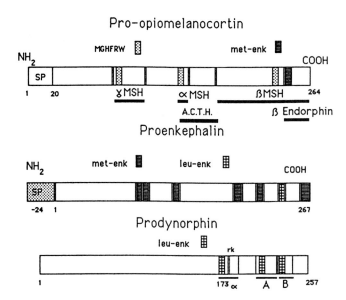

Figure 6.9. Prohormone precursors of opiate peptides. Bar diagrams of the three principal proproteins in mammals. The basic amino acid sequences that are the cleavage sites for processing are indicated as single or double vertical lines. Pro-opiomelanocortin and proenkephalin have consensus signal sequences at their N termini. From Cooper JR, Bloom FE, Roth RH. *The Biochemical Basis of Neuropharmacology.* 8th ed. New York, NY: Oxford University Press; 1986.

is particularly rich in neurons that contain neuropeptides. However, neuropeptides are also found in all other parts of the nervous system. In cortical structures, the neuropeptide-containing cells often seem to be a class of interneuron.

Colocalization

Individual synaptic terminals may contain more than one neurotransmitter. This is particularly true of the neuropeptide systems. Sometimes, neuropeptides are colocalized with other types of neurotransmitters such as GABA, ACh, dopamine, norepinephrine, or serotonin. In other situations, different neuropeptides are colocalized. Colocalization implies that the substances may be released simultaneously as cotransmitters. In this regard, it is of interest that the neuropeptides are capable of modulating the effect of nonpeptide neurotransmitters.

Other Neurotransmitter Candidates

A number of other molecules have been proposed as possible neurotransmitters, including adenosine, epinephrine, and histamine. These substances are unquestionably neuroactive in that they affect neural activity,

however, the other criteria for a neurotransmitter role have not yet been met.

Delivery of Neurotransmitters and Neurotransmitter Enzymes

Delivery of neurotransmitters or their enzymes to synaptic terminals depends on rapid transport mechanisms. For the neurotransmitters that are accumulated within terminals (the amino acids), or for transmitters that are synthesized locally within terminals, delivery is not a problem. The molecules are already in place within the terminal. However, the transporters and synthetic enzymes (both of which are proteins) must be delivered from the neuronal cell body. Because the transporters are membrane proteins, it is probable that they are transported via the rapid axonal transport systems. Similarly, the synthetic enzymes that are localized in vesicles are usually transported via rapid transport systems. Secretory granules containing neuropeptides are delivered to terminals via rapid transport systems.

Release of Neurotransmitters From Presynaptic Terminals

The aspect of release that is special in neurons is that it occurs at *presynaptic terminals* that are structurally specialized for the purpose. In neurons that transmit via action potentials *(spiking neurons),* neurotransmitter release is triggered by the invasion of the action potential. In *nonspiking neurons* (those that do not support an action potential, such as sensory receptors and neurons in the retina), the release varies in relation to the membrane potential. Even so, the cellular mechanisms governing the release process are similar. In what follows, we focus on the more common form of release of neurotransmitters in response to the action potential.

The Invasion of the Action Potential Triggers Release Through a Process of Excitation-Secretion Coupling

When the action potential invades the presynaptic terminal, the terminal is depolarized to near the equilibrium potential for Na^+. This depolarization sets into motion a cascade of events that leads to neurotransmitter release. This cascade is termed *excitation-coupled secretion.* This is an example of *regulated release,* as distinct from *constitutive release* processes in other cell types that are not triggered by depolarization.

In the same way that an appropriate model system (the squid giant axon) made possible the initial characterization of the mechanisms of the action potential, another model system (the squid giant synapse) made possible the initial studies of the mechanisms of release. This work, performed by Bernard Katz and his colleagues, formed the basis for our current understanding of the release process.

Release Is Measured Experimentally by Measuring the Effect of the Neurotransmitter on the Postsynaptic Cell

The easiest and most accurate measure of neurotransmitter release is the response that is generated in the postsynaptic cell. At the squid giant synapse, neurotransmitter release results in a strong depolarization on the postsynaptic side, which is termed the *excitatory postsynaptic potential* (EPSP). The ionic

basis of the EPSP is explained further in Chapter 7. For now, you need to know that the EPSP is a graded response, the amplitude of which varies in direct relationship with the amount of neurotransmitter that is released. Thus, the postsynaptic cell represents a superb bio-sensor for the amount of neurotransmitter released at any one time.

The key discoveries made by Katz and his colleagues were that (1) release is dependent on Ca^{2+} flux, and (2) release occurs in quantal units that we now know to be multimolecular packets of neurotransmitter contained within synaptic vesicles.

Excitation-Coupled Secretion Depends on Calcium

The release of neurotransmitter from presynaptic terminals in response to depolarization depends critically on extracellular Ca^{2+}. The lower the concentration of Ca^{2+} in the extracellular medium, the less neurotransmitter is released. When Ca^{2+} channels are blocked (for example, by increasing extracellular Mg^{2+}), release is blocked. This and other work led to the formulation of the *calcium hypothesis*—that neurotransmitter release is triggered by Ca^{2+} flux across the presynaptic terminal membrane. Subsequent studies by R. Llinas and his colleagues demonstrated that neurotransmitter release was directly related to the magnitude of the Ca^{2+} flux.

Release Depends on Ca^{2+} Channels in the Presynaptic Plasma Membrane

The pivotal role of Ca^{2+} flux was established before there was a clear understanding of the molecular nature of ion channels. As knowledge about ion channels increased, it became possible to describe release in terms of ion channel function. Thus, it is now known that when the action potential invades the presynaptic terminal, the resulting depolarization (produced by the transient Na^+ flux) triggers the opening of voltage-gated Ca^{2+} channels in the presynaptic membrane. Imaging studies using Ca^{2+}-sensitive fluorescent probes have indicated that the calcium flux occurs predominantly at the portion of the presynaptic membrane that is apposed to the postsynaptic membrane *(the release site)*. The calcium influx then triggers the release of neurotransmitter by promoting the fusion of vesicles with the plasma membrane.

Each step in the process requires a few tenths of a millisecond. At the squid giant synapse, the Ca^{2+} current that is induced by the presynaptic action potential occurs about 0.5 ms after the onset of the action potential; neurotransmitter release is delayed by an additional 0.2 ms. In warm-blooded animals, the release process is somewhat faster. The time required for the action potential to trigger release and for the neurotransmitter to diffuse to the postsynaptic cell determines the *synaptic delay.*

Neurotransmitter Is Released in Quantal Units That Correspond to Individual Synaptic Vesicles

The other major discovery made by Katz and his colleagues was the quantal nature of neurotransmitter release. The key studies were carried out using the neuromuscular junction. Stimulation of a motor axon produces an excitatory postsynaptic potential in the muscle fiber that is termed the *end plate potential* (EPP). The amplitude of the EPP provides a convenient bioassay for the amount of neurotransmitter released at any one time.

During their analysis of synaptic transmission at the neuromuscular junction, P. Fatt and Katz made a key discovery: in the absence of presynaptic stimulation, there were small spontaneous depolarizations that had features that resembled the EPP elicited by presynaptic stimulation. In particular, the spontaneous depolarizations were affected by the same drugs that affected the EPP, indicating that both were the result of the action of the neurotransmitter (acetylcholine) used by the motor axon. These small potentials were termed *miniature end plate potentials* (commonly referred to as *MEPPs,* or in lab jargon, *minis*).

Katz and his colleagues subsequently discovered several important facts about MEPPs that were the key to the model for release that was subsequently developed: (1) MEPPs had a characteristic size (about 0.5 mV), (2) MEPP frequency increased when the presynaptic terminal was depolarized, and (3) careful calculations of the extent of depolarization produced by applied acetylcholine indicated that MEPPs produced a depolarization equivalent to that produced by about 5000 molecules of ACh. These observations suggested that MEPPs were the result of a spontaneous release of packets, or quanta, of neurotransmitter from the presynaptic terminal.

The final piece in the puzzle came with the discovery that the EPP was a result of the simultaneous release of a number of quanta. The key evidence came from recordings of neuromuscular transmission when extracellular Ca^{2+} concentrations were very low. This can be achieved experimentally by reducing extracellular Ca^{2+} concentrations. In this setting, neurotransmitter release is decreased, leading to a decrease in the size of the EPP. Del Castillo and Katz found that when Ca^{2+} concentrations were very low, the amplitude of the EPP evoked by presynaptic stimulation varied. Sometimes there was no response at all (termed a *failure*); sometimes, there was a response that was approximately the same amplitude as a MEPP; and sometimes there was a response that appeared to be an integral multiple of the MEPP amplitude (Figure 6.10).

As extracellular Ca^{2+} concentrations were increased from the minimal levels, the amplitude of the integral units did not change; however, there were fewer failures, and the probability of observing responses that were multiples of the unitary response increased. These changes could be accurately modeled by the Poisson theorum, assuming that neurotransmitter release occurred in packets of the size of MEPPs, and that the release of any individual packet was a probabilistic event. These observations led Katz and his colleagues to propose the quantal hypothesis for synaptic transmission, which holds that the EPP is generated as the result of the release of

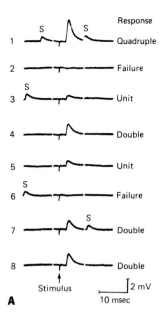

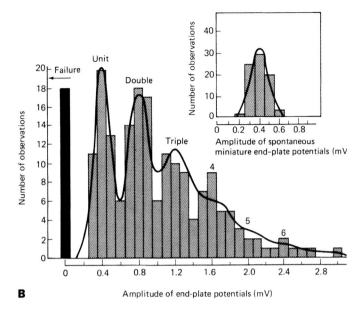

Figure 6.10. The amplitude distributions of MEPPs indicate that stimulus-coupled release involves the release of multimolecular packets of neurotransmitter. From Kandel ER, Schwartz JH, Jessel TM. *Principles of Neural Science.* 3rd ed. New York, NY: Elsevier; 1991. **A** is adapted from Liley AW. The quantal components of the mammalian end-plate potential. *J Physiol (Lond).* 1956;133:571–587. **B** is adapted from Boyd IA, Martin AR. The end-plate potential in mammalian muscle. *J Physiol. (Lond).* 1956;132:74–91.

a number of quanta, each of which carries a constant amount of neurotransmitter.

An important technique arising from the quantal hypothesis of synaptic transmission is the technique of *quantal analysis* (see Box), which has provided a way to define how release parameters change in different situations.

Synaptic Vesicles Are the Repository for the Packets of Neurotransmitter That Undergo Quantal Release

The recognition of the quantal nature of neurotransmitter release provided the context for interpreting the results of the initial electron microscopic analyses of synapses. When early electron microscopic studies by S. Palay revealed the existence of *synaptic vesicles,* the obvious conclusion was drawn that the vesicles were the repository for the quanta of neurotransmitter, and that quantal release reflected the release of the contents of the vesicle into the synaptic cleft. This is the *vesicle hypothesis* of neurotransmitter re-

Quantal Analysis

Neurotransmitter release is a regulated process that can be modified. It is important to define what aspect of neurotransmission is actually altered by particular manipulations, that is, whether changes occur presynaptically or postsynaptically. Because neurotransmitter is released in quanta, and the release is a probabilistic event, the process can be defined mathematically. The factors in the equation are (1) the probability of release *(p)*, (2) the number of quanta released *(n)*, and (3) the amplitude of the response produced in the postsynaptic cell *(m)*. Thus, $m = np$. By determining the values of *p, n,* and *m,* it is possible to determine the elements of release that are modified.

lease. This hypothesis has largely stood the test of time and has been expanded upon and refined so that release is now explained on the basis of vesicle fusion with the presynaptic membrane.

Release Occurs as a Result of the Fusion of Synaptic Vesicles With the Presynaptic Plasma Membrane

The vesicle hypothesis of neurotransmission holds that release involves *exocytosis,* in which the synaptic vesicles fuse with the external membrane of the terminal so that the contents of the vesicle are released into the synaptic cleft. Important evidence in support of this hypothesis was provided by a series of careful studies of release at the neuromuscular junction by T. Reese and J. Heuser. The key to these studies was the use of a rapid freezing technique that allowed presynaptic terminals to be quickly frozen as they were releasing neurotransmitter. When terminals were frozen during periods of massive release, *omega profiles* appeared along the portion of the presynaptic membrane abutting the synaptic junction. These omega profiles represent vesicles that had partially fused with the presynaptic plasma membrane (Figure 6.11).

Careful quantitative studies of the surface membrane area of presynaptic terminals and the number of vesicles in stimulated and nonstimulated presynaptic terminals indicated that, under conditions of massive release, there were decreases in the number of vesicles and corresponding increases in the area of the plasma membrane of the presynaptic terminal. As the presynaptic terminals recovered from the period of massive release, the number of vesicles increased, and the area of the plasma membrane decreased. These observations were the basis of the *vesicle recycling hypothesis,* which holds that vesicles fuse with the plasma membrane, and this membrane is then recovered during the reformation of the vesicles.

Dramatic visual documentation of the vesicle fusion process has been provided by studies of the presynaptic plasma membrane of the neuromuscular junction using *freeze–fracture* techniques (Figure 6.12). Freeze–

Figure 6.11. The vesicle fusion process during stimulated release. Electron micrographs of the presynaptic component of the frog neuromuscular junction: **A.** nonstimulated synaptic terminals; **B.** synaptic terminals that were massively activated. Note the omega profiles suggesting vesicle fusion. Reproduced with permission from Heuser JE, Reese TS. *The Journal of Cell Biology* 1973;73: 366–381, by copyright permission of the Rockefeller University Press.

Figure 6.12. Freeze–fracture views of the release process. An *en face* view of the internal leaflet of the presynaptic plasma membrane of nonstimulated **(A)** versus stimulated terminals **(B)**. When presynaptic terminals are massively stimulated, pitlike depressions appear on the cytoplasmic leaflet of the presynaptic membrane (Figure 6.11). These pits mark the site of fusion of the vesicles, and the pit itself represents the aperture into the fused vesicle. Reproduced with permission from Heuser JE, Reese TS. *The Journal of Cell Biology* 1981;88:564–580, by copyright permission of the Rockefeller University Press.

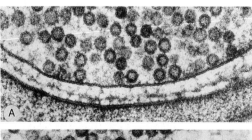

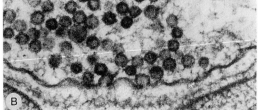

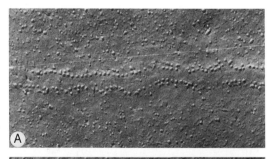

fracture techniques yield *en face* views of large areas of the internal leaflets of plasma membranes. When presynaptic terminals are massively stimulated, pitlike depressions appear on the cytoplasmic leaflet of the presynaptic membrane (Figure 6.12). These pits mark the site of fusion of the vesicles, and the pit itself represents the aperture into the fused vesicle.

Vesicle Fusion Occurs Selectively on the Portion of the Presynaptic Membrane Apposed to the Postsynaptic Site

To maximally affect the postsynaptic site, vesicles must be released into the synaptic cleft. Because the entire terminal is depolarized during the action potential, there must be a mechanism for directing the vesicles to the appropriate site on the terminal membrane. Freeze–fracture studies of presynaptic terminals at neuromuscular junctions revealed that vesicles fused at particular locations termed *active zones* that were characterized by arrays of intramembranous particles (Figure 6.12). In synapses in the CNS, active zones are located in the *presynaptic grid* that is visible using specialized staining techniques.

The entire process of neurotransmitter release as it is currently understood is summarized in Figure 6.13. Considerable effort is currently being directed toward defining the molecules that mediate the various stages in the vesicle docking and release process.

Nondirected Release Occurs Via a Different Mechanism Than Directed Release

The general model for release outlined here is thought to apply to all situations in which neurotransmitter is released from small vesicles into a synaptic cleft. In some synapses, however, release appears to be nondirectional, so that release occurs essentially anywhere along the plasma membrane. This appears to be true of the release of transmitters from certain dense core vesicles.

There are differences in the molecular composition of the membranes of small vesicles and dense core vesicles that may account

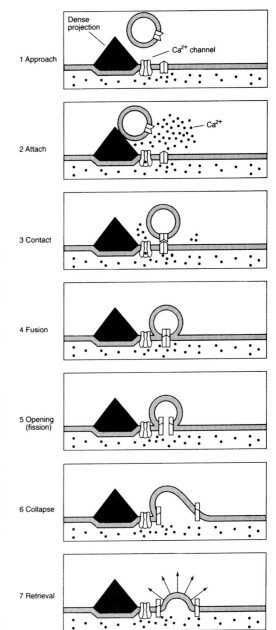

Figure 6.13. Model for vesicle docking and release. Schematic illustration of the steps that are thought to occur during vesicle docking and release. Adapted from Kandel ER, Schwartz JH, Jessel TM. *Principles of Neural Science.* 3rd ed. New York, NY: Elsevier; 1991. From Llinas RR, Heuser JE. Depolarization-release coupling systems in neurons. *Neuroscience Research Program Bulletin.* 1977;15:555–687.

for the different modes of release. In particular, small vesicles contain a number of unique proteins that are thought to be involved in localizing the vesicles near the release site, and docking the vesicles with the presynaptic membrane at the time of release. These proteins are apparently absent from the membranes of large, dense core vesicles, accounting for the differences in their behavior. In synapses that contain both types of vesicles, it is thought that small vesicles fuse selectively at the active zone, whereas large, dense core vesicles fuse nonselectively.

The Amount of Neurotransmitter Released Is Regulated by Processes Within the Presynaptic Terminal

The total amount of neurotransmitter that is released into the synaptic cleft depends on the number of vesicles that fuse during the action potential. The probability of vesicle fusion can be modified in at least three ways: (1) by altering presynaptic Ca^{2+} levels, (2) by altering the voltage-gated Ca^{2+} channels that are responsible for Ca^{2+} flux; or (3) by regulating the amplitude or duration of the presynaptic action potential. Each mechanism is used in physiologically important ways.

Increasing Intraterminal Ca^{2+} Levels Facilitates Neurotransmitter Release

Intracellular Ca^{2+} levels in the presynaptic terminal are closely regulated by cytoplasmic buffering mechanisms. However, under certain circumstances, these buffering mechanisms are overridden. For example, when a train of action potentials is generated in the presynaptic terminal, the multiple action potentials produce so much Ca^{2+} influx that the intracellular Ca^{2+} buffering mechanisms in the presynaptic terminal are temporarily saturated. As a result, there is a residual high concentration of Ca^{2+} near the release site that promotes vesicle fusion in response to subsequent action potentials.

For historical reasons, the enhanced release after a train of action potentials is called *posttetanic potentiation*. The early studies of this phenomenon were carried out at the

neuromuscular junction, where high frequency stimulation of the presynaptic axon produced muscle tetany. For this reason, the stimulation was termed *tetanic stimulation*. This term is now applied by physiologists to any high-frequency stimulation paradigm. The increases in the probability of neurotransmitter release during posttetanic potentiation are typically short-lasting (minutes to hours). Thus, this is a type of *short-term neuronal plasticity* in which the probability of neurotransmitter release varies depending on the history of activity of the synapse.

Ca^{2+} levels within presynaptic terminals can also be regulated by processes that affect the membrane potential in the presynaptic terminal. Ca^{2+} channels are voltage dependent, so that any persistent change in the presynaptic membrane potential will lead to changes in Ca^{2+} concentrations. In the same way that residual calcium at the release site leads to increased release, changes in cytoplasmic Ca^{2+} as a result of prolonged depolarization or hyperpolarization affect release for as long as cytoplasmic Ca^{2+} levels are altered.

Neurotransmitter Release Can Be Altered by Modifying Voltage-Gated Ca^{2+} Channels

There are a number of different types of voltage-dependent calcium channels. For some of these, the voltage sensitivity of the channel can be modified, usually by enzymatic means. A common means for regulating channels is via enzymes that phosphorylate (kinases) or dephosphorylate (phosphates) the protein subunits of the channel. These protein kinases and phosphatases are regulated by a variety of signal transduction cascades.

One important way that Ca^{2+} channels can be modified is via receptors for neurotransmitters released by other terminals. As noted in Chapter 4, some presynaptic terminals are contacted by other presynaptic terminals. Such *axo-axonic synapses* allow one synapse to regulate another. When the neurotransmitter released by one terminal onto another enhances neurotransmitter release in the second terminal, the result is termed *presynaptic facilitation*. Conversely, when the neurotransmitter released by one terminal onto another depresses neurotransmitter release in the second terminal, the result is termed *presynaptic inhibition*. Presynaptic inhibition and facilitation play an important role in regulating information flow along certain pathways.

Another way that neurotransmitter release is modulated is by negative feedback mechanisms. Many synapses possess *autoreceptors* for the neurotransmitter released at the particular synapse. If there are already high levels of the neurotransmitter extracellularly as a result of prior release, subsequent release will be downregulated.

Neurotransmitter Release Can Be Altered by Modifying the Amplitude or Duration of the Presynaptic Action Potential

Because Ca^{2+} flux across the presynaptic membrane occurs through voltage-gated Ca^{2+} channels that are gated by the presynaptic action potential, any change in the amplitude or duration of the presynaptic action potential will alter the amount of neurotransmitter that is released.

Increases in the *duration* of the action potential increase release; decreases in the duration decrease release. The duration of the action potential depends on two processes: (1) the inactivation of Na^+ channels with continued depolarization, and (2) the opening of voltage-dependent K^+ channels that contribute to the repolarizing phase. Normally, repolarization occurs as a result of the K^+ channel opening before there is significant Na^+-channel inactivation. Thus, one way to alter the *duration* of the action potential is to inactivate the voltage-sensitive K^+ channels.

The voltage sensitivity of K^+ channels is regulated by enzymes that phosphorylate or dephosphorylate the protein subunits of the channels. Phosphorylation leads to channel *inactivation* (i.e., the channels do not open in response to depolarization during the action potential and thus do not contribute to membrane repolarization). As a result, the duration of the action potential increases. Because the action potential lasts for a longer period of time, voltage-gated Ca^{2+} channels remain open for longer, increasing Ca^{2+} flux across the presynaptic membrane. The increase in Ca^{2+} flux in turn leads to increased neurotransmitter release.

The activity of the relevant kinases and phosphatases that regulate channel phosphorylation are regulated by signal transduction cascades. One of the most important ways that the intracellular signaling cascades are regulated is via the action of neurotransmitters (see Chapter 7). This can occur via autoreceptors that are activated by the neurotransmitter that is released by the terminal, or via axo-axonic synapses. The distance over which an intracellular signaling process can operate depends on the extent to which the intracellular signals that are generated are *compartmentalized*.

SUMMARY OF KEY POINTS

Neurotransmitter Synthesis, Accumulation, and Storage

1. Different neurotransmitters are synthesized in different ways. ACh, the chatecholamines, 5-HT, and GABA are synthesized by particular enzymes. Amino acid neurotransmitters are selectively taken up into vesicles within presynaptic terminals. Peptide neurotransmitters are synthesized like other proteins—on rough endoplasmic reticulum in the neuronal cell body.

2. The enzymes that synthesize nonpeptide neurotransmitter molecules such as acetylcholine, GABA, norepinephrine, dopamine, and serotonin are present within nerve terminals. However, the synthetic enzymes themselves, being proteins, must be synthesized in the neuronal cell body and transported to the presynaptic terminals.

3. The enzymes responsible for neurotransmitter synthesis may be localized either in the nerve terminal cytoplasm or in vesicles. Choline acetyl transferase is localized in the terminal cytoplasm; after synthesis, ACh is rapidly concentrated in synaptic vesicles. The synthetic enzymes for the catecholamines are localized in *dense core vesicles*.

4. Amino acid neurotransmitters are accumulated in terminals and then in vesicles by selective *transporters* located in the plasma membrane.

5. After release, the action of neurotransmitters is terminated either by degrading the neurotransmitter molecule or by removing the transmitter from the synaptic cleft through reuptake.

Neurotransmitter Release

1. When the action potential invades the presynaptic terminal, the resulting depolarization triggers the opening of voltage-gated Ca^{2+} channels in the presynaptic membrane.

2. The Ca^{2+} flux through the membrane triggers the fusion of synaptic vesicles with the terminal membrane, resulting in the release of neurotransmitter into the synaptic cleft.

3. After fusion, the membrane of the vesicle is recovered through endocytosis.

4. The probability of vesicle fusion (and thus the extent of release) can be modified by varying intraterminal Ca^{2+} concentrations, the properties of the voltage-sensitive Ca^{2+} channels, or the amplitude or duration of the presynaptic action potential.

7

Postsynaptic Mechanisms

Neurotransmitter Receptors and Signal Transduction Cascades

The previous chapter describes how action potentials in axons lead to the release of neurotransmitters at presynaptic terminals. Once released, the neurotransmitters diffuse across the synaptic cleft and interact with receptors on the postsynaptic target cells. Given the great diversity of neurotransmitter types, it is not surprising that neurotransmitters exert a wide variety of effects.

To understand the variety of actions that neurotransmitters elicit, an important first principal to understand is that the action of a neurotransmitter is determined by the properties of the receptor with which it interacts. The most common effect of a neurotransmitter on its target cell is to induce a *synaptic potential* in the postsynaptic cell (either hyperpolarizing or depolarizing). On the basis of the electrical response that neurotransmitters induce in their target cells, neurotransmitters are classified as *excitatory* or *inhibitory,* and the corresponding synaptic potentials are termed *excitatory* or *inhibitory postsynaptic potentials* (EPSPs and IPSPs, respectively). Neurons integrate the summed inputs over time, and the overall level of depolarization determines the neuron's moment-by-moment

firing. The integration process is discussed in Chapter 8. Here we focus on the nature of the response mediated by the postsynaptic receptors.

It is important to understand that although most neurotransmitters have characteristic actions, the neurotransmitter molecules themselves are not inherently excitatory or inhibitory. For this reason, certain neurotransmitters can exert excitatory actions in some situations and inhibitory actions in others. The nature of the response in the postsynaptic cell to a particular neurotransmitter depends on which receptors are present at the synapse. We will see specific examples of this in the case of photoreceptors in the retina, where the neurotransmitter released at the photoreceptor terminal is depolarizing for some types of bipolar neurons and hyperpolarizing for others (see Chapter 18).

TERMINOLOGY FOR RECEPTORS

Historically, receptors were named on the basis of the neurotransmitters that activate them, and then on the basis of their pharmacological characteristics. Thus, there are different receptors for each individual neurotransmitter, and the receptors are further distinguished on the basis of the drugs that interfere with or mimic the action of the neurotransmitter. This is neuropharmacological nomenclature (see Box). More recently, as the protein subunits of receptors have been cloned, the protein subunits of the receptors have been named according to the gene families to which they belong (a genetic nomenclature).

THERE ARE TWO PRINCIPAL CLASSES OF NEUROTRANSMITTER RECEPTORS: IONOTROPIC AND METABOTROPIC

Ionotropic receptors are ion channels that are gated by ligand binding. *Metabotropic receptors* on the other hand are indirectly coupled (through G proteins) to signal transduction cascades. Hence, metabotropic receptors exert their action by regulating the levels of intracellular *second messengers*. These second messengers regulate the activity of enzymes (especially protein kinases), which in turn modify other proteins. It is important to understand the differences in the mode of action of the two classes of receptor in order to understand the nature of the responses that are induced by receptor activation. We begin by considering the ionotropic receptors.

Ionotropic Receptors

Binding of the neurotransmitter to an ionotropic receptor modulates channel conductance. Ionotropic receptors are similar to other plasma membrane ion channels in that they are selectively permeable to particular ions. Two important features of ionotropic receptors are that (1) the channel is gated, and (2) this gating is modulated by the binding of the appropriate neurotransmitter to a particular site on the receptor. To understand the mode of action of an ionotropic receptor, it is helpful to consider an example: the nicotinic ACh receptor that mediates synaptic transmission at the *neuromuscular junction.*

The neuromuscular junction is the synapse between a motor neuron and a striated muscle fiber. This synapse has particular advantages for experimental investigation because of its size and accessibility (see Figure 7.1).

The basic strategy for evaluating synaptic transmission at the neuromuscular junction is to position recording electrodes within the muscle fiber and stimulate the incoming nerve electrically. Nerve stimulation triggers an action potential in the motor axon; when the action potential invades the presynaptic terminal, ACh is released, diffuses across the synaptic cleft, and interacts with the receptors present on the muscle.

Binding of ACh to the receptors at the neuromuscular junction produces an inward-going current (synaptic current) and a depolarizing potential in the muscle fiber termed the *end plate potential* (EPP) (see Figure 7.1). Normally, the depolarization produced by activation of the motor axon is sufficient to induce an action potential in the muscle fiber. However, for experimental studies, the typical strategy is to partially block the EPP using curare (a selective antagonist for nicotinic ACh receptors). In this case, the amplitude of the EPP is below the threshold for action potential generation in the muscle fiber.

The EPP Is Generated as a Result of the Opening of a Channel That Is Permeable to Both the Monovalent Cations Na+ and K+

The ionic basis of the EPP at the neuromuscular junction was established by determining the *reversal potential* of the EPP using voltage and current clamp techniques (Figure 7.2). When the membrane potential of the muscle is held at a particular value and an EPP is generated, the polarity and amplitude of the synaptic current and EPP vary as a function of the membrane potential. At the resting potential, there is an inward-going current causing depolarization. As the membrane is depolarized, the amplitude of the current decreases and eventually reverses to an outward-going current, which causes hyperpolarization (Figure 7.2). This reversal

Neuropharmacological Terminology

Drugs that interfere with the action of the neurotransmitter are termed *antagonists*. Many antagonists act by blocking the binding of the neurotransmitter to the receptor. These are termed *competitive antagonists* because they compete with the neurotransmitter for the *ligand binding site* (the site on the receptor that recognizes the neurotransmitter). Antagonists that interfere with receptor function in some way other than blocking the binding of the neurotransmitter are termed *noncompetitive antagonists*.

Drugs that activate receptors are called *agonists*. Agonists bind to the same site on the receptor as the natural neurotransmitter and often have a chemical structure similar to the neurotransmitter. However, the kinetics of activation by an agonist may be quite different from those for the natural neurotransmitter.

The terminology for receptors can best be appreciated by considering some examples. Thus, there are two classes of receptors for acetylcholine (ACh receptors), termed *nicotinic* and *muscarinic,* based on the fact that the different receptor types are selectively activated by the agonists nicotine and muscarine, respectively. Nicotinic and muscarinic receptors also have different antagonists. Curare (the toxin used by South American Indians to poison arrows) and certain snake toxins (cobra toxin and the toxin of the banded krait, which is called *bungarotoxin*) block the action of ACh at nicotinic receptors. Atropine and scopolamine (two common constituents of cold medicines) block the action of ACh at muscarinic receptors. Muscarinic receptors are further subdivided (M1, M2, etc.) based on their pharmacology.

Excitatory amino acid receptors are distinguished primarily on the basis of differences in their response to agonists. All excitatory amino acid receptors are activated by glutamate and are thus called *glutamate receptors* (even though the receptors are also activated by aspartate). However, there are different receptor types that are activated preferentially by different agonists. For example, two classes of glutamate receptors are preferentially activated by the amino acid derivatives kainic acid and quisqualate. These have been called *kainate receptors* and *quisqualate-A receptors,* respectively. But both kainate and quisqualate-A receptors are also activated by another glutamate agonist called alpha-amino-3-hydroxy-5-methyl-4 isoxazole proprionic acid (AMPA), and so the two are also called *AMPA receptors.* Another receptor type is preferentially activated by the drug *n*-methyl-D-aspartate (NMDA), and these are termed *NMDA receptors* (see later).

The basic theme continues in the case of receptors for other neurotransmitters. Thus, there are GABA-A and GABA-B receptors, α- and β-adrenergic receptors, and so forth. The key point to understand is how receptors are named.

It is also important to note that, to some extent, the terminology used to refer to receptors is a matter of fashion. As new and more selective agonists or antagonists are discovered, the terminology used to refer to the receptor types changes. In addition, there is increasing use of the genetic nomenclature in place of the neuropharmacological nomenclature. For example, there are a number of closely related protein molecules that make up glutamate receptors. Molecular biologists refer to these different protein subunits as GluR1, GluR2, and so forth. The changing terminology is a constant challenge for anyone not in the field.

---

occurs at a membrane potential of about 0 mV, and 0 mV is thus termed the *reversal potential.* There is no potential change when the membrane potential is at 0 mV because the driving force for ion flux (the concentration gradient) is exactly balanced by the membrane potential.

The reversal potential identifies the *equilibrium potential* for the ion species that generate the potential. If the potential is generated by a single ion, then the reversal potential will be at the equilibrium potential for that ion (about -100 mV for K^+ and $+55$ mV for Na^+). However, the reversal potential of 0 mV does not correspond to the equilibrium potential of any of the major ions. This observation suggests that the EPP results from a simultaneous flux of more than one ion. Subsequent studies have demonstrated that the ionic currents during the EPP are the result of a simultaneous increase in permeability to both Na^+ and K^+.

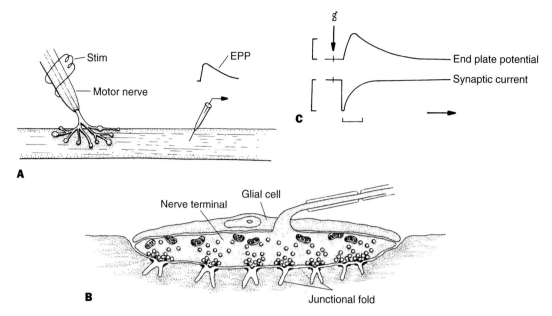

Figure 7.1. Synaptic transmission at the neuromuscular junction. A, The neuromuscular junction is the synapse between a motor neuron and a striated muscle fiber. The motor nerve breaks up into multiple presynaptic terminals, which form synaptic contacts with the muscle **(B)**. Studies of synaptic transmission at the junction involve placing recording electrodes within the muscle fiber and stimulating the incoming nerve electrically. Nerve stimulation triggers an action potential in the motor axon, causing the release of ACh. The ACh interacts with the receptors present on the muscle and generates a synaptic current and an end plate potential (EPP). **C** illustrates an example of the EPP and the synaptic current measured under voltage clamp. The EPP is slower than the underlying synaptic current because the current must first charge the capacitor of the membrane before a potential change is produced. **B** is from Dowling JE. *Neurons and Networks: An Introduction to Neuroscience.* Cambridge, Mass: Harvard University Press; 1992.

It is important to emphasize the differences in the ionic currents generated during the EPP versus during the action potential. Both the EPP and the action potential involve ionic currents generated by changes in transmembrane flux of Na^+ and K^+; however, the nature of the ionic currents is different. The action potential is due to a sudden increase in permeability for Na^+ (which is responsible for the depolarizing phase) and a subsequent increase in permeability for K^+ (which is responsible for the hyperpolarizing phase). We now know that these respective currents are generated as a result of the opening of an ion channel that is selective for Na^+, followed by the opening of a different ion channel that is selective for K^+. In contrast, the synaptic current at the neuro-muscular junction is the result of simultaneous Na^+ and K^+ currents. As a result, the equilibrium potential (0 mV) is a weighted average of the currents produced by the simultaneous changes in transmembrane flux of these two ions.

The Synaptic Current at the Neuromuscular Junction Is Due to the Opening of a Cation Channel That Is Permeable to Both Na^+ and K^+

A very powerful tool for characterizing how ligand-gated channels operate is the patch clamp technique developed by Ernst Nehrer and Bert Sackman (Figure 7.3). Using glass micropipettes with specially polished tips, a small piece of membrane containing a

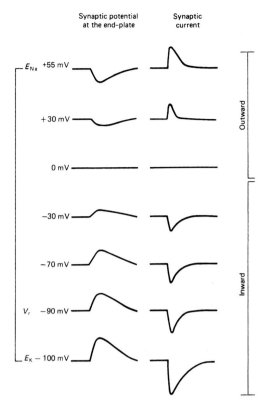

Synaptic potential at the end-plate | Synaptic current

E_{Na} +55 mV

+30 mV

0 mV

−30 mV

−70 mV

V_r −90 mV

E_K −100 mV

Outward / Inward

Figure 7.2. Ionic basis of the EPP at the neuromuscular junction. The reversal potential identifies the *equilibrium potential* for the ion species that generate the EPP. To determine the reversal potential, an EPP is induced while the postsynaptic membrane potential is "clamped" at different values. At the resting potential, there is an inward-going synaptic current and a depolarizing EPP. As the membrane is depolarized, the amplitude of the synaptic current decreases and eventually reverses to an outward-going current. The point of reversal is termed the *reversal potential.* If the potential is generated by a single ion, then the reversal potential will be at the equilibrium potential for that ion (about −100 mV for K^+ and +55 mV for Na^+). However, the reversal potential of 0 mV does not correspond to the equilibrium potential of any of the major ions. This observation suggests that the EPP results from a simultaneous flux of more than one ion (in the case of the neuromuscular synapse, both Na^+ and K^+). From Kandel ER, Schwartz JH, Jessel TM. *Principles of Neural Science.* 3rd ed. New York, NY: Elsevier; 1991.

single channel is plucked from the postsynaptic membrane. The barrel of the pipette can be filled with any solution so that the isolated receptors can be exposed to neurotransmitters or other molecules of interest. When the pipette contains the appropriate neurotransmitter (ACh in the case of receptors from muscle), the channel is activated. Opening of the channel leads to a stepwise increase in current flow across the membrane (Figure 7.3). When the channel closes, this current ceases. The current resulting from the opening of a single channel is termed the *elementary current.* When more than one channel is present in the patch of membrane, the elementary currents summate (Figure 7.3).

The reversal potential for the elementary currents is identical to the reversal potential for the synaptic response, indicating that the current carried by the individual channel is in fact the same as that responsible for the EPSP. Different types of channels have different elementary conductances. Thus, the patch clamp technique has yielded a wealth of information about the biophysics of channel function that could not have been obtained in any other way. Nehrer and Sackman shared the Nobel Prize in Physiology or Medicine in 1992 for this work.

The Structure of the ACh Receptor

Our understanding of the physical structure of receptors is based on studies of the ACh receptor. Progress was made possible by two fortunate circumstances: (1) the ACh receptor is present in extremely high concentrations in some tissues (the neuromuscular junction and the electric organs of certain fish), thus providing a rich source for receptor isolation; and (2) a nearly irreversible antagonist was available that could be used to purify the receptor protein (the toxin of a snake—the banded krait. *Bungaris multicinctus*). Thus, it was possible to actually visualize the native receptor in the postsynaptic membrane using electron microscopy, purify the protein (using bungarotoxin as a tool), define the molecular structure of the channel,

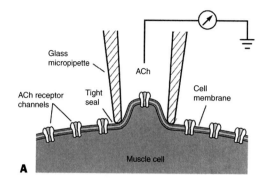

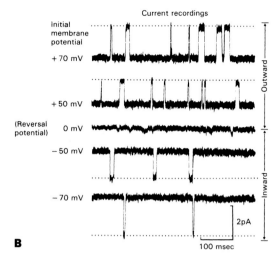

Figure 7.3. Recording the currents generated by individual channels using the patch clamp technique. A illustrates the basic patch clamp technique. To record the currents generated by single channels, a micropipette filled with saline and a low concentration of ACh is placed against the surface membrane of the muscle. Gentle suction is then applied so as to form a tight seal. **B** illustrates an example of the single-channel currents generated by 2 μM ACh. The opening of the channel causes a transient increase in conductance and generates a current of about 3 pA (picoamps). In this experiment, the currents were recorded as the membrane potential was held at different holding potentials, in order to determine the reversal potential of the ionic current generated by the opening of the single channel. Note that the reversal potential is about 0 mV (the same as the EPP). **A** is from Kandel ER, Schwartz JH, Jessel TM. *Principles of Neural Science.* 3rd ed. New York, NY: Elsevier; 1991, after Alberts B, Bray D, Lewis J, Raff M, Roberts K, Watson JD. *Molecular Biology of the Cell.* 2nd ed. New York, NY: Garland Publishing; 1989. **B** is from Kandel ER, Schwartz JH, Jessel TM. *Principles of Neural Science.* 3rd ed. New York, NY: Elsevier; 1991.

and clone the genes that encode the protein constituents.

Electron microscopic studies of the external surface of receptor-rich membranes such as the neuromuscular junction reveal dense clusters of irregularly shaped globular masses that have a hollow center (Figure 7.4). The hollow center is thought to represent the ion channel. Binding of the neurotransmitter to the receptor is thought to induce a rapid transient conformational change in the protein, leading to channel opening, thus allowing the transmembrane flux of Na^+ and K^+.

The cloning of the genes that encode the protein subunits of the ACh receptor made it possible to define in detail the molecular structure of the receptor (Figure 7.4). The native receptor is a pentameric structure made up of four different proteins (two alpha, one beta, one gamma, and one delta) that are encoded by different genes. Each protein subunit is a membrane-spanning protein, and the protein subunits are assembled into a single molecular conglomerate that mediates both receptor and channel functions (that is, has the binding sites for the neurotransmitter and forms the pore in the membrane that is the ion-selective channel). These conglomerates are termed *heteromeric assemblies* because they are made up of different protein subunits. The internal walls of the ion channel are formed by portions of the membrane-spanning domains of each of the subunits.

The ligand binding site that recognizes ACh is present on the alpha subunits in the heteromeric structure. There are two alpha subunits in each heteromeric assembly, and

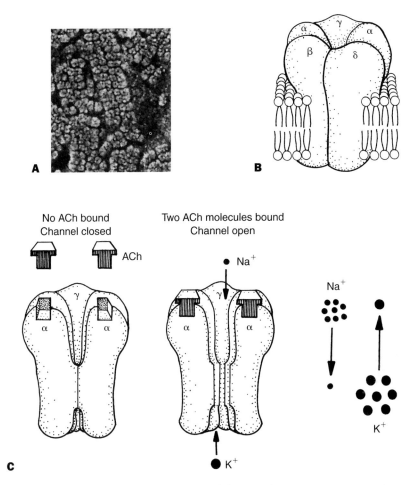

Figure 7.4. Structure of the ACh receptor. High-resolution electron micrograph of ACh receptors in the postsynaptic membrane of the electric organ in *Torpedo californica,* an electric fish. This is a view of the external surface of the postsynaptic membrane (as it would be seen from the presynaptic terminal). **B** illustrates how the channel would appear in the membrane. The channels are made up of a protein complex composed of two alpha, one beta, one gamma, and one delta subunit. The binding domain for ACh is on the alpha subunit. **C**, binding of ACh to both sites causes a conformational change in the receptor that leads to the opening of the channel. Modified from Kandel ER, Schwartz JH, Jessel TM. *Principles of Neural Science.* 3rd ed. New York, NY: Elsevier; 1991. The photomicrograph is courtesy of J. E. Heuser and S. R. Salpeter.

activation of the receptor requires that ACh bind to both sites. This simple fact accounts for the stoichiometry of receptor activation (that two molecules of ACh are required for channel opening). In the case of the ACh receptor, it is thought that each protein subunit forming the walls of the pore (channel) has negatively charged amino acids positioned so as to form three rings. These rings determine the *ion selectivity* of the channel (i.e., the fact that the channel is selective for *cations*) and thus produce a depolarizing potential when the channel opens.

Ionotropic Receptors on Neurons Have Properties That Are Very Similar to the Nicotinic ACh Receptor of Muscles

EPSPs generated by excitatory synapses onto neurons have the same ionic basis as the EPP—that is, the EPSP is due to the opening of a monovalent cation channel. The ionic basis of the EPSP was first established in studies carried out by John Eccles and colleagues who evaluated synaptic activation of spinal motor neurons by sensory axons projecting into the spinal cord via the dorsal roots (Ia afferents). For a complete description of this circuit, see Chapter 14.

Eccles and his colleagues established that the reversal potential for the EPSP was at about 0 mV (Figure 7.5). The time course of the synaptic currents produced by Ia afferents was also comparable to that at the neuromuscular junction, suggesting the operation of similar types of receptors. Subsequent physiological studies of excitatory synaptic connections on other types of CNS neurons revealed similar properties. It remained for neuropharmacology to define which neuro-

transmitters actually generated these synaptic potentials.

Eccles was awarded the Nobel Prize in Physiology or Medicine in 1963 for this work. His corecipients were Alan Hodgkin and Andrew Huxley, for their work on the ionic basis of the action potential.

The Most Common Form of Excitatory Synaptic Communication Between Neurons in the CNS Involves Excitatory Amino Acids Acting at Ionotropic Receptors

Identification of the neurotransmitters used at central synapses has been based on a combination of approaches beginning with neuropharmacology. The general strategy is to record from postsynaptic cells, activate particular inputs, and evaluate the effects of selective antagonists on the resulting synap-

Figure 7.5. Ionic basis for the EPSP at Ia afferent synapses onto motoneurons. Ia afferents were activated by stimulating electrodes positioned either on the dorsal root itself or on the incoming peripheral nerve. Recording microelectrodes were positioned intracellularly within the motor neurons. Activation of the Ia afferents produced depolarizing EPSPs within the motor neurons that were similar in form to the EPP at the neuromuscular junction. At the resting potential, there is an inward-going current causing depolarization (the synaptic currents are not illustrated but are essentially identical to the currents associated with the EPP, see Figure 7.2). As the membrane is depolarized, the amplitude of the EPSP decreases and then reverses. The reversal potential is at about 0 mV.

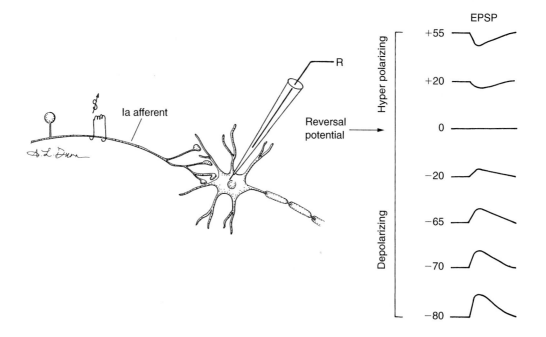

tic potentials. Once a candidate neurotransmitter has been identified, other criteria are applied, including (1) whether the action of the neurotransmitter can be mimicked by application of the neurotransmitter candidate; (2) whether the appropriate synthetic and degradative enzymes are present (if appropriate); and (3) whether the appropriate high-affinity uptake systems are in place, and so forth (see the criteria for identifying a substance as a neurotransmitter outlined in Chapter 6).

Based on accumulated evidence of this sort, it is thought that most excitatory synapses in the CNS are glutamatergic and operate via ionotropic receptors similar to the ACh receptor.

The native receptor complex that is present in the plasma membrane is made up of heteromeric subunits that together form the ligand-gated ion channels. Each protein subunit is encoded by a different mRNA; hence, the expression of the different subunits can be differentially regulated. When the protein subunits are present in different combinations in the heteromeric complex, they form receptors with somewhat different physiological properties. Thus, receptor diversity in vivo arises in part as a result of constructing heteromeric complexes with different combinations of the protein subunits.

There are two principal types of ionotropic glutamate receptors with distinct pharmacological characteristics: the AMPA/kainate receptor (Figure 7.6) and the NMDA receptor (Figure 7.7). AMPA receptors are ionotropic receptors that have channel properties similar to those of the ACh receptor. The AMPA/kainate-sensitive glutamate receptors are thought to mediate "fast" glutamatergic transmission such as is seen at the synapse between the IA afferent and the motoneuron. It is also the most common mode of transmission at other excitatory CNS synapses.

The NMDA receptor has unique properties that distinguish it from other glutamate receptors. It is an important member of the glutamate receptor family and is so named because the drug n-methyl-D-aspartate (NMDA) is a selective agonist. This receptor has quite different properties from the remainder of the ionotropic receptor family. The NMDA receptor is a ligand-gated channel that is activated by glutamate and is similar to the other ionotropic receptors in that sense. However, when activated, the channel is permeable to both monovalent cations

Figure 7.6. Ionotropic glutamate receptors. AMPA/kainate receptors are ionotropic receptors that have channel properties similar to those of the ACh receptor. Binding of the neurotransmitter causes a conformational change that leads to channel opening. The channels are permeable to both Na$^+$ and K$^+$ and produce synaptic currents with a reversal potential near 0 mV. The AMPA/kainate class of glutamate receptors mediate the "fast" glutamatergic transmission that occurs at most excitatory synapses in the CNS. Adapted after Kandel ER, Schwartz JH, Jessel TM. *Principles of Neural Science*. 3rd ed. New York, NY: Elsevier; 1991.

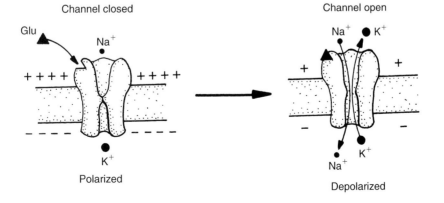

Channel closed

Channel open

Polarized

Depolarized

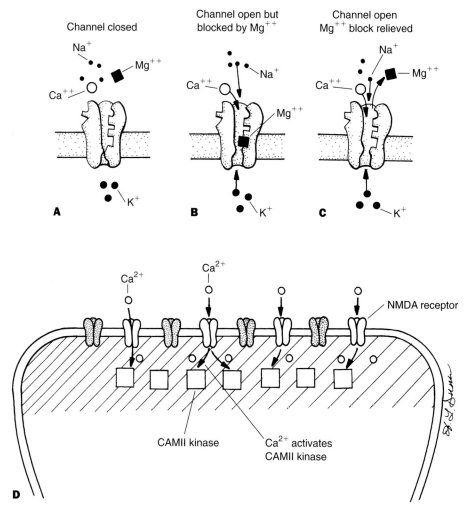

Figure 7.7. The NMDA receptor. The NMDA receptor is also an ionotropic receptor; its channel is permeable to both monovalent cations (Na^+, K^+) and the divalent cations Ca^{2+} and Mg^{2+}. **A** illustrates the channel in its closed state. **B,** Binding of the neurotransmitter causes the channel to open, but unless the neuron is already depolarized, Mg^{2+} enters and blocks the channel. **C,** If the neuron is depolarized prior to receptor activation, Mg^{2+} is displaced, allowing the flux of Na^+, K^+, and Ca^{2+}. **D,** The Ca^{2+} influx increases the Ca^{2+} concentration in the subsynaptic cytoplasm and can activate the calcium/calmodulin-dependent kinase (CAMII kinase), which is highly concentrated in the postsynaptic density. The NMDA channel is modulated by glycine, and the channel has binding sites for Zn^{2+} and also certain important drugs (for example, phencyclidine, or PCP). **A, B,** and **C** are adapted from Kandel, ER, Schwartz, JH, Jessel, TM. *Principles of Neural Science.* 3rd ed. New York, NY: Elsevier; 1991.

(N^+, K^+) and divalent cations (Ca^{2+} and Mg^{2+}).

There are several unusual and important features of the NMDA receptor: (1) the channel function of the NMDA receptor is dependent on the membrane potential, (2) the NMDA receptor can activate second messenger cascades because its activation leads to influx of Ca^{2+} into the postsynaptic cytoplasm, and (3) the NMDA channel is modulated by glycine.

The channel function of the NMDA receptor depends on the level of polarization of the postsynaptic membrane (that is, channel function is voltage dependent). The voltage dependency comes about because the receptor is blocked in a voltage-dependent manner by magnesium ions. If the membrane potential is near the resting potential when glutamate interacts with the receptor, the channel opens but is immediately blocked by Mg^{2+}. As a result, there is little or no current flow. However, if the membrane is depolarized when the channel opens, Mg^{2+} is displaced, and there is a flux of Na^+, K^+, and Ca^{2+}. Complete relief from Mg^{2+} blockade occurs when the membrane is depolarized by about 30 mV from the resting potential (Figure 7.7).

The voltage dependency of the NMDA receptor gives it unique signaling characteristics. If the membrane is near the resting potential when glutamate activates the receptor, there is no current flow because of Mg^{2+} blockade. In this case, signaling is mediated entirely by non-NMDA receptors, which produce a rapid, transient depolarizing current. Alternatively, if the membrane is somewhat depolarized when the receptor is activated, then there is no Mg^{2+} block; thus, opening of the channel leads to Na^+, K^+, and Ca^{2+} flux. Together these cationic currents cause a longer-lasting depolarization and lead to Ca^{2+} influx into the postsynaptic cytoplasm (Figure 7.7). Whether the neuron is depolarized when the neurotransmitter activates the receptor depends on concurrent activity at other synapses *(spatial summation),* and the recent history of activity at the synapse that is activated *(temporal summation).*

The operation of the NMDA receptor is modulated by glycine. In the presence of glycine, the NMDA receptor is efficiently gated by glutamate; in the absence of glycine, glutamate is much less effective. An important question is the extent to which NMDA receptors are actually regulated by glycine in physiological settings.

The NMDA receptor can activate second messenger cascades because its activation leads to influx of Ca^{2+} into the postsynaptic cytoplasm. This Ca^{2+} pulse can activate a number of calcium-dependent processes (for example, the enzyme calcium/calmodulin protein kinase II, calcium-activated proteases, etc.). The calcium-activated enzyme systems then modify other cellular processes within the postsynaptic cell. In this way, NMDA receptors can regulate some of the same enzyme systems that are regulated by metabotropic receptors (see later).

Inhibitory Neurotransmission Is Mediated Via Receptors That Are Similar in Structure to ACh and EAA Receptors But Form Anion-Selective Channels

The receptors for inhibitory neurotransmitters are also ligand-gated ion channels. However, the channel is selectively permeable to Cl^-, so that the ionic current shifts the membrane potential toward the equilibrium potential for Cl^- (about -70 mV). Channel opening and the resulting increase in flux of Cl^- produce the inhibitory postsynaptic potential (IPSP) at inhibitory synapses (Figure 7.8). Converging evidence from pharmacological and immunocytochemical studies indicated that the principal inhibitory neurotransmitters in the CNS are GABA and glycine.

Ionotropic Receptors Are Members of Families of Closely Related Supramolecular Structures

The cloning of the genes for the different ionotropic receptors allowed direct comparisons of their protein sequences and three-dimensional structures predicted by the sequence information. Substantial sequence homologies were found between the protein subunits of the different types of non-NMDA glutamate receptors as well as the GABA and glycine receptors. These results led to the hypothesis that the different receptors are members of receptor *super families* that include both anionic and cationic channels. Structure–function analyses indicated that the ion selectivity of the different channels is determined by the distribution of charged amino acids in the portions of the molecule forming the walls of the channel.

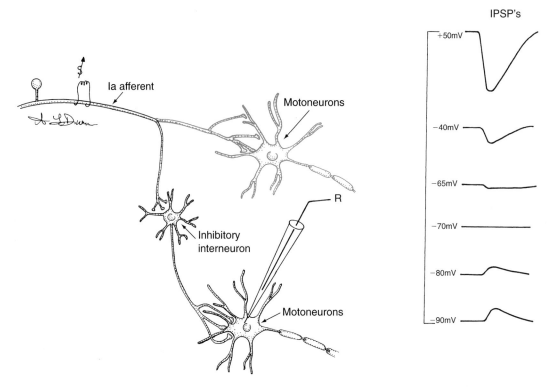

Figure 7.8. Ionic basis of the IPSP. The foundation of our knowledge of inhibitory synaptic transmission in the CNS came from studies of inhibitory inputs to motor neurons by Eccles and colleagues. In addition to forming excitatory synaptic connections on motor neurons, Ia afferents synapse on *inhibitory interneurons*, which then synapse on motor neurons. Hence, stimulation of the Ia afferent generates monosynaptic excitation in one population of motoneurons and disynaptic inhibition in another. The reversal potential for IPSPs on motor neurons is about −70 mV, which is the equilibrium potential for Cl⁻. Based on this and other evidence, it was surmised that the synaptic current produced by the inhibitory neurotransmitter was a Cl⁻ current generated as a result of the opening of an anion-selective channel.

The similarity in the structure of different ionotropic receptors makes sense because certain aspects of channel function are common. For example, the receptor–channel complex must span the membrane; the pore of the channel must be hydrophilic and permeable to ions; and the receptor must bind a ligand and undergo changes in confirmation to alter channel conductances. The fact that structurally similar channels are associated with binding sites for different neurotransmitters suggests that the two functions of the receptor–ion channel complex are discrete and separable.

In summary, ionotropic receptors are molecular assemblies that possess ligand binding sites and also form ion channels that are selectively permeable to particular ion species. Binding of the neurotransmitter leads to a transient opening of the channel, which in turn leads to a transient ionic current. Because of their channel functions, ionotropic receptors have some features in common with the voltage-gated ion channels that are responsible for the action potential.

Metabotropic Receptors

Metabotropic receptors are not ion channels; instead, they regulate the production of intracellular second messengers that in turn regulate particular enzymes. Receptors can be conceived of as having ligand-binding domains and effector domains. In ionotropic receptors, the two domains are located on the same complex. For metabotropic receptors, the receptor contains the ligand-binding domains, whereas the effector is an enzyme that is often not even uniquely related to the particular receptor in question. The effector enzymes can be regulated in a number of different ways, only some of which are related to receptor activation.

Metabotropic receptors exert their action through GTP-binding proteins *(G proteins)*. The G proteins act as "go-betweens" for the receptor and the enzyme systems that produce the second messengers. Each receptor type activates a specific G protein. Thus, the specificity of action of particular neurotransmitters depends on the G protein that is activated.

The enzyme systems associated with metabotropic receptors are constitutively active and can be either upregulated or downregulated. For example, activation of a receptor that leads to the production of a particular second messenger upregulates the activity of the enzyme that produces this messenger, but the enzyme operates at some basal level in the absence of receptor activation. In fact, it is generally true that the activity of enzymes that produce second messengers is upregulated by activation of some receptors and downregulated by others (see later). The upregulation and downregulation are mediated by different G proteins.

Because of the profound differences in the characteristics of different metabotropic receptor systems, it is more convenient to classify the receptors according to the enzyme systems that they regulate. With this in mind, examples of the principal metabotropic receptor systems can be considered. It is important to emphasize that these examples are not exhaustive. Because of the nature of metabotropic receptor systems, a great deal of diversity is possible (by pairing different receptors with different effector systems).

Metabotropic Receptors That Regulate the Levels of Cyclic Nucleotides

One class of metabotropic receptors regulates the levels of cyclic nucleotides (*cAMP* and *cGMP*) by modulating the activity of *adenylate cyclase* or *guanidylate clyclase* (the synthetic enzymes for cAMP and cGMP, respectively). Receptor activation leads to activation of a G protein that then regulates the activity of the cyclase. Different receptors activate stimulatory and inhibitory G proteins (G_s and G_i) to either up- or downregulate the activity of the cyclases (Figure 7.9). The cyclic nucleotides in turn alter the activity of *cAMP-* or *cGMP-dependent protein kinases*, which phosphorylate specific protein substrates.

Neurotransmitters that exert their effects by modifying adenylate cyclase include dopamine, serotonin, norepinephrine, histamine, octopamine, and many hormones. Neurotransmitters that operate by regulating guanidylate cyclase include acetylcholine (operating through muscarinic receptors), histamine, norepinephrine, and glutamate.

Metabotropic Receptors That Regulate the Metabolism of Phosphatidyl Inositol

Another important second messenger system induces the hydrolysis of membrane *phosphatidyl inositol* (Figure 7.10). Receptor activation leads to activation of a G protein that upregulates the activity of the enzyme *phosphoinositidase* or *phospholipase C*. Phospholipase C then hydrolyzes membrane phosphatidyl inositol into *inositol 1,4,5-trisphosphate* (IP3) and *diacylglycerol* (DG). Both IP3 and DG then act as second messengers. IP3 is released from the membrane into the cytoplasm, where it induces the release of calcium from nonmitochondrial stores within the cell (especially the endoplasmic reticulum). Once in the cytoplasm, the calcium can act as an intracellular messenger in several ways, for example by activating *calcium/calmodulin-dependent protein kinase*, which then phosphorylates its particular set of substrate proteins, and *calcium-activated protease*, which selectively degrades certain

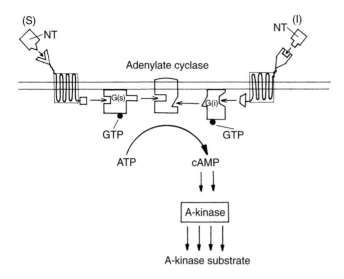

Figure 7.9. Metabotropic receptors that regulate nucleotide cyclases. There are several metabotropic receptors that operate by modulating the activity of nucleotide cyclases. Typically, the cyclases are regulated both positively and negatively. For example, one type of neurotransmitter might activate a receptor that is coupled to a GTP-binding protein (G protein) that stimulates the cyclase (G$_s$); another might be coupled to a G protein that inhibits the cyclase (G$_i$). Modulating the activity of the kinase will alter the cytoplasmic concentration of the cyclic nucleotide. The cyclic nucleotides in turn regulate the activity of cAMP- or cGMP-dependent protein kinases, which phosphorylate specific protein substrates.

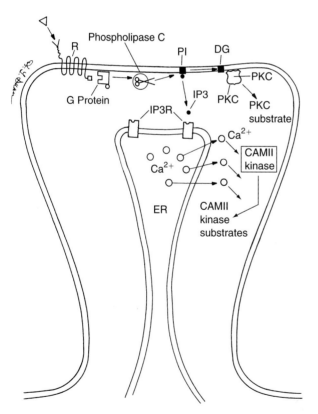

Figure 7.10. Metabotropic receptors that regulate the metabolism of phosphatidyl inositol. Receptor activation leads to activation of a G protein that upregulates the activity of phospholipase C (PLC). PLC then hydrolyzes membrane phosphatidyl inositol (PI) into inositol 1,4,5-trisphosphate (IP3) and diacylglycerol (DG). IP3 is released from the membrane into the cytoplasm, where it interacts with an IP3 receptor (IP3R) that is present in the membrane of the endoplasmic reticulum. The IP3R then mediates the release of calcium from the endoplasmic reticulum. DG remains in the membrane and activates protein kinase C, which phosphorylates a different set of substrate proteins.

elements of the cytoskeleton. DG remains in the membrane and activates another enzyme, *protein kinase C,* which phosphorylates a different set of substrate proteins. In this way, receptor-mediated hydrolysis of phosphatidyl inositol generates multiple signals, each of which modulates different cellular processes.

Some Metabotropic Receptors Activate Enzymes That Degrade Second Messengers

For any molecule that serves a regulatory function, both increases and decreases in its concentration are important. Changes in concentration can be brought about either by altering the activity of the enzymes responsible for producing the molecule, or by altering degradative processes. These principles also apply to molecules that operate as second messengers. Thus, it is not surprising that some receptor-mediated signaling processes lead to activation of enzymes that degrade second messengers.

An important example of this type of signaling occurs in the photoreceptors of the retina. Activation by light bleaches photopigments that are a type of metabotropic receptor. The activation of these metabotropic receptors activates a *phosphodiesterase,* an enzyme that degrades cyclic nucleotides. When phosphodiesterase is activated, there is a rapid decrease in cGMP within the photoreceptor. These decreases in cGMP trigger the physiological changes that are responsible for sensory transduction (see Chapter 18).

Different Signaling Pathways Converge on Particular Enzyme Systems

The IP3 produced as a result of hydrolysis of membrane phosphatidyl inositol exerts its effect through the release of calcium from intracellular stores. However, intracellular calcium levels can also be increased by activating voltage-dependent calcium channels and the NMDA receptor. Thus, three different signaling pathways can converge to modulate intracellular calcium levels. In the same way, the intracellular concentrations of other second messengers can be regulated through the converging action of many multienzyme cascades that either increase or decrease the concentration of the messenger.

Functional Consequences of Metabotropic Receptor Activation

Metabotropic receptors regulate a wide range of processes within neurons. In terms of neurotransmission, one of the most important consequences of receptor activation is that the physiological activity of the target cell is modified. As seen in the following examples, second messengers can modulate physiological activity in a number of ways. Here we consider only a few examples.

Second Messengers Modulate Ion Channels and Ionotropic Receptors

Second messengers produced by metabotropic receptors modulate physiological processes within postsynaptic cells by modulating ion channels and ionotropic receptors. Generally, this occurs as a result of the activation of protein kinases that phosphorylate the channels and receptors. An example of this type of modulation is the slow EPSP that is produced in sympathetic ganglia as a result of activation of muscarinic ACh receptors.

Preganglionic fibers of the autonomic nervous system release ACh at their synaptic terminals. Release of ACh activates nicotinic receptors on the postsynaptic neurons, producing fast EPSPs that are typical of ionotropic receptors (which last a few milliseconds). There is, however, a second form of transmission involving muscarinic ACh receptors. When preganglionic fibers are stimulated repetitively, a slow EPSP is generated in the ganglion cells, which persists for many seconds (Figure 7.11). This slow EPSP can be blocked by antagonists of muscarinic

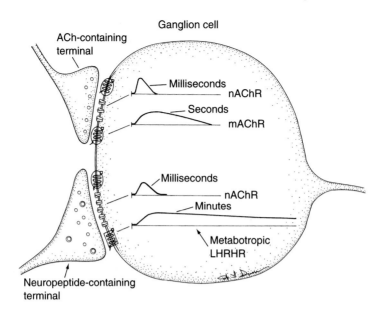

Figure 7.11. Slow EPSPs in sympathetic ganglia.
Preganglionic fibers of the autonomic nervous system release ACh at their synaptic terminals. Release of ACh activates both nicotinic receptors, which mediate fast EPSPs (a few milliseconds in duration), and muscarinic receptors, which mediate slow EPSPs (tens of seconds in duration). Some synaptic terminals also contain the neuropeptide LHRH (luteinizing hormone-releasing hormone), which activates a metabotropic receptor that mediates an ultraslow EPSP (minutes in duration). Adapted from Kandel ER, Schwartz JH, Jessel TM. *Principles of Neural Science.* 3rd ed. New York, NY: Elsevier; 1991.

receptors, and hence the response is mediated by muscarinic ACh receptors.

The ionic basis of this slow EPSP was found to be quite different from that of the fast EPSP produced by nicotinic receptors. Rather than being due to an *increase* in ionic conductance, the slow EPSP is produced as a result of a *decrease* in a resting K^+ current. The resting K^+ current is responsible for the resting potential; hence, the inactivation of the resting K^+ current produces a depolarization.

The inactivation of the K^+ current occurs as a result of the phosphorylation of proteins that compose the K^+ channel. Thus, activation of muscarinic ACh receptors leads to the activation of a protein kinase that phosphorylates K^+ channels.

Because the K^+ channels in sympathetic neurons are regulated by muscarinic ACh receptors, they are termed *M-type channels.* Another neurotransmitter that operates in a similar way is serotonin, and the K^+ channels that are modulated by it are termed *S-type channels.* Norepinephrine (acting through a cAMP pathway) exerts a similar effect on K^+ channels in certain cortical neurons, again producing slow EPSPs. Some of the actions of neuropeptides are also the result of modulation of K^+ channels.

An important consequence of the fact that metabotropic receptors operate by regulating second messenger cascades is that the physiological events that they induce often have a slow onset and are and long lasting. For example, in the cases already discussed, the depolarization resulting from inactivation of resting K^+ flux (so-called slow EPSPs) develop slowly over the course of several seconds and persist for tens of seconds. Slow potentials produced by neuropeptides last even longer. For example, the potentials produced in sympathetic neurons by the hormone LHRH persist for many minutes, and these are termed ultraslow EPSPs.

At the same time, metabotropic receptors can produce very rapid physiological effects.

For example, the physiological events in retinal photoreceptors that result from activation of phosphodiesterase (see earlier) occur in milliseconds (see Chapter 18).

Second messengers can also modulate channel function in presynaptic terminals, leading to changes in neurotransmitter release (Figure 7.12). Phosphorylation of K^+ channels reduces their voltage sensitivity.

As a result, K^+ channels do not open as readily while the membrane is being depolarized during the action potential, so the contribution of K^+ currents to the repolarizing phase of the action potential is reduced. The result is an increase in the duration of the action potential. Increases in action potential duration in turn result in an enhanced release of neurotransmitter. Second

Figure 7.12. An overview of the signal transduction processes that occur at CNS synapses. The cartoon summarizes some of the ways that second messengers operate to regulate presynaptic and postsynaptic function. Presynaptically, second messengers can modulate K^+ channel function in presynaptic terminals, increasing the duration of the presynaptic action potential (1); modulate presynaptic Ca^{2+} channels (2); and alter the phosphorylation state of proteins associated with synaptic vesicles (3). All of these modulations alter neurotransmitter release. Postsynaptically, second messengers can alter the phosphorylation state of ionotropic receptors (4); modulate steps in the signal transduction cascades induced by metabotropic receptors (5); regulate the function of a host of enzyme systems (6); and regulate gene expression (7).

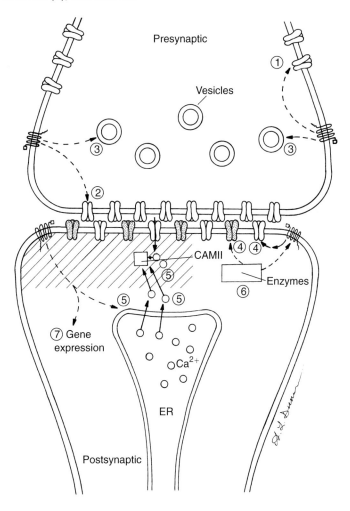

messengers can also alter the phosphorylation state of presynaptic Ca^{2+} channels, which alters channel operation and modifies neurotransmitter release. Finally, second messengers can alter the phosphorylation state of certain proteins associated with synaptic vesicles; these changes also alter neurotransmitter release.

Second messengers can also alter the functional properties of receptors (both ionotropic and metabotropic). For example, the subunits of the ACh receptor can be phosphorylated by a number of different second messenger–dependent protein kinases. Phosphorylation by cAMP-dependent kinase increases the rate at which the receptor is desensitized when continuously exposed to ACh. Similarly, the β-adrenergic receptor can be phosphorylated by a cAMP-dependent kinase, and this phosphorylation inhibits the interaction of the receptor with G_s. These changes may play a role in regulating receptor sensitivity.

Second Messengers Also Regulate Other Processes Within Target Cells

In addition to regulating receptor and channel function, and thus the electrical activity of the neuron, metabotropic receptors can regulate a broad range of cellular processes. Activation of metabotropic receptors can even modulate gene expression in postsynaptic cells. By regulating gene expression, synaptic inputs can modulate cellular function over very long time periods. There is currently a tremendous interest in determining how these types of regulatory mechanisms operate.

Different Receptors Are Colocalized at Individual Postsynaptic Sites

The postsynaptic membrane of glutamatergic synapses contains a blend of different receptors, including members of both the ionotropic and metabotropic family (Figure 7.12). The blend of receptor types determines the overall functional characteristics of the synapse. Exactly how signaling is regulated at such synapses is just now being elucidated.

KEY POINTS

1. Neurotransmitters are not inherently excitatory or inhibitory; their action depends on the receptors with which the neurotransmitter interacts.

2. There are two principal types of neurotransmitter receptors: ionotropic receptors are ligand-gated ion channels; metabotropic receptors modulate enzyme cascades in the postsynaptic cell.

3. Ionotropic receptors are ion channels that open as a result of the binding of the neurotransmitter to the receptor; channel opening causes a rapid, transient ionic current.

4. Ionotropic receptors have two principal functional domains. The portion of the molecule that binds the neurotransmitter is termed the *ligand binding site;* this site determines the neurotransmitter to which the receptor responds. The portion of the molecule that forms the channel through the membrane is termed the *channel domain.*

5. The properties of the channel domain (i.e., distribution of charged amino acids in the parts of the molecule that form the walls of the pore) determine the *ion selectivity* of the channel (i.e., the particular ions to which the channel is permeable).

6. The physiological consequences of activating ionotropic receptors depend on the ion selectivity of the channel. Most *excitatory* neurotransmitters activate ionotropic receptors that are permeable to cations (usually K^+ and Na^+). Activation of these receptors leads to a simultaneous increase in permeability to both cations, which results in *inward* current flow, leading to membrane depolarization. Most inhibitory neurotransmitters activate ionotropic receptors that are permeable to anions (Cl^-). Opening of the channel leads to an increase in permeability to Cl^-, which results in membrane hyperpolarization.

7. Metabotropic receptors operate through enzymes to produce intracellular *second messengers*. These second messengers produce long-lasting metabolic effects in their postsynaptic targets, often by activating specific *protein kinases* that alter the phosphorylation state of other proteins (channels, receptors, or enzymes).

8

Neuronal Integration

Our goal in this chapter is to outline how input signals modulate the activity of individual neurons. It is useful to begin by recalling the concepts that underlie the principle of *dynamic polarization.* Essentially all neurons *receive* input signals, *integrate* those signals, *transmit* information over distances, and *communicate* with target cells via release of neurotransmitters. This information flow is unidirectional because neurons are *functionally polarized:* different parts of the neuron (functional compartments) are specialized to carry out different functions. The receiving and integrating compartment in most neurons is made up of dendrites and cell bodies. Based on the integrated input, action potentials are generated in the axon so that information can be transmitted over distances and communicated to target cells.

In virtually all neurons, input signals produce a graded change in membrane potential in the receiving compartment. This input signal can be in the form of a receptor potential (in the case of a sensory neuron) or a synaptic potential (in the case of other neurons). The graded input signal (an analog signal) is transformed into patterns of action potentials (a digital signal). This is the process of integration.

The simplest forms of integration are seen at sensory receptors. The receptors transduce physical energy into graded changes in membrane potential; the changes in membrane potential are transformed into changes in firing frequency (see Figure 8.1). In the case of many CNS neurons, however, the integration process can be quite com-

plex. In the first place, many CNS neurons receive tens of thousands and even hundreds of thousands of individual synaptic inputs. As described in the previous chapter, different types of synapses produce different postsynaptic responses. Some inputs are excitatory, others are inhibitory, some produce short-acting effects, others produce prolonged effects, and so forth. Hence, neuronal firing is regulated on a moment-by-moment basis through the integration of these different converging inputs.

We begin by considering how graded depolarization is transformed into action potential frequency at a prototypical sensory receptor. We then consider the signals that are generated by synapses. Of particular importance are how different input signals are integrated in space (spatial summation) and time (temporal summation), and how the physical form of the neuron affects this spatiotemporal integration process. Finally, we consider the interactions between depolarizing (excitatory) and hyperpolarizing (inhibitory) process.

SIGNAL TRANSFORMATION AT A SENSORY RECEPTOR

The basic process of signal transformation can be seen at sensory receptors. Sensory receptors are specialized to transduce physical energy into graded changes in membrane potential. Most often, the transduction process involves the opening or closing of an ion

145

channel in the receptor membrane, which generates a change in membrane potential called the *receptor potential.*

A simple example of signal transformation occurs at sensory nerve endings in the skin; this mediates somatic sensation (our sense of touch). The sensory nerve endings are the distal processes of sensory neurons whose cell bodies lie in the dorsal root ganglia (Fig-ure 8.1). The plasma membrane of these distal processes contains mechanosensitive ion channels. Mechanical stimulation causes the channels to open, leading to ionic currents that are similar to those produced by iono-tropic neurotransmitter receptors.

An important feature of the distal nerve ending is that Na^+ channels are not present at sufficient density to make the membrane

Figure 8.1. Signal transformation at mechanosensitive nerve endings. The nerve endings are the distal processes of sensory neurons whose cell bodies lie in the dorsal root ganglia. The plasma membrane of these distal processes contains mechanosensitive ion channels. Mechanical stimulation causes the channels to open, leading to ionic currents that are similar to those produced by ionotropic neurotransmitter receptors. **A,** Activation of mechanosensitive ion channels produces a receptor potential, which is a graded response that is determined by the intensity of the mechanical stimulus. When the spike-generating zone at the junction between the nerve ending and the axonlike portion of the process becomes sufficiently depolarized, action potentials are triggered in the axon. The graded depolarization at the spike-generating zone is called the generator potential. **B,** A stronger stimulus produces a larger generator potential and a greater number of action potentials. **C,** A stronger and longer duration stimulus produces a still larger and longer-lasting generator potential and a higher-frequency and larger number of action potentials.

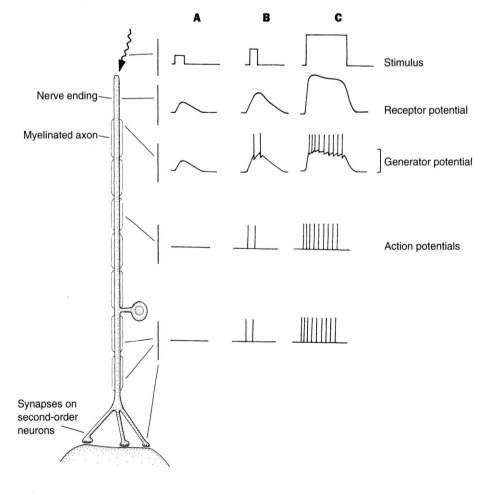

"excitable." Hence, the activation of mechano-sensitive ion channels produces a receptor potential that is a graded response that is determined by the intensity of the mechanical stimulus. The stronger the mechanical stimulus, the greater the number of individual channels that are activated and the larger the resulting receptor potential. In this way, a graded mechanical stimulus is transformed into a graded depolarization of the nerve ending.

The somatosensory nerve ending is an extension of the peripheral process of the sensory neuron. The portion of the process that lies proximal to the nerve ending functions as an axon, in that it contains Na$^+$ channels and supports action potentials. Indeed, the peripheral process of most sensory neurons is myelinated. At the junction between the nerve ending (which does not support action potentials) and the axonlike portion of the process is a *spike-generating zone*. When this region becomes sufficiently depolarized as a result of receptor activation, action potentials are triggered in the axonlike process and propagate centrally. The graded depolarization at the spike-generating zone is called the *generator potential* because it generates action potentials. The larger the generator potential, the greater the frequency of action potential generation.

SIGNAL TRANSFORMATION IN CNS NEURONS

The same basic process of signal transformation occurs in the case of the input signals generated by synapses onto neurons (Figure 8.2). Each individual synapse produces a synaptic current of a given amplitude because the synapse fires in an all-or-none fashion. The synaptic currents depolarize or hyperpolarize the postsynaptic neuron, depending on the nature of the receptor, that is, they produce *excitatory* or *inhibitory postsynaptic potentials* (EPSPs or IPSPs). The EPSPs and IPSPs modulate the probability of action potential generation at the spike-generating zone at the axon hillock.

The overall integration process is complicated, however. The first complication is due to the large numbers of individual synapses

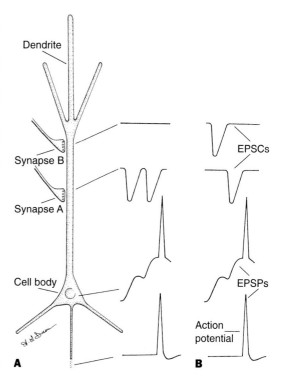

Figure 8.2. Spatial–temporal summation in CNS neurons. There are two basic processes that underlie neuronal integration: temporal summation and spatial summation. Temporal summation is the process through which two input signals that occur sequentially summate to produce a larger depolarization at the spike-generating site than would be generated by a single signal. In the example illustrated in **A,** Synapse A fires twice in rapid succession, which produces two excitatory postsynaptic currents (EPSCs) in the postsynaptic neuron. The resulting EPSPs summate so as to produce a larger net depolarization at the cell body. As a result, an action potential occurs in response to the second EPSP. Spatial summation is the process through which the signals generated by different inputs summate. In the example illustrated in **B,** Synapse A and B fire sequentially. Each generates its own synaptic current. The resulting EPSPs summate so as to produce a larger net depolarization at the cell body. As a result, an action potential occurs in response to the second EPSP.

that a neuron receives; the second is due to the physical form of neuronal receptive compartments (dendrites), which causes synapses to lie at varying distances from the spike-generating site; the third results from the diverse nature of the signals that synapses can generate. We consider each of these points in turn.

Interactions Between Synapses Depend on Spatiotemporal Summation

Most CNS neurons receive a large number of synapses, and it is rare that an individual synapse is powerful enough to depolarize the postsynaptic neuron sufficiently to generate an action potential. Instead, each time a synapse is activated, it produces a small depolarization or hyperpolarization; this signal then sums with the signals being generated by other synapses at that point in time. In this way, the membrane potential of the postsynaptic cell at any given point in time is determined by the integrated sum of the excitatory and inhibitory potentials being generated at that point in time.

There are two basic processes that underlie the summation process that we call neuronal integration: temporal summation and spatial summation.

Temporal summation is the process through which two input signals that occur sequentially can summate to produce a larger depolarization at the spike-generating site than would be generated by a single signal (Figure 8.2). Activation of an individual excitatory synapse produces a short-duration synaptic current and a longer-duration EPSP (the actual duration of the EPSP depends on the time constant of the postsynaptic cell; see later discussion). If this synapse fires once, a given level of depolarization is produced. If the synapse fires twice in rapid succession, the synaptic currents produced by each firing are the same and are separable in time, but the longer-duration EPSPs overlap in time. As a result, the second EPSP occurs during the depolarization induced by the first. In this way, the depolarizations produced during the two EPSPs are additive and cause a sufficient amount of depolarization to trigger an action potential.

The duration of the potential produced by a given synaptic current depends on the time constant of the cell, which in turn is determined by shape parameters (see later). The time constant is defined as the time required for a potential to decay to $1/e$ (about 37%) of its initial value.

Spatial summation is the process through which the signals generated by different inputs can summate. Consider, for example, a situation in which two inputs to a given neuron produce small EPSPs. If these two inputs fire simultaneously, the EPSPs will summate. Again, the depolarizations produced during the two EPSPs are additive and cause a sufficient amount of depolarization to trigger an action potential. In this way, two inputs that are individually too weak to discharge the postsynaptic cell can induce a postsynaptic action potential when activated simultaneously (Figure 8.2).

The processes of temporal and spatial summation are interdependent. In particular, as the example in Figure 8.2 illustrates, one cannot have spatial summation unless there is also temporal summation (that is, the two input signals must come close enough together in time to allow the depolarizations produced to summate). The term *spatiotemporal summation* applies to the combination of the two processes.

Input signals sum in a nonlinear fashion. Synaptic currents are generated by the opening of ion channels. When the channel opens, the membrane potential shifts to the equilibrium potential for the ions to which the channel is permeable. In the case of excitatory synapses, inward synaptic current produces a depolarizing EPSP. But the maximal depolarization that can be achieved during an EPSP is the equilibrium potential for the ions carrying the current (0 mV for excitatory synapses). Once a sufficient number of channels have opened to bring the membrane potential near the equilibrium potential for the EPSP, the opening of additional ion channels will have no additional effect on the membrane potential. For this reason, different inputs exhibit nonlinear summation.

Consider again the simple circuit illustrated in Figure 8.2. When activated individually, each of the two excitatory inputs produces an EPSP that is subthreshold for action potential generation. However, when activated together, the two inputs generate an EPSP that brings the neuron to threshold. If a third excitatory input were simultaneously activated, it would create only a small additional depolarization because the first two EPSPs have already brought the membrane potential near the equilibrium potential (0 mV).

Spatiotemporal summation occurs in the receptive and integrative elements of neurons (dendrites and cell bodies). Usually, these portions of the neuron are not electrically excitable because they do not possess voltage-gated Na^+ channels. For this reason, spatiotemporal summation depends on passive or *electrotonic* properties of the receptive compartments. These electrotonic properties

are determined by the physical size and shape of the compartment, that is, on dendritic form.

Electrotonic Properties of Dendrites Are Determined by Their Form

When the plasma membrane of a process (a dendrite, for example) is not electrically excitable, depolarization at one site spreads passively rather than actively. This is termed *electrotonic conduction.* The way that the current spreads is determined by the *cable properties* of the dendrite (Figure 8.3).

A dendrite can be thought of as a cable with a low-resistance core and possessing an insulating element with high resistance and

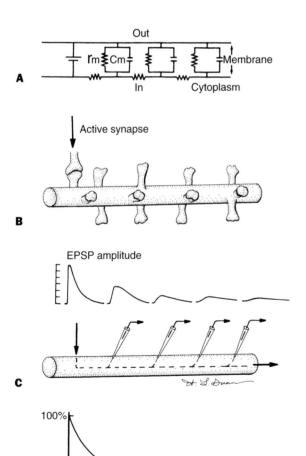

Figure 8.3. Cable properties of dendrites. Because they are not electrically excitable, dendrites behave as *core conductors.* **A,** The electrotonic spread of current in a core conductor depends on the relative amount of current flowing through the cytoplasm (core) and the leak current flowing across the membrane. These current flows are determined by the ratio between the cytoplasmic resistance (R_c) and membrane resistance (R_m). **B,** When a synaptic current is generated by activating a particular synapse (arrow), the resulting current flow occurs along the core of the dendrite and across the membrane, depending on the ratio of R_c and R_m, and the membrane capacitor becomes charged. **C,** The net result is that with increasing distance from the active synapse, the amplitude of the synaptic potential decreases. **D,** The rate of decay of the potential can be defined mathematically based on the length constant (λ) of the dendrite.

capacitance (the plasma membrane). Thus, dendrites behave as *core conductors*. The electrotonic spread of current in a core conductor depends on the relative amount of current flowing along the longitudinal axis (core) and the leak current flowing across the limiting membrane (Figure 8.3). When a synaptic current occurs at one location, the pattern of current flow within the dendrite determines the spread of the postsynaptic potential. Current flow occurs along the core of the dendrite and across the membrane, depending on the ratio of cytoplasmic resistance (R_c) and membrane resistance (R_m), and the membrane capacitor becomes charged. The net result is that with increasing distance along the dendrite, the amplitude of the synaptic *potential* decreases (Figure 8.3).

Because of the Cable Properties of Dendrites, the Closer an Excitatory Input Is to the Spike-Initiating Site, the More Effective It Will Be in Discharging the Postsynaptic Cell

For an excitatory synaptic current of a given magnitude and duration, the resulting depolarization at the spike-initiating site depends on the electronic distance between the active synapse and the spike-initiating site. A synapse that produces a synaptic current of a given amplitude produces a smaller depolarization at the cell body if it terminates on a distal dendrite than a similar synapse that terminates proximally (Figure 8.4).

The other consequence of electrotonic conduction is that the time course of electrotonic potentials is modified as a function

Figure 8.4. The efficacy of proximal versus distal synapses. A–C illustrate the excitatory postsynaptic currents (EPSCs) and EPSPs at the cell body produced by three synapses situated at different proximodistal locations. Note that for a given size EPSC, the EPSP produced in the cell body is smaller for distal than for proximal synapses. Also illustrated is the fact that a synaptic current in a distal dendrite generates an EPSP in the cell body that has a slower onset and longer time to peak than the voltage transient generated by a synapse on a proximal dendrite. The relationship between the shape of the EPSP and the site of termination of the active synapse provides a shape index for intracellularly recorded EPSPs.

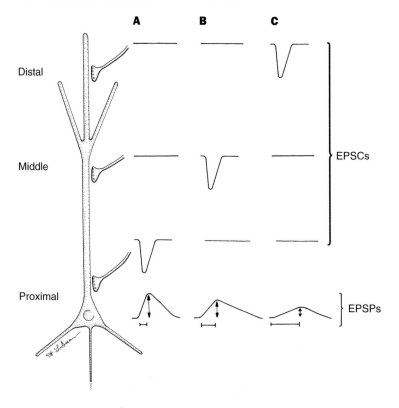

of distance from the initiating site. This is because the dendrite acts as a resistance–capacitance circuit. The plasma membrane acts as a resistor and capacitor in parallel; in other words, it operates as a high-pass filter. The high-frequency components of the voltage transient charge the membrane capacitance, producing what amounts to a low-pass filter for the intracellular voltage transient. For this reason, a synaptic current in a distal dendrite generates a voltage transient in the cell body that has a *slower onset* and *longer time to peak* than the voltage transient generated by a synapse on a proximal dendrite (Figure 8.4). The greater the electrotonic distance, the greater this effect. The relationship between the shape of the EPSP and the site of termination of the active synapse provides a *shape index* for intracellularly recorded EPSPs, which can be used to compare sites of termination of different afferents to a given cell.

Cable Properties Are Determined by Dendritic Diameter

When the resistance of the core is low with respect to the resistance of the membrane, most of the current flows along the core. Conversely, when the resistance of the core is high with respect to the resistance of the membrane, less current flows along the core. These properties determine *core conductance* (the relative amount of current flowing along the longitudinal axis of the dendritic cylinder).

The key variables determining the core conductance are the *specific resistance of the cytoplasm* (R_i), the *resistance of the plasma membrane* (R_m), the *capacitance of the membrane* (C_m), and the *diameter of the dendrite*. Because R_i, R_m, and C_m are similar in different dendrites, differences in cable properties are determined primarily by differences in dendritic diameter. The greater the diameter of a dendrite, the greater the core conductance.

Differences in the cable properties of dendrites are expressed according to differences in *characteristic length,* which is measured in *length constants.* A length constant is the distance over which an electrotonic potential falls to $1/e$ (which equals approximately 0.37) of its initial value (V_o). If the postsynaptic potential that is induced by a particular synapse falls to 0.37 of its initial value over the length of the dendrite, that dendrite is said to have a characteristic length of 1.0. Given two dendrites of different diameters with a comparable *physical length* (in millimeters), the characteristic length of the small-bore dendrite will be greater than that of the large-bore dendrite. Thus, a synapse terminating at a given *physical distance* from the cell body will be at a greater *electrotonic distance* from the cell body on the smaller-bore dendrite (Figure 8.5).

Dendritic Spikes

In general, dendrites conduct electrotonically, not via action potentials. Nevertheless, the dendrites of certain neurons possess "trigger zones" that contain clusters of voltage-gated ion channels, usually at dendritic branch points (Figure 8.6). In some cases, voltage-dependent calcium channels are present; in other cases sodium channels may be present. These clusters of voltage-gated ion channels create "hot spots" in the dendritic membrane in which *regenerative potentials* can be produced (either Ca^{2+} spikes or Na^+ "sparks"). It is thought that these trigger zones serve to amplify the signals generated in distal dendrites.

Inhibitory Synapses Operate by Producing Current Shunts That Effectively Clamp the Postsynaptic Membrane at a Level That Is Below Threshold for Action Potential Initiation

When inhibitory connections are added into the picture, the situation becomes somewhat more complex. The IPSPs produced by inhibitory inputs are opposite in sign to EPSPs. However the more important consequence of activating inhibitory synapses is that by opening an anion channel, a current shunt is produced that will affect current flow produced by excitatory synapses.

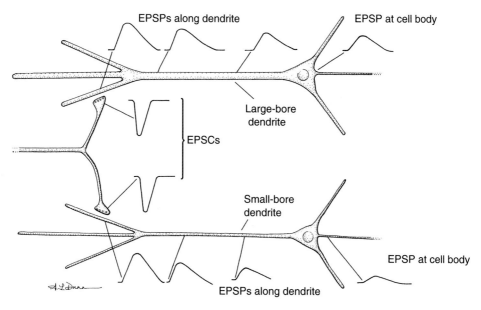

Figure 8.5. Characteristic length is determined by dendritic diameter. Two neurons, one with a large-diameter dendrite (above), the other with a smaller-diameter dendrite (below). An incoming afferent fiber forms excitatory synapses on both dendrites. An action potential in this fiber produces excitatory synaptic currents of similar magnitude in the postsynaptic cell. However, in the neuron with the larger diameter dendrite, the net depolarization at the cell body is greater.

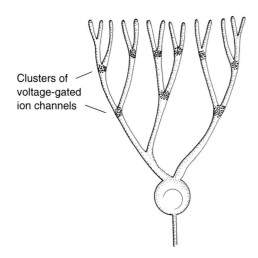

Figure 8.6. Dendritic trigger zones trigger regenerative potentials in dendrites. The dendrites of certain neurons possess "trigger zones" that contain clusters of voltage-gated ion channels, usually at dendritic branch points. These clusters of voltage-gated ion channels create "hot spots" in the dendritic membrane in which *regenerative potentials* can be produced (either Ca^{2+} spikes or Na$^+$ "sparks"). It is thought that these trigger zones serve to amplify the signals generated in distal dendrites.

Consider the simple circuit illustrated in Figure 8.7. Activation of the synapse causes a synaptic current that produces an EPSP. Activation of the inhibitory synapse leads to the opening of ligand-gated Cl$^-$ channels. The equilibrium potential for Cl$^-$ is −70 mV, so that the IPSP produced by the inhibitory synapse is −5 mV in amplitude. However, if the inhibitory synapse is activated simultaneously with the excitatory synapse, the resulting membrane potential is predicted not by an arithmetic summation, but rather by the Goldman equation. The opening of the Cl$^-$ channel effectively shunts the depolarizing current produced by the excitatory synapses. Thus, even though the potential change produced by inhibitory synapses is small, the current shunt is more than sufficient to negate the effect of the excitatory input.

Because inhibitory synapses work primarily by shunting the current produced by excitatory synapses, inhibitory synapses that open Cl$^-$ channels are most effective when situated near the spike-initiating zone (the cell body or the axon initial segment). In-

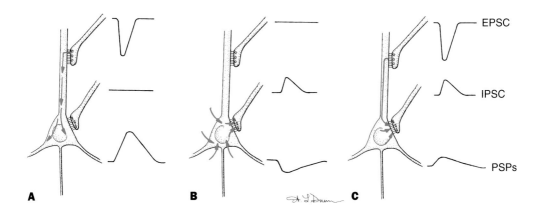

		EPSC
		IPSC
		PSPs

A   **B**   **C**

Figure 8.7. Shunting by inhibitory synapses. A, Activation of an excitatory synapse causes an excitatory postsynaptic current (EPSC) that produces a depolarizing potential (EPSP). **B,** Activation of an inhibitory synapse leads to the opening of ligand-gated Cl⁻ channels, producing an inhibitory postsynaptic current and hyperpolarizing potential (IPSP). The equilibrium potential for Cl⁻ is -70 mV, so that the IPSP produced by the inhibitory synapse would be about -5 mV in amplitude. **C,** If the inhibitory synapse is activated simultaneously with the excitatory synapse, the Cl⁻ conductance effectively shunts the depolarizing current produced by the excitatory synapses.

hibitory synapses on dendrites have a much more local effect, regulating the effectiveness of nearby excitatory synapses, but having little effect on excitatory synapses at other locations on the neuron.

The Relationship Between Presynaptic and Postsynaptic Activity Can Be Characterized Based on Input–Output Functions

The relationship between the firing of presynaptic fibers and the firing of the postsynaptic neuron can be defined on the basis of *input–output functions.* (see Figure 8.8). In terms of excitatory synapses, input–output functions can be characterized as (1) *one-to-one* (one presynaptic action potential gives rise to one postsynaptic action potential), (2) *one-to-many* (one presynaptic action potential leads to burst firing of the postsynaptic neuron), or (3) *many-to-one* (many presynaptic inputs must fire simultaneously to induce postsynaptic discharge).

The phraseology is less convenient for inhibitory inputs, but the principles are the same. Thus, inhibitory inputs can totally block postsynaptic discharge in response to excitatory synapses, can reduce the probabil-

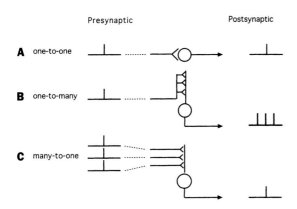

Presynaptic   Postsynaptic

A one-to-one

B one-to-many

C many-to-one

Figure 8.8. Input–output functions. In terms of excitatory synapses, input–output functions can be characterized as **A,** one-to-one (one presynaptic action potential gives rise to one postsynaptic action potential), **B,** one-to-many (one presynaptic action potential leads to burst firing of the postsynaptic neuron), **C,** many-to-one (many presynaptic inputs must fire simultaneously to induce postsynaptic discharge).

ity of postsynaptic firing in response to other inputs, or can reduce the firing rate if the neuron fires spontaneously.

The neuromuscular junction is an example of a one-to-one synapse. Neuron-to-neuron synapses are not usually one-to-one in the strictest sense, although some neurons are innervated by a small number of very large and powerful synaptic contacts. Examples include neurons in the ciliary ganglion and neurons in certain sensory relay nuclei (for example, the cochlear nucleus). Each time one of the synapses fires, there is a high probability that the postsynaptic cell will fire. Synapses that have a high probability of discharging the postsynaptic cell are termed *detonator synapses* or *command inputs,* and the synapse is said to have a high *safety factor.*

The best-known example of a one-to-many synapse in the CNS is the climbing fiber input to Purkinje cells of the cerebellum (see Chapter 17). Each Purkinje cell is innervated by one climbing fiber, and the climbing fiber forms numerous powerful synaptic contacts along the Purkinje cell's dendrite. When this climbing fiber discharges, a brief train of action potentials is triggered in the Purkinje cell.

Most synapses in the CNS are of the many-to-one variety, where individual neurons are innervated by a very large number of inputs that are too weak to fire the neuron independently. In this situation, neuronal firing occurs when the summed activity in a large number of excitatory inputs produces a sufficient level of depolarization at the spike-initiating site.

Spontaneous Activity of Neurons

Some neurons generate action potentials only when they are sufficiently depolarized by their inputs; others have a *spontaneous activity* that is *modulated* by synaptic inputs. Spontaneously active neurons possess intrinsic mechanisms that lead to a steady-state depolarization that is sufficient to maintain a certain level of activity. The general rules for synaptic action apply equivalently whether

synapses directly drive neurons or modulate their "spontaneous" activity. However, neurons with intrinsically determined spontaneous activity may also be modulated nonsynaptically (for example, by circulating hormones). Neurons with intrinsically regulated programs for spontaneous activity play an important role in regulating a number of important biological processes (circadian rhythms, hormonal surges, etc.).

Examples of Simple Neural Circuits

A neural circuit is a series of interconnected neurons. These circuits can be put together in various ways.

Excitatory Neural Circuits

In an excitatory neural circuit, all synaptic relays are excitatory (Figure 8.9A). A simple example is the relay that begins at the specialized stretch receptor in a muscle spindle (the Ia afferent). Ia afferents form excitatory connections with motoneurons. Hence, the sensory neuron, motor neuron, and the muscles that the motor neurons supply are components of a simple circuit in which there are two excitatory synaptic relays. Action potentials in the afferent axon trigger an EPSP and then an action potential in the motor neuron; the action potential in the motor neuron then triggers an end plate potential in the muscle. In excitatory neural circuits, the transformation function can be one-to-one, one-to-many, or many-to-one, but in each case, activity at the input stage *increases* the probability of activity at the end stage (that is, the circuit mediates *net excitation*).

Neural Circuits That Involve Recurrent Inhibition

An example of a circuit that mediates recurrent inhibition is illustrated in Figure 8.9B. This is the same circuit illustrated in Figure 8.9A, involving sensory afferents, motoneurons, and the muscles that the mo-

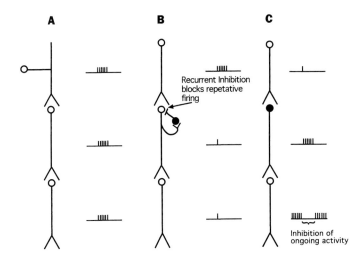

Figure 8.9. Examples of simple neural circuits. A illustrates an excitatory neural circuit in which all relays are excitatory. **B** illustrates a circuit that mediates recurrent inhibition. **C** illustrates an example of an inhibitory neural circuit in which one or more of the synaptic relays are inhibitory. This example represents the situation in the cerebellar relay involving climbing fibers, Purkinje neurons, and neurons of the deep cerebellar nuclei. Climbing fibers are excitatory; Purkinje neurons are inhibitory. Action potentials in the climbing fiber trigger an EPSP and a burst of action potentials in the Purkinje neuron; the train of action potentials in the Purkinje neuron then induce a series of IPSPs in the neurons of the deep cerebellar nuclei. The IPSPs will strongly inhibit the target neurons, blocking their spontaneous activity and preventing their activation by other excitatory inputs. This is an example of projected inhibition, in which activity at the input stage (the climbing fiber) causes a decrease in activity at the end stage (that is, the circuit mediates net inhibition).

toneurons supply, but adding an interneuron. In addition to the projection to the muscle, motoneurons give rise to collaterals that terminate on interneurons that use GABA as their neurotransmitter (see Chapter 7). These *inhibitory interneurons* then project back to the motor neurons. Hence, an action potential in the motor neuron triggers the firing of the inhibitory interneuron, which then produces IPSPs in the motoneuron (termed *recurrent inhibition*). Recurrent inhibitory circuits are common throughout the nervous system and are thought to play a role in limiting multiple spiking.

Neural Circuits That Involve Feedforward Inhibition

The same basic circuit also forms the substrate for feedforward inhibition. In addition to receiving projections from motor neurons, inhibitory interneurons are directly innervated by sensory afferents and then project on to a different group of motor neurons that receive direct excitatory projections. Hence, activity in the sensory afferent triggers the firing of the inhibitory interneuron, which then produces IPSPs in the motoneuron. This is termed *feedforward inhibition.*

Inhibitory Neural Circuits

In an inhibitory neural circuit, one or more of the synaptic relays are inhibitory. An example of a simple inhibitory circuit is the cerebellar relay involving climbing fibers, Purkinje neurons, and neurons of the deep cerebellar nuclei. Climbing fibers are excitatory, but Purkinje neurons use GABA as their neurotransmitter and are thus inhibitory. Action potentials in the climbing fiber trigger an EPSP and a burst of action potentials in the Purkinje neuron; the train of action potentials in the Purkinje neuron then induces a series of IPSPs in the neurons of the deep cerebellar nuclei. The IPSPs will strongly inhibit the target neurons, blocking their spontaneous activity and preventing their activation by other excitatory inputs. This is an example of *projected inhibition,* in which activity at the input stage (the climbing fiber) causes a decrease in activity at the end stage (that is, the circuit mediates *net inhibition*).

There are also neural circuits with two inhibitory connections in series. An example is the relay involving neurons in the cortex, striatum (caudate nucleus and putamen), globus pallidus, and thalamus (Figure 8.9C). The projections from the cortex to the striatum are excitatory; however both the striatal neurons and the neurons of the globus pallidus use GABA as their neurotransmitter. Activation of striatal neurons inhibits neurons in the globus pallidus. This reduces their level of ongoing activity, which otherwise causes a *tonic inhibition* of neurons in the thalamus. Hence, when the ongoing activity of pallidal neurons is inhibited, the tonic inhibition of neurons in the thalamus is blocked (a process called *disinhibition*).

Neural Circuits Involving Nonspiking Neurons

There are also neural circuits involving interconnections between neurons that do not generate action potentials. One important example of such a circuit is found in the retina. The photoreceptors form synapses with bipolar neurons and horizontal neurons, which in turn form synapses on ganglion cells. None of these neuron types generates action potentials. Instead, the neurotransmitters produce graded responses that modulate neurotransmitter release in a graded fashion. We will learn more about the details of this important circuit in Chapter 18.

3

The Somatosensory System

The neural systems responsible for somatic sensation begin in the periphery where various types of receptors give rise to the *general somatic afferent pathways*. It is useful to begin by setting out some of the key concepts for physicians.

1. *The somatosensory system transmits information from peripheral receptors to the CNS via several parallel pathways* (Figure III.1). The parallel nature of the somatosensory system derives from two facts: (1) different classes of peripheral receptor are specialized for different submodalities of somatosensory information (pain, temperature, discriminative touch, vibration sensitivity, and position sense), and (2) different receptors give rise to axons that travel through the spinal cord and brain over

different pathways that cross at different locations along the neuraxis. Injuries at different levels of the nervous system produce different types of somatosensory symptoms because of the existence of these parallel pathways. An understanding of the organization of the different components of the somatosensory system forms the basis for localizing lesions that affect the somatosensory pathways.

2. *Conscious perception of somatic sensation depends on inputs to the cerebral cortex* . Thus, interruption of the pathways for somatic sensation anywhere along their course from the periphery to the cortex leads to a loss of conscious sensation. An exception to this generalization is pain perception, which persists after cortical injuries

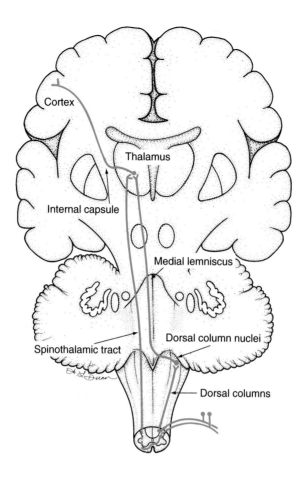

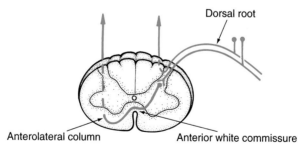

Figure III.1. Overview of somatosensory system organization. The basic route of information flow in the somatosensory is indicated in the schematic. Sensory axons from one side of the body enter the cord in the dorsal roots. At this point, the pathways related to the different submodalities of somatic sensation diverge. The components conveying information about discriminative touch, vibration sensitivity, and position sense ascend ipsilaterally in the dorsal columns (DC). The primary afferent fibers synapse at the spinomedullary junction in the dorsal column nuclei. Axons from second-order neurons in the dorsal column nuclei then decussate in the lower medulla and ascend through the brainstem in the medial lemniscus to terminate in the thalamus. Axons of thalamic relay neurons then project on to the somatosensory cortex (Cx) through the internal capsule. The component conveying pain and temperature sensitivity enters the cord and synapses on neurons in the dorsal horn of the spinal cord. The axons of second-order neurons decussate near the level of entry in the ventral white commissure and ascend in the anterolateral column of the spinal cord on the contralateral side. These fibers travel through the lateral brainstem in the spinothalamic tract and terminate in the thalamus. Throughout much of their course, the two components remain quite separate. Not shown are the somatosensory inputs from the face, which arrive via the trigeminal nerve.

and so is thought to be mediated by subcortical systems.

3. *Pathways carrying somatosensory information decussate.* Hence, one side of the cortex receives somatosensory information from the opposite side of the body. As a result, cortical injuries produce sensory symptoms on the contralateral side.

4. *The pathways carrying different submodalities of somatosensory information decussate at different levels.* The pathways conveying pain and temperature sensation decussate near their level of entry into the spinal cord (at the segmental level) and travel up the spinal cord on the side contralateral to their point of entry. The pathways that convey the sensations for discriminative touch, vibration sensitivity, and position sense travel up the spinal cord ipsilateral to their point of entry and decussate at the spinomedullary junction. These facts lead to important diagnostic signs that allow the physician to determine the level within the spinal cord of a lesion that affects the somatosensory systems.

5. *The pathways carrying different submodalities of somatosensory information travel through different white matter tracts within the spinal cord and brainstem.* Information related to discriminative touch, vibration sensitivity, and position sense is carried by fibers that travel through the dorsal columns and medial lemniscus. This system is thus called the dorsal column–medial lemniscal system. Information related to pain and temperature sensitivity travels through the anterolateral columns of the spinal cord and spinothalamic tract in the brainstem. This system is thus called the spinothalamic system. Lesions can affect one or the other of these systems selectively.

Through an understanding of the precise routes taken by the various pathways conveying somatosensory information, it is possible to deduce the location of lesions affecting the somatosensory pathways.

9

Somatic Sensation I

Peripheral Receptors and Primary Afferent Neurons

Principles of Sensory Systems

This chapter is the first to deal with one of the sensory systems, and so we begin with some generalizations that apply to all sensory systems.

1. Sensory systems detect four attributes of stimuli: modality, intensity, duration, and location.

    Modalities refer to the different senses: somatic sensation, sight, hearing, taste, and smell. The sensory receptors that are the first stage of the different sensory systems are specialized to transduce particular types of physical energy: mechanical, thermal, light, sound, and chemical. Receptors of the somatosensory system detect mechanical or thermal energy, receptors in the visual system detect light, and so forth. However, the use of the term *modalities* is not entirely consistent in the literature. Some texts also refer to different "modalities" of somatic sensation: pain, temperature, vibration sensitivity, discriminative touch, and position sense. Other texts refer to these as *submodalities* or *qualities* of somatic sensation.

    The convention in this text is to use the term *modalities* to refer to the five classical senses (somatic sensation, sight, hearing, taste, and smell). The term *submodalities* is used to refer to situations where specialized components of the pathway are dedicated to detect particular types of stimuli. For example, in the somatosensory system, anatomically distinct components of the system are dedicated to convey information related to discriminative touch, vibration sensitivity, position sense, pain, and temperature. Other pathways are dedicated to visceral sensation. These will be referred to respectively as the submodalities of somatic and visceral sensation. To avoid confusion, it is important for students to remember that other texts may use these terms somewhat differently.

    Intensity refers to stimulus strength. The *dynamic range* of a receptor is the range of intensities over which the receptor varies its response. The lowest intensity that produces a response is the *receptor threshold*. The lowest intensity at which an individual can detect a stimulus (that is, perceive it consciously) is termed a *sensory threshold*. Sensory thresholds are regulated in part by descending influences, providing a mechanism for the brain to focus attention on particular inputs and ignore irrelevant stimuli.

Duration refers to the time during which the stimulus is applied, and sensory receptors are classified according to the way that they respond to stimuli of different durations. In particular, *adaptation* refers to the way that a receptor (or a neuron along the sensory pathway) responds to an enduring stimulus. *Rapidly adapting receptors* respond to changes in stimulus intensity (for example, at stimulus onset) and then rapidly return to their baseline firing rate. They may also respond at stimulus offset. In this way, rapidly adapting receptors encode dynamic aspects of the stimulus. *Slowly adapting receptors* continue to respond throughout the duration of the stimulus, and for this reason, slowly adapting receptors are better able to encode the intensity of stimuli.

Location has two components: (1) the ability to locate the site of an individual stimulus, and (2) the ability to distinguish between closely spaced stimuli. Both of these components are aspects of *spatial resolution.*

2. All sensory systems have *receptor cells* that *transduce* physical energy into ion fluxes in nerve cells. The ion fluxes in turn are integrated so as to produce changes in action potential frequency in *first-order neurons* that send their axon from the periphery into the CNS.

   In some systems, the receptor cells are separate from the first-order neurons. For example, in the auditory system, vibration causes a depolarization of receptors called *hair cells,* which release an excitatory neurotransmitter onto the peripheral process of a cochlear ganglion cell. In the visual system, the receptors are rods and cones in the retina. These *photoreceptors* release a neurotransmitter onto nonspiking neurons in the retina that integrate input from receptors and then activate retinal ganglion cells, which send an axon to the brain.

   In the somatosensory system, the axon terminals of the first-order neurons function as the receptor elements. As described later, the transduction of physical energy occurs at these terminals.

3. In all sensory systems, the transduction process leads to changes in ion fluxes that generate changes in membrane potential within the receptor cell. The changes in membrane potential within receptor cells in response to a stimulus is termed the *receptor potential.* The receptor potential is a *graded response* that varies as a function of stimulus intensity (that is, the receptor potential is not an all-or-none action potential). As described in this and later chapters, the nature of the receptor potential differs in different sensory systems.

4. Different receptor types are specialized to transduce particular types of physical energy. The receptors of the somatosensory system are *mechanoreceptors* and *thermoreceptors* that are specialized to transduce mechanical or thermal energy, respectively.

5. The stimulus that produces a response in a particular receptor is termed an *adequate stimulus* for that receptor. Many receptors are *tuned* to be most sensitive to a narrow range of energy (for example, a particular frequency of vibration or a particular wavelength of light). For receptors that are tuned, it is possible to define a *tuning curve* that defines the range of energy to which the receptor is most sensitive.

6. Either directly or indirectly, the receptor potential produces graded changes in membrane potential within the first-order neuron. These are integrated so as to produce changes in the frequency of action potentials within the axons of the first-order neurons. When the changes in membrane potential within first-order neurons are depolarizing, they generate action potentials. In this case, the depolarizing potentials are termed *generator potentials.*

7. As a result of the characteristics of the receptor and the integrative properties of the first-order neuron, stimulus features are encoded into a pattern of action potentials termed the *neural code.*

8. There are two types of neural code. The fact that different modalities and submodalities are carried by particular pathways means that activity in each pathway (regardless of how is induced) is interpreted in terms of its modality. If the optic nerve is electrically stimulated, light will be perceived; if a peripheral nerve is electrically stimulated, the

perception will be of a somatic sensation. This is termed a *labeled line code*. Other aspects of a stimulus are encoded by the frequency or pattern of action potentials generated termed *rate codes* and *pattern codes*, respectively.

With this brief, general overview of sensory processing in mind, we can begin to consider processing within the somatosensory system.

THE CELLULAR ELEMENTS OF THE SOMATOSENSORY SYSTEM

Somatosensory submodalities are conveyed via separate "channels" that begin at the re-ceptor level; the pathways conveying different submodalities take different routes through the spinal cord and brainstem. But all channels are made up of the same cellular elements:

1. Somatic sensation begins at peripheral receptors in skin, muscles, tendons, and joints. The peripheral receptors are the nerve endings of first-order neurons.

2. First-order neurons have their cell bodies in the dorsal root ganglion and, in the case of sensory axons from the head and face, the trigeminal ganglion. These neurons are *unipolar* and give rise to a single process that bifurcates, forming a peripherally directed and a centrally directed axon (Figure 9.1). The peripheral branch of the axon joins the peripheral nerves. At the end of the

Figure 9.1. First-order neurons. The cell bodies of the first-order neurons of the somatosensory system are found in the dorsal root ganglia, trigeminal ganglion, and mesencephalic nucleus of the trigeminal nerve. The peripheral branch of the axon of the first-order neurons projects into the periphery. The sensory receptors themselves are the terminal specializations of these axons. The central branch of the axon travels into the CNS, where it synapses upon second-order neurons.

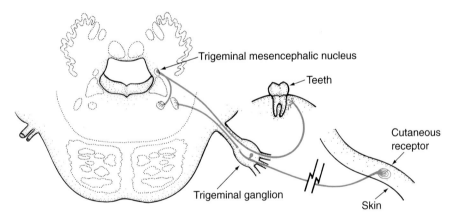

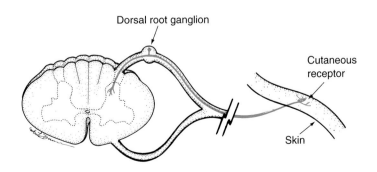

peripheral branch is the receptor apparatus that transduces physical energy. The other branch of the axon projects into the spinal cord via the dorsal root and terminates on second order neurons. The axons of the dorsal root are termed *primary afferents.*

Somatosensory information from the head and face travels via cranial nerve V (the trigeminal nerve). Most of the somatosensory input is via first-order neurons with cell bodies in the trigeminal ganglion (Figure 9.1). Thus, the trigeminal ganglion is equivalent to the dorsal root ganglia. However, one group of fibers that arrives via the trigeminal nerve has its cell bodies in a nucleus within the brainstem itself—the *trigeminal mesencephalic nucleus* or *mesencephalic nucleus of cranial nerve V.* This nucleus is the only example of a first-order neuron in the somatosensory system that has its cell body in the CNS.

3. Second-order neurons receive input from axons arriving via dorsal roots and cranial nerves. The synapse between first- and second-order neurons thus represents the *first synaptic relay* of the somatosensory pathway. The cell bodies of second-order neurons of the somatosensory system are located in the spinal cord, dorsal column nuclei, and in the brainstem (in the case of somatosensory inputs from the head and face that arrive via cranial nerves). Second-order neurons send their axons to third-order neurons in the thalamus. The synapses along the pathway from the spinal cord to the cortex are excitatory. However, at each synaptic station, there are important inhibitory interactions that help to shape the *receptive field* of the neurons (see Chapter 11).

Different Types of First-Order Neurons Respond to Different Types of Physical Energy

The receptors of the somatosensory pathways are specializations of the terminals of the peripheral axons of the first-order neurons. Different receptors are specialized to transduce different types of stimuli. As a result of receptor specialization, the axons deriving from the different receptors carry information related to the submodalities of pain, temperature, discriminative touch, vibration sensitivity, and position sense. With regard to somatosensory receptors, the following terms are used:

- *Mechanoreceptors* include all receptors that are specialized to respond to physical displacement (touch, pressure, and movement).

- *Thermoreceptors* are specialized to sense changes in temperature.

- *Nociceptors* respond to stimuli that are potentially damaging (that is, they are pain receptors).

- *Proprioceptors* provide information about body position and movement. These are located in joints and muscle.

- *Chemoreceptors* respond to particular chemical substances.

- *Cutaneous receptors* are the receptors that are present in the skin.

- *Deep receptors* are the receptors that are present in muscles, tendons, joints, and viscera.

- *Superficial sensation* refers to the sensation mediated by *cutaneous receptors* for touch, superficial pressure, temperature, and pain.

- *Tactile sensation* refers to the sensation of touch and superficial (light) pressure.

- *Deep sensation* refers to the sensation mediated by receptors in joints and muscle including the receptors for position, vibration, and deep pressure.

- *Visceral sensation* refers to the sensation mediated by receptors in the viscera, including visceral pain as well as the general sensations of hunger and nausea, which are mediated by chemoreceptors.

Different Qualities of Somatic Sensation Are Mediated by Different Receptors

Although a consideration of somatic sensation can be organized in a variety of ways, based on the above categories and definitions, a useful way to begin is to consider sensation from the point of view of what is perceived (that is, the submodality of sensation). Thus, the following summarizes somatic sensation by submodality.

It is important to emphasize that the different submodalities are conveyed by anatomically distinct components. Thus, different peripheral receptors are specialized to respond to certain types of stimuli, and the axons from these receptors project to distinct sites in the CNS. The earliest evidence for the fact that different submodalities are conveyed by anatomically distinct systems came from clinical observations, specifically, that certain diseases or injuries selectively abolished particular submodalities (the ability to recognize skin contact, heat and cold, or pain, for example).

Tactile Perception

Tactile perception is mediated by cutaneous and subcutaneous mechanoreceptors in the skin. There are a number of different types of receptors distributed in different types of skin (Figure 9.2). The receptors are specializations of the peripheral terminal of the dorsal root ganglion neuron, which can be either a bare nerve ending or an end organ consisting of the nerve ending surrounded by a specialized nonneuronal capsule. The differences in receptor morphology correspond to differences in physiological properties and functions. These are briefly summarized in Table 9.1 and are considered in more detail in Chapter 11.

There are some facts about cutaneous mechanoreceptors that are important to know (some of which are easy to figure out by self-examination). For example, hair re-

ceptors respond to slight bending of the hair and are sometimes directionally selective. Examine this yourself with the hairs on your arm. Touch receptors respond to low-intensity mechanical deflection of the skin. These are termed *low-threshold mechanoreceptors.* There are slowly adapting and rapidly adapting proprioceptors that are sensitive to absolute position and change in position (movement), respectively.

Vibration Sensitivity

Vibration sensitivity is a special mode of tactile perception that is mediated by specialized mechanoreceptors that adapt very rapidly (*Pacinian corpuscles*). Pacinian corpuscles are composed of accesssory structures (a capsule) surrounding the axon terminal. Vibration sensitivity is important for the physician because the central pathways for vibration sensitivity, along with those for position sense and discriminative touch, travel via the dorsal column system. In fact, vibration sensitivity is conveyed exclusively via the dorsal column system, whereas some touch sensitivity remains after complete interruption of the dorsal column system. For this reason,

Figure 9.2. Types of cutaneous receptors.
Different types of cutaneous mechanoreceptors in skin: left, hairy skin; right, glabrous skin. Adapted from Light and Perl, Peripheral sensory system.

In: Dyck PJ, Thomas PK, Lambert EH, Burge R, eds. *Peripheral Neuropathy.* 2nd ed. Vol 1. Philadelphia, Pa: WB Saunders; 1984:210–230.

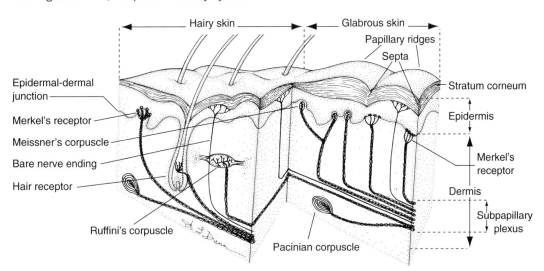

Table 9.1. Cutaneous and subcutaneous receptors

Location	Type	Adaptation	Function	Fiber group
cutaneous	Hair receptors (hairy skin)	SA/RA	Hair deflection	A-beta
	Meissner's corpuscles (glabrous skin)	RA	Touch (dynamic)	A-beta
	Merkel's receptors	SA	Touch (steady)	A-beta
	Bare nerve endings	SA	Nociceptors	A-delta, C
			Thermoreceptors	A-delta, C
subcutaneous	Pacinian corpuscle	VRA	Vibration	A-beta
	Ruffini's corpuscles	SA	Touch (steady)	A-beta

SA = slowly adapting; RA = rapidly adapting, responds when the velocity of skin deformation changes; VRA = very rapidly adapting; signals onset and offset of a stimulus to skin

tests of vibration sensitivity are more definitive for diagnostic purposes than tests of touch sensitivity. Tests for vibration sensitivity are a routine part of the neurological exam because they are simple to carry out using a tuning fork.

Pain Perception

Pain perception is thought to be mediated by free nerve endings in the skin. There seem to be several different types of pain receptors that respond to different forms of noxious stimulation. Some are high-threshold mechanoreceptors that respond to pinch or pressure but not to light touch. Some respond to high-intensity mechanical pressure as well as high temperatures (temperatures above 45°C). Some, termed *polymodal nociceptors,* respond to high-intensity mechanical stimuli, thermal stimuli, and also the application of certain chemical substances that are typically released at sites of tissue damage (bradykinin, histamine, and potassium). It is thought that different types of nociceptors convey different qualities of pain—sharp, pricking pain versus slow, burning pain, for example.

Temperature Sensation

Temperature sensation is mediated by specialized cutaneous thermoreceptors. Neuro-

physiological studies have demonstrated that there are both *warm receptors* and *cold receptors.* Most mild changes in skin temperature will cause increases in firing in the axons from one type of receptor and decreases in firing in axons from another type of receptor.

Position Sense

Position sense (proprioception) is mediated by joint receptors, tendon receptors, and receptors in the muscles. Tendon receptors are termed *Golgi tendon organs.* Muscle receptors are sensory nerve endings found in specialized structures termed *muscle spindles.* Sensory endings have a somewhat different form in the two types of muscle spindles. One form is termed the *annulospiral ending;* the other is termed the *flower spray ending.* The muscle spindles do not play a role in conscious perception but are extremely important for regulating muscle function. For this reason, these receptors are described in more detail in the chapters on the motor system.

Visceral Sensation

Visceral sensation is mediated by sensory nerve endings in the viscera. Most of these sensory nerve endings are *regulatory interoreceptors* that play a role in maintaining homeostasis. Visceral chemoreceptors contribute to the general visceral sensations that mediate ingestive behaviors (for example, hunger). Visceral pain is thought to be mediated by pain receptors that are sensitive primarily to stretch or distension of the viscera.

The generalizations about the relationship between submodality and receptor type are useful but by no means absolute. In particular, free nerve endings can probably respond to a broad range of stimuli. For example, the cornea has only free nerve endings, yet we can perceive touch, temperature, and pain from the cornea.

Different Types of Sensory Fibers Are Myelinated to a Different Extent

An important fact for physicians is that the different submodalities of somatic sensation are carried by axons with different conduction velocities. The differences in conduction velocity are due to differences in the degree of myelination. For example, when a peripheral nerve is stimulated electrically at one site, and recordings are made a few centimeters away, the compound action potential exhibits several peaks that reflect fibers with different conduction velocities (Figure 9.3). There are actually two classification schemes

(groups I–IV, and types A, B, A-ß, A-δ, and C). These two schemes are summarized in Table 9.2 to help students generalize between reference sources.

Axons of cutaneous mechanoreceptors are termed A-ß (group II); these have a conduction velocity ranging from 25 to 70 m/s. There are two groups of nociceptors, A-δ and C (groups III and IV), which have conduction velocities of 10 to 30m/s, and 2.5 m/s or less, respectively. The A-δ fibers respond to sharp, pricking pain; the C fibers respond to slow, burning pain.

Tendon, muscle, and joint receptors are described using the groups I–IV terminology. Muscle spindles and Golgi tendon organs give rise to axons that conduct at the highest velocity (group I), and these are termed Ia and Ib afferents, respectively. There are also lightly myelinated and unmyelinated axons (groups III and IV) in muscles and joints, many of which are pain receptors.

From the point of view of somatic sensation, a key point is that fibers carrying discriminitive touch and vibration sensitivity are fast conducting; those carrying pain and temperature are slow conducting.

Some disease processes can selectively affect the peripheral receptors or peripheral axons carrying particular submodalities. Peripheral neuropathies (for example, diabetic

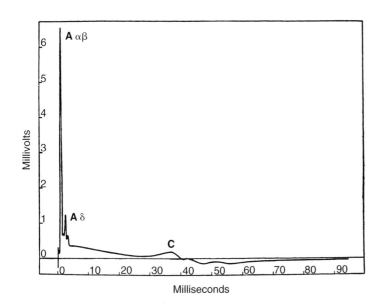

Figure 9.3. Multiple components in a compound action potential recorded from a cutaneous nerve (saphenous nerve). The drawing is a trace of a compound action potential recorded from a peripheral nerve. The distance between stimulating and recording electrodes was approximately 37 mm. Reproduced with permission from Gasser HS. The classification of nerve fibers. *Ohio Journal of Science.* 1941;41:145.

Table 9.2. Somatosensory receptors

Fiber type	Submodality	Receptor type	Myelination	Conduction velocity (m/s)
Group I (A)	Muscle stretch	Annulospiral Muscle spindle	Yes	70–120
Group I (B)	Tendon stretch	Golgi tendon	Yes	
Group II (A-β)	Discriminative touch, vibration	Cutaneous Mechanoreceptors	Yes	25–70
Group III (A-δ)	Pain, temperature	Free nerve?	Yes	10–30
Group IV (C)	Pain, temperature	Free nerve?	No	<2.5

neuropathy) may selectively affect unmyelinated C fibers, leading to a loss of pain and temperature sensation. The selective loss of particular classes of axons can be detected in the compound action potential generated by electrical stimulation of peripheral nerves. The loss of particular classes of fibers is reflected by the disappearance of particular components of the compound action potential.

Localization of a Stimulus

The somatosensory system allows a person to determine the part of the body surface that is activated by a particular stimulus. This is termed *localization*. Simply stated, localization is based on the selective activation of the receptors that are present in the particular area of skin. For example, if you touch your forearm with a pencil, you perceive the stimulus as a touch to the forearm because the terminals of a particular set of first-order neurons are activated. A touch to the back of the hand activates the terminals from a different set of first-order neurons. The brain distinguishes between the stimuli because the first-order neurons supplying the different sites project in a *somatotopic* fashion to different populations of second-order neurons. In this way, the somatotopic mapping that begins in the periphery is continued in the central projections.

The ability to distinguish between two sites is termed *spatial resolution*. Spatial resolution is tested clinically by using *two-point discrimination*. For example, when testing a human subject, two light stimuli are applied to nearby sites on the skin, and the subject is asked whether he or she perceives one stimulus or two. When the two stimuli are applied to areas of the skin that are close to one another, they are perceived as a single stimulus. As the spatial separation increases, two stimuli are perceived. The distance at which an individual is just able to distinguish two stimuli is termed the *two-point discrimination threshold*.

Receptor Density Differs in Different Areas of the Skin

Spatial resolution can occur only when the sensory nerve endings from different first-order neurons are activated (Figure 9.4). Conversely, activating different branches of the axon from a single first-order neuron generates an equivalent signal for the CNS. Thus, spatial resolution depends on the spacing of receptors in the skin (termed *receptor density*). The more closely spaced the receptors (the higher the receptor density), the better the spatial localization.

Receptor density varies among different parts of the body surface. For example, the

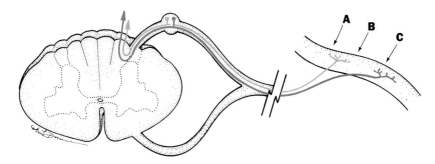

Figure 9.4. Two-point discrimination requires activation of separate axons. When two stimuli are applied at points that activate the terminals of only one sensory axon, only one stimulus is perceived (**A** and **B**). However, when the two stimuli are widely spaced enough to activate the terminals of two axons, two stimuli are perceived (**B** and **C**).

lips and finger tips have a very high receptor density; the fingers have a moderately high receptor density; the palms have a medium density; and the skin of the thigh and back have a low density (Figure 9.5). You can demonstrate these facts to yourself by evaluating the two-point discrimination threshold in the different sites.

Collection of Primary Afferent Axons in Peripheral Nerves and Dorsal Roots

The axons arising from the terminals of different first-order neurons serving particular areas of skin collect into cutaneous nerves and then into *peripheral nerves*. Each part of the body contains receptors for the different submodalities, and so the sensory axons from the different receptor subtypes comingle in cutaneous and peripheral nerves. For this reason, damage to a peripheral nerve will lead to a loss of all submodalities of cutaneous sensation in the area of skin that the nerve supplies.

Cutaneous nerves and the most distal branches of peripheral nerves contain the axons of primary afferents that supply sharply delimited areas, and there is little overlap between areas supplied by different distal branches. However, each of the distal branches of peripheral nerves contains axons deriving from several dorsal root ganglia (see Figure 9.6). As a result, there is considerable overlap between the areas of skin supplied by different dorsal roots.

The area supplied by an individual dorsal root is termed a *dermatome* (Figures 9.7 and 9.8). It is helpful to remember a few key landmarks. The neck is supplied by C2 and C3. The forearm and thumb arm are supplied by C6, the ulnar border of the hand and the little finger are supplied by C8. The nipple is roughly the boundary between the dermatomes supplied by T4 and T5. The umbilicus is the boundary between the dermatomes supplied by T10 and T11. The groin is supplied by L1. The dorsal side of the foot and great toe are supplied by L5; the lateral side of the foot and little toe are supplied by S1. The genitoanal region is supplied by S2 through S4.

The understanding of the differences between dermatomes and peripheral nerve distribution is the key to distinguishing between disease processes affecting peripheral nerves and those affecting dorsal roots or the spinal cord. In particular, a disease affecting the dorsal roots or spinal cord will lead to sensory loss with a dermatomal distribution. In contrast, a disease affecting the peripheral nerves will lead to a sensory loss with a distribution matching that of the peripheral nerves. The differences in the expected pattern of sensory loss can be appreciated by comparing the territories supplied by dorsal roots (dermatomes) versus particular peripheral nerves (Figures 9.7 and 9.8). Many dis-

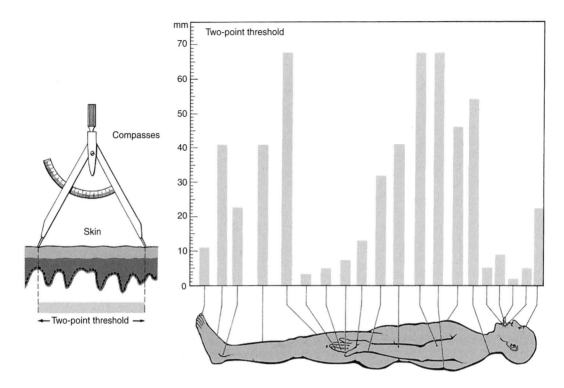

Figure 9.5. Differences in receptor density in different parts of the skin. As a result of substantial differences in the density of receptors in different parts of the body, the two-point discrimination threshold varies widely over the body surface. The two-point threshold is the smallest distance between two stimuli for which the subject can detect the existence of separate stimuli. The hands and face have the lowest thresholds for two-point discrimination; the back and thigh have the highest. From Zimmerman M, The somatovisceral sensory system. In: Schmidt RF, Thews G, eds. *Human Physiology.* 2nd ed. Berlin, Germany: Springer-Verlag; 1989:196–222. Presentation adapted from Ruch T, Patton HD, eds. *Physiology and Biophysics. The Brain and Neural Function.* Philadelphia, Pa: WB Saunders; 1979. Data from Weber EH. *Archiv für Anatomie, Physiologie und wissenschaftliche Medizin.* 1835:152.

eases of peripheral nerves primarily affect distal branches, leading to a disruption of sensation in the distal limbs. This is termed a *stocking–glove distribution* and is distinguished from a *dermatomal distribution* indicative of injury to dorsal roots or the spinal cord.

There is considerable overlap between individual dermatomes (Figure 9.9; see also Figure 9.6). The reason for the overlap is the regrouping of fibers that occurs in peripheral nerves. As a result of the regrouping, stimulation within any particular dermatome will activate fibers that distribute to two to four dorsal roots. For the same reason, *damage to a single dorsal root usually does not lead to complete anesthesia throughout the dermatome.*

With regard to dorsal roots and dermatomes, there are several important points to remember:

1. Damage or irritation of a dorsal root will lead to sensory symptoms that have a dermatomal distribution. Damage or irritation of a peripheral nerve will produce sensory symptoms with a stocking–glove distribution.

2. Because of the considerable overlap between individual dermatomes, damage to a single dorsal root will usually not lead to complete anesthesia within the dermatome supplied by that root. In fact, there may be minimal

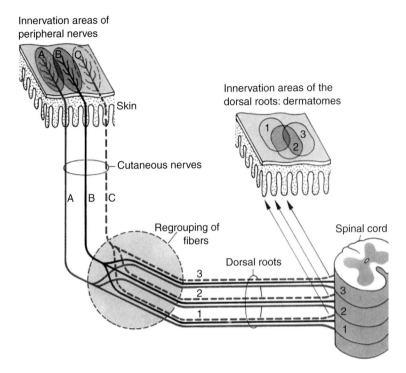

Figure 9.6. Innervation areas of peripheral nerves and dorsal roots. Cutaneous nerves and the most distal branches of peripheral nerves contain the axons of primary afferents that supply sharply delimited areas between which there is little overlap (see, for example, A, B, and C). However, the distal branches of peripheral nerves contain axons deriving from several dorsal root ganglia. As a result, there is considerable overlap between the innervation areas of different dorsal roots (see, for example, 1, 2, and 3). The innervation area of a particular dorsal root is termed a *dermatome* (see Figure 9.7). From Schmidt RF, Thews G, eds. *Human Physiology.* 2nd ed. Berlin, Germany: Springer-Verlag; 1989:196–222.

sensory loss. In contrast, damage to a peripheral nerve leads to complete anesthesia in the area supplied by that nerve.

3. Irritation of a single root will result in the perception of pain in the appropriate dermatomal zone. Such irritation is usually caused by compression of the root as it leaves the spinal canal.

The Cells of Origin of Fibers Subserving Different Submodalities Can Be Distinguished Within the Dorsal Root Ganglion

Fibers subserving different somatosensory submodalities originate from dorsal root ganglion cells that can be distinguished on the basis of size and other characteristics. Differences in size are directly related to differences in axon diameter and myelination. Thus, unmyelinated (C) and thinly myelinated (A-δ) fibers originate from small cell bodies; faster-conducting axons originate from large cell bodies.

The cells conveying different submodalities also contain different neuropeptides—a fact that provides a means for identification of cell types through immunocytochemistry. For example, nociceptors contain a mix of neuropeptides including substance P (the first to be identified), calcitonin gene–related peptide, vasoactive intestinal peptide, and somatostatin, as well as other neuropeptides. These neuropeptides are also

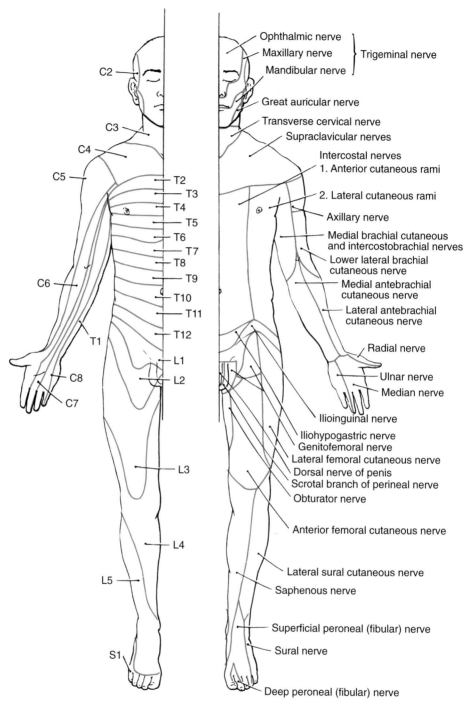

Figure 9.7. Dermatomes and peripheral nerves: anterior body surface. Each dorsal root supplies a particular area of the body, termed a *dermatome*. Dermatomes are labeled according to the dorsal root that supplies them. The figure illustrates the sensory innervation of the anterior body surface via dorsal roots (left) and peripheral nerves (right). Some key landmarks: the neck is supplied by C2 through C3. The forearm and thumb arm is supplied by C6, the ulnar border of the hand and the little finger are supplied by C8. The nipple is roughly the boundary between the dermatomes supplied by T4 through T5. The umbilicus is the boundary between the dermatomes supplied by T10 and T11. The groin is supplied by L1. The dorsal side of the foot and great toe are supplied by L5; the lateral side of the foot and little toe are supplied by S1. The genitoanal region is supplied by S2 through S4. Reproduced with permission from Gilman S, Newman SW. *Essentials of Clinical Neuroanatomy and Neurophysiology.* Philadelphia, Pa: F.A. Davis Co; 1996.

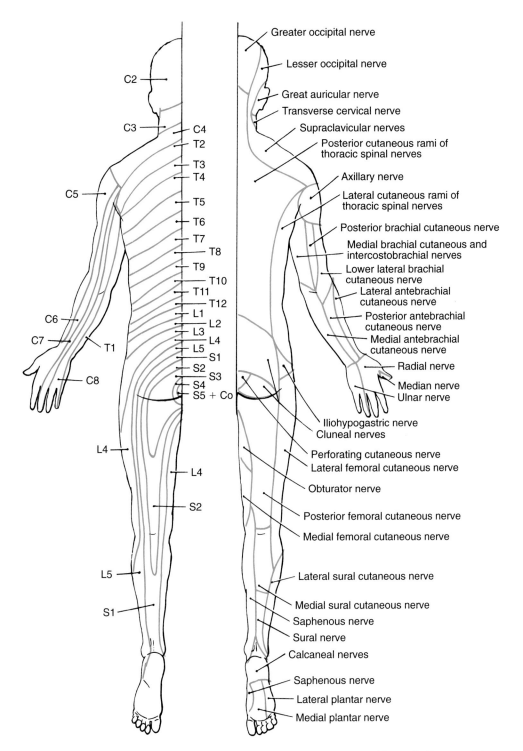

Figure 9.8. Dermatomes and peripheral nerves: posterior body surface. The sensory innervation of the posterior body surface via dorsal roots (left) and peripheral nerves (right). Reproduced with permission from Gilman S, Newman SW. *Essentials of Clinical Neuroanatomy and Neurophysiology.* Philadelphia, Pa: F.A. Davis Co; 1996.

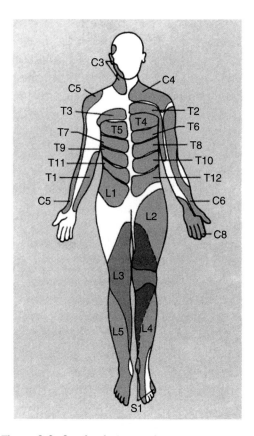

Figure 9.9. Overlap between dermatomes. The complete distribution of the cutaneous regions supplied by different dorsal roots; note the substantial overlap between the territories supplied by different dorsal roots. From Zimmerman M. the somatovisceral sensory system. In: Schmidt RF, Thews G, eds. *Human Physiology.* 2nd ed. Berlin, Germany: Springer-Verlag; 1989:196–222. Presentation adapted from Lewis T. Anatomical basis of pain. In: Lewis T, ed. *Pain.* New York, NY: Macmillan; 1942:11–31. As adapted from data from Foerster O. Dermatomes in man. *Brain* 1933;56:1.

present in the terminals of nociceptors within the spinal cord and so serve as a useful marker for the sites of termination of nociceptive afferents.

The different neuropeptides play an important role as neurotransmitters and neuromodulators at the central terminals of primary afferents. Also, the neuropeptides are transported into the peripheral axon, and it is thought that they are released by the sensory axon terminals. This peripheral release of neuropeptides is believed to play an effector role, for example, altering blood flow in a particular area of skin following noxious stimulation, and regulating the sensitivity of the receptor. These processes may play an important role in increasing sensitivity after injury to an area of skin (*hyperalgesia,* see Chapter 12).

Somatosensory fibers conveying information for the different submodalities travel together in the dorsal roots as they enter the spinal cord. However, after entering the spinal cord, the fibers of the different submodalities diverge and take a different course. This fact is of extreme importance to the physician because it provides the basis for localizing lesions. In the following chapter, the central course of the different pathways is described in detail.

10

Somatic Sensation II

Ascending Pathways From the Spinal Cord and Brainstem to the Thalamus

Upon entering the spinal cord and brainstem, the axons carrying information related to different submodalities diverge into separate pathways. Some axons enter and synapse in the dorsal horn at the level of entry; some axons branch and ascend or descend several segments, dropping off collaterals into the dorsal horn along the way; some axons join the dorsal columns and travel up the spinal cord to terminate in the dorsal column nuclei. The key point is that axons conveying different types of information take different courses, and this fact forms the basis for differential diagnosis of lesions affecting the spinal cord.

In this chapter we define the course of the somatosensory pathways from the spinal cord to the cortex and the sites of synaptic relays. Students should be familiar with the location of the different pathways in each of the regions that are traversed and be able to identify tracts in cross sections through each level of the neuraxis. Students should also be able to state at which levels the different pathways decussate. In this chapter we also define the nature of the deficits that result from damage to ascending somatosensory pathways. Students should be able to describe the deficits that result from selective lesions at different levels

of the neuraxis and localize the sites of lesions based on a defined set of signs and symptoms.

THE MAJOR ASCENDING PATHWAYS THAT CONVEY SOMATOSENSORY INFORMATION

In the spinal cord, the two somatosensory pathways that are most important for the physician are the *dorsal column–medial lemniscal* (DC-ML) *pathway* and the *anterolateral quadrant* (ALQ) *system* (which is the *spinothalamic tract*). These are the focus of the following discussion. Other pathways include the *spinocervicothalamic pathway* and the *spinocerebellar pathways.*

The spinocerebellar pathway conveys important proprioceptive information to the cerebellum and thus plays an important role in coordinating motor function. However, it is thought that the spinocerebellar pathway does not convey information that reaches conscious perception. For this reason, the spinocerebellar pathways is considered separately from the pathways mediating conscious perception.

The basic organization of the primary afferents of the ascending sensory pathways that contribute to conscious perception is illustrated in Figure 10.1.

The DC-ML System

The DC-ML system contains axons from receptors specialized for discriminative touch, vibration sense, and position sense. Upon entering the spinal cord via the dorsal roots, the axons of dorsal root ganglion cells conveying these submodalities enter the ipsilateral dorsal column directly (Figure 10.2) and also give off collaterals that project into the dorsal horn, where they terminate on neurons in the gray matter (see Figure 10.4).

Most of the axons in the dorsal column are from first-order neurons (dorsal root ganglion neurons). However, there are a few axons of second-order neurons in the dorsal horn of the spinal cord that receive input from primary afferents that convey information regarding discriminative touch, vibration sense, and position sense. For the physician, the most important points are the submodalities that the fibers of the dorsal column convey, and that dorsal columns convey this information from the ipsilateral side of the body.

In cervical and upper thoracic levels, the dorsal column is divided into two distinct columns of fibers, the *cuneate* and *gracile fasciculi*. The gracile fasciculus is the most medial and contains axons from dorsal root ganglion cells in the lumbar and low thoracic region that convey information from the lower body. The cuneate fasciculus lies laterally and contains axons from dorsal root ganglion cells in upper thoracic and cervical ganglia.

The axons within the dorsal columns terminate in the cuneate and gracile nuclei at the spinomedullary junction (Figure 10.2). These two nuclei are termed the *dorsal column nuclei*. The neurons in the cuneate and gracile nuclei are thus considered second-order neurons. However, as already noted, some of the axons of the dorsal column actually originate from second-order neurons in the dorsal horn, and so some of the neurons in the dorsal column could be considered third-order neurons.

The neurons within the dorsal column nuclei give rise to axons that leave the nuclei ventrally and curve toward the midline as the *internal arcuate fibers*. The fibers cross the midline in a loose collection of fibers (the *decussation of the medial lemniscus*) and then collect into the medial lemniscus (Figure 10.2). In the medulla, the medial lemniscus lies immediately adjacent to the midline. As it ascends through the pons and midbrain, the

Figure 10.1. Ascending somatosensory pathways within the spinal cord. The course of axons of the DC-ML system (left); the course of axons of the ALQ system, which carry pain and temperature information (right). Some axons conveying pain and temperature ascend or descend for a few segments in the tract of Lissauer before entering the dorsal horn.

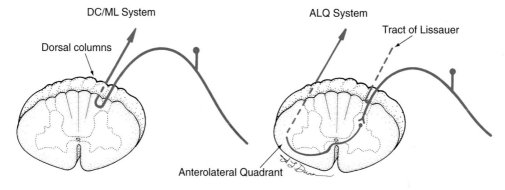

DC/ML System

Dorsal columns

ALQ System

Tract of Lissauer

Anterolateral Quadrant

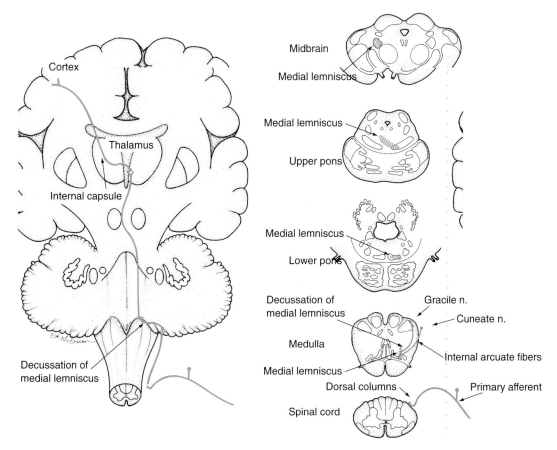

Figure 10.2. The DC-ML system. The axons of dorsal root ganglion cells that project via the DC-ML system enter the dorsal columns soon after entering the cord and also give off collaterals that project into the dorsal horn, where they terminate on neurons in the gray matter (see also Figure 10.4). The axons terminate on second-order neurons in the ipsilateral dorsal column nuclei (the cuneate and gracile nuclei) at the spinomedullary junction. The neurons within the dorsal column nuclei give rise to axons that leave the nuclei ventrally and curve toward the midline as the internal arcuate fibers and decussate in a loose collection of fibers (the decussation of the medial lemniscus) and then collect into the medial lemniscus on the contralateral side. The axons within the medial lemniscus ascend to the ventral posterior lateral nucleus of the thalamus. The neurons in the thalamus then send their axons through the posterior limb of the internal capsule to terminate in the somatosensory division of the cerebral cortex.

tract is positioned more and more laterally (Figure 10.2). The axons within the medial lemniscus ascend to the ventral posterior lateral (VPL) nucleus of the thalamus. The neurons in the thalamus then send their axons through the *posterior limb of the internal capsule* to terminate in the somatosensory division of the cerebral cortex.

The neurons in both the dorsal column nuclei and the thalamus are sometimes termed *relay neurons* because they represent the synaptic stations in the relay from the periphery to the cortex.

It is very important for students to commit the information in Figures 10.1 and 10.2 to memory. In particular, it is important to know (1) the location of the tracts containing the axons of the DC-ML system at each level of the neuraxis, and (2) that the axons conveying information via the DC-ML system decussate at the level of the spinomedullary junction. Thus, a unilateral lesion *below* the level of the spinomedullary junction that involves the DC-ML system will lead to a loss of discriminative touch, vibration sense, and position sense on the *ipsilateral* side of the body below the level of the lesion. A unilateral lesion *above* the level of the spinomedullary junction that involves the DC-ML system will lead to a loss of discriminative touch, vibration sense, and position sense on the *contralateral* side of the body.

One other point of clinical importance is that there is a *somatotopic organization* of axons within the tract and of neurons within the dorsal column nuclei and the thalamus. This is due to the fact that as the fibers from given segments join the tract, they layer onto the fibers that are ascending from lower levels. For example, in the case of the dorsal columns, the fiber from the cervical levels (representing the arm) layer onto the lateral side of the column of fibers carrying ascending information from the legs. This is the basis of the somatotopic organization. Hence, within dorsal columns, the legs are represented medially (in the gracile fasciculus); within the dorsal column nuclei the legs are represented medially (in the gracile nucleus).

There is also somatotopy at higher levels. Within the thalamus, the legs are represented laterally; and as described later, within the cortex, the legs are represented medially (Figure 10.3). It is important for the physician to know this information because small lesions can involve the tracts or nuclei and so lead to partial loss of somatic sensation in the portion of the body represented by that area.

The concept just outlined is that the dorsal column system carries axons that convey discriminative touch, vibration sensitivity, and position sense. There are, however, some complications to this simple story. If the dorsal column system was the only route for ascending information related to these submodalities, then complete dorsal column lesions should eliminate discriminative touch, vibration sensitivity, and position sense below the level of the lesion. Clinically, the situation is not this simple.

First, some aspects of touch are preserved following complete dorsal column lesions. In particular, patients are able to detect light touch or pressure—that is, know that they have been touched. They are deficient in localizing the touch and in making complicated discriminations, for example, the ability to recognize an object by touch—what we call *stereognosis*. The reason for the sparing of crude touch is that some touch information is conveyed via the spinothalamic system. In contrast, position sense and vibration sensitivity are represented almost exclusively in the DC-ML system, so that these submodalities are lost below the level of a lesion affecting the dorsal columns. Hence, tests of position sense and vibration sensitivity are more useful for detecting selective lesions involving the dorsal columns than are tests of discriminative touch.

Second, complete dorsal column lesions at the thoracic or lumbar levels reduce but do not eliminate position sense in the leg. The reason is that some fibers conveying position sense from the leg ascend in the dorsolateral funiculus. These fibers terminate in a small cell group that is closely related to the gracile nucleus (termed *nucleus z*). This pathway is referred to as the *spinomedullary tract*. The axons from the second-order neurons in nucleus z cross the midline with axons from nucleus gracilis neurons and continue in the medial lemniscus.

Collaterals of Dorsal Column Axons Also Terminate at Segmental Levels

In addition to conveying information to the brain, axons that enter the dorsal columns also give rise to collaterals that terminate in the dorsal horn of the spinal cord near their segment of entry. These collaterals play a role in modulating simple motor behaviors that are mediated by segmental circuitry (for example, the scratch reflex). Interestingly, the pattern of termination of afferents conveying different submodalities is quite distinctive, indicating that the different afferents innervate different populations

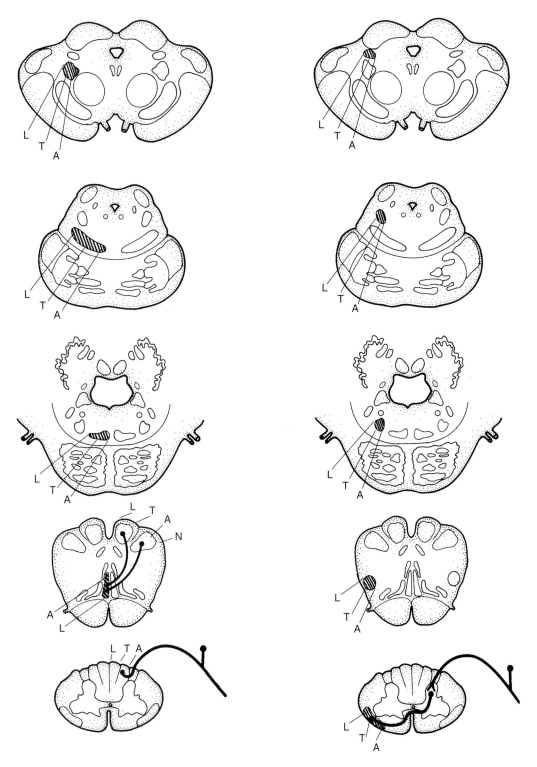

Figure 10.3. Somatotopy of ascending pathways. The approximate position of axons conveying information from the leg (L), trunk (T), arm (A), and neck (N) are indicated for the DC-ML system (left) and the ALQ/spinothalamic system (right).

of neurons in the gray matter of the spinal cord (Figure 10.4).

The ALQ, or Spinothalamic, System

The ALQ, or spinothalamic, system contains axons from receptors specialized for pain and temperature. These axons also mediate a crude form of touch sensitivity. The primary afferents conveying these submodalities enter via the dorsal roots and then either enter the dorsal horn directly (Figure 10.5) or project up or down the cord for a few segments in the *tract of Lissauer* before entering the dorsal horn. The primary axons synapse upon neurons in the dorsal horn, especially neurons in the *substantia gelatinosa,* and also send collaterals to deeper laminae within the dorsal horn. The axons that ascend in the anterolateral quadrant originate from at least second-order or higher-order neurons in the gray matter of the cord.

Ascending axons carrying pain and temperature information ascend for one to two segments on the ipsilateral side and then decussate in the *anterior white commissure* of the spinal cord, which is a thin band of white matter located just below the central canal (Figures 10.1 and 10.5). After crossing the midline, the axons carrying pain and temperature information join the white matter in the *anterolateral column* in the anterior (ventral) funiculus and ventral portion of the lateral funiculus. Many of the ascending fibers of second-order neurons in the ALQ-spinothalamic system project directly to the thalamus. This component is the spinothalamic tract.

The fact that the fibers ascend for one or two segments before they cross the midline is of clinical significance. For this reason, a unilateral lesion at a particular level of the cord will lead to a loss of pain and temperature sensitivity at dermatomal levels beginning about one or two segments below the injured segment (Figure 10.6). This is in contrast to the situation with the dorsal column system in which a unilateral lesion at a particular level leads to a loss of position sense and vibration sensitivity at and at all levels below the site of injury.

The fact that fibers carrying pain and temperature information decussate in the anterior white column is also of clinical significance. Congenital disorders and trauma can lead to a condition in which a cystic cavity forms in the center of the spinal cord. The cavity is called a *syrinx,* and the condition is called *syringomyelia.* The expansion of the syrinx can interrupt the crossing fibers of the anterior white columns, leading to a bilateral loss of pain and temperature sensitivity at the level of the interruption (more on this later).

The spinothalamic fibers in the anterolateral column also have a somatotopic organization (Figure 10.3). Again, this is due to the fact that as the fibers from given segments join the tract, they layer onto the fibers that ascend from lower levels. In the case of the anterolateral column, fibers layer onto the medial side of the tract (because they arise from the contralateral side). Hence, in cervical segments, which contain fibers from all levels of the body, fibers from lower levels of the cord are found laterally within the tract; fibers from higher levels are found medially in the tract.

So, in the spinal cord, the somatotopic organization is as follows:

Dorsal column	Legs medial, arms lateral
ALQ	Legs lateral, arms medial

The somatotopic organization of the spinothalamic tract is of clinical importance. One treatment for the relief of intractable pain is cordotomy—the transection of the fibers of the spinothalamic tract. The somatotopic organization of the tract allows selective lesions to be made. For example, chronic pain arising from the lower portion of the body can be relieved by sectioning laterally located fibers, sparing pain and temperature sensation arising from higher levels.

The fibers of the spinothalamic tract travel through the medulla just dorsal to the inferior olivary nucleus in the lateral part of the tegmentum. At this level, the spinothalamic tract is well separated from the medial lem-

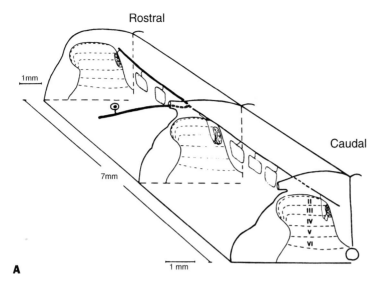

A

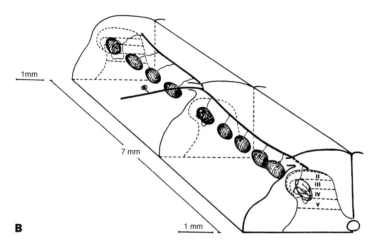

B

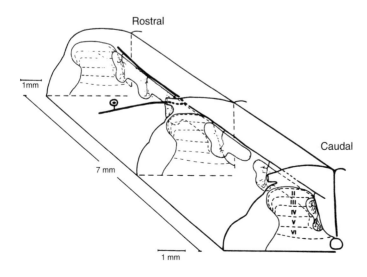

C

Figure 10.4. Collateral projections of the dorsal column to the spinal cord. Termination patterns of primary afferent fibers in the cat spinal cord were characterized physiologically and then labeled by injecting a tracer (horseradish peroxidase). **A,** Termination pattern of a rapidly adapting mechanoreceptor in the lumbosacral spinal cord. **B,** Termination pattern of a slowly adapting mechanoreceptor. **C,** Termination pattern of the primary afferent axons from a Pacinian corpuscle. The ascending axon of each type of axon enters the dorsal column, and collaterals enter the gray matter of the dorsal horn. **A** and **C** are from Brown AG, Fyffe REW, and Noble R. Projections from Pacinian corpuscles and rapidly adapting mechanoreceptors of glabrous skin to the cat's spinal cord. *J Physiol.* 1980;307:385–400. **B** is from Brown AG, Rose PK, Snow PS. Morphology and organization of axon collaterals from afferent fibers of slowly adapting type I units in cat spinal cord. *J Physiol.* 1978;277:15–27.

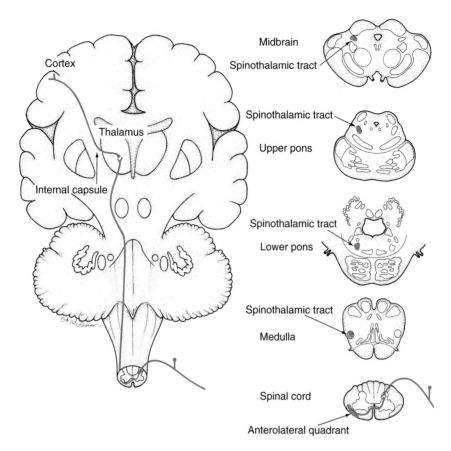

Figure 10.5. The ALQ, or spinothalamic, system. The axons of dorsal root ganglion cells that convey information related to the submodalities of pain, temperature, and a crude form of touch enter via the dorsal roots and then either enter the dorsal horn directly or project up or down the cord for a few segments in the tract of Lissauer before entering the dorsal horn (see also Figure 10.1). The primary axons synapse upon second-order neurons in the dorsal horn, especially neurons in the substantia gelatinosa. The axons of second-order neurons cross the midline in the ventral white commissure and join the spinothalamic tract in the anterolateral quadrant. The spinothalamic tract ascends in the ventrolateral brainstem and midbrain and terminates in the ventrobasal thalamus.

niscus (Figure 10.5). In the pons, the medial lemniscus travels in a more and more lateral location as it ascends, so that the medial lemniscus and spinothalamic tract are adjacent at pontine levels and higher. These facts are of considerable diagnostic significance. A lesion that affects one or the other tract selectively can occur only where the two tracts are physically separated (that is, below the pontine level).

It is of clinical importance that the vascular supply to the medial and lateral medulla differs. As noted in Chapter 3, the medial–ventral medulla (where the medial lemniscus is located) is supplied by perforating vessels from the vertebral and basilar artery. In contrast, the lateral medulla (where the spinothalamic fibers are located) is supplied by circumferential vessels. Thus, vascular insults can affect either the medial lemniscus or the spinothalamic tract selectively, leading to a sensory dissociation syndrome (the loss of particular submodalities of sensation).

The axons of the spinothalamic tract ascend to the ventral VPL nucleus of the thalamus where they terminate upon *relay neu-*

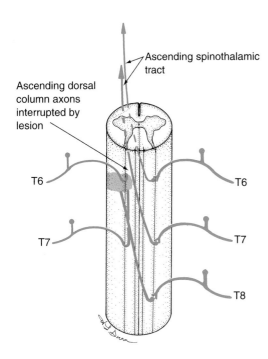

Ascending spinothalamic tract

Ascending dorsal column axons interrupted by lesion

T6

T7

T6

T7

T8

Figure 10.6. Decussation of spinothalamic axons. The decussation of ascending axons carrying pain and temperature information occurs over one or two segments. For this reason, a unilateral lesion that interrupts the anterolateral funiculus will lead to a loss of pain and temperature sensitivity that begins at dermatomal levels about one or two segments *below* the injured segment. In this example, the lesion interrupts ascending axons of dorsal root ganglion cells that give rise to dorsal column projections at T6 and below. At the same time, spinothalamic axons carrying information from T6 and T7 are spared, whereas those from T8 are interrupted. Thus, in this example, there is a loss of vibration sensitivity from T6 and below on the left, and a loss of pain and temperature from T8 and below on the right.

rons in the VPL. Some axons also terminate in nuclei within the posterior thalamus and in intralaminar nuclei.

In addition to the projections to the thalamus, some of the axons of the ALQ system terminate in the reticular formation and in the periaqueductal gray of the midbrain. These are termed *spinoreticular* and *spinomesencephalic projections;* some older texts separate the ALQ tract in the spinal cord into spinothalamic, spinoreticular, and spinomes-

encephalic components. In fact, however, there are not separate populations of axons in the ALQ, and so the separate terminology refers to sites of termination and not to collections of axons that can be separately identified.

The Spinocervicothalamic System

A minor component of ascending somatosensory fibers travels in the so-called spinocervicothalamic system. In humans, this system is small and is therefore mentioned only in passing. The primary afferent fibers convey information primarily from hair receptors. The primary afferents synapse upon neurons in the dorsal horn. The axons of the second-order neurons then ascend in the dorsolateral funiculus. The spinocervicothalamic tract terminates in a special nucleus in the upper cervical cord (levels C1–C3), termed the *lateral cervical nucleus.* The axons from neurons in the lateral cervical nucleus decussate and join the medial lemniscus on the contralateral side, where they project onto the VPL nucleus of the thalamus.

In summary, the DC-ML and ALQ/spinothalamic systems differ in several key respects. These differences are summarized in Table 10.1.

SOMATOSENSORY INFORMATION CONVEYED VIA THE TRIGEMINAL NERVE (CRANIAL NERVE V)

Somatosensory information from the face and from the oral and nasal cavities is carried by the *trigeminal nerve.* The nerve has three divisions: opthalmic, maxillary, and mandibular. These divisions supply different areas of the skin and mucous membranes of the head. The cell bodies of most of the primary afferents are in the *trigeminal ganglion* (also known as the *Gasserian ganglion* or *semilunar ganglion*). The exception to this generalization is the *trigeminal mesencephalic nucleus,* which is the only example of a first-order neuron with its cell body in the CNS.

Table 10.1. Important facts regarding the DC-MLS and ALQ/spinothalamic systems

System	Decussation	Location of first synapse	Vascular supply in medulla	Sub-modalities
DC-MLS	Lower medulla	Dorsal column n.	Medial	Discriminative touch Vibration Position sense
ALQ/Spinothalamic	1–2 segments above point of entry	Dorsal horn	Lateral	Pain Temperature

The axons of neurons in the trigeminal ganglion enter the brainstem and synapse in the *trigeminal nuclear complex* (Figure 10.7). This nuclear complex is composed of three individual nuclei: the *principal sensory nucleus* (also called the *main sensory nucleus* or *chief nucleus* in some texts), the *spinal trigeminal nucleus,* and the *mesencephalic nucleus of cranial nerve V.* The principal sensory nucleus and the rostral two-thirds of the spinal trigeminal nucleus are analogous to the DC-ML system. Primary afferents conveying information pertaining to discriminative touch terminate on neurons in the principal nucleus. There is a somatotopic organization within the nucleus such that the lower face (mandible) is represented dorsally, the cheek is represented in an intermediate location, and the nose and forehead are represented ventrally.

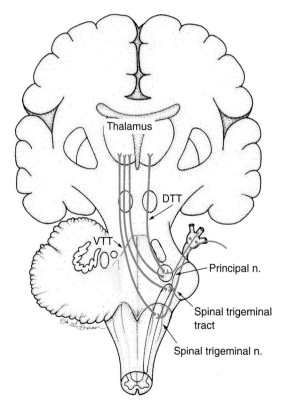

Figure 10.7. Trigeminal afferents that convey information related to discriminative touch. The sensory components of the trigeminal nerve in a dorsal view of the brainstem. Primary afferents that convey information pertaining to discriminative touch terminate on neurons in the principal nucleus and in the rostral two-thirds of the spinal trigeminal nucleus. Axons of the second-order neurons in the principal sensory and rostral spinal trigeminal nucleus cross the midline, collect in the ventral trigeminothalamic tract, and ascend to terminate in the VPM nucleus of the thalamus. There is also a component of fibers that projects ipsilaterally to the VPM nucleus via a tract termed the dorsal trigeminothalamic tract.

Axons of the second-order neurons in the principal sensory and rostral spinal trigeminal nucleus cross the midline and collect in a tract termed the *ventral trigeminothalamic tract* (VTT) (see Figure 10.7). These axons terminate in the VPM nucleus of the thalamus. There is also a component of fibers that projects ipsilaterally to the VPM nucleus via a tract termed the *dorsal trigeminothalamic tract*.

The caudal one-third of the spinal trigeminal nucleus (termed the *subnucleus caudalis*) is analogous to the ALQ/spinothalamic system. Primary afferents that convey the submodalities of pain and temperature enter at the level of the pons and then project caudally in a well-defined tract called the *spinal trigeminal tract*. These fibers terminate on neurons in the subnucleus caudalis. The axons of neurons in subnucleus caudalis project across the midline and ascend in the VTT. The fibers of the VTT then join the spinothalamic fibers and project onto the VPM nucleus of the thalamus (Figure 10.8).

It is of clinical significance that pain and temperature are represented in the spinal trigeminal nucleus, whereas discriminative touch is represented in the more rostrally located principal nucleus, because injuries can affect components selectively. For example, occlusion of the vascular supply to the lateral medulla can damage the spinal trigeminal nucleus (disrupting pain and temperature sensitivity for the face). The principal nucleus in the pons may be spared, however, and so discriminative touch in the face is spared.

Lesions of the lateral medulla or posterior pons can damage the fibers of the spinal tract of cranial nerve V as well as the ascending spinothalamic fibers. In this case, there is a loss of pain and temperature sensitivity in the ipsilateral face and contralateral body. However, lesions in the upper pons or midbrain interrupt trigeminothalamic fibers after they have crossed and thus produce sensory deficits in the contralateral face and body.

The mesencephalic nucleus is unique in that it contains the cell bodies of first-order

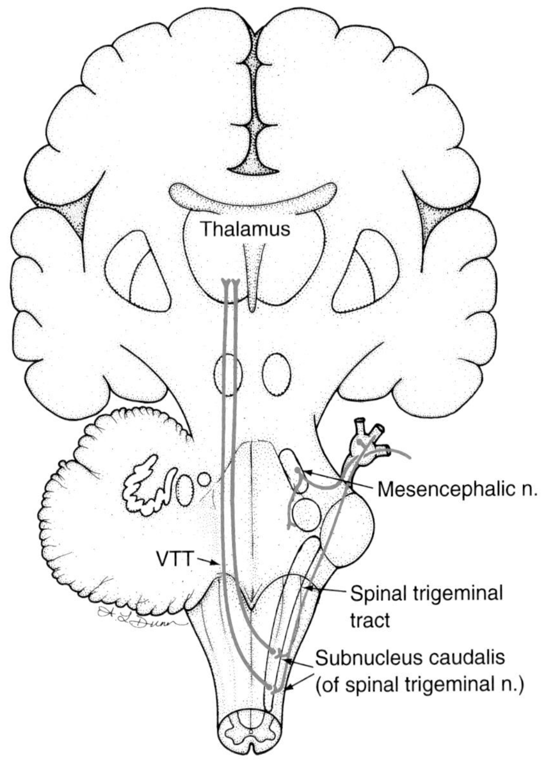

Figure 10.8. Trigeminal afferents that convey pain and temperature sensitivity, and the special case of the mesencephalic nucleus. Primary afferents that convey pain and temperature sensitivity enter at the level of the pons and project caudally in the spinal trigeminal tract to terminate in the caudal one-third of the spinal trigeminal nucleus. The axons of neurons in subnucleus caudalis decussate, join the VTT, and ascend to terminate in the VPM nucleus of the thalamus. The first-order neurons of the mesencephalic nucleus are also illustrated. The afferents convey sensory information from periodontal ligaments and gingiva as well as proprioceptive information from extraocular muscles. The central axons of these neurons project to the motor nuclei of cranial nerves III, IV, and VI, where they play a role in reflex control of eye movement.

neurons that give rise to sensory axons that project into the periphery (Figure 10.8). The receptors for some of the peripheral axons are in periodontal ligaments and gingiva and are sensitive to pressure on the teeth. The central axons of these cells project bilaterally to the motor nucleus of cranial nerve V, controlling the muscles of mastication. This circuit is thought to play an important role in reflex control of chewing.

Other axons arising from neurons in the mesencephalic nucleus of cranial nerve V convey proprioceptive information from extraocular muscles. The central axons of these neurons project to the motor nuclei of cranial nerves III, IV, and VI, where they play a role in reflex control of eye movement.

The Spinocerebellar Pathway

The spinocerebellar pathway does not play a role in conscious perception because it does not relay information to the cerebral cortex. Instead the pathway conveys proprioceptive information to the cerebellum. The primary afferents of the pathway enter via the dorsal roots and join the dorsal columns on the ipsilateral side. The axons from dorsal root ganglion cells in sacral and lumbar regions, as well as some dorsal root ganglion cells in lower thoracic regions, ascend for several segments and then enter the gray matter of the spinal cord to terminate in a special nucleus termed *Clarke's nucleus* (Figure 10.9). The second-order neurons in Clarke's nucleus then give rise to projections that ascend through the spinal cord in the dorsolateral funiculus as the dorsal spinocerebellar tract. These axons enter the inferior cerebellar peduncle and terminate in the ipsilateral cerebellum.

Spinocerebellar axons from dorsal root ganglion cells in upper lumbar, thoracic, and cervical regions ascend in the dorsal columns and terminate in the *lateral* or *external cuneate nucleus* in the medulla. Projections from the external cuneate nucleus join the fibers of the dorsal spinocerebellar tract and enter the cerebellum.

It is important to emphasize the fact that the spinocerebellar pathways are exclusively ipsilateral. In this way, one side of the cerebellum receives proprioceptive information from the ipsilateral side of the body. As we see in the chapter pertaining to the cerebellum, one side of the cerebellum also influences motor function on the ipsilateral side.

Collaterals of Primary Afferents Mediate Segmental Reflexes

So far we have focused on the somatosensory pathways that contribute to conscious perception. But it is important to understand that primary afferents conveying information related to all of the different submodalities also terminate within the gray matter of the spinal cord. These segmental circuits play an important role in reflex functions.

A *reflex* is a motor response to a sensory stimulus. The simplest reflex, the stretch reflex, involves only two neurons—the afferent (sensory) neuron and the efferent (motor) neuron—and is mediated by a monosynaptic connection between the two (see Chapter 14). Somewhat more complex reflexes involve polysynaptic connections at the segmental (spinal) level (for example, the flexion–withdrawal reflex). The afferent component of these reflexes is conveyed by the same first-order neurons that convey the information that reaches consciousness. Indeed, many of the segmental connections arise as branches of axons that project centrally. These segmental connections of primary afferents will be considered in more detail in the section pertaining to motor control.

The Somatosensory Thalamus

Ascending axons in the different somatosensory pathways terminate in the VPL and VPM nuclei of the thalamus. These nuclei

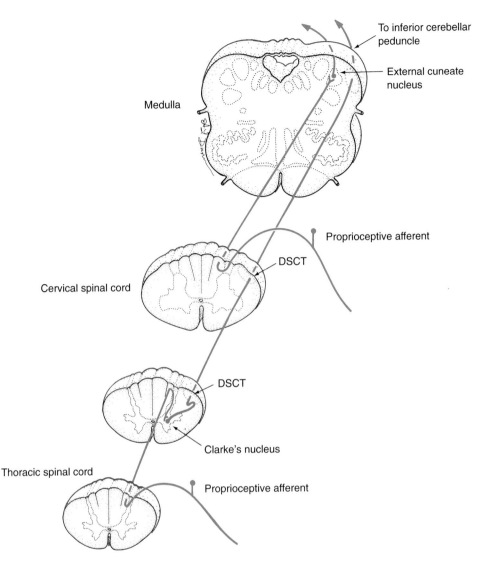

Figure 10.9. Central projections of proprioceptive afferents. Proprioceptive afferents from T1 and below enter the spinal cord in the dorsal roots, ascend for a few segments in the dorsal column, and then enter the gray matter of the spinal cord to synapse on neurons in Clarke's nucleus. Neurons in Clarke's nucleus then give rise to axons that enter the dorsolateral column and ascend in the dorsal spinocerebellar tract (DSCT). Proprioceptive afferents from the cervical cord enter via the dorsal roots and ascend in the dorsal columns to the spinomedullary junction where they synapse on neurons in the lateral or external cuneate nucleus. The neurons in the lateral cuneate nucleus then give rise to axons that enter the inferior cerebellar peduncle, along with the axons of the dorsal spinocerebellar tract.

are collectively termed the *ventral tier* of nuclei or the *ventrobasal complex*). These represent the *thalamic relay nuclei* of the somatosensory system. These nuclei are also termed *specific nuclei* in recognition of the fact that they project to particular cortical regions. This is in contrast to *nonspecific nuclei* that receive input from the reticular formation and project more widely within the cerebral cortex (more on this in Chapter 30).

There is an important topographic organization of the inputs to the ventrobasal complex. First, there is a somatotopographic organization. In general, fibers conveying information from the legs terminate laterally within the nuclei; fibers conveying information from the arms and face terminate medially (Figure 10.10). Second, there is at least some degree of segregation of inputs conveying information from different types of receptors (cutaneous versus deep). Axons conveying information from cutaneous receptors terminate within the "core" of the ventrobasal complex; axons conveying information from deep receptors terminate within the "shell." The different subregions then project to different subdivisions of the somatosensory cortex. This forms the anatomical basis for the fact that neurons in different subdivisions of the somatosensory

Figure 10.10. Topographic organization of the somatosensory thalamus. The figurines illustrate the representation of the body surface in the ventrobasal complex of the monkey, as revealed by physiological mapping techniques. This is a coronal section through the nucleus. In general, legs are represented laterally, head and face are represented medially. This figure also illustrates the microelectrode mapping technique that has played a key role in defining the organization of the somatosensory system. In this technique, a recording electrode is advanced through the region of interest. Recordings are made of the responses at different locations (inset in upper right) in response to tactile stimulation of different sites on the body surface. Similar techniques have been used to map all of the relays of the somatosensory system including the cerebral cortex. VBex and VBarc indicate external and arcuate divisions of the ventrobasal complex; GLD, lateral geniculate nucleus. Reproduced with permission from Mountcastle VB, Henneman E. The representation of tactile sensibility in the thalamus of the monkey. *J Comp Neurol.* 1953;97:409–440, reprinted by permission of Wiley-Liss, Inc., a subsidiary of John Wiley & Sons, Inc.

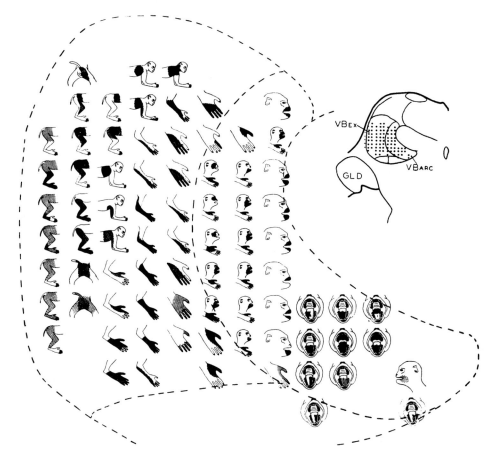

cortex are involved in the processing of information from different types of receptor (more on this in Chapter 11).

THALAMOCORTICAL PROJECTIONS AND THE ORGANIZATION OF THE SOMATOSENSORY CORTEX

Relay neurons in the VPL and VPM nuclei of the thalamus give rise to axons that project to the cerebral cortex via the posterior limb of the internal capsule. The axons have a somatotopic organization so that axons conveying information from the legs are located toward the posterior portion of the internal capsule; axons from the arms and head are located more rostrally near the genu of the internal capsule.

On reaching the cortex, the fibers distribute in a somatotopic fashion to several different cortical regions, the most important being the cortical areas on the *postcentral gyrus (Brodmann's areas 1 through 3)*. This area is termed the *primary somatosensory cortex*, or *SI* (Figure 10.11).

The different cytoarchitectonic areas (Brodmann's areas 1 through 3) are actually composed of four functional areas (1, 2, 3a, and 3b) that play a somewhat different role in somatic sensation (more on this in Chapter 11). There is also a secondary somatosensory region (SII) in the lower margin of the postcentral gyrus in the upper bank of the lateral sulcus. SII receives input from SI and then projects to sites in the insular cortex that play a role in somatosensory discrimination. Finally, the posterior parietal lobe receives input from the somatosensory system and other modalities and plays a role in spatial discriminations.

There is a detailed topographic map of somatosensory representation in each of the subregions the somatosensory cortex. The most important map to remember is the one for the primary somatosensory cortex (Figure 10.11). Fibers conveying information from the pharynx, tongue, and teeth are represented in order on the ventral-most portion of the postcentral gyrus; the lips and

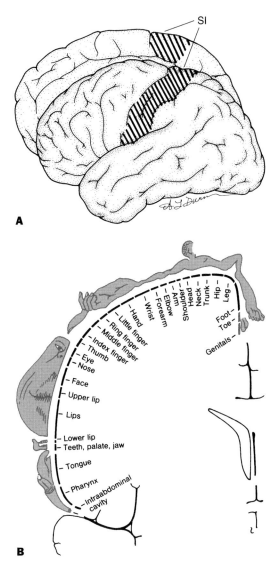

Figure 10.11. The sensory homunculus in the primary somatosensory cortex. A, Location of the primary somatosensory cortex, or SI. The medial aspect of the hemisphere has been projected as a mirror image. **B,** "Homunculus" illustrating the somatotopic organization of the primary sensory cortex. Reproduced with permission from Penfield W, Rassmussen T. *The Cerebral Cortex of Man.* New York, NY: Macmillan; 1950.

face are represented in the next-most dorsal regions, followed by the digits of the hand and the upper limb. The trunk and upper limb are represented in the dorsal-most portion of the gyrus.

The representation for the foot, toes, and genitals extends onto the medial surface of the cerebral hemisphere into the *paracentral lobule*. In this location, the regions of the cerebral cortex representing the feet, on the two sides, are adjacent to one another across the midline. It is for this reason that injury to the medial walls of the cortex in the area of the paracentral lobule (for example, as a result of a meningeal tumor) can result in bilateral sensory impairments involving the feet and toes.

The somatotopic map is often illustrated schematically by a caricature termed a *homunculus* (Latin for "little man") (Figure 10.11). In the case of the somatosensory system, this schematic map is termed the *sensory homunculus*. A similar representation is commonly used for the motor system (termed the *motor homunculus*). Similar caricatures are also sometimes used to illustrate the somatotopic organization in the thalamus and spinal cord.

There Is a Disproportionate Representation of Different Parts of the Body on the Cortical Surface

An important feature of the sensory homunculus is that the caricature appears distorted. Specifically, relatively large regions of the cortex are devoted to particular regions of the body. For example, a large portion of the somatosensory cortex is devoted to the digits of the hand and the lower face and lips. This reflects an important fact: the amount of brain devoted to the representation of any region of the body is not proportional to the relative size of that part of the body, but instead is proportional to the receptor density in that region of the body. As described in later sections, the same principle applies in other sensory systems, and a similar principle applies in the motor system.

Simple Diagnostic Techniques for Assessing Deficits in Somatosensory Submodalities

Cutaneous mechanoreception (light touch) is tested clinically by lightly touching the skin with a piece of cotton or similar material. The patient is asked whether he or she feels the touch. In a routine test, the physician evaluates only detection. In more-detailed tests, the physician also asks the patient to identify where on the skin the touch occurred (to test for accurate localization). Pain sensitivity is tested by touching areas of the skin in a random fashion with either the point or the rounded end of a safety pin. The patient is asked to identify sharp or blunt.

Position sense is assessed by passively moving the patient's extremities (for example, fingers or toes) when the patient's eyes are closed. Deficits in position sense are reflected by an inability to determine the direction of movement or the position of the limb. Deficits in proprioception also produce deficits in motor behaviors because of the importance of proprioceptive feedback for motor function. Such deficits are collectively termed *ataxias* and involve difficulty in fine motor control (for example, walking with eyes closed or in the dark, or manipulating objects with the fingers).

Vibration sense is assessed by striking a tuning fork and placing it on a bony protuberance (a knuckle, wrist, elbow, ankle bone, knee cap, etc.). If there is damage to the DC-ML system, the patient will be unable to detect the vibration. It is important to be sure that the patient is discriminating between vibration and touch. This can be done by asking the patient to discriminate between a tuning fork that has been struck and one in which vibration has been stopped.

Spatial resolution can be assessed in several ways. One way is to determine the two-point discrimination threshold. An even easier way is to draw letters or numbers on the patient's skin or to ask the patient to identify objects by touch (with eyes closed). The ability to identify objects is termed *stereognosis*. The ability to identify letters or numbers that are traced on the skin surface is termed *graphesthesia*. Another measure of spatial resolution is to ask the patient to identify the direction of movement if the skin is lightly stroked, or to determine the texture of objects (smooth or rough). These discriminations require integration of information from adjacent regions of the skin and thus test spatial resolution. Together, these discriminations are termed *epicritic* sensitivity (fine touch). Epicritic touch is disrupted by damage to the DC-ML system, although simple recognition of tactile stimulation (termed *protopathic* touch) is preserved as long as spinothalamic pathways are intact.

Pain sensation can be tested clinically by applying a sharp object to the skin (a pin prick) or by pinching an area of skin (for example, with serrated forceps).

Temperature sensation is tested by warming or cooling a tuning fork and applying it to the skin surface.

INJURY TO DIFFERENT COMPONENTS OF THE SOMATOSENSORY PATHWAY PRODUCES CHARACTERISTIC SYMPTOMS

And so we return to the topic of sensory deficits that follow lesions, but this time with a much more specific agenda—to outline the specific symptoms resulting from injury to different parts of the system.

Lesions or Disorders of Peripheral Nerves

Lesions or disorders of peripheral nerves lead to decreased sensation in the areas of the skin that are served by that nerve. Many peripheral nerve disorders are a result of metabolic disorders or toxins (e.g., diabetes, alcohol, thiamine deficiencies, hyperthyroidism, and uremia). Such disorders tend to produce loss of sensation that is often bilateral and most severe distally (i.e., most severe in feet and/or hands). This pattern of sensory loss is termed a *stocking–glove distribution,* to distinguish it from a *dermatomal distribution,* which results from injuries to dorsal roots.

Following injury, peripheral nerves are capable of regeneration and sprouting, and areas of anesthesia that result from trauma to a peripheral nerve may disappear with time. Regeneration is common following crush injuries and is less successful following major injuries that transect the axon. When regeneration is unsuccessful (for example,

when the injury substantially disrupts the tissue terrain through which the sensory axons must grow), the axons may become arrested and form a tangle of nerve endings termed a *neuroma.* Abnormal sensory perceptions including pain often originate from neuromas.

Lesions of a Dorsal Root

Lesions of an individual dorsal root may not produce areas of complete anesthesia because of the overlap of dermatomes. However injuries to several dorsal roots will produce areas of complete anesthesia corresponding to the completely denervated areas of the skin.

Spinal Cord Lesions

The symptoms following lesions of the spinal cord depend on the sites of the damage. Of course complete transection leads to a complete loss of sensation below the level of the lesion.

Lesions of the Dorsal Columns

Lesions of the dorsal columns produce deficits in discriminative touch, vibration sensitivity, and position sense. This is termed the *posterior column syndrome.* One setting in which this occurs is tertiary syphilis, and in this case, the posterior column syndrome is referred to as *tabes dorsalis.* If the lesions involve the dorsal columns on both sides, the deficits are bilateral. If the lesions are unilateral, the sensory deficits are *ipsilateral* to the lesion.

Lesions of the Lateral Columns

Lesions that involve the lateral columns produce deficits in pain and temperature sensitivity on the *contralateral* side of the body beginning one to two segments below the level of the lesion. It is rare for a physical injury to involve the lateral columns on both sides of the cord because they are physically separated. However, certain disease processes, for example, demyelinating disease (multiple sclerosis), can affect these pathways.

Lesions That Affect the Dorsal Column and the Spinothalamic System

Unilateral injuries of the spinal cord that affect both the dorsal column and the spinothalamic system produce injuries that are mixed with regard to laterality. Unilateral lesions produce deficits in discriminative touch, vibration sensitivity, and position sense on the *ipsilateral* side, and deficits in pain and temperature discrimination on the *contralateral* side. These are the sensory symptoms that are part of the *Brown Sequard syndrome* that occurs following *hemisection* of the spinal cord (a unilateral transection). There are also motor symptoms (paralysis on the side ipsilateral to the lesion). These are described in more detail in later sections.

Lesions at the Ventral Midline

Lesions at the *ventral midline* of the spinal cord can disrupt the *anterior white commissure*, which carries the crossing fibers of second-order axons of the spinothalamic tract. The result is a *bilateral loss of pain and temperature sensitivity* in the areas of the body that are supplied by the involved segments. Other submodalities are preserved. Such a lesion can develop as a result of the development of a cystic cavity in the center of the cord (termed *syringomyelia*). The cystic space formed as a result of the expansion of the central canal is termed a *syrinx*. The resulting deficit is termed a *suspended–dissociated* loss of sensation. The loss is "*suspended*" because areas above and below the dermatomes exhibiting loss of pain and temperature exhibit normal sensation. It is "*dissociated*" because pain and temperature sensation is affected selectively (sparing other submodalities). Syringomyelia is rare except in cervical segments, and so the suspended–dissociated symptom complex usually involves the arms.

Lesions Above the Level of the Spinomedullary Junction

The localization rules change in key ways when the lesion occurs above the level of the spinomedullary junction. In particular, unilateral lesions produce deficits in sensory discrimination on the *contralateral* side of the *body*. But if the lesion occurs in the medulla or lower pons, there may also be deficits in somatic sensation involving the face and head on the side *ipsilateral* to the injury, due to damage of the primary afferents of the trigeminal nerve or the trigeminal nuclei. This is an example of the phenomenon of *crossed signs* or *alternating signs* (ipsilateral head and face; contralateral body) that are characteristic of lesions that affect the brainstem, pons, and midbrain. The reason is that the lesions affect ascending and descending pathways serving the contralateral side of the body, along with cranial nerves or their central projections serving the ipsilateral head and face. Other examples of crossed signs are seen in the case of motor pathways.

Lesions That Involve Thalamocortical Projections

Lesions that involve thalamocortical projections (a lesion of the internal capsule) or the cortex itself produce somewhat different symptoms than injuries at lower levels. In particular, pain perception is preserved even when the lesions completely destroy the portion of the somatosensory cortex representing a particular area of the body. It is thought that conscious perception of pain is mediated at subcortical levels. The exact identity of the pathways involved is not known.

Abnormal Sensations

Abnormal sensations (as distinct from lost sensation) can occur after damage to peripheral nerves. These are termed *dysesthesias*. Abnormal sensations from neuromas have already been discussed. Another example is the *phantom limb* phenomenon. When a limb is amputated, patients often continue to "feel" the limb. Some patients perceive that they can move the phantom limb; others feel the limb in a fixed position. Unfortunately, many patients experience persistent pain that is thought to be due to abnormal neuronal activity in central circuits as a result of the loss of primary afferent input. Presumably because the pain arises from abnormal activity within the CNS, it is very difficult to treat.

Another important abnormal sensation is *trigeminal neuralgia* or *tic douloureux*—attacks of severe pain in the area supplied by branches of the trigeminal nucleus that last for a few seconds. The attacks of pain are sometimes set off by stimulation of a *trigger zone* (a particular site on the skin) by temperature changes or as a result of facial movement.

11

Somatic Sensation III

The Physiology of Somatic Sensation

The preceding chapter describes the functional anatomy of the somatosensory system, and that coverage provides the basis for much of what physicians need to know. However there are some clinical phenomena that require an understanding of the physiology of the somatosensory system. This chapter summarizes the physiological principles that are important for physicians.

FIRST-ORDER NEURONS AND THE RECEPTOR APPARATUS

Transduction of physical energy into ionic currents in nerve cells occurs at the sensory nerve ending. As noted earlier, different types of receptor specializations at the nerve ending respond preferentially to particular types of stimuli. But all the first-order neurons of the somatosensory system have important features in common.

The terminal portion of the axon of the first-order neuron is the part that functions as the receptor (Figure 11.1). This terminal portion has electrical properties that are similar to dendrites, that is, the terminal portion does not possess voltage-gated sodium channels and therefore can be depolarized in a graded fashion. An *adequate stimulus* for the particular receptor causes a depolarization *(receptor potential)* in this portion of the axon, the amplitude of which is determined by the intensity of the stimulus.

In the somatosensory system, the receptor potential is also the *generator potential* that initiates action potentials in the myelinated portion of the axon. The greater the intensity of the stimulus, the greater the amplitude of the depolarizing receptor potential. The amplitude of the generator potential then determines the frequency of the action potentials that are generated in response to the stimulus (Figure 11.1). In this way, stimulus intensity is encoded by action potential frequency.

Another way of saying this is that the graded receptor potential/generator potential is an analog signal that is directly related to stimulus strength. This analog signal is then converted to a digital signal (the train of action potentials at a particular frequency). Thus, the receptor actually carries out two conversions: (1) it transduces physical energy into a receptor potential/generator potential; and (2) it integrates the receptor potential, producing trains of action potentials. Hence, the peripheral sensory axon functions as an integrator in a way that is similar

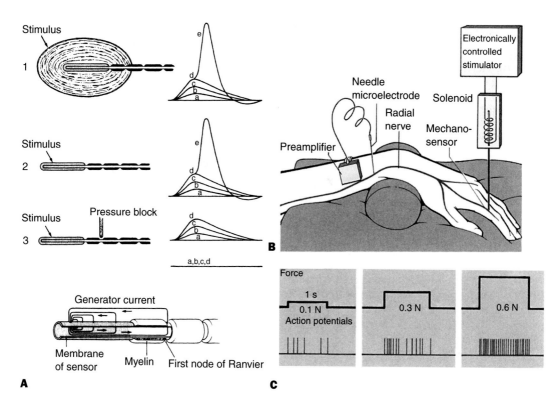

Figure 11.1. The receptor apparatus. A, Mechanical stimulation of the receptor apparatus (in this case, a Pacinian corpuscle) produces a graded depolarization of the terminal portion of the axon, depending on the intensity of the stimulation. (The response to different intensities of stimulation is illustrated in a–e). This is the receptor potential. In the case of the mechanoreceptors in the somatosensory system, the receptor potential is also the generator potential, which triggers an action potential in the primary afferent when the depolarization is sufficient (e). The axon itself is actually mechanosensitive because it exhibits graded depolarizations even when the capsule is removed. Pressure applied to the myelinated portion of the nerve blocks action potentials without affecting the receptor potential. The diagram in the lower portion of the figure illustrates the path of current flow that leads to action potential generation. **B,** Setup for assessing sensory function in humans. The tip of a microelectrode is inserted into a cutaneous nerve so that action potentials can be recorded. **C,** The greater the amplitude of the generator potential, the higher the frequency of the action potentials that are generated in response to the stimulus. In this way, stimulus intensity is encoded by action potential frequency. From Zimmerman M. The somatovisceral sensory system. In: Schmidt RF, Thews G, eds. *Human Physiology*. Berlin, Germany: Springer-Verlag; 1989:176–195.

to the somatodendritic compartment of a typical neuron.

Different Features of Tactile Stimuli Are Decoded by Receptors With Different Response Dynamics

It is useful to know the response properties of different types of receptors and how these response properties relate to function. In this regard, receptors are classified according to their sensitivity and response dynamics. In terms of sensitivity, receptors can be categorized as low threshold (very sensitive) versus high threshold (sensitive only to strong stimuli). In terms of response dymanics, receptors are classified as rapidly adapting, slowly adapting, and very rapidly adapting. These properties can be measured experimentally using an electromagnetic stimulator in which a stylus can be moved so as to indent the skin and hold the indentation (see Figure 11.1).

Slowly adapting receptors respond to a mechanical displacement as it occurs, but also continue to respond if the stimulus is maintained (Figure 11.2). The action potential frequency generated depends on the strength of the stimulus, and so these receptors enable a determination of stimulus magnitude. Rapidly adapting receptors respond to stimulus onset and offset (i.e., when skin is displaced and when the displacement is relieved). If the displacement is held, the receptor ceases to fire. Very rapidly adapting receptors respond only at the beginning of the stimulus and when the dynamic phase of the stimulus ceases.

These characteristic response patterns are exhibited by different types of receptors. For example, Merkel's disk receptors and

Figure 11.2. Response dynamics of different types of mechanoreceptors. The responses of different types of receptors to the application of a ramp deformation of the skin produced with an electromechanical stimulator (Figure 11.1). The dynamics of the skin deformation are illustrated in the lower trace. From Zimmerman M. The somatovisceral sensory system. In: Schmidt RF, Thews G, eds. *Human Physiology.* Berlin, Germany: Springer-Verlag; 1989:176–195.

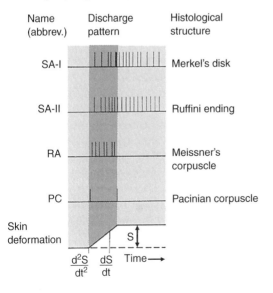

Ruffini endings are slowly adapting, Meissner's corpuscles are rapidly adapting, and Pacinian corpuscles are very rapidly adapting. It is important to know that there are different response characteristics and to understand what these response features encode.

The differences in the response dynamics of rapidly adapting versus slowly adapting receptors are seen following application of a constant stimulus of long duration (Figure 11.2). Slowly adapting mechanoreceptors increase their firing rate when the stimulus is applied and maintain a higher rate throughout the application of the stimulus. Rapidly adapting mechanoreceptors show a transient increase in activity at the onset of the stimulus and another transient increase in activity when the stimulus ceases. Slowly adapting mechanoreceptors are well suited to encode stimulus intensity. The greater the amplitude of the stimulus, the higher the frequency of the action potentials that are generated. Rapidly adapting mechanoreceptors are particularly well suited to encode *stimulus dynamics* (stimulus onset and offset).

The receptors that encode particular submodalities have response characteristics that are predictable based on the type of information that they convey. For example, Pacinian corpuscles are transducers for vibration sensitivity and so are extreme examples of rapidly adapting receptors. Nociceptors are slowly adapting, with the ones encoding slow, burning pain being more slowly adapting than the ones encoding fast pricking pain. And thermoreceptors are very slowly adapting.

The response characteristics of receptors and first-order neurons determine the pattern of action potentials (neural code) that is conveyed to the CNS by the primary afferent. Each type of receptor detects and accentuates certain features of the stimulus (submodality, stimulus onset and offset, and stimulus intensity). This is an example of what is termed *feature extraction*—a process that is manifested in the receptors of all sensory systems.

Receptive Fields

An important concept in sensory neurophysiology is the concept of the *receptive field*. In the somatosensory system, the receptive field for a particular neuron is the area of the body that, when stimulated, affects the firing of that neuron (either increasing or decreasing its rate of discharge).

First-Order Neurons

The receptive field of a first-order somatosensory neuron is determined by the distribution of the axon terminals of that neuron (see Figures 9.6 and 11.3). When the terminals of a particular neuron are activated by an adequate stimulus, generator potentials are produced in the sensory axon terminal. If the generator potential in any axon terminal is above threshold for action potential generation, then an action potential is conveyed to the CNS. Above-threshold activation anywhere within the distribution of the terminals of the sensory neuron can produce action potentials, and it is this area that is the neuron's receptive field.

The size of the receptive field of first-order neurons varies across the body surface (see, for example, Figure 11.4). This is directly related to (and indeed a consequence of) differences in receptor density across the skin surface (see Figure 11.5 and Chapter 9). The greater the density of receptors, the more restricted the distribution of the sensory axon terminals. The distribution of sensory axon terminals in turn determines the size of the receptive field.

The receptive fields of first-order neurons are *simple receptive fields* because the response of the first-order neuron (dorsal root ganglion cells) is either an increase in action potentials (excitation) when the stimulus occurs within the innervation territory of the peripheral axon, or no response at all when the stimulus occurs outside of the innervation territory (see Figure 11.4). There is no inhibition at the peripheral terminals of the somatosensory system, and so there are no inhibitory components of the receptive fields of dorsal root ganglion cells.

Receptive Fields of Cutaneous Mechanoreceptors Have a Central Area of Maximal Excitability

As already noted, the receptive field of a first-order sensory neuron is determined by the area of the body (for example, a particular patch of skin) that contains the axon terminals of that neuron. The discharge of the first-order neuron is highest when a maximal number of individual branches (receptor specializations) are activated. This is because the generator potentials in the different branches can summate to produce a greater net depolarization. The discharge of a first-order neuron is highest when the stimulus is applied to the center of a receptive field and weaker when a stimulus is applied near the edge of the receptive field. Thus, the receptive field of a first-order cutaneous mechanoreceptor has a gradient of excitability from the center to the edge.

Receptive Fields of Second-Order and Higher-Order Neurons Have Excitatory and Inhibitory Domains

The receptive fields of second-order neurons are determined by the inputs the neurons receive from first-order neurons and by input from interneurons. An important feature of the receptive fields of second- and higher-order neurons is that they have *center-surround* organization. Stimulation of an area of skin that lies in the center of the receptive field leads to excitation. Stimulation of a nearby area of skin actually causes a decrease in firing (inhibition) of the second-order neuron. This center-surround receptive field organization is based on *lateral inhibition*, which is mediated by inhibitory interneurons.

The way that receptive fields of second-order neurons are determined is illustrated in Figure 11.3. For example, the inputs from different first-order neurons that serve a par-

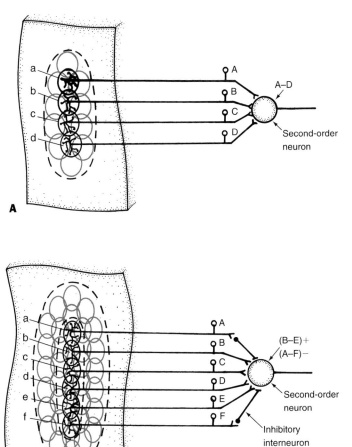

Figure 11.3. Convergence of inputs from first-order neurons onto second-order neurons, and the shaping of receptive fields by inhibitory inputs. A, The receptive field of a first-order neuron in the somatosensory system is determined by the distribution of its terminals in the periphery (small circles). The receptive field of a second-order neuron is determined by the summed inputs it receives from first-order neurons (A + B + C + D). **B,** The receptive fields of second- and higher-order sensory neurons can have inhibitory components that are generated via inhibitory interneurons located in the region of the second-order neurons. In this case, the second-order neuron receives direct input from first-order neurons B–E, and indirect inhibition from first-order neurons A and F. Modified from Kandel ER, Schwartz JH, Jessel TM. *Principles of Neural Science.* 3rd ed. New York, NY: Elsevier; 1991.

ticular surface area may converge on a particular second-order neuron (Figure 11.3). Primary afferents are exclusively excitatory, and so the information that is conveyed to the second-order neuron is a pattern of excitatory postsynaptic potentials that is determined by the pattern of action potentials in the primary afferent. In this case, the receptive field of the second-order neuron is the sum of the receptive fields of the different primary afferents (Figure 11.3A).

Second-order neurons also receive inputs from inhibitory interneurons that are activated by other first-order neurons (Figure 11.3B). These synaptic interactions lead to *complex receptive fields* in which there are in-

creases in the activity of the neuron in response to stimulation of one area of the body (excitation), and decreases in response to stimulation of another area (inhibition). A common receptive field structure is one with an excitatory center and an inhibitory surround. Because complex receptive fields (with both excitatory and inhibitory regions) can be generated only through synaptic interactions, they are observed only in second- or higher-order neurons (that is, after the first synapse in the somatosensory pathway).

An example of a center-surround receptive field at the level of the dorsal column nucleus is illustrated in Figure 11.4. In this case, a second-order neuron in the dorsal

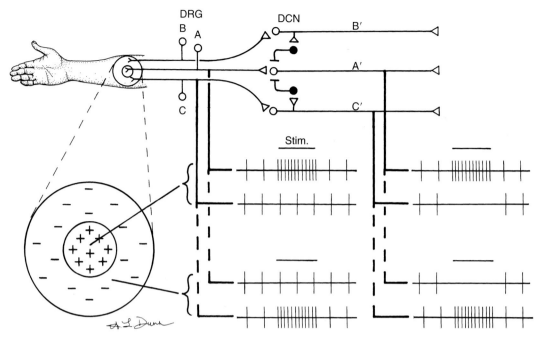

Figure 11.4. Center-surround antagonistic receptive fields of second-order neurons of the somatosensory system. In this example of a center-surround receptive field at the level of the dorsal column nucleus (DCN), first-order neurons (A–C) innervate a patch of skin, with A innervating the center of the patch. The cell bodies of the first-order neurons are located in the dorsal root ganglia (DRG). When the center of the patch is stimulated, first-order neuron A is strongly activated; first-order neurons B and C are unaffected. Second-order neuron A′ receives excitatory input from first-order neuron A. When the center of the patch of skin is stimulated, neuron A′ is strongly activated. Stimulation of the surrounding region (for example, in the distribution of the terminals of neuron C) leads to inhibition of the ongoing discharge of second-order neuron A′ (see lower traces). This inhibition is mediated by inhibitory interneurons in the nucleus (see diagram). Adapted from Kandel ER, Schwartz JH, Jessel TM. *Principles of Neural Science.* 3rd ed. New York, NY: Elsevier; 1991.

column nucleus (neuron A) receives input from a dorsal column axon that innervates a particular patch of skin on the arm. When the center of the patch of skin is stimulated, the second-order neuron is strongly activated. Lateral inhibition comes into play when an adjacent area of skin is activated (just outside the receptive field of the primary neuron that provides the excitatory input to the secondary neuron). In this case, the primary neuron supplying the adjacent area of skin projects to and excites its own second-order neuron in the dorsal column nucleus (neuron B). But that second-order neuron gives rise to a collateral axon that innervates an inhibitory interneuron, which in turn projects to neuron A. Thus, when the

patch of skin innervated by neuron B is stimulated, the result is *lateral inhibition* of neuron A.

Lateral inhibition is the substrate of *surround inhibition.* Consider again the patch of skin innervated by neuron A in Figure 11.4. All around this patch of skin are patches of skin innervated by other first-order neurons (B, C, and so forth, and see Figure 11.3). Each of these innervates second-order neurons in the dorsal column nucleus, which in turn give rise to collateral projections to inhibitory interneurons that feed back upon second-order neurons supplying adjacent areas of skin (in particular, neuron A). As a consequence of this circuitry, stimulation of patches of skin surrounding the patch innervated by neuron A produces lateral inhibition of neuron A. The collection of lateral inhibitory circuits is the substrate for surround inhibition.

Summary

The key points concerning receptive fields are that (1) primary neurons have receptive fields with excitatory centers and no inhibitory surrounds, and (2) second-order neurons and neurons in thalamic relay nuclei have receptive fields with antagonistic center-surround organization (often with excitatory centers and inhibitory surrounds).

Surround inhibition provides the mechanism for simple *feature extraction*. The circuit is set up to enhance discrimination between adjacent areas of skin. Thus the circuit provides the substrate for the determination that one patch of skin is activated at the same time that an adjacent area is not (otherwise there would be lateral inhibition). In this way, the spatial boundaries of a somatosensory stimulus is encoded by the pattern of activity in the entire population of second-order neurons. In this case, the feature that is extracted is an edge that represents the boundary between activated and nonactivated regions.

PROCESSING OF SOMATOSENSORY INFORMATION WITHIN THE CORTEX

Somatosensory information that underlies conscious perception is conveyed to the cortex by thalamic inputs originating from neurons in the ventrobasal complex. The information is then processed sequentially along intracortical circuitry. The initial stages of processing occur in the primary somatosensory areas, and processing continues as information is conveyed to sensory association areas.

Information processing within the cortex is determined by the interconnections between cortical neurons. To review the basic cellular organization of the somatosensory cortex, thalamic inputs from the "specific thalamic relay nuclei" (in this case, the neurons of the ventrobasal thalamus) terminate primarily in layer IV in tuftlike endings a few hundred micrometers in diameter (Figure 11.5). In this way, the terminals from a given thalamic neuron terminate on a spatially distinct cluster of cortical neurons.

Neurons in layer IV interconnect with neurons in other cortical layers via *intralaminar* (vertical) connections (Figure 11.5). The intralaminar connections create a column of interconnected neurons (a functional module) that sequentially processes whatever information is delivered into the column. This is the cellular basis of the *columnar organization* of the cerebral cortex. Following are some key points concerning this organization:

1. The thalamic recipient layers are occupied by small neurons that have a stellate form. In general, stellate neurons are interneurons whose axons do not leave the cortex gray matter of the cortical area in which the neurons reside. They can be either excitatory or inhibitory on their target cells. Although stellate neurons are especially numerous in layer IV (where they represent the majority of the neurons in the layer), they are also found in layers II and III. They give rise to axons that project locally within the laminae and across the laminae.

2. In addition to the interlaminar projections, stellate neurons also give rise to important lateral projections within the cortex (not represented in Figure 11.5). These lateral projections form the substrate for communication between columns.

3. Pyramidal neurons give rise to the axons that project to nearby cortical areas (for example, sensory association areas) as well as to the contralateral cerebral cortex. The former are termed *associational projections,* the latter, *commissural projections.* Pyramidal neurons are found primarily in layers III and V.

4. Neurons in layer VI give rise to projections back to the thalamus as well as to projections to the superior colliculus (not represented in Figure 11.5).

As described in later chapters, the organization of the cortex into functional columns is an organizational principal that is reiterated in different cortical regions.

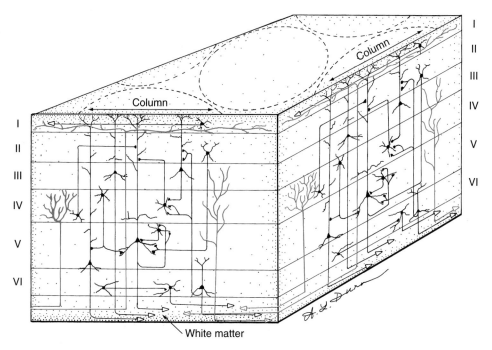

Figure 11.5. Intralaminar interconnections between neurons in the cortex form the basis for functional columns. The figure illustrates an idealized scheme of the translaminar interconnections between different neuron types in the somatosensory cortex. The inputs from thalamic relay nuclei terminate in discrete patches, primarily in layer IV. The thalamic recipient neurons (mostly stellate cells, although thalamic inputs also terminate on pyramidal neurons) give rise to projections to both superficial and deep laminae. Interneurons in other layers also give rise to translaminar projections that interconnect neurons in the different laminae, forming a functional module (a column of interconnected neurons that process a similar type of information). This circuitry forms the basis for columnar processing.

Information processing in the somatosensory system occurs in two ways: (1) by *parallel processing,* and (2) by *hierarchical processing.*

As already noted, parallel processing is reflected at subcortical levels by the existence of discrete pathways that carry certain types of information (pain and temperature versus cutaneous mechanosensation and vibration sensitivity). Parallel processing continues in the cortex, in that particular subdivisions of the somatosensory cortex are dominated by inputs from particular classes of receptors.

Hierarchical processing is reflected at subcortical levels by the increasing complexity of receptive fields, progressing from primary to secondary neurons. In the cortex, hierarchical processing is reflected by an increasing complexity of receptive field organization of individual neurons, so that individual neurons begin to respond selectively to even more complex patterns of somatosensory input.

Parallel Processing: Different Subdivisions of the Somatosensory Cortex Are Dominated by Inputs From Different Classes of Receptors

Preceding sections have noted how information related to different submodalities is carried by different pathways (labeled lines). As described in Chapter 10, this segregation is maintained up to the level of the thalamus. Specifically, axons conveying information from different types of receptors (cutaneous versus deep) terminate in different locations within the ventrobasal complex (core versus shell, respectively).

The segregation of information is still seen at the level of the somatosensory cortex. There are two aspects of this organization. First, there are different subdivisions (termed *Brodmann's areas 1, 2, 3a,* and *3b*) within the primary somatosensory cortex (SI). Each subdivision is dominated by inputs from particular types of receptors. This is because each area receives input preferentially from a different subdivision of the ventrobasal complex (see Figure 11.6). Second, within

Figure 11.6. Different subdivisions of the somatosensory cortex are dominated by inputs from different classes of receptors. Within SI there are four subdivisions termed *Brodmann's areas 1, 2, 3a,* and *3b,* which are parallel strips of cortex that border the postcentral gyrus. Each subdivision is dominated by inputs from particular types of receptors. This is because each area receives input preferentially from different subdivisions of the ventrobasal complex. Each subdivision receives input from all parts of the body, and the somatotopic map is similar in each. Area 1 is dominated by input from rapidly adapting cutaneous receptors; area 2 is dominated by input from deep pressure receptors; area 3a is dominated by input from muscle stretch receptors; and area 3b is dominated input from both rapidly adapting and slowly adapting cutaneous mechanoreceptors. LP indicates lateral posterior nucleus; VL, ventrolateral nucleus; VPL, ventroposterior lateral nucleus. Adapted from Jones EG, Friedman DP. Projection pattern of functional components of thalamic ventrobasal complex on monkey somatosensory cortex. *Journal of Neurophysiology* 1982;48:521–544.

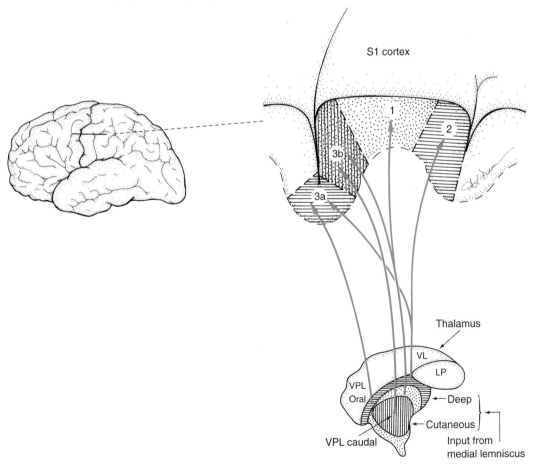

each area, inputs conveying information from particular types of receptors terminate in discrete cortical columns.

The four subdivisions of SI are found as parallel strips of cortex along the postcentral gyrus. Each subdivision receives input from all parts of the body, and the somatotopic map is similar in each. But the thalamic inputs to each subdivision convey information that is dominated by particular receptor types. For example, area 1 is dominated by input from rapidly adapting cutaneous receptors; area 2 is dominated by input from deep pressure receptors; area 3a is dominated by input from muscle stretch receptors; and area 3b is dominated by input from both rapidly adapting and slowly adapting cutaneous mechanoreceptors (Figure 11.6). But domination does not mean exclusivity, and so each area also receives some input from fibers carrying some other receptor types.

Within each area, inputs conveying information from different types of receptors (e.g., rapidly adapting, slowly adapting) distribute to different cortical columns. Because of vertical interconnections between neurons in different layers of the cortex, the neurons in a particular cortical column are primarily involved in processing whatever information arrives via the thalamic projection to thalamic recipient neuron in layer IV. For example, if the thalamic recipient neuron receives input from a fiber conveying information from rapidly-adapting receptors, then the response properties of neurons in the entire column of cells reflect that input.

A remarkable aspect of the columnar organization of the cortex is that columns are arranged in a highly regular fashion, alternating systematically from one receptor type to another. Figure 11.7 illustrates one example of this; in area 3b, columns that are specialized for rapidly adapting and slowly adapting cutaneous receptors in each digit lie adjacent to one another. Other digits have similar pairs of modules. Thus, the hand representation in the cortex is a stripelike mosaic. Similar columns exist for other receptor types. For example, there are alternating stripes of cortex in which cells are devoted to superficial tactile stimuli versus deep pressure in a particular area of skin.

Hierarchical Processing: As Information Is Processed in the Cortex, the Receptive Fields of Neurons Become More Complex

Within the circuitry of the somatosensory cortex, information processing occurs so that neurons begin to respond selectively to particular features of a somatosensory stimulus. For example, some neurons respond better to movement than to touch (that is, they are motion sensitive); some neurons respond selectively to movement in a particular direction (direction sensitive); and some neurons respond selectively to movement along a particular axis (orientation sensitive). In this way, activity in particular populations of neurons "represents" the detailed features of particular physical stimuli.

Extrapolating from these results, one might imagine that there would be neurons further and further along the hierarchy that respond selectively to ever more complex stimulus configurations. For example, humans can easily distinguish between items on the basis of touch alone (for example, coins in one's pocket). This is termed *stereognosis* (knowing based on three-dimensional form). One could imagine the existence of neurons somewhere in the brain that respond selectively to the stimuli generated by a quarter. Just how far this response specificity actually goes is not certain.

It is noteworthy that receptive fields tend to become larger as one progresses along the hierarchy, at the same time that the characteristics of the receptive field become more complex. Thus, for example, the simple receptive fields of relay neurons supplying the digits are quite small and represent only a small patch of skin on the digit. However, a cortical neuron may respond in a directionally selective manner to a stimulus on any of several digits. In this way, neurons respond to common stimulus features even when the stimuli are delivered to different receptive surfaces.

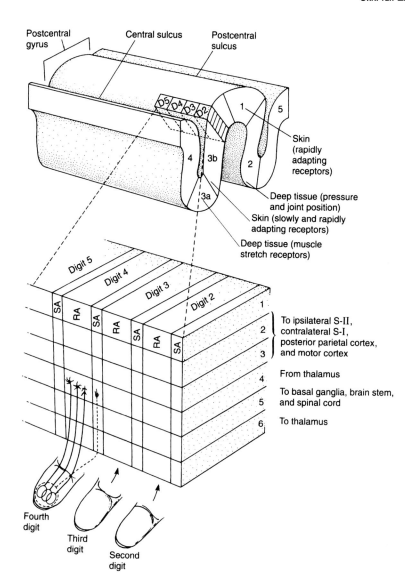

Figure 11.7. Cortical modules. In the somatosensory cortex, the inputs carrying information from different receptor types terminate in columns that alternate in regular fashion. This figure illustrates an example of this in area 3. Thus, each cortical module that represents a single digit contains separate modules that receive information from rapidly adapting (RA) and slowly adapting (SA) receptors. This organization is reiterated for each digit. From Kandel ER, Schwartz JH, Jessel TM. *Principles of Neural Science.* 3rd ed. New York, NY: Elsevier; 1991.

Somatosensory Cortex Projections

Neurons in the SI project to a number of other cortical areas. There are extensive interconnections between SI and the secondary somatosensory cortex (SII), which is at the basal end of the postcentral gyrus. SII also receives input from neurons in the thalamus. There are also extensive projections to the posterior parietal cortex—a sensory association area responsible for integrating information from different sensory modalities. Finally, there are extensive projections to the motor cortex, and descending projections to the motor centers in the brainstem and spinal cord. Indeed, in the human, about one-third of the axons in the descending corticospinal tract originate from the somatosensory cortex.

The Somatosensory Cortex Plays a Key Role in Motor Control

In addition to providing the substrate for conscious perception of somatic sensation, the somatosensory cortex plays an important role in motor control. Fine motor skills require continuous sensory feedback. You can appreciate this when you try to manipulate a door key when your hands are numb from the cold. This feedback is provided by sensory–motor interconnections at each level of the neuraxis from the spinal cord to the cerebral cortex. Because fine motor control requires sensory input, damage to the dorsal column–medial lemniscal system leads to *ataxia* (a lack of coordination in fine motor skills).

In addition, higher-order somatic sensation often requires the participation of the motor system. This can be appreciated immediately if one asks a person to identify an object by touch alone; the object must be manipulated to be identified (try it with a coin in your pocket).

THE POSTERIOR PARIETAL CORTEX INTEGRATES INFORMATION FROM SOMATOSENSORY AND OTHER MODALITIES

One important site of integration of information from different sensory systems is the posterior parietal area. The posterior parietal area lies immediately posterior to the dorsal part of the postcentral gyrus. It receives a major projection from the primary somatosensory cortex as well as projections from cortical areas subserving other sensory modalities (vision, audition). The posterior parietal cortex integrates this multimodal information in order to allow a determination of the position of the body with respect to the immediate environment (an *egocentric spatial map*).

Injury to the posterior parietal cortex produces an interesting symptom termed *sensory neglect,* in which stimuli on the opposite side of the body are ignored. So too are stimuli in the contralateral visual field, or sound stimuli originating from sources that lie on the side contralateral to the lesion. It is as if the patients do not notice these stimuli even though their sensory systems are demonstrably intact. This and other interesting symptoms resulting from parietal lobe injury are discussed in the Chapter 31.

Descending Projections Provide Feedback to Lower Levels to "Gate" the Flow of Sensory Information

An important feature of sensory processing is *sensory gating.* Much of the activity generated by the somatosensory system does not reach our consciousness. This is easy to appreciate. As you sit and read this book, you are not conscious of points of contact with your clothes or contact with your chair unless you direct your attention to those points of contact. Then they can be felt quite easily. In the somatosensory system, as in other systems, we do not always notice particular stimuli unless we "pay attention" to them. This is termed *selective attention,* and it is thought that part of the mechanism for selective attention is sensory gating by the cerebral cortex.

Sensory gating is mediated by pathways that project from the cortex to lower relay nuclei (Figure 11.8). These projections terminate on inhibitory interneurons within the relay nuclei that in turn project onto the relay neurons themselves. In this way, cells in the cortex can control the flow of information over ascending sensory pathways. This is an example of a mechanism termed *distal inhibition,* in which a long-distance excitatory projection produces inhibition via an inhibitory interneuron in the target area.

Feedback pathways that mediate distal inhibition provide another example of a cellular mechanism that enables *contrast enhancement.* It is thought that distal inhibition is part of the mechanism for *selective perception* (when we perceive one stimulus and not a host of others). Distal inhibition and sensory gating are especially important in the case of painful stimuli, as described in the following chapter.

Functional Reorganization of Somatosensory Representation Following Injury

Damage at any level of the somatosensory system leads to a loss of function that depends on the site of the injury. However, an important feature of the somatosensory sys-

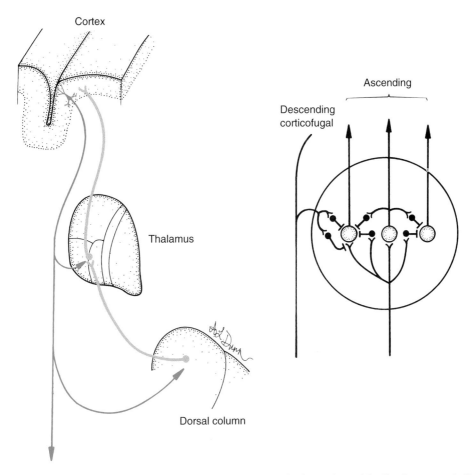

Figure 11.8. Descending projections control transmission through sensory relay nuclei. Transmission over sensory pathways can be modulated by descending inputs. Neurons in the cerebral cortex give rise to descending projections to relay nuclei in the thalamus and the dorsal column nuclei. These inputs modulate transmission through the nucleus. For example, the circuit shown is an idealized representation of connections within the dorsal column nuclei. Descending inputs terminate on inhibitory interneurons, which in turn modulate the output from the relay neurons. The diagram of dorsal column circuitry is adapted from Kandel ER, Schwartz JH, Jessel TM. *Principles of Neural Science.* 3rd ed. New York, NY: Elsevier; 1991.

tem is that it is capable of some reorganization; it is believed that this reorganization is one of the mechanisms through which the brain compensates following injury.

One well-studied example of functional reorganization occurs following injury to peripheral nerves. When peripheral sensory nerves are cut, the areas of the somatosensory cortex that these sensory nerves supply are no longer activated. If regeneration is prevented or unsuccessful, the cortex remains disconnected from its normal peripheral inputs. Over a period of weeks to months, however, the areas of the cortex that lose their normal inputs become responsive to stimulation of sites on the skin in which peripheral innervation is intact. An example of this phenomenon is illustrated in Figure 11.9. In this case, the median and ulnar nerves supplying the hand of a squirrel monkey were transected and ligated (in order to prevent regeneration). In the squirrel monkey, these nerves supply the glabrous skin on the volar surface of the hand. Microelectrode recordings carried out several months after nerve transection revealed that

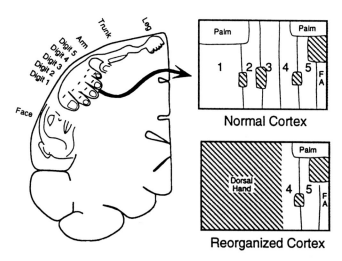

Figure 11.9. Functional reorganization in the somatosensory cortex after peripheral nerve injuries. The median and ulnar nerves normally innervate the glabrous skin on the volar surface of the fingers and hand and are represented in the normal cortex as indicated in the upper schematic. When the median and ulnar nerves are transected and ligated, there is a reorganization of the somatosensory map in the cortex. The region of cortex that would normally represent the glabrous skin becomes responsive to stimulation of the hairy skin on the dorsal surface of the hand. Reproduced with permission from Kaas J, Florence SL. Brain reorganization and experience. *Peabody Journal of Education.* 1996;71:152–167.

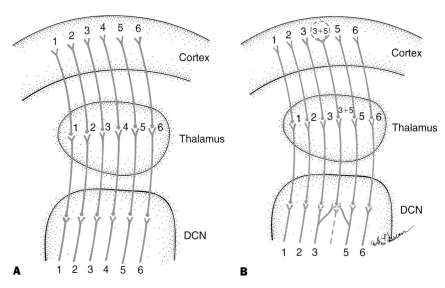

Figure 11.10. A possible mechanism underlying functional reorganization. The cartoon in **A** schematizes the normal topography of somatosensory relays through the dorsal column nuclei, ventroposterior nucleus of the thalamus, and the somatosensory cortex. 1–6 represent different primary axons carrying information from the periphery. **B** illustrates what might happen when the peripheral branch of primary neuron 4 is damaged. In this case, the central projections of primary neuron 4 are replaced as a result of the expansion of the terminal field of neurons 3 and 5. As a result, the postsynaptic target of neuron 4 in the dorsal column nuclei (DCN) is now innervated by primary axons 3 and 5. Site 4 is no longer represented in the cortex, but the part of the cortex that previously represented site 4 now represents sites 3 and 5. In this way, the part of the cortex that normally represented site 4 has taken over the function of nearby parts of cortex.

the region of cortex that would normally represent the glabrous skin could now be activated by stimulation of the hairy skin on the dorsal surface of the hand. Thus, the hand representation in the cortex was reorganized so that no part of the cortex was unresponsive.

A reorganization of cortical representation may play a role in functional recovery following peripheral nerve injury. Damaged peripheral nerves are capable of regeneration; however, they often regenerate in a disordered fashion and so establish an abnormal pattern of innervation in the periphery. As a result, the somatotopic map is scrambled, and stimuli are mislocalized. Deficits in the ability to localize stim-uli sometimes recover over prolonged periods, however. One hypothesis is that the recovery occurs as a result of a reorganization that specifically compensates for peripheral reinnervation errors.

The mechanisms that underlie the reorganization of somatosensory maps are not yet known. One possible mechanism is illustrated in Figure 11.10. The idea is that, in response to damage to peripheral nerves, there is a reorganization of the central projections of primary afferents, in this case in the dorsal column nuclei. In this way, the central projections from undamaged primary afferents would come to activate neurons that were normally supplied by the primary afferents that had been damaged.

12

Somatic Sensation IV

Pain

Pain is usually considered as a separate topic because it has unusual features in comparison with other submodalities of somatic sensation and because it is so important to the physician. From the clinical standpoint, pain is significant in two respects. First, it can be an important symptom of pathology. Second, persistent pain can be extremely debilitating, affecting virtually every aspect of a patient's life. This is reflected by the enormous amount of money that is spent on pain medication.

Physicians (as well as patients) recognize different types of pain. Terms that are used include aching, burning, stinging, and soreness. These terms describe the patient's perception of the sensory input. To some extent, these different perceptions arise from activation of different types of *nociceptors*. But even when nociceptors are activated, patient's perceptions vary. An important feature of pain is the psychological "spin" that patients put on the interpretation of the sensory input. As result, the response to pain is highly variable among individuals and under different circumstances in the same individual. Thus, pain is considered to have *sensory–discriminative components* that are directly related to receptor activation, as well as *motivational–affective components* that are less directly linked to the stimulus that

produces the sensation. In the following sections, we first consider the transmission of information from nociceptors (the sensory–discriminative component) and then the more complex aspects of pain that are related to motivational–affective components, rather than sensation per se.

NOCICEPTORS ARE HIGH THRESHOLD RECEPTORS THAT RESPOND TO POTENTIALLY INJURIOUS STIMULI

As noted in Chapter 9, nociceptors are free nerve endings that are specialized to respond to intense stimuli that have the potential of producing tissue damage. Several different types of nociceptors are thought to exist, including receptors that are preferentially sensitive to mechanical, thermal, or chemical stimuli. There are also *polymodal nociceptors,* which have a broad range of sensitivity. The mechanisms of nociceptor activation are similar to those of other mechanoreceptors in that an *adequate stimulus* produces a depolarizing receptor potential that leads to the generation of action potentials in the primary afferent fiber. An important difference is that nociceptors have a higher threshold for activation, so that under normal circumstances, they respond only to intense stimuli.

Nociceptors Become Sensitized Following Injury

An important aspect of nociceptor function is that nociceptors become sensitized following tissue damage. For example, following a mechanical or burn injury to a localized area, innocuous stimulation of that area or surrounding areas is perceived as painful. The increased sensitivity is termed *hyperalgesia,* and a distinction is made between *primary hyperalgesia* (at the site of injury) and *secondary hyperalgesia* (in areas surrounding the site of injury). Both of these can be readily appreciated through personal experience.

It is thought that hyperalgesia occurs in part because the threshold for nociceptor activation decreases. The mechanism underlying the decrease in receptor threshold is believed to involve a combination of factors including (1) the release of histamine, bradykinin, and prostaglandin by damaged cells; (2) the release of neuroactive peptides by nociceptor nerve endings; and (3) the release of vasodilatory peptides (for example, substance P), which increase blood flow and cause edema, which in turn causes additional release of bradykinin (Figure 12.1). There may also be other factors that contribute to hyperalgesia, including facilitation of transmission at central synapses of pain pathways.

Recent neurophysiological studies suggest that there may be a population of *silent nociceptors* that are normally unresponsive to high-intensity mechanical stimuli. It is thought that these receptors become sensitized (active) following injury and play a role in the development of hyperalgesia.

Figure 12.1. Nociceptor sensitization. Peripheral nociceptors become sensitized at sites of tissue damage as a result of several factors.
(1) Bradykinin and prostaglandin are released at sites of tissue damage; these substances activate and sensitize nociceptors. (2) Neuroactive peptides are released by nerve endings, especially substance P. Substance P stimulates degranulation of mast cells and histamine release. The histamine directly excites nociceptors. Substance P also acts on blood vessels to increase blood flow, causing edema, which in turn causes additional release of bradykinin.

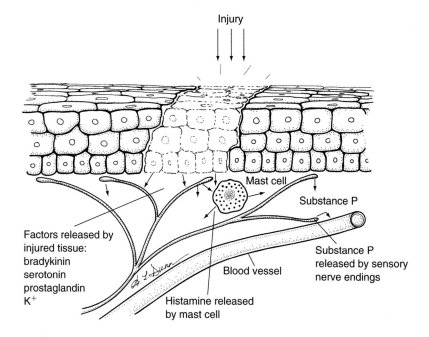

Damage to Nociceptive Pathways Can Lead to Pain

One paradoxical feature of nociceptor pathways is that injury to the pathways themselves often leads to pain. In particular, a primary complaint in disease states that damage peripheral nociceptor axons (for example, peripheral neuropathies) may be an uncomfortable burning sensation. The loss of pain sensation is then detected upon clinical evaluation. Similarly, following injuries that extensively damage peripheral nerves, patients may feel pain (burning or electric pain) in areas exhibiting complete anesthesia. Phantom limb pain is another example of a situation in which pain is perceived in the absence of nociceptors.

The mechanisms underlying pain perception following peripheral nerve injury are not well understood. One hypothesis is that loss of afferent input leads to abnormal activity of deafferented second-order neurons, which is perceived as pain by the patient. In this case, pain is a *positive symptom* that results from *abnormal activity* in surviving circuitry, rather than a *negative symptom* due to a loss of function of the circuit. As discussed in more detail later, abnormal activity in surviving circuits following injury is thought to underlie the clinically important phenomenon that damage to central pain pathways may *cause* pain syndromes.

NEUROTRANSMISSION ALONG CENTRAL PAIN PATHWAYS

The sensory axons subserving nociceptors are A-delta and C fibers. These are lightly myelinated and unmyelinated fibers and therefore relatively slow conducting. Upon entering the spinal cord, the fibers bifurcate and project both up and down the cord in the *tract of Lissauer,* dropping off collaterals into the dorsal horn throughout their course. Nociceptor axons terminate primarily in the superficial portion of the dorsal horn (the *substantia gelatinosa*), although some A-delta fibers pass through the superficial laminae to

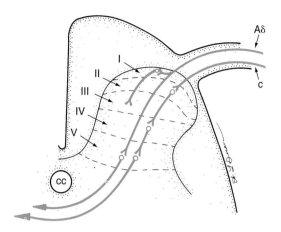

Figure 12.2. Terminations of primary afferents conveying nociceptive information. Primary afferents of nociceptors include C fibers (unmyelinated) and A-delta fibers (myelinated). C fibers terminate in lamina II; A-delta fibers terminate in laminae I and V. Nociceptive information is conveyed through polysynaptic relays to projection neurons in lamina V. These give rise to axons that cross the midline in the ventral white commissure and ascend in the anterolateral quadrant on the contralateral side.

terminate in deeper laminae. These pathways are illustrated in Figure 12.2.

The primary afferents of nociceptors terminate on two types of neurons: (1) neurons that are dedicated to nociception (that is, they receive input only from high-threshold nociceptors), and (2) neurons that receive input from high-threshold nociceptors as well as low-threshold mechanoreceptors. The former cells are termed *nociceptive specific*; the latter are termed *wide dynamic range* neurons because they respond to low-intensity stimuli but continue to increase their response as stimulus intensity increases to noxious levels.

An important feature of the primary afferents of nociceptors is that they release both excitatory amino acids and neuropeptides onto their target cells in the spinal cord. The excitatory amino acids produce fast excitatory postsynaptic potentials; the neuropeptides produce slow excitatory postsynaptic potentials that generate long-lasting activation of the second-order neurons. For this reason, the activity that is produced over central pathways by activation of nociceptors may be of long duration.

Modulation of Transmission Between Primary Nociceptive Afferents and Second-Order Neurons

The degree of activation of second-order neurons by primary afferents carrying nociceptive information depends on the context in which the activation occurs. This fact was first demonstrated in neurophysiological studies of experimental animals. Specifically, it was found that the response of second-order neurons to nociceptor afferents was *decreased* if low-threshold mechanoreceptors were simultaneously activated. These observations formed the basis for the *gate control theory of pain*, which holds that the flow of information along pain pathways is gated by activity in pathways carrying other submodalities. In this way, central transmission of nociceptive information depends on the balance of activity in nociceptive versus other submodalities.

One possible cellular mechanism that could underlie the gating of transmission along pain pathways is illustrated in Figure 12.3. In this circuit, a nociceptive afferent terminates on a particular second-order neuron. Mechanoreceptor afferents from the same dermatome innervate their own second-order neurons but also terminate on inhibitory interneurons that in turn project to the second-order neuron of the nociceptive pathway. Activation of the mechanoreceptor afferent would activate the inhibitory interneuron so as to inhibit transmission along the nociceptive pathway.

The gate control theory of pain was readily accepted by physicians because it explained some otherwise puzzling facts; the theory also led to simple therapies that proved to be efficacious in relieving pain. For example, one treatment that was developed for chronic pain on the basis of the gate theory involves *transcutaneous electrical stimulation* of peripheral nerves in the areas in which pain is perceived. Such stimulation al-

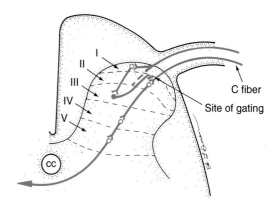

Figure 12.3. Gate theory of pain. A possible circuit that could underlie the gating of transmission along pain pathways is illustrated. In this circuit, a nociceptive afferent terminates on a particular second-order neuron. Mechanoreceptor afferents from the same dermatome innervate their own second-order neurons but also terminate on inhibitory interneurons that project to the second-order neuron of the nociceptive pathway. Activation of the mechanoreceptor afferent activates the inhibitory interneuron inhibiting transmission along the nociceptive pathway.

leviates pain, theoretically by gating transmission along the pain pathways. The gate control theory also explains an interesting natural response to a painful stimulus—the tendency to massage a site of injury immediately after the injury occurs. It is common experience that this simple action helps to alleviate the pain. Again, the reason may be that massaging activates mechanoreceptors in the area, which turn gate transmission over pain pathways. Whatever the reason, the perception of pain can be modulated by activating other receptor subtypes supplying the same part of the body surface.

ASCENDING NOCICEPTIVE PATHWAYS

Ascending fibers carrying nociceptive information arise from second-order neurons that receive direct input from primary afferents (especially the A-delta fibers) and from third- and higher-order neurons that receive indirect (polysynaptic) input from nociceptive afferents via interneurons in the dorsal horn (the route of transmission for information conveyed by C fibers; see Figure 12.2). The fact that nociceptive information is conveyed via polysynaptic pathways is important because transmission along polysynaptic pathways is sensitive to anesthetics. It is for this reason that nociceptive pathways are more sensitive to anesthesia than the pathways carrying other submodalities.

Ascending Fibers Carrying Nociceptive Information Terminate in Several Locations in the Brain

The ascending fibers of the anterolateral quadrant that carry nociceptive information terminate in several locations within the brain (Figure 12.4). In addition to the principal projection to the thalamus (spinothalamic projections), nociceptive fibers project to the *reticular formation* and the *periaqueductal gray of the mesencephalon* (a collection of neurons that surround the cerebral aqueduct in the midbrain).

Some textbooks name the different components of the ascending nociceptive system according to their sites of termination. For example, the fibers projecting to the reticular formation are termed the *spinoreticular tract;* the fibers projecting to the periaqueductal gray of the mesencephalon are termed the *spinomesencephalic tract.* The different projection systems mediate somewhat different functions.

Spinothalamic projections. The nociceptive fibers of the spinothalamic tract terminate within two nuclear groups in the thalamus: (1) a lateral group that includes the ventrobasal and posterior nuclei, which also receive input from the dorsal column–medial lemniscus system; and (2) the *medial nuclear group,* which includes the central lateral nucleus and the intralaminar nuclei. The circuit involving the lateral group of nuclei conveys information regarding pain to the cortex. The circuit involving the medial group is thought to play a role in general arousal.

Spinoreticular projections. The nociceptive fibers that terminate in the reticular formation innervate neurons that project to the thalamus (especially the medial nuclei), forming a polysynaptic relay for nociceptive information that complements the more direct projection via the spinothalamic tract. These polysynaptic relays through the reticular formation, like the direct spinothalamic projections to the medial nuclear group of the thalamus, are thought to play a role in general arousal.

Spinomesencephalic projections. The nociceptive fibers that terminate in the periaqueductal gray of the mesencephalon innervate neurons that give rise to descending projections to the medulla and ascending projections to the hypothalamus. The descending projections are thought to play an important role in gating the transmission of nociceptive information (see later discussion). The ascending projections are thought to represent one of the ways that nociceptive information reaches the structures that make up the *limbic system.*

LOCALIZATION OF PAIN DEPENDS ON THE DORSAL COLUMN/MEDIAL LEMNISCAL SYSTEM

Returning to the theme that pain is a complex perception, it is important to note that the dorsal column/medial lemniscal system also plays a role. In particular, damage to the dorsal columns leads to an inability to properly localize pain. Patients with such injuries may report that they feel pain "somewhere" in the affected region but cannot say exactly where.

Pain Perception Does Not Depend on the Cerebral Cortex

Although the spinothalamic tract innervates thalamic neurons that project to the somatosensory cortex, damage to the cortex

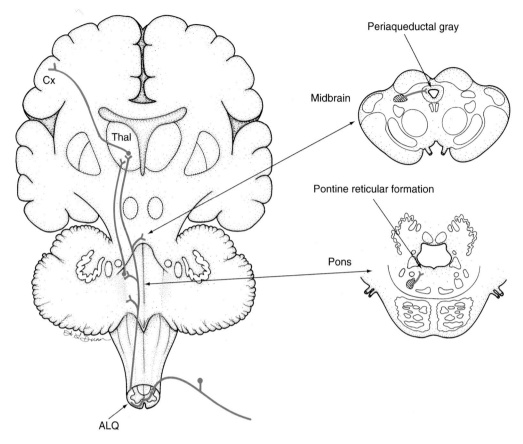

Figure 12.4. Ascending nociceptive pathways. Ascending fibers carrying nociceptive information project centrally via the anterolateral quadrant of the spinal cord. The axons terminate in the thalamus, reticular formation, and periaqueductal gray of the mesencephalon. The nociceptive fibers that terminate in the reticular formation innervate neurons that project onto the thalamus, forming a polysynaptic relay for nociceptive information that complements the direct projection via the spinothalamic tract. The nociceptive fibers that terminate in the periaqueductal gray innervate neurons that give rise to descending projections to the medulla and ascending projections to the hypothalamus (see Figure 12.7).

does not lead to a loss of pain perception. For this reason, it is said that pain perception is represented subcortically. However, this is little more than a restatement of a clinical fact. Exactly how pain is perceived in the absence of the cerebral cortex is not well understood. Cortical injury does disrupt pain localization in much the same way as injuries to the dorsal columns. For the physician, the key point is that injury to the somatosensory cortex or to thalamocortical projections (the internal capsule) leads to a loss of discriminative touch, position sense, and vibration sensitivity and, to some extent,

disrupts pain localization on the side contralateral to the injury; however, pain sensitivity is spared.

VISCERAL PAIN

Pain from the viscera is of special importance for the physician as a diagnostic clue to visceral dysfunction. Visceral pain has characteristics that are different in important ways from the pain arising from cutaneous and joint nociceptors.

Visceral pain is conveyed via nociceptors located in the viscera. The characteristics of visceral nociceptors are not well understood. It is thought that some visceral nociceptors sense stretch (distension of the viscera) whereas others are thermoreceptors or chemoreceptors. The axons of visceral nociceptors travel from the viscera to the spinal cord with autonomic axons and traverse prevertebral and paravertebral autonomic ganglia. En route through prevertebral parasympathetic ganglia, they give rise to collaterals that innervate postganglionic neurons (Figure 12.5). These connections form the substrate for peripheral autonomic reflexes. The primary afferents then continue on via the dorsal roots to synapse on neurons in the dorsal horn.

Pain Arising From Viscera Is Often "Referred" to Other Locations

An important aspect of visceral pain is that the pain is difficult to localize and is often not perceived as arising from viscera. Instead, the pain is perceived as originating from a particular cutaneous or subcutaneous site.

This is the phenomenon of *referred pain*. Referred pain is of extreme importance for the physician as a localizing symptom, because the pain from a particular organ is referred to a characteristic dermatomal location. Thus the source of the pain can be inferred by the location of the referred pain.

The most common areas to which visceral pain is referred are as follows. Pain from myocardial infarction or angina is referred to the chest and/or the left arm. Pain from the esophagus is referred to the chest in a pattern that is similar to pain arising from the heart. It is for this reason that pain resulting from myocardial infarction may be mistaken for pain resulting from indigestion (and vice versa). Pain from distension of the colon is referred to the periumbilical region.

It is important to note that in some patients, visceral pain is referred in an atypical fashion. For example, pain arising from the heart may sometimes be referred to the jaw. Physicians must be alert for these atypical referents.

The cellular mechanisms underlying referred pain are not well understood. The most widely accepted hypothesis is that primary afferents carrying visceral pain activate

Figure 12.5. Visceral afferents travel through autonomic nerves and ganglia. Axons of visceral nociceptors travel from the viscera to the spinal cord with autonomic axons and traverse prevertebral and paravertebral autonomic ganglia. En route through prevertebral parasympathetic ganglia, they give rise to collaterals that innervate postganglionic neurons, which forms the substrate for peripheral autonomic reflexes. The primary afferents enter the spinal cord via the dorsal roots to synapse on neurons in the dorsal horn.

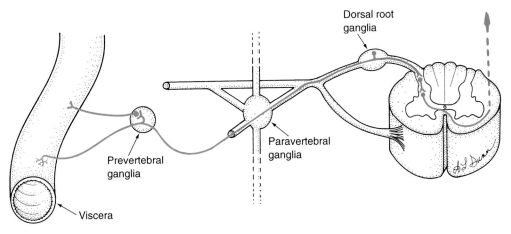

some of the same populations of neurons in the spinal cord as primary afferents from cutaneous locations. This is termed the *convergence–projection hypothesis* of referred pain. Thus, activity over visceral nociceptor afferents would drive some of the same second-order neurons that give rise to central pathways conveying information from cutaneous receptors (Figure 12.6). For this reason, activity over visceral nociceptors is perceived as originating from the cutaneous receptors.

TRANSMISSION OVER PAIN PATHWAYS IS MODULATED BY DESCENDING INFLUENCES

An important aspect of nociception is that transmission along pain pathways is modulated by descending systems. This was discovered through experimental studies that revealed that stimulation of the periaqueductal gray matter led to a loss of sensitivity to painful stimuli. The mechanism for this effect is an inhibition of transmission along no-

ciceptive pathways at the spinal level. This finding led to the development of therapies for intractable pain that involved stimulating the periaqueductal gray with implanted electrodes. Such stimulation alleviates pain, but patients do not lose tactile sensitivity.

Descending modulation of pain sensitivity is mediated by the descending projections from the periaqueductal gray to two nuclear groups in the medulla: the *raphe nucleus,* which lies at the midline, and neurons in the lateral portion of the reticular formation (Figure 12.7). These neurons in turn give rise to descending projections to the spinal cord that travel in the *dorsolateral funiculus* and terminate in the dorsal horn. The axons from the raphe nucleus contain the neurotransmitter serotonin (5-HT); the descending axons from the reticular formation contain norepinephrine.

The Circuit Responsible for Descending Modulation Involves Neurons That Use Endogenous Opioid Peptides

An important feature of nociceptive circuitry is that certain of its component neurons use *opioid peptides* as their neurotransmitter. The descending circuits that modulate transmission along nociceptive pathways are thought to operate in part by activating these neurons.

One important group of opiate-containing neurons is localized in the dorsal horn (Figure 12.8). These neurons contain the neuropeptide *enkephalin*. When the enkephalin-containing interneurons are activated, they release enkephalin near the synapses between the primary afferent and the second-order neuron. Enkephalin then inhibits transmission between the primary afferent and the second-order neuron. It is thought that enkephalin acts both presynaptically and postsynaptically, that is, by inhibiting release from primary afferent terminals and by exerting a postsynaptic inhibitory action on the second-order neurons of the nociceptive pathway.

There are also opiate-containing neurons in the periaqueductal gray and medulla, some

Figure 12.6. The convergence–projection hypothesis of referred pain. One possible circuit is shown in which primary afferents carrying visceral pain terminate on some of the same neurons in the spinal cord that receive input from cutaneous receptors. In this circuit, activity over visceral nociceptor afferents activate some of the same second-order neurons that give rise to central pathways conveying information from cutaneous receptors. The brain interprets this activation as arising from the cutaneous site that is represented by the second-order neuron.

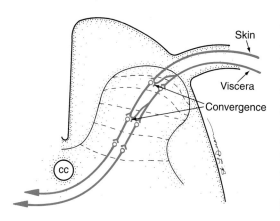

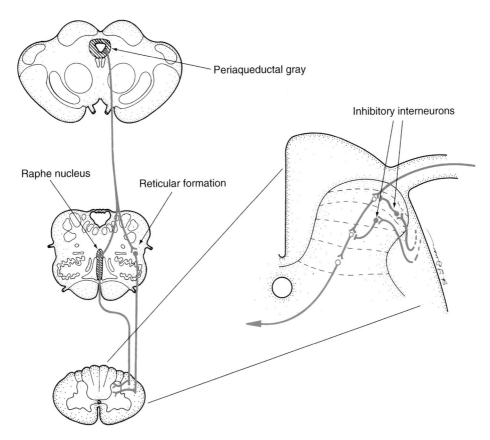

Figure 12.7. Descending modulation of transmission along nociceptive pathways. Descending modulation of pain sensitivity is mediated by projections from the periaqueductal gray to the raphe nucleus and the reticular formation. Neurons in the raphe nucleus and reticular formation give rise to projections to the spinal cord that descend in the dorsolateral funiculus and terminate on inhibitory interneurons in the dorsal horn.

of which contain another opiate-like peptide called *dynorphin.* The opiate-containing neurons in the periaqueductal gray and the medulla activate the neurons that give rise to the descending pathways.

Morphine acts on opiate receptors by mimicking the action of the opiate peptide neurotransmitters. Systemic opiates such as morphine act in two ways: (1) by inhibiting transmission at the level of the synapse between the primary nociceptive afferent and the second-order neuron, and (2) by activating the descending modulatory pathways from the raphe nucleus and the reticular formation. Physicians take advantage of the multiple sites of opiate action. For example, morphine can be injected intrathecally (into

the cerebrospinal fluid of the spinal cord) where it blocks nociceptive transmission at the level of the spinal cord without the side effects of systemic delivery of morphine (respiratory depression). This is the reason intrathecal injections of opiates are used to relieve the pain of labor.

Descending Modulatory Pathways Provide a Mechanism for Controlling Pain

The perception of pain depends on the context in which the pain occurs. For example, there are a host of anecdotal accounts of individuals who suffer injuries under conditions of high emotion but do not notice pain until

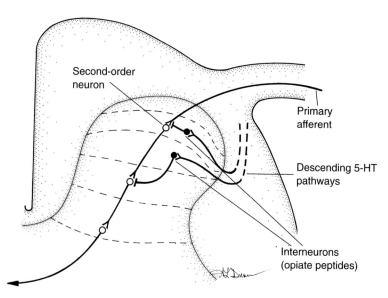

Second-order neuron

Primary afferent

Descending 5-HT pathways

Interneurons (opiate peptides)

Figure 12.8. Neurotransmitter systems that modulate transmission along nociceptive pathways. The projections from the raphe nucleus contain the neurotransmitter serotonin (5-HT); the descending axons from the reticular formation contain norepinephrine. Transmission along nociceptive pathways is modulated by interneurons that use *opioid peptides* as their neurotransmitter. Especially important are the enkephalin-containing interneurons in the dorsal horn. When activated, these interneurons release enkephalin near the synapses between the primary afferent and the second-order neuron. It is thought that enkephalin acts both presynaptically and postsynaptically by inhibiting release from primary afferent terminals and by exerting an inhibitory action on the second-order neurons of the nociceptive pathway.

the emotion subsides. Most people can recall such events from their own experience.

Analgesia (decreased pain sensitivity) resulting from situations that generate strong emotions is termed *stress-induced analgesia.* The mechanisms that contribute to diminished pain perception are thought to play an important role in allowing an adaptive response to emergency situations. It is believed that the reduction of pain perception is mediated by the descending pathways in which endogenous opiate-like peptides play an important role.

The descending pathways may also play a role in the motivational–affective component of pain perception. The degree of discomfort and distress that a patient feels as a result of nociceptive input depends to a great extent on the emotional state of the patient. For example, there are enormous individual differences in the response to surgical trauma that surgeons must be prepared to address.

The mechanisms through which motivation and emotion affect pain perception are poorly understood. What is known is that the descending modulatory pathways receive input from areas that are part of the limbic system that are thought to play an important role in regulating motivation and emotion. Thus, motivational–affective variables may modulate pain perception in part via the descending modulatory pathways.

Positive Symptoms Can Result From Damage to Pain Pathways

The concept of a positive symptom has already been discussed with respect to the consequences of peripheral nerve injury. Often, destruction of peripheral nociceptors has the paradoxical effect of producing a *chronic pain syndrome.* The same is true of injuries that disrupt central pain pathways. For example, small lesions in the thalamus can lead to intractable pain (termed the *thalamic pain syndrome*). The reason is likely due to the fact noted earlier—that damage in particular parts of the pain circuit leads to abnormal activity in surviving components of the circuit, and that activity is perceived as painful.

The fact that damage to central pain pathways sometimes has the paradoxical effect of producing chronic pain has important clinical implications. Because pain is carried by specific pathways, an obvious strategy to treat intractable pain is to surgically transect pain pathways. For example, intractable pain in cancer patients is sometimes treated by transecting the ascending spinothalamic tract in the spinal cord *(tractotomy)*. Although surgical treatments are used when all else fails, such surgeries generally yield disappointing results. The intractable pain that brought the patient to the physician may be temporarily relieved but is often replaced by unusual sensations that are perceived as extremely unpleasant. Again, the explanation is abnormal activity in surviving components of the pain circuit.

4

The Motor System

Motor systems are the efferent pathways through which the nervous system controls the movement of the body and the operation of internal systems (viscera). The systems that control movement of the body (soma) are called the *somatic motor pathways* (also called *general somatic efferents*). The systems that regulate the viscera are termed *viscero-motor pathways* (also called *general visceral efferents*) and represent the motor component of the *autonomic nervous system*. The next several chapters concentrate on the somatic motor pathways. Visceromotor pathways will be considered further in the chapter on the autonomic nervous system (Chapter 27).

The components of the somatic motor system include structures found at various levels of the neuraxis. In general, structures at higher levels control structures at lower levels. In this way, the motor system is said to be organized hierarchically.

The components of the somatic motor system include (1) the muscles and the lower motoneurons that directly innervate muscles; (2) various structures that contain upper motoneurons (found in areas of the brainstem, midbrain, and motor cortex) that control the discharge of lower motoneurons; and (3) ancillary structures that control impulse traffic along the motor pathways including the cerebellum and the basal ganglia (Figure IV.1).

In the following chapters, we begin by considering the muscles and lower motoneurons and then work our way up the hierarchy of the primary motor pathways from the cerebral cortex. Then we consider the ancillary systems (the cerebellum and basal ganglia) that control impulse traffic along primary motor pathways.

There are several key concepts for clinicians:

1. *Pathways from the cerebral cortex mediate voluntary motor control.* Thus, interruption of the descending motor pathways anywhere along their course from the cortex to the periphery leads to loss of voluntary motor control (paralysis) in the portion of the body that is disconnected from cortical input. Some definitions:

   Paraplegia is paralysis affecting only the lower limbs.

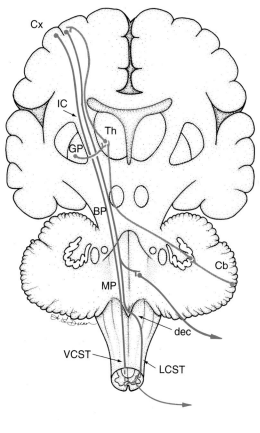

Figure IV.1. Overview of motor system organization. The basic motor circuit is formed by axons of the corticofugal pathways that originate in the cortex (Cx), descend in the internal capsule (IC), and travel through the midbrain in the basis pedunculi (BP) and then through the medulla in the medullary pyramid (MP). Most of the axons of the corticospinal tract decussate at the level of the spinomedullary junction in the pyramidal decussation (dec). The axons descend through the spinal cord on the side contralateral to their point of origin in the lateral corticospinal tract (LCST). A smaller component of fibers (about 10%) do not decussate and instead travel down the cord on the side ipsilateral to their point of origin in the ventral corticospinal tract (VCST). The output of the motor cortex is controlled by the thalamus (Th), which in turn is controlled by the basal ganglia, via the globus pallidus (GP), and by the contralateral cerebellum (Cb).

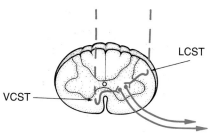

Quadriplegia is paralysis affecting all four limbs.

Hemiplegia is paralysis affecting only one side of the body.

Monoplegia is paralysis affecting only one limb.

Paresis is weakness resulting from partial loss of motor function. The same prefix modifiers (*para-, quadri-, hemi-,* and *mono-*) apply.

2. *Corticofugal pathways decussate.* In this way, one side of the cerebral cortex controls the opposite side of the body (see Figure IV.1). The axons of the corticofugal pathways descend in the internal capsule, travel through the midbrain and pons, and then collect in the medullary pyramid. Most (about 90%) decussate at the level of the spinomedullary junction in the pyramidal decussation and descend through the spinal cord on the side contralateral to their point of origin in the brain. The remaining 10% descend on the ipsilateral side and decussate near their point of termination in the spinal cord.

3. *Descending pathways originating in the brainstem mediate postural control and balance.* Injuries above the spinomedullary junction often damage corticofugal pathways and spare brainstem pathways. The result is a loss of voluntary control (paralysis) coupled with characteristic *fixed postures* that reflect the continued operation of the brainstem pathways that mediate postural control.

4. *Activity in the motor cortex is regulated by thalamocortical projections.* Thalamocortical projection neurons are in turn regulated via inputs from the basal ganglia and the cerebellum. Thus, damage to the basal ganglia or cerebellum results in a dysregulation of motor cortical output.

5. *The motor cortex receives input from other cortical areas including the supplemental motor and premotor cortices.* The supplemental and premotor regions are important for the planning and programming of complex movements and exert their effects by playing upon the primary motor cortex.

Obviously, the circuitry responsible for motor control is complex. Through an understanding of the precise routes taken by the various motor pathways, it is possible to quite precisely deduce the location of lesions affecting the motor system. In the following chapters, we begin by describing the organization of motor circuitry at each level of the neuraxis, beginning at the segmental level (the site of origin of the final common path). Then we consider the role of the basal ganglia and the cerebellum in regulating the principal motor circuit.

13

Motor System I

The Anatomical Organization of the Motor System

The somatic motor system includes the neurons in the spinal cord and cranial nerve nuclei within the brainstem that directly innervate striated muscles, the segmental circuitry that controls these motor neurons, and structures in the brainstem, midbrain, and cerebral cortex that provide descending input to the motor neurons. These components are organized in a hierarchy such that higher levels control lower levels. But there is also a tremendous amount of local control at the different levels that mediates sensory–motor integration.

The goal in this chapter is to provide an overview of the anatomical organization of the motor system. It is important to become familiar with the topographic organization of motoneurons within the spinal cord, with the names and locations of the cranial nerve nuclei with somatomotor functions, and with the areas of the brain that give rise to descending motor pathways, and to identify these regions in histological sections. Students should also be familiar with the names, locations, and topographic organization of the principal descending motor pathways from the cortex and brainstem. This information provides the key to localizing lesions that produce particular motor symptoms.

We begin by considering the neurons that directly innervate striated muscles. Then, we work our way up the neuraxis, considering the organization of motor pathways at each level.

MOTONEURONS AND THE MOTOR UNIT

Muscles are innervated by *motor neurons,* also termed *motoneurons,* that have their cell bodies in the ventral horn of the spinal cord and in cranial nerve nuclei in the brainstem. The motoneurons send their axons to muscles via the ventral roots and cranial nerves. The output from the motoneurons to the muscles is called the *final common path* because all descending pathways operate through the motoneurons.

There are two types of motoneurons that are intermixed in the ventral horn: *alpha motoneurons,* which are directly responsible for muscle contraction, and *gamma motoneurons,* which control the *intrafusal muscle fibers* that are an intregral component of the specialized sensory apparatus termed the *muscle spindle.* The role of gamma motor neurons in motor control will be discussed further in Chapter 14.

An important point for the physician is the concept of upper versus lower motoneurons. *Lower motoneurons* include all neurons that directly innervate muscles. They are found in both the spinal cord and brainstem (cranial nerve nuclei). *Upper motoneurons* provide the descending inputs to the lower motoneurons, for example, the neurons in the motor cortex and brainstem that give rise to corticofugal and brainstem pathways, respectively. The distinction between upper and lower motoneurons is very important for the physician because the symptoms resulting from damage to the two are quite different, as described later.

The Basic Functional Unit of the Final Common Path Is the Motor Unit

Each individual muscle fiber receives input from one and only one motor axon. However, the converse is not true; instead, each motoneuron projects to a group of muscle fibers. The basic functional unit of the final common path is the *motor unit*, defined as *an individual motoneuron and all the muscle fibers that it contacts*. The collection of motoneurons that projects to a particular muscle or muscle group is termed the *motor neuron pool* for that muscle or muscle group. As described later, these two concepts are important for understanding diseases that affect muscles *(myogenic disease)* or the motoneurons or their axons *(neurogenic disease)*.

There Is an Important Topographic Organization of the Motoneurons in the Ventral Horn

The motoneurons within a spinal segment are topographically organized according to the muscle groups that they innervate. There are two axes of organization. One axis is based on proximal versus distal muscles; the other axis is based on flexors versus extensors. These are summarized by two anatomi-cal/functional rules: the *proximodistal rule* and the *flexor–extensor rule*.

The proximodistal rule is that motoneurons innervating proximal musculature are located medially in the ventral horn; motoneurons innervating distal musculature are located laterally. Motoneurons innervating axial musculature are found most medially, and extend all along the extent of the spinal cord. Motoneurons innervating limbs are collected in enlargements (the *cervical* and *lumbar enlargements*). In the enlargements, the motoneurons innervating the limbs tend to lie lateral to the motoneurons innervating the axial musculature of the particular spinal segment.

The medial motoneurons serving proximal musculature are involved primarily in maintaining equilibrium and posture. Lateral motoneurons serving distal musculature are involved in fine motor movements. This functional difference is matched by differences in the patterns of descending input to lateral and medial motoneurons (see later).

The flexor–extensor rule is that motoneurons innervating extensor muscles lie ventral (anterior) to those innervating flexors.

A useful mnemonic for recalling the proximodistal rule and the flexor–extensor rule together is to picture the torso of a body builder flexing his upper arm, superimposed on a spinal cord section (Figure 13.1). The forearm (distal limb) lies laterally, the upper arm lies medially, the bicep (the principal flexor) lies dorsally, and the triceps (the principal extensor) lies ventrally.

Visceromotor Neurons in the Spinal Cord Have Properties That Are Similar to Somatomotor Neurons

Related to the somatic motor neurons are the *visceromotor neurons* that give rise to *preganglionic autonomic fibers*. The preganglionic autonomic motoneurons of the sympathetic nervous system are found in a distinctive *intermediolateral column* that is present from spinal levels T1 through L2. Preganglionic

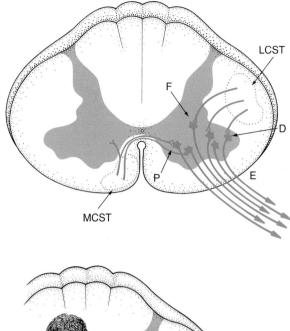

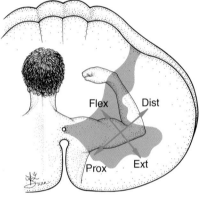

Figure 13.1. The proximodistal and flexor–extensor rules define the topographic organization of motoneurons within the ventral horn. P indicates motoneurons that innervate proximal musculature; F, motoneurons that innervate flexors; D, motoneurons that innervate distal musculature; E, motoneurons that innervate extensors; LCST, lateral corticospinal tract; MCST, medial corticospinal tract.

motoneurons of the parasympathetic nervous system are found in *nucleus X* in the sacral spinal cord and in the nuclei supplying certain cranial nerves.

Motoneurons Give Rise to the Motor Component of Cranial Nerves

The motoneurons supplying the head and face are found in a series of discrete nuclei within the brainstem and midbrain (Figure 13.2). The cranial nerve nuclei that have somatomotor functions (i.e., innervate striated muscle) include the following:

1. *Nucleus ambiguus* is found at the level of the olive in the caudal medulla. Neurons within the nucleus give rise to axons that enter the glossopharyngeal nerve (IX), the vagus nerve (X), and the cranial root of the accessory nerve (XI), which innervate the striated muscles of the pharynx and larynx. The other motoneurons that contribute to the accessory nerve are in the ventral horn of the cervical spinal cord.

2. *The hypoglossal nucleus* lies adjacent to the midline just below the fourth ventricle at the level of the olive. The hypoglossal nucleus innervates the muscles of the tongue via the hypoglossal nerve (XII).

3. *The facial nucleus* is located at the midpontine level in a ventral and lateral location. Motoneurons of the facial nucleus control what are called collectively the *muscles of facial expression* (lips, cheeks, and forehead). Their axons travel via the facial nerve (VII).

4. *The abducens nucleus* is also located at the midpontine level. The nucleus lies adjacent

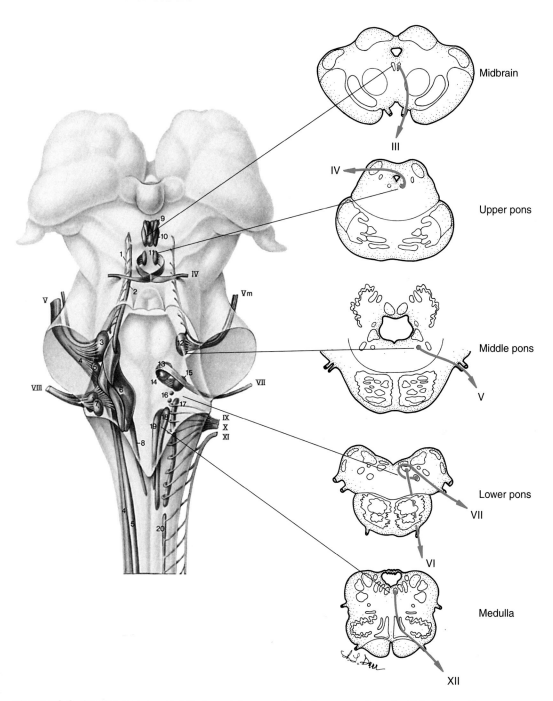

Figure 13.2. Cranial nerve nuclei that contain motoneurons supplying voluntary musculature.
Motor nuclei are represented on the right-hand side of the drawing (including both the motor nuclei innervating voluntary musculature and parasympathetic motor nuclei). Drawings on the right illustrate representative sections through the brainstem and midbrain. III indicates the oculomotor nerve; IV, trochlear nerve; V, trigeminal motor nerve; VI, abducens nerve; VII facial nerve; XII, hypoglossal nerve. The drawing on the left is from Nieuwenhuys R, Voogd J, van Huijzen C. *The Human Nervous System.* Berlin: Springer-Verlag; 1985.

to the midline just beneath the fourth ventricle. Axons from the facial nucleus enter the abducens nerve.

5. *The motor nucleus of the trigeminal nerve* is located within the lateral pons. The motoneurons in the trigeminal nucleus control the muscles of mastication. Their axons travel via the trigeminal nerve (V).

6. *The trochlear nucleus* is located in the mesencephalon just beneath the boundary between the inferior and superior colliculus. The motoneurons control extrinsic eye muscles. Their axons travel via the trochlear nerve.

7. *The oculomotor nucleus* is also located in the mesencephalon, but beneath the superior colliculus. There are two divisions of the oculomotor nucleus: a somatomotor division that supplies the extraocular muscles (voluntary eye movement), and a visceromotor division (the Edinger–Westphal nucleus) that is part of the parasympathetic nervous system. Motor axons from the oculomotor nucleus project to the eye via the oculomotor nerve (III).

Inputs to Lower Motoneurons

Spinal motoneurons receive direct input from the following sources:

1. Descending pathways (However, many of the descending inputs operate through interneurons.)

2. Sensory afferents

3. Interneurons within the spinal segment (*intrasegmental connections*)

4. Interneurons in nearby segments (*propriospinal connections*).

PATHWAYS AND TRACTS

Local Segmental Circuitry

The sensory afferent input to motoneurons and the interneurons within the spinal cord form the basis for the segmental control over motor output—the subject of Chapter 14. The anatomical substrates for this control are the intrasegmental connections between interneurons and motoneurons within a particular segment, and the intrasegmental connections (also called propriospinal con-

nections) between interneurons and motoneurons in nearby segments.

The *propriospinal pathways* are topographically organized so as to interconnect motor nuclei that control similar muscle groups (axial muscles versus distal muscles). The medially located propriospinal pathways that interconnect motoneurons supplying axial muscles project across a number of segments. Lateral propriospinal pathways have a more limited longitudinal distribution, being more closely limited to the segments associated with the limbs.

Descending Pathways

Descending pathways terminate directly on motoneurons and also on interneurons. These descending pathways originate from the cerebral cortex *(corticofugal pathways)* and the brainstem *(brainstem pathways)* (see Table 13.1). There are medial and lateral components of each that control motoneurons in the medial versus lateral ventral horn (and so control axial versus distal musculature, respectively).

Corticofugal Pathways

Following are the most important components of the corticofugal pathways:

1. The *corticospinal tract* (CST), which projects to motoneurons in the spinal cord

2. The *corticobulbar tract*, which projects to motoneurons in the medulla, pons, and midbrain (in older terminology, collectively termed the *bulb*)

There are also important cortical projections to nuclei in the midbrain and brainstem that in turn give rise to descending pathways:

1. The *corticorubral tract*, which projects to the red nucleus, which in turn gives rise to the rubrospinal tract

2. *Corticoreticular fibers*, which project to the portions of the reticular formation that give rise to *medial brainstem* pathways

Table 13.1. Summary of descending motor pathways of the spinal cord

Pathway	Type	Location	Controls	Function
LCST	Lateral	Lateral column	Distal muscles; flexors	Fine motor control
MCST	Medial	Ventral column	Proximal muscles; extensors	Trunk muscles, body orientation
RuST	Lateral	Lateral column	Distal muscles; flexors	Fine motor control
TST	Medial	Ventral column (cervical cord only)	Head movements	Coordinates head movement with eye movement
MRST	Medial	Ventral column	Proximal muscles; extensors	Postural control; facilitates motoneurons; controls extensors
LRST	Medial	Ventral column	Proximal muscles; extensors	Postural control; inhibits motoneurons; controls extensors
VST	Medial	Ventral column	Proximal muscles; extensors	Postural control/balance; facilitates motoneurons; controls extensors

In this way, the cerebral cortex controls motor output via direct and polysynaptic pathways.

The CST originates from the large pyramidal neurons *(Betz cells)* of the *primary motor cortex* (Brodmann's area 4) and also from pyramidal neurons in Brodmann's areas 1, 2, and 3 of the somatosensory cortex. The fibers of the CST travel through the *posterior limb* of the *internal capsule* and then into the *basis pedunculi* in the midbrain. Upon entering the pons, the tract breaks up into small fascicles. When the axons emerge from the pons, they come together to form the *medullary pyramid* on the basal surface of the brainstem. Because the fibers of the CST travel through the medullary pyramid, the pathway is also termed the *pyramidal tract.* This name has nothing to do with the fact that the fibers of the CST originate from pyramidal cells.

The corticospinal system separates into lateral and medial components termed the *lateral* and *medial corticospinal tracts,* respectively, at the level of the pyramidal decussation (Figures 13.3 and 13.4). The fibers of the lateral CST cross the midline at the *pyramidal decussation* at the spinomedullary junction and continue into the cord in the *lateral column.* They leave the lateral column at the level of termination and innervate motoneurons and interneurons in the lateral portion of the ventral horn in the area that contains motoneurons that supply distal muscles and flexors (Figure 13.3).

The medial, or ventral, CST is formed by a small component of fibers that do not decussate and instead continue in the ipsilateral ventral column. The axons of the ventral CST terminate on motoneurons and interneurons in the medial and ventral portion of the ventral horn in the area that contains motoneurons that supply proximal muscles and extensors (Figures 13.3 and 13.4).

It is important for physicians to know the location of the descending CST at different levels of the neuraxis to understand the symptoms that result from lesions in different locations. This is summarized in Figure 13.4.

Direct projections from the cortex to spinal motoneurons are especially important in humans and other animals that exercise

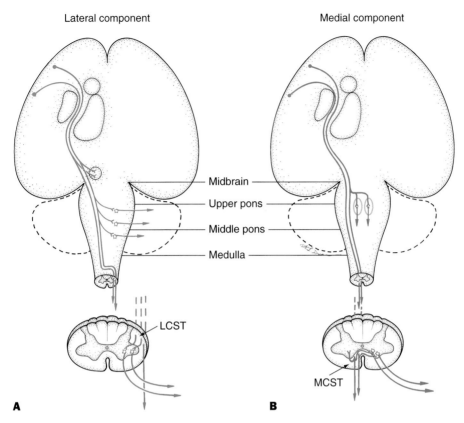

Figure 13.3. Corticofugal pathways. A illustrates the lateral corticofugal pathway (the lateral corticospinal tract, LCST). **B** illustrates the medial corticofugal pathways (the medial or ventral CST).

fine motor control of their distal extremities. These direct connections between cortex and spinal motoneurons allow for a *fractionation of movement* of distal musculature (fine control of movement, especially of the digits).

Corticobulbar Projections Control Muscles of the Head and Face

The corticobulbar tract can be considered as being functionally identical to the CST. These pathways control the lower motor neurons in cranial nerve nuclei. Axons of the *corticobulbar pathway* travel with the axons of the CST into the brainstem. These fibers leave the tract near the level at which they terminate and cross the midline to innervate the lower motoneurons in the cranial nerve nuclei in the brainstem (Figure 13.3).

As in the case of the projections to the spinal cord, corticobulbar projections to the cranial nerve nuclei are crossed (such that the cerebral cortex on one side controls muscles on the contralateral side). The projections to the hypoglossal nucleus appear to be exclusively crossed. The other cranial nerve nuclei receive predominantly crossed projections, but the trigeminal nucleus, nucleus ambiguus, and the facial nucleus also receive input from the ipsilateral cerebral cortex. This fact is especially important in the case of the facial nucleus. The motoneurons that supply the musculature of the lower face receive input predominantly from the contralateral cortex, whereas motoneurons supplying the upper face receive bilateral cortical input.

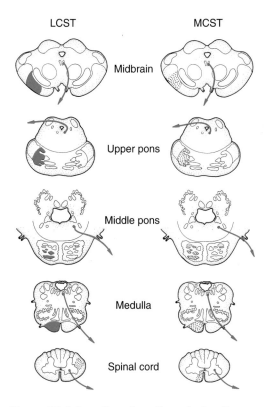

Figure 13.4. Location of corticospinal pathways at different levels of the neuraxis.

There is controversy about whether there are direct cortical projections to the motoneurons on the ipsilateral side or whether the input is relayed from the contralateral side via an interneuron. The important clinical fact is that unilateral damage to the motor cortex or to its descending projections produces a contralateral paralysis of the tongue and the muscles of the lower face but spare motor function in the upper face, the pharynx, and the muscles of mastication that are supplied by the trigeminal nerve.

The other cranial nerve nuclei that contain motoneurons are the nuclei that innervate extrinsic eye muscles (the oculomotor, trochlear, and abducens nuclei). Eye movement is a special motor function that depends critically on visual as well as vestibular input. It has both voluntary and reflex components, the latter controlling conjugate gaze, which brings images in the two eyes into register. The voluntary component de-pends on input from a special part of the cortex termed the *frontal eye field*. Motoneurons controlling eye movement are not directly innervated by axons from the cerebral cortex. Instead, cortical control is mediated via interneurons that coordinate the movement of the two eyes. Because of the special nature of eye movement control, it will be considered as a separate topic (see Chapter 21).

The Corticospinal Tract Is Somatotopically Organized

All levels of the motor pathways have a somatotopic organization but the somatotopic organization of the CST is especially important to physicians. The somatotopic organization of motoneurons is summarized by the proximodistal and flexor–extensor rules. The somatotopic organization of the lateral CST in the spinal cord is leg lateral and arm medial. The same somatotopy is present in the CST, in the medulla and pons. In the basis pedunculi, the somatotopy of the axons controlling the arms and legs is the same; axons controlling the face lie medial to those controlling the arm. In the internal capsule, the organization is legs posterior, arms next, and face anterior.

The primary motor cortex lies in the precentral gyrus (Brodmann's area 4, see Figure 13.5). The somatotopic organization of the primary motor cortex mirrors that of the somatosensory cortex. Indeed, the somatomotor and somatosensory maps lie in register to one another across the central sulcus (Figure 13.5). Thus the so-called motor homunculus is organized so that motor representation of the head and face is in the ventral-most portion of the precentral gyrus; the digits are represented in the next most dorsal regions of the precentral gyrus, and the trunk and upper portion of the leg are represented in the dorsal-most portion of the gyrus.

The representation for the lower leg and foot is in the portion of Brodmann's area 4 that extends onto the medial surface of the cerebral hemisphere into the *paracentral lob-ule*. In this location, the regions of the cere-

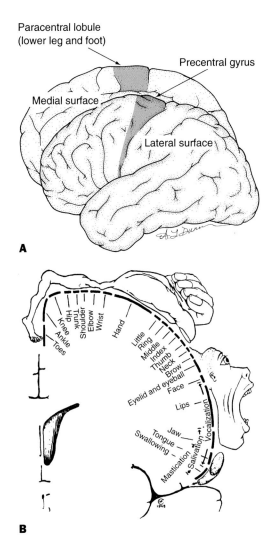

Paracentral lobule
(lower leg and foot)

Precentral gyrus

Medial surface

Lateral surface

A

B

Figure 13.5. Somatotopic organization of the primary motor cortex (the motor homunculus). A, Location of the primary motor cortex. The medial aspect of the hemisphere has been projected as a mirror image. **B,** Caricature illustrating the somatotopic organization of the primary motor cortex. Reproduced with permission from Heimer L. *The Human Brain and Spinal Cord.* New York, NY: Springer-Verlag; 1995.

bral cortex representing the two lower limbs are adjacent to one another across the midline. It is for this reason that injury to the medial walls of the cortex in the area of the paracentral lobule (for example, as a result of a meningeal tumor) can result in bilateral motor impairments involving the lower limbs.

Brainstem Pathways

The Medial Pathways

There are medial and lateral brainstem pathways. Medial brainstem pathways originate from the vestibular nuclei, the reticular formation, and the tectum (Figure 13.6) and include the *vestibulospinal tract* (VST), the *reticulospinal tract,* and the *tectospinal tract.* These pathways project to the spinal cord via the ventral columns and terminate on medially located interneurons and motoneurons that innervate axial musculature. These pathways are especially important for controlling musculature that maintains posture and balance.

The VST originates from neurons in the vestibular nuclei (also called *Deiters' nucleus*) and projects ipsilaterally in the ventral column of the spinal cord. Neurons in the vestibular nucleus receive input from the vestibular labyrinth, and the pathway is important for vestibular control of posture, balance, and equilibrium.

The VST is made up of lateral and medial components. The *lateral vestibulospinal tract* originates from the lateral vestibular nucleus. It terminates on interneurons in such a way as to facilitate extensor muscles and inhibit flexor muscles. The *medial vestibulospinal tract* originates from the medial vestibular nucleus. The initial course of the tract is along the *medial longitudinal fasciculus.* In the spinal cord, the tract continues in the ventral funiculus. The medial VST innervates interneurons in cervical levels and serves to control head movement to maintain balance and equilibrium.

The reticulospinal tract originates from several nuclei in the reticular formation and projects ipsilaterally in the ventral column. The neurons giving rise to the descending projections receive input from the motor cortex, thus forming a *corticoreticulospinal pathway.* This pathway is important for cortical regulation of spinal reflexes.

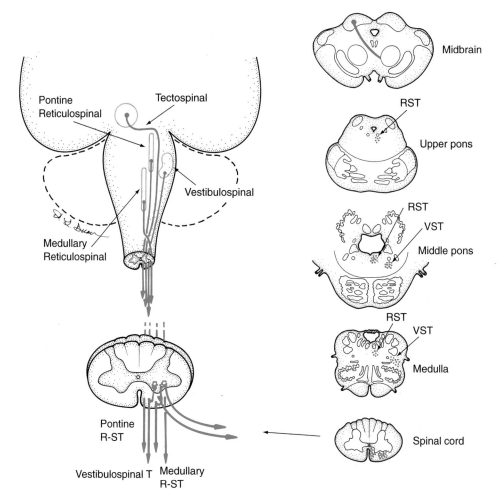

Figure 13.6. Medial brainstem pathways. Medial brainstem pathways include the reticulospinal tract (RST) and VST. The tectospinal tract is also considered a medial brainstem pathway. Its axons run in the medial longitudinal fasciculus medial to the reticulospinal system.

The tectospinal tract originates from neurons in the superior colliculus. The axons cross the midline in the midbrain and project to cervical segments of the spinal cord. The neurons giving rise to the descending projections receive input from the motor cortex, thus forming a *corticotectospinal pathway*. This pathway is important for coordinating head movements with eye movements during visual tracking.

The Lateral Pathway

The lateral brainstem pathway originates from the red nucleus and projects to the spinal cord via the lateral column (Figure 13.7). This pathway is termed the *rubrospinal tract.* The fibers of the rubrospinal tract decussate below the nucleus, travel through the ventrolateral tegmentum, and descend in the dorsal portion of the lateral column, leaving the lateral column at the level of termination to innervate interneurons and motoneurons in the lateral portions of the ventral horn. The cells of origin of the rubrospinal pathway receive input from the motor cortex and also the cerebellum.

Clinical Aspects

It is of considerable clinical relevance that the medial and lateral brainstem pathways are well separated from the CST throughout

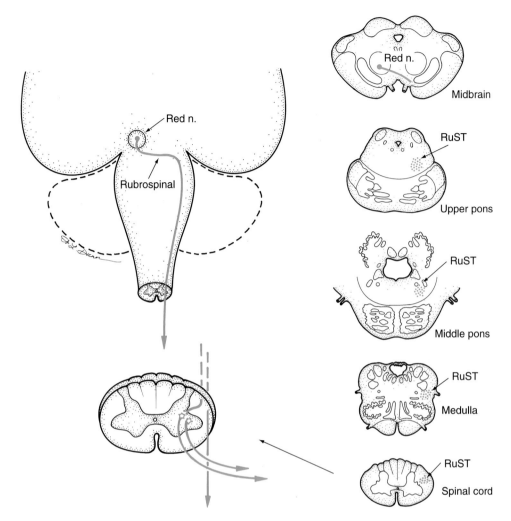

Figure 13.7. The lateral brainstem pathway. The sole lateral brainstem pathway is the rubrospinal tract (RuST).

their course in the medulla. As a result, vascular or other injuries can damage the CST and spare the brainstem pathways, resulting in *rigidity syndromes* (see Chapter 15).

Proprioceptive Pathways Subserve Sensory–Motor Integration at Different Levels of the Neuraxis

The operation of the somatic motor system depends critically on information from the somatosensory system. Indeed, any motor act is a coordinated undertaking that requires sensory–motor integration. Especially impor-

tant are the proprioceptive inputs that provide information on body position and movement (muscle, tendon, and joint receptors). Thus, an understanding of motor system function requires an understanding of the interconnections between sensory and motor circuits at several levels of the neuraxis.

Sensory–motor integration is mediated by sensory afferents carrying proprioceptive information. At the segmental level, there are direct connections between proprioceptive afferents and motoneurons. These direct segmental connections mediate spinal reflexes (see Chapter 14). Primary afferents carrying proprioceptive information also give rise to important ascending projections to the cerebellum. These are the spinocerebellar pathways described in Chapter 10.

Proprioceptive information is integrated within the cerebellum, which then projects onto the motor thalamus to neurons that modulate the output of the motor cortex. These projections are discussed in more detail in Chapter 17.

Injury to the Motor Pathways Produces Different Symptoms Depending on the Level of the Injury

Lower motor neurons can be injured as a result of direct trauma to the brainstem or spinal cord, as a result of an interruption of the blood supply, or as a result of disease processes (for example, amyotrophic lateral sclerosis). Upper motor neurons are most often damaged as a result of stroke, for example, because of the interruption of the blood supply to the motor cortex (via the middle cerebral artery), a loss of blood supply to the posterior limb of the internal capsule, or a loss of blood supply to the medullary pyramid. Understanding of the signs and symptoms of the two types of injury requires an understanding of how the different components of the motor system operate physiologically, which is the topic of subsequent chapters.

14

Motor System II

Segmental Control of Movement: The Motor Unit and Segmental Reflexes

The complex motor activities of higher organisms are mediated through the activation of muscles by their respective motoneurons (the *final common path*). Motor responses are categorized according to their complexity, and the degree to which they are under conscious control.

1. *Reflex responses* are the simplest. They are rapid, stereotyped, and involuntary. Examples include the stretch reflex of muscles, the withdrawal reflex to escape a painful stimulus, and the eye blink reflex that protects the eye from damage. Many reflexes can be mediated entirely at the segmental level. Reflexes typically involve only a few synapses. Indeed, the simplest reflex (the stretch reflex) is monosynaptic and requires only the local circuitry that is present at the spinal segment that contains the motor neuron.

2. *Rhythmic motor patterns* are voluntary repetitive movements in which part of the pattern of behavior is handled unconsciously. Examples include walking and chewing. These behaviors are mediated by rhythmic pattern generators in the spinal cord and brainstem, in which the patterned activity is generated by the circuit properties and the overall level of activity of the circuit is controlled by descending influences.

3. *Voluntary movements* are any motor activity that is under conscious control. These range from simple acts, for example, deliberately contracting a particular muscle, to highly complex undertakings that require coordination of systems throughout the body, such as playing the piano, catching a ball, and so forth. Complex voluntary movements require not only integration between somatosensory and motor systems, but also feedback from other sensory systems, especially the vestibular system and visual system (or the auditory system for animals that use audition for navigation).

Thus motor control requires a complex integration of information at several levels of the neuraxis. In this chapter we begin by outlining the key functional properties of the final common path, and then describe how motor output is regulated at the segmental level through sensory–motor interactions.

THE MOTOR UNIT

The basic functional unit of the final common path is the *motor unit,* defined as an individual motoneuron and all the muscle fibers that it innervates. Motor units are distinguished on the basis of their size, that is, the number of individual muscle fibers innervated by a single motoneuron. Differences in size are expressed in terms of the *innervation ratio* (the number of muscle fibers innervated by an individual motoneuron).

The average size of the motor unit varies, depending on the system, from about 10 in the case of extraocular muscles (innervation ratio = 1:10) to about 2000 (innervation ratio = 1:2000) in the case of the largest muscles of the leg. In general, higher innervation ratios are found in larger muscles. The smaller the motor unit, the finer the degree of control. Larger motor units provide greater power, but less control.

The size of the motoneuron cell body is related to the size of its associated motor unit: small motor units are served by motoneurons with small cell bodies; large motor units are served by motoneurons with large cell bodies. The diameter of the axon is closely related to the size of the *axon arbor* (number of collaterals). Because conduction velocity is related to axon diameter, axonal conduction velocity tends to be faster in large motor units.

There Are Different Types of Motor Units in Which the Motoneurons and Muscle Fibers Have Different Physiological Characteristics

Muscles are differentiated according to type as *fast twitch* or *slow twitch,* based on their contraction time, and according to whether they are *fatigable* or *fatigue resistant,* based on

the speed at which they fatigue. Thus three classes of muscle fibers are recognized: fast fatigable, fast fatigue-resistant, and slow-twitch.

1. Fast-fatigable fibers tend to be of larger diameter and are part of larger motor units. They exhibit brief (phasic) rapid contractions. Fast-fatigable fibers are used for intermittent activity requiring large force (explosive movements).

2. Fast fatigue-resistant fibers contract almost as rapidly as the fast fatigable fibers but are almost as resistant to fatigue as slow fibers.

3. Slow-twitch fibers tend to be of smaller diameter, are part of smaller motor units, and exhibit sustained (tonic) slow contractions that are highly resistant to fatigue. Slow-twitch fibers are used for sustained activities (e.g., maintenance of posture).

There are important biochemical differences between the different muscle types that account for the different physiological properties:

1. Fast-fatigable muscles are poor in myoglobin (the oxygen-binding pigment that results in red coloration), rich in glycolytic enzymes for anaerobic metabolism, and poor in oxidative enzymes; they have high myofibrillar ATPase activity.

2. Fast fatigue-resistant muscles have properties of both fast and slow muscles. They are rich in oxidative enzymes but have high ATPase activity.

3. Slow-twitch muscles are rich in myoglobin (and are thus termed *red muscles*), rich in oxidative enzymes for aerobic metabolism, poor in glycolytic enzymes, and have low ATPase activity.

Differences in Muscle Properties Are Determined by the Pattern of Activity in Motor Axons

Motoneurons innervating fast-twitch muscles fire in brief (phasic) bursts, whereas motoneurons innervating slow-twitch muscles fire at sustained rates. The characteristic patterns of activity of the motor axons determine the muscle fiber type. This fact was demonstrated by cross-innervation experiments. For example, when a nerve that carries axons supplying fast-twitch fibers is transected and allowed to regenerate into a muscle composed of mostly slow-twitch fibers, the muscle fibers convert to slow-twitch fibers. The opposite conversion occurs when a nerve supplying slow-twitch fibers is allowed to regenerate into a muscle composed of fast-twitch fibers. Thus, with cross-innervation, the contractile properties of the muscle are converted to the type appropriate to the innervating nerve.

Fiber-Type Conversion Occurs in Diseases That Affect Motoneurons

Diseases that affect the motor unit are of two types: *myogenic* (diseases of the muscle) or *neurogenic* (diseases of the motoneuron or its axons). An example of a neurogenic disease is amyotrophic lateral sclerosis (also called Lou Gehrig's disease), in which there is a slow degeneration of lower motoneurons, leading to weakness and then paralysis. As motoneurons die, the muscle fibers they serve are denervated, and the affected muscle fibers atrophy. In the early stages of the disease, when some motoneurons have died and others still survive, there is a characteristic change in the distribution of fast and slow fibers in the muscle.

The explanation for this fact is that as some motoneurons die, surviving motoneurons sprout new terminal arbors to reinnervate muscle fibers that have lost their normal input. Because muscle fiber type is controlled by the nerve, the muscle that is reinnervated is converted to the type appropriate for that nerve (Figure 14.1). This results in a cluster of fibers of similar type defined on the basis of the pattern of termination of sprouting fibers, which is distinctly different from the mosaic pattern of fiber types in normal muscle. The pattern is called *muscle fiber–type grouping* and is diagnostic for neurogenic diseases of the motor unit.

THE PHYSIOLOGY OF MUSCLE ACTIVATION

When a motoneuron fires an action potential, all the muscle fibers that are contacted by its axons contract simultaneously. However, muscles are made up of collections of fibers that belong to different motor units. Thus, the different motor units can contract singly or collectively. Increases in the force of contraction of the total muscle mass are achieved by increasing the number of motor units that are simultaneously active. This process is termed *recruitment.*

Muscle fibers within an individual muscle are recruited in order, based on motor unit size. Small motoneurons have the lowest threshold, hence small motor units are recruited first. Large motoneurons are recruited as the overall level of descending activation increases. This is termed the *size principle.* Also, slow fibers, which are resistant to fatigue, are recruited before fast-fatigable fibers. Maximal force is generated when all the motor units represented in a given muscle are activated simultaneously.

The overall extent of contraction of a muscle fiber depends on the frequency of action potentials (APs) in the motor axon. Each AP in the motor axon produces an AP in the muscle that leads to a muscle twitch 10 to 100 ms in duration. If the frequency of APs is low, each one elicits a twitch. But if AP frequency increases, individual twitches will begin to summate (Figure 14.2). This is a form of temporal summation termed *rate modulation.* When the frequency of APs is sufficient to produce a sustained contraction of the muscle, the muscle fiber is said to be in *tetanus.* The stimulation required to produce such contraction is termed *tetanic stimulation.* Notice that this summation does not involve the excitatory postsynaptic potential; it comes about because of the time course of the contraction–relaxation cycle of the muscle.

Thus, the overall contraction of a muscle depends on two variables: (1) the number of individual motor units that are simultaneously active, and (2) the frequency of APs over motor axons.

Normal Innervation Pattern

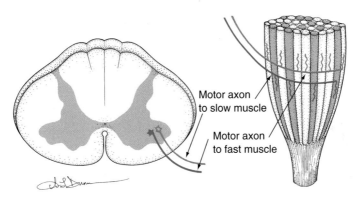

Motor axon
to slow muscle

Motor axon
to fast muscle

Innervation Pattern in Neurogenic Disease

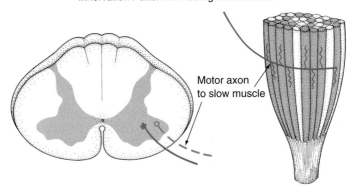

Motor axon
to slow muscle

Figure 14.1. Histochemical differences between fast- and slow-twitch muscles, and the diagnosis of motoneuron disease by means of fiber typing. Fast and slow fibers are distinguished by their content of myofibrillar ATPase, which can be revealed using histochemistry. In normal muscles, the fiber types are distributed in a random-appearing patchwork that is determined by the pattern of activity in the motoneuron. In muscles in which there has been a partial loss of motoneuron innervation, the different muscle types are found in groups. This is termed muscle-type grouping. The reason for the type grouping is that when some motoneurons die, surviving axons of a particular type sprout from one muscle fiber to reinnervate nearby fibers that have lost their normal input. The reinnervated fibers are then converted to the type determined by the reinnervating nerve. For example, the upper panel of the schematic illustrates a muscle that is innervated by four nerves, two fast and two slow. In the lower panel, one fast and one slow motoneuron have died (dashed lines) and the remaining motoneurons have sprouted to reinnervate the muscle fibers that have lost their normal input. The resulting fiber-type grouping is illustrated in the bottom panel.

The electrical activity generated by a single motor unit is termed the *motor unit response.* The motor unit response can be recorded using needle electrodes positioned in the muscle. This procedure is called *electromyography.* An electromyogram (EMG) can also be recorded using surface electrodes on the skin, but in this case, the EMG represents the summed activity of many motor unit responses. Electromyography provides an important diagnostic tool for neurologists to assess motor unit dysfunction.

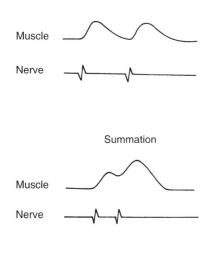

Summation

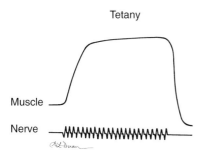

Tetany

Figure 14.2. Rate modulation of muscle fiber contraction. An action potential in a motor axon induces a single contraction (twitch) of a muscle fiber. Single action potentials that occur at low frequency induce single twitches (upper trace). Because of the long duration of the twitch response, a second action potential coming soon after the first will induce another twitch that summates with the first (second trace). The summation increases the overall extent of the contraction (that is, the total amount of force generated). High-frequency stimulation produces a sustained contraction (tetany) of the muscle (third trace). Adapted after Kandel ER, Schwartz JH, Jessel TM. *Principles of Neural Science.* New York, NY: Elsevier; 1991.

PROPRIOCEPTORS AND PROPRIOCEPTIVE PATHWAYS

Motor control depends critically on proprioceptive input. This is intuitively obvious. Close your eyes and bring your finger to your nose. Proprioceptive input provides in-formation about the starting position of your arm and information about its position throughout its trajectory to your nose. Without proprioceptive input, you would either miss your nose entirely or perhaps break it because you did not decelerate in time. It is for this reason that patients who have local nerve block for surgical procedures involving the hand are told not to try to scratch their nose!

Proprioceptors provide several key pieces of information that are required for initiating a movement, including (1) information about the starting position of the limb (joint receptors), (2) information about the load on the muscle (stretch receptors), and (3) feedback about the progress of the motor act.

There are several types of proprioceptive inputs that derive from different specialized sensory receptors. These are summarized in Table 14.1. In the following sections we briefly consider the sensory receptors themselves and the organization of proprioceptive pathways within the spinal cord, and then outline the role of the different receptors in mediating segmental reflexes.

Proprioceptors

Muscle Spindles

Muscle spindles are sensory organs embedded within the muscle mass itself (Figure 14.3). The muscle spindle is made up of a connective tissue sheath surrounding 10 to 12 very small muscle fibers. Embedded within the sheath is a sensory nerve ending that functions as a stretch receptor. The sheath is larger in the center, with tapered ends (*fusiform* or *spindle-shaped*), giving rise to the two root terms used with reference to the spindle *fusi-* and *spindle*. The small muscles within the sheath are termed *intrafusal muscles*. The remainder of the fibers comprising the muscle are termed *extrafusal*.

The sensitivity of the stretch receptor within the sheath depends on the tension (tone) of the intrafusal muscle fibers. The

Table 14.1. Proprioceptors*

Group name	Diameter	Conduction velocity (m/s)	End organ	Function
Ia (A-alpha)	12–20	70–120	Primary muscle spindle	Proprioception
Ib (A-alpha)	12–20	70–120	Golgi tendon organ	Proprioception
II (A-beta and gamma)	5–12	30–70	Secondary muscle spindle	Proprioception
			Joint receptors	Proprioception

*From Lindsley DF, Holmes JE, *Basic Human Neurophysiology.* New York, NY: Elsevier; 1984:104.

tone of the intrafusal muscle fibers is regulated by input from *gamma motoneurons.* These motoneurons innervate the muscle fibers within muscle spindles and, for this reason, are termed *fusimotor.* In contrast, the alpha motoneurons that innervate extrafusal muscle fibers are termed *skeletomotor.* Intrafusal and extrafusal muscle fibers are organized in parallel within the muscle mass (Figure 14.3).

The sensory component of the muscle spindle is a stretch receptor. There are two types of stretch receptor that differ in their response dynanics. The stretch receptor of *primary muscle spindles* provides information about both dynamic and static phases of muscle stretch. As the stretch is applied, the rate of firing of the receptor is highest. The rate decreases during the sustained phase of the stretch. These receptors have response characteristics that are similar to other rapidly adapting receptors (see Chapter 11). In this way, primary spindles signal both the rate of change and the length of the muscle. In contrast, the firing of *secondary spindles* depends only on the amount of stretch applied (not the rate of application). As stretch is applied, the firing rate increases and remains high throughout the period of stretch application. Secondary spindles have response characteristics that are similar to other slowly adapting receptors. Thus, secondary spindles provide static information about the change in fiber length.

The sensory axons of primary and secondary muscle spindles enter the spinal cord via the dorsal roots as Ia and II (beta) afferents, respectively. The afferents form important intrasegmental connections with neurons in the gray matter of the spinal cord and also give rise to the ascending propriospinal pathways to the cerebellum (the spinocerebellar tract) described in Chapter 13. As described later, Ia afferents form important intrasegmental connections with both motoneurons and interneurons that mediate segmental reflexes.

Golgi Tendon Organs

Golgi tendon organs are localized at the junction between the muscle fibers and the tendon (Figure 14.3). This receptor is a stretch receptor that gives rise to an Ib afferent fiber. The Golgi tendon organ provides information that is different from that provided by the muscle spindle because it is in series rather than in parallel with the muscle. Hence, the tendon organ directly senses the amount of stretch (tension) in the muscle, which in turn is regulated by extrafusal muscle fibers.

Golgi tendon organs are innervated by Ib afferent fibers. The central projections of Ib afferents are similar to those of Ia afferents. One difference, however, is that Ib afferents do not form direct monosynaptic connections with alpha motoneurons. Instead, their

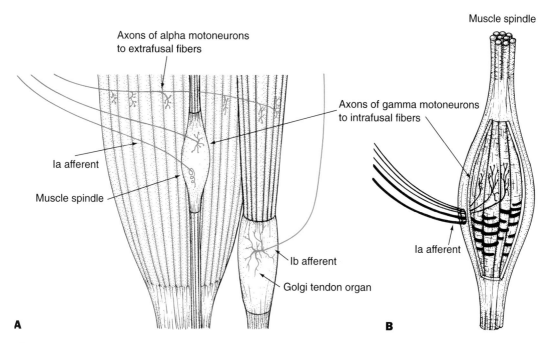

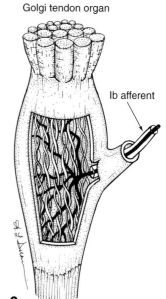

Figure 14.3. Muscle spindles and Golgi tendon organs.
Proprioceptive information about muscle and tendon stretch is provided by two types of receptors in muscles and tendons: muscle spindles and the Golgi tendon organ. Muscle spindles are stretch receptors that are embedded in a fusiform-shaped sheath that surrounds a group of specialized muscle fibers termed *intrafusal fibers.* The axon from the stretch receptor is a large-diameter (Ia) afferent. The intrafusal muscles are innervated by axons from gamma motoneurons. The Golgi tendon organ is a stretch receptor that is embedded in the tendon. The axon from the Golgi tendon organ is an Ib afferent. **A** is after Houk JC, Crago PE, Rymer WZ. Functional properties of the Golgi tendon organs. In: Desmedt JE, ed. *Spinal and Supraspinal Mechanisms of Voluntary Motor Control and Locomotion.* Vol 8. Basel, Switzerland: Karger; 1981:33–43. *Progress in Clinical Neurophysiology.* **B** is after Schmidt RF. Motor systems. In: Schmidt RF, Thews G, eds. Biederman-Thorson MA, trans. *Human Physiology.* Berlin, Germany: Springer-Verlag; 1983:81–110. **C** is after Hulliger M. The mammalian muscle spindle and its central control. *Rev Physiol Biochem Pharmacol.* 1984;101:1–110.

influence on motoneurons is mediated via interneurons.

Spinal Reflexes: Sensory–Motor Integration at the Segmental Level

Sensory input plays a key role in regulating *spinal reflexes,* also called *segmental reflexes.*

Reflexes are involuntary and stereotyped responses to sensory stimuli; thus all reflexes have afferent and efferent components. The efferent component is the output of the motoneurons to the muscle fibers. The afferent component is the input from the sensory receptors at the segmental level that comes into the spinal cord via dorsal root fibers.

Reflexes can be thought of as being hard-wired in that activation of particular sensory receptors leads to a reflex response in a particular muscle group, with the strength of the sensory stimulus determining the amplitude of the motor response. It is in this respect that reflexes are *stereotyped.*

There are three principal segmental reflexes: the *stretch reflex,* the *Golgi tendon organ reflex,* and the *flexion–withdrawal reflex.* These originate from different sensory receptors and have different intraspinal circuitry.

The Stretch Reflex

Ia afferents from primary spindles form direct excitatory synaptic connections with the alpha motoneurons innervating the same *(homonymous)* muscle and also innervate other muscles that act synergistically (that is, exert a similar action; Figure 14.4). Moreover, the Ia afferents innervate inhibitory interneurons that project to motoneurons that in turn supply *antagonist muscles.* Any stimulus that rapidly stretches the muscle activates the Ia afferent. The Ia afferents then activate the motoneurons supplying that muscle and other muscles that act synergistically and produce inhibition (via the interneuron) of motoneurons supplying antagonist muscles. This is termed the *stretch reflex.*

Physicians refer to the stretch reflex as the *deep tendon reflex* because it is elicited by

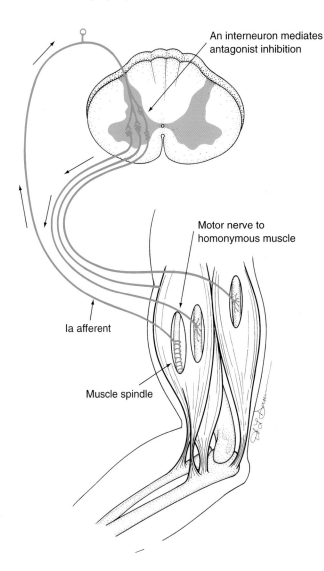

An interneuron mediates antagonist inhibition

Motor nerve to homonymous muscle

Ia afferent

Muscle spindle

Figure 14.4. The stretch reflex. When the muscle spindle receptor is activated by stretch, activation is conveyed centrally via Ia afferents. Ia afferents form excitatory monosynaptic connections with motoneurons innervating the same (homonymous) muscle and with motoneurons innervating synergistic muscles. Thus, the activity over the Ia afferent results in contraction of the muscle in which the stretch receptor was activated. Activation of Ia afferents also inhibits antagonist muscles via inhibitory interneurons (antagonist inhibition). Adapted from Kandel ER, Schwartz JH, Jessel TM. *Principles of Neural Science.* New York, NY: Elsevier; 1991.

tapping the muscle tendon. This leads to rapid muscle stretch, which activates the Ia afferent and initiates the reflex. Unfortunately, this term is confusing because the reflex has nothing to do with the receptor that actually is in the tendon—the *Golgi tendon organ* (see later). Other terms for the stretch reflex are the *myotactic reflex* and the *muscle spindle reflex*. The term *myotatic reflex* derives from the important role that the reflex plays in regulating normal muscle tone (see later).

The stretch reflex provides a measure of the integrity of both the segmental circuitry and descending input. Obviously, damage to either afferent or efferent arms of the reflex (sensory input or the motoneuron or its axon) disrupts the reflex. Partial damage of the circuitry that mediates the stretch reflex leads to *hyporeflexia;* complete destruction of either the afferent or efferent arms leads to *areflexia* at that segmental level.

An important fact for physicians is that the reflex becomes hyperactive when descending pathways are disrupted (hyperreflexia). The reason is that descending systems inhibit segmental reflexes via inhibitory interneurons. When the descending activation is removed, the reflexes are disinhibited.

The Stretch Reflex Plays an Important Role in Normal Muscle Function

Gamma motoneurons control intrafusal muscle fibers within the muscle spindle. Contraction of these muscles increases the sensitivity of the stretch receptor, leading to increased activation of the muscle spindle and the Ia afferent. Descending systems can control the level of activity of gamma motoneurons and thus the level of activity over the Ia afferent. This in turn controls the degree of tonic activation of alpha motoneurons by Ia afferents. In this way, regulation of the gamma system by descending pathways determines muscle tone.

The activation of alpha motoneurons via the circuit from gamma motoneurons to intrafusal muscles to Ia afferents to alpha mo-

toneurons is termed the *gamma loop* (Figure 14.5).

Muscle tone is assessed clinically by passively extending and flexing the limbs. Abnormal activity of the gamma system is manifested by increased "stiffness" (that is, resistance to movement). This condition is termed *hypertonus.* Hyperactivity of the stretch reflex also results in an abnormal response to rapid stretch—a prolonged oscillation of contraction and relaxation of the limb termed *clonus.* The term *spasticity* refers to the combination of hyperreflexia and hypertonus. And the term *spastic paralysis* refers to the loss of voluntary motor function in conjunction with hyperreflexia and hypertonus. All of these conditions are typically the result of disruption of descending influences on the segmental circuitry.

Alpha–Gamma Coactivation

During a movement, the alpha motoneurons innervating extrafusal muscles and the gamma motoneurons activating the intrafusal fibers are concurrently activated by descending pathways. This is termed *alpha–gamma coactivation* (Figure 14.6). If this did not occur, intrafusal fibers would not contract simultaneously with the extrafusal fibers, and so the muscle spindle would be unloaded as the muscle shortens. Because the sensitivity of the receptor depends on the tension on the intrafusal muscles, unloading would result in a loss of receptor sensitivity to stretch and decrease the level of ongoing activity in the Ia afferent. Instead, recordings from human subjects have revealed that Ia afferent activity actually increases during a purposive movement. Thus, alpha–gamma coactivation ensures that the muscle spindle maintains its sensitivity to stretch as the muscle contracts.

The Golgi Tendon Organ Reflex

Golgi tendon organs are considered to be in series with the muscle. When the muscle is stretched, the tension on the Golgi tendon

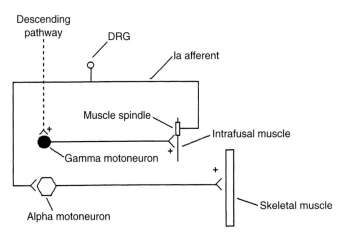

Figure 14.5. The gamma loop. (See text for description.)

organ increases, resulting in increases in the firing of Ib afferents. Ib afferents innervate interneurons in the spinal cord that inhibit the homonymous and synergistic muscles and excite the motoneurons of antagonistic muscles (Figure 14.7). These connections form the basis for the Golgi tendon organ reflex, also known as the *inverse myotatic reflex*, the *clasp-knife reflex*, or the *lengthening reaction*.

The Golgi tendon organ reflex is present in normal individuals but is difficult to demonstrate unless it is hyperactive as a result of disease or injury. For this reason, it is a useful diagnostic sign. For example, if the lower leg of a patient with an injury to descending systems is passively extended, there is considerable resistance to stretch (because of the hypertonus already described). If the pressure to extend the leg is increased, the Golgi tendon organ reflex is activated, leading to inhibition of the muscles resisting the passive extension, and activation of the antagonist muscles. The result is a sudden release of the resistance to extension, allowing the muscle to lengthen to a new position where resistance to further extension will be reestablished. Because of the sudden release of resistance, the limb behaves like the blade of a pocket knife that will snap into either an open or closed position as the blade is moved. The term *clasp-knife reflex* derives from the old name for a pocket knife (a clasp knife).

The Flexion–Withdrawal Reflex

The flexion-withdrawal reflex involves the withdrawal of a limb that experiences a painful stimulus. Afferent fibers from cutaneous pain receptors project to the spinal cord and synapse on interneurons that form polysynaptic connections with motoneurons. Activation of pain receptors leads to polysynaptic activation of the flexor motoneurons supplying the limb affected by the painful stimulation (Figure 14.8). This leads to flexion of the limb (withdrawal from the stimulus).

Strong noxious stimuli also lead to a polysynaptic activation of extensor motoneurons on the contralateral side. The result is a compensatory extension of the contralateral limb. This is termed the *crossed extensor reflex.* In the case of the upper limbs, activation of the crossed extensor reflex leads to a pushing response such that the body is pushed away from whatever produced the painful stimulus. In the case of the lower limb, activation of the crossed extensor reflex helps the organism to maintain balance. When one foot experiences a painful stimulus, that foot is raised; the contralateral limb is extended so as to support the added weight.

Unlike the stretch reflex, the flexion–withdrawal reflex can be suppressed by conscious effort; you can hold your hand in a fire if you choose to. Presumably the suppression of the reflex is possible because

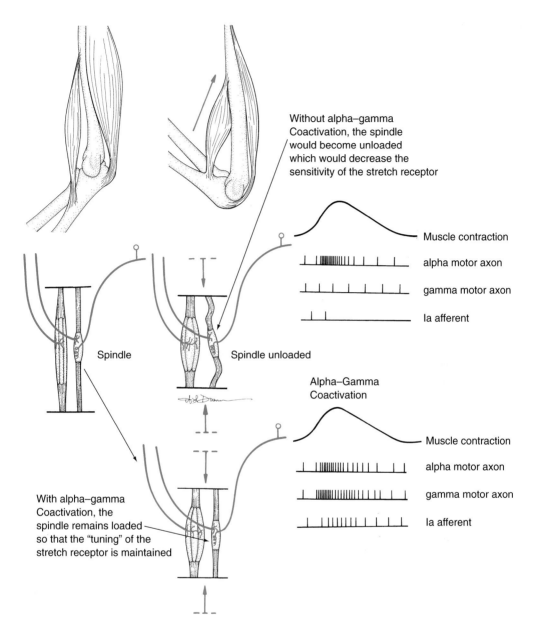

Without alpha–gamma Coactivation, the spindle would become unloaded which would decrease the sensitivity of the stretch receptor

Muscle contraction

alpha motor axon

gamma motor axon

Ia afferent

Spindle

Spindle unloaded

Alpha–Gamma Coactivation

Muscle contraction

alpha motor axon

gamma motor axon

Ia afferent

With alpha–gamma Coactivation, the spindle remains loaded so that the "tuning" of the stretch receptor is maintained

Figure 14.6. Alpha–gamma coactivation during a purposive movement. During a muscle contraction, alpha and gamma motoneurons are normally coactivated by descending pathways. If this did not occur, intrafusal fibers would not contract simultaneously with the extrafusal fibers, and so the muscle spindle would be unloaded as the muscle shortens (middle panel above). Because the sensitivity of the receptor depends on the tension on the intrafusal muscles, unloading would result in a loss of receptor sensitivity to stretch, resulting in a decrease in the level of ongoing activity in the Ia afferent. Thus, alpha–gamma coactivation ensures that the muscle spindle maintains its sensitivity to stretch as the muscle contracts. In fact, recordings from humans indicate that there is actually an increase in activity over the Ia afferent during the dynamic phase of the movement. Adapted from Kandel ER, Schwartz JH, Jessel TM. *Principles of Neural Science.* New York, NY: Elsevier; 1991.

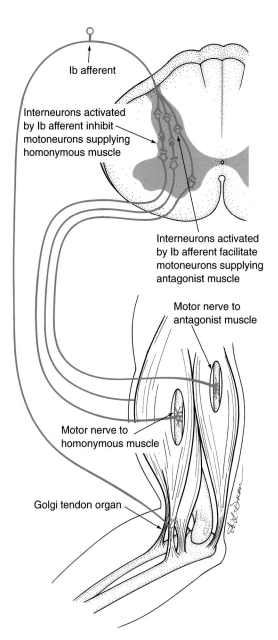

Ib afferent

Interneurons activated
by Ib afferent inhibit
motoneurons supplying
homonymous muscle

Interneurons activated
by Ib afferent facilitate
motoneurons supplying
antagonist muscle

Motor nerve to
antagonist muscle

Motor nerve to
homonymous muscle

Golgi tendon organ

Figure 14.7. The Golgi tendon organ reflex.
When tension is applied to the muscle, the stretch receptor that is part of the Golgi tendon organ is activated, leading to discharge of Ib afferents. Ib afferents innervate interneurons in the dorsal horn that in turn activate inhibitory interneurons that project to the motoneurons supplying the same (homonymous) muscle from which the Ib afferent originates. The Ib afferent also innervates excitatory interneurons that project to antagonist muscles. Thus, activation of Ib afferents results in inhibition of the motoneurons supplying the muscle that experiences the stretch, leading to relaxation of the muscle and simultaneous contraction of antagonist muscles. Adapted from Kandel ER, Schwartz JH, Jessel TM. *Principles of Neural Science.* New York, NY: Elsevier; 1991.

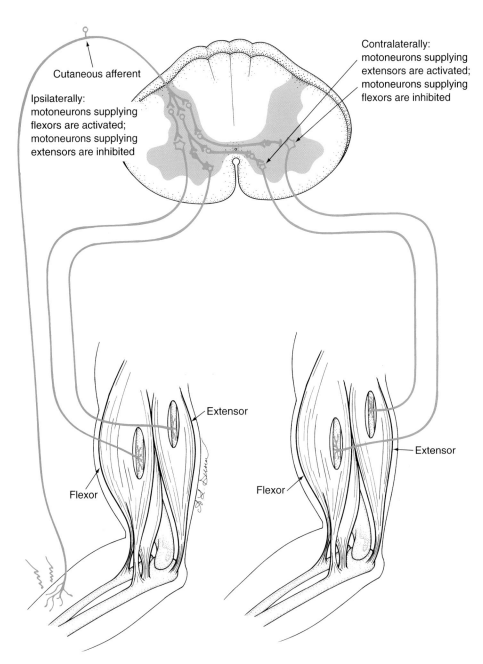

Figure 14.8. The flexion–withdrawal and crossed extensor reflexes. The flexion–withdrawal reflex allows rapid escape from a noxious stimulus. When a painful stimulus occurs, nociceptors are activated. The primary nociceptive afferent projects into the gray matter, where it terminates on interneurons in the dorsal horn. Through a chain of excitatory connections, motoneurons supplying flexor muscles are activated; motoneurons supplying extensors are inhibited on the side ipsilateral to the stimulation. The result is flexion of the limb. On the contralateral side, the polysynaptic pathways produce excitation of extensors and inhibition of flexors. Thus, the contralateral limb is extended. In the case of the arms, the extension of the contralateral limb serves the function of pushing the individual away from the source of the painful stimulus. In the case of the leg, extension on the side contralateral to the noxious stimulus helps to support the body as the leg that experiences the painful stimulus is withdrawn. Adapted from Kandel ER, Schwartz JH, Jessel TM. *Principles of Neural Science*. Elsevier; 1991.

Image labels:
- Cutaneous afferent
- Ipsilaterally: motoneurons supplying flexors are activated; motoneurons supplying extensors are inhibited
- Contralaterally: motoneurons supplying extensors are activated; motoneurons supplying flexors are inhibited
- Extensor
- Extensor
- Flexor
- Flexor

descending pathways control inhibitory interneurons that block transmission along the polysynaptic circuit.

When descending influences are interrupted, the circuitry that mediates the flexion–withdrawal reflex is still present in decentralized portions of the spinal cord. This circuitry can still mediate the flexion–withdrawal response to a painful stimulus. Thus, individuals with a spinal cord injury that interrupts both ascending and descending fibers do not consciously experience pain when a painful stimulus is applied to the limbs, but do exhibit the flexion–withdrawal reflex.

Differentiation of Spinal Reflexes

Spinal reflexes are differentiated according to the number of intervening neurons between afferent and efferent components. *Monosynaptic reflexes* involve a direct synaptic connection between sensory afferents and motoneurons (the stretch reflex). *Polysynaptic reflexes* involve one or more intervening interneurons (the Golgi tendon organ reflex, flexion–withdrawal, and crossed extensor reflexes).

Because of the difference in the number of intervening synapses, polysynaptic reflexes have a longer and more variable latency than monosynaptic reflexes. Polysynaptic reflexes also have a longer duration response that persists after the sensory stimulus that initiates the response has ceased. Finally, polysynaptic reflexes are more susceptible to descending control. For example, as already noted, the flexion–withdrawal reflex can be inhibited by conscious effort.

Interneurons Provide the Substrate for Coordination Between Muscle Groups

Although it is possible to voluntarily contract almost any muscle or set of muscles, voluntary movement involves highly coordinated cycles of facilitation and inhibition of muscle groups. This is intuitively obvious. If you flex your arm, extensor muscles in the arm must be inactive. When movements involve an entire limb (as in reaching and

walking), there must be coordination between muscle groups throughout the limb. Much of the coordination is mediated by interneurons at the segmental level.

One important substrate for coordination is the interneurons that mediate *reciprocal inhibition* in the circuitry responsible for the stretch reflex. As noted earlier, muscle stretch activates homonymous muscles and inhibits antagonistic muscles via inhibitory interneurons. These interneurons are termed *Ia inhibitory interneurons* because they are innervated by Ia afferents. The same circuit operates during activation of motoneurons by descending pathways. Descending projections that activate particular motor neurons also activate Ia inhibitory interneurons that project to antagonistic muscles (Figure 14.9). This is an example of *projected inhibition,* in which long-distance excitatory projections produce inhibition via an inhibitory interneuron. In this way, excitation of particular muscles by descending inputs simultaneously inhibits antagonistic muscles.

A similar group of interneurons mediates reciprocal inhibition in the Golgi tendon organ reflex, termed *Ib inhibitory interneurons.* These too are innervated by descending pathways and play an important role in voluntary motor control by inhibiting particular sets of motoneurons.

Another substrate for coordination is provided by inhibitory interneurons named *Renshaw cells* by Sir John Eccles, who won the Nobel Prize for his studies of synaptic physiology. Renshaw cells provide the substrate for *recurrent inhibition* of motoneurons (Figure 14.10). Recurrent collaterals from motoneurons synapse on Renshaw cells, which in turn project back upon the motoneurons, forming inhibitory synaptic connections.

Renshaw cells also innervate Ia inhibitory neurons, thus providing a substrate for disinhibition. Specifically, activation of the Renshaw cells inhibits the Ia inhibitory interneurons, which in turn disinhibits the motoneurons on which the Ia inhibitory interneurons terminate. In addition to the recurrent excitation from motoneurons, Renshaw cells also receive input from descending

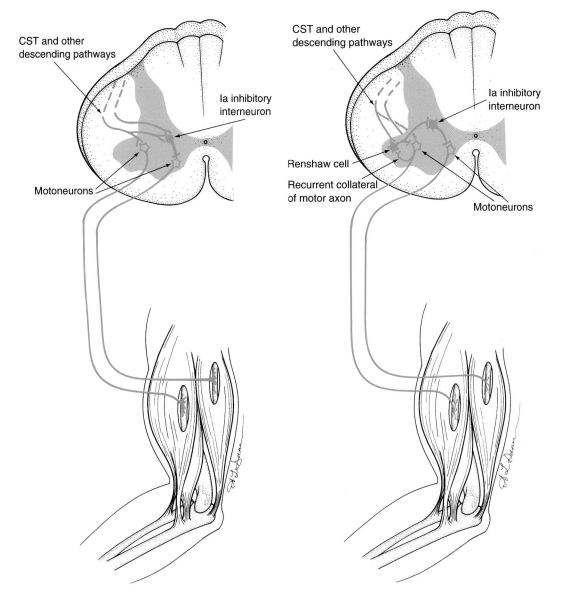

Figure 14.9. Inhibitory interneurons provide a substrate for projected inhibition by descending pathways. An important mechanism for coordination is provided by the population of inhibitory interneurons that mediate reciprocal inhibition (see, for example, the Ia inhibitory interneurons illustrated in Figure 14.5). Descending projections that activate particular motor neurons (flexors in this example) also activate Ia inhibitory interneurons that project to antagonistic muscles (extensors). This is an example of projected inhibition, in which long-distance excitatory projections produce inhibition via an inhibitory interneuron. Coordination is accomplished as a result of the simultaneous activation of one muscle group and the inhibition of antagonistic muscles by descending inputs.

Figure 14.10. Renshaw cells provide a substrate for projected inhibition and also disinhibition. Renshaw cells provide the substrate for recurrent inhibition of motoneurons. Recurrent collaterals from motoneurons synapse on Renshaw cells, which in turn project back upon the motoneurons, forming inhibitory synaptic connections. Renshaw cells also innervate Ia inhibitory neurons, thus providing a substrate for disinhibition of other motoneurons. Activation of the Renshaw cells inhibits the Ia inhibitory interneurons, which in turn disinhibits the motoneurons on which the Ia inhibitory interneurons terminate. In this way, descending pathways can produce both inhibition and disinhibition.

pathways. In this way, descending pathways can produce both inhibition and disinhibition.

Descending Inputs Provide a Tonic Excitation of Segmental Circuitry

Descending pathways are tonically active and so provide a tonic excitation of motoneurons as well as inhibitory interneurons. When descending inputs are damaged, this tonic excitation is eliminated. As a result, motoneurons lose their excitatory inputs but are also disinhibited because of the loss of excitatory input to inhibitory interneurons. This is part of the explanation for the fact that damage to descending pathways produces not just loss of motor function (paralysis) but also hyperreflexia and spasticity.

Segmental Circuitry Mediates Simple Motor Programs

Locomotion depends on rhythmic stepping movements. The stepping pattern involves alternating extension and flexion of one leg at the same time the other leg is performing the opposite movement. That is, while one leg extends, the opposite leg flexes. This is termed a *motor pattern*. Such motor patterns are mediated by somewhat more complex segmental circuitry.

Studies in experimental animals carried out primarily by Sir Charles Sherrington demonstrated that cats with complete spinal transections still exhibited alternating stepping movements when placed on a treadmill. This finding demonstrated the existence of circuits in the spinal cord that can operate autonomously. However, voluntary control of the segmental circuitry that mediates simple motor programs depends on input from supraspinal levels.

Spinal Cord Injury

Spinal cord injuries damage motoneurons directly in the injured segment (a lower motoneuron injury) and also disconnect segments from descending inputs (an upper motoneuron injury).

The signs and symptoms of spinal cord injury evolve over time after the injury.

Immediately after a traumatic injury that completely transects the spinal cord, motor symptoms include *paralysis* and *areflexia* or *hyporeflexia* in all segments at or below the level of the injury. This condition is termed *flaccid paralysis*. Autonomic reflexes are suppressed. There is also a loss of somatic sensation in segments below the level of the injury. This condition of flaccid paralysis and hyporeflexia is termed *spinal shock*. Spinal shock persists for one or more weeks after the injury.

Over the course of several weeks, there is a gradual return of segmental reflexes in segments below the level of the injury. Then the reflexes become hyperactive *(hyperreflexia)*. Segmental reflexes at the level of the injury never return because the circuitry responsible for the reflex is destroyed.

Over the course of several months, hyperreflexia continues to increase to the point that *spasms* are very large in amplitude and quite disturbing to the patient. Another symptom is muscle *clonus* (alternating flexion and contraction), which is triggered by rapid displacements of the limb. The collection of symptoms (paralysis with hyperreflexia) is termed *spastic paralysis*.

Also over the course of weeks to months, affected muscles undergo atrophy. Atrophy is especially prominent in the muscles supplied by the injured segments because those muscles have lost their motor innervation completely. Denervated muscles undergo a complex series of changes that lead to alterations in membrane properties, muscle atrophy, and changes in ion channels and neurotransmitter receptors in the muscle membrane. As a result of these changes, individual muscle fibers exhibit spontaneous twitching *(fibrillation)*.

Hyperactivity of segmental reflexes is another example of positive symptoms following injury to the nervous system. (Positive symptoms were first described in the section on the somatosensory system.) Hughlings

Jackson introduced the terms that are now in common use. *Negative signs* are those that result from a loss of function of particular neurons. In the case of the motor system, this would be muscle weakness or paralysis. *Positive signs* are those in which an abnormal response occurs because of the emergence of an abnormal pattern or level of activity in some group of neurons as a result of loss of their input. Thus, for example, segmental reflexes that depend on local circuitry within the spinal cord become hyperactive after the removal of descending input, which normally inhibits the reflex via inhibitory interneurons. The appearance of abnormal positive signs or symptoms is termed a *release phenomenon.*

One extremely important example of such a release phenomenon is the *inverted plantar reflex* (the *Babinski sign*). In normal individuals, when the sole of the foot is briskly stroked moving from the heel to the toes, the foot and toes flex. This is the *normal plantar reflex.* However, if there is injury to the lateral corticospinal tract supplying the foot, there is an abnormal dorsiflexion of the great toe, and the remaining toes fan out. The responses are opposite, and so the collection of responses is termed an *inverted plantar reflex* or, more commonly, a *Babinski sign.* A Babinski sign is considered to be one of the best indicators of injury to the corticospinal tract.

In summary, the long-term signs and symptoms reflecting the injury to lower motor neurons are the following:

1. Flaccid paralysis of muscles
2. Loss of segmental reflexes
3. Muscle atrophy
4. Muscle fibrillation (twitching)

The long-term signs and symptoms reflecting the upper motoneuron injury are the following:

1. Muscle paralysis without atrophy
2. Exaggerated (hyperactive) stretch reflexes
3. Abnormal reflexes, spasticity, and maintained flexion or extension
4. Clonus (alternating flexion and contraction) following rapid displacements of the limb

15

Motor System III

Supraspinal Control of Movement

The spinal reflexes described in the preceding chapter are the basic building blocks of motor behavior, but clearly these represent only a minor component of the overall motor repertoire of higher animals. Coordinated functions such as postural control, locomotion, and voluntary motor activities depend on higher levels of the nervous system.

It is useful to consider the three primary types of motor activities separately to understand the contribution of different components of the motor system.

1. *Postural control* and balance depend on circuits that regulate antigravity muscles; important components of this circuitry include the reticulospinal and vestibulospinal tracts; these in turn are regulated by descending inputs from the cerebral cortex and cerebellum.
2. *Locomotion* depends on spinal pattern generators that produce stepping or walking movements. Spinal pattern generators are local circuits that can operate autonomously to bring about coordinated movements of pairs of limbs. The activity of these circuits is regulated by descending inputs.
3. *Voluntary motor activities* are mediated by descending inputs from the cerebral cortex that travel via the corticospinal, or pyramidal tract.

An important principle to understand is that different descending pathways control different aspects of motor behavior. This is important for the physician because lesions often damage some components of the descending pathways and spare others. Of particular importance are the paralyses with fixed postures that result when corticospinal pathways are damaged but brainstem pathways survive. An understanding of the symptoms produced by injuries requires an understanding of which components are damaged and which survive.

POSTURAL CONTROL IS MEDIATED PRIMARILY BY THE RETICULOSPINAL AND VESTIBULOSPINAL TRACTS

The muscles that are important for postural control are primarily the muscles of the trunk and the extensor muscles of the limbs. This is intuitively obvious if you think about which muscles you use for routine maintenance of posture and balance. For example, imagine yourself standing on the deck of a boat or on a moving train. You are able to stand in such an unstable environment by

257

continuously adjusting your stance by extending one leg or the other, or through adjustments of the position of your trunk. Most of these adjustments are carried out without conscious effort or even awareness.

As we have seen in previous chapters, the descending pathways that innervate motoneurons supplying proximal (trunk) and extensor muscles are the *medial brainstem pathways:* the *reticulospinal, vestibulospinal,* and *tectospinal tracts.* In what follows, the organization and operation of these pathways is described in more detail.

Two Components of the Reticulospinal Pathway Exert Opposite Actions on Antigravity Muscles

The way that reticulospinal pathways regulate the tone of extensor musculature was defined in experiments carried out in the 1950s by Horace Magoun and Ruth Rhines. These investigators demonstrated that stimulation of the *pontine reticular formation* activated the extensor musculature. This region was termed the *extensor facilitatory area.* In contrast, stimulation of the *medullary reticular formation* had the net effect of inhibiting extensor muscles. This region was therefore termed the *extensor inhibitory area.* The two principal components of the reticulospinal tract arise from the extensor inhibitory and facilitatory areas, respectively (Figure 15.1). These physiological studies formed the basis for our understanding of how the descending reticulospinal pathways operate. The pontine reticular formation gives rise to the medial reticulospinal tract, which facilitates extensors. The medullary reticular formation gives rise to the lateral reticulospinal tract, which inhibits extensors. Hence, these two components of the reticulospinal pathways exert opposite actions on the antigravity (extensor) muscles.

This summary is an oversimplification. In particular, the descending projections of the medullary reticular formation have a variety of actions. For example, projections from the medullary reticular formation also inhibit motoneurons supplying neck and back musculature, and some of the axons also excite extensors. And of course because of the organization of segmental circuitry (see Chapter 14), both pathways also produce reciprocal actions (that is, facilitation of extensors is accompanied by reciprocal inhibition of flexors and vice versa). Nevertheless, these generalizations are useful to summarize the principal actions.

Reticulospinal Pathways Regulate the Tone of Antigravity Muscles by Controlling Gamma Motoneurons

An important fact is that the descending facilitatory input from the reticulospinal tract acts primarily by activating *gamma motoneurons.* As a result of the increase in gamma motoneuron activity, the sensitivity of the muscle spindle increases, leading to an increase in activity over the associated Ia afferents. The result is a tonic increase in activation of extensor motoneurons, leading to extensor facilitation. In this way, the reticulospinal pathway controls muscle tone via the *gamma loop.*

The Cells of Origin of the Reticulospinal Tract Integrate Inputs From the Cerebellum, Cerebral Cortex, and Ascending Somatosensory Pathways

Activity over reticulospinal pathways is controlled by inputs from the cerebellum, cerebral cortex, and ascending somatosensory pathways. The inputs from the cerebral cortex are especially important for postural adjustments in anticipation of and during the execution of a voluntary movement. For example, any voluntary movement of the limbs must be accompanied by postural adjustments to maintain balance. These adjustments are mediated by the descending projections from the cortex to the reticulospinal centers, operating in conjunction with corticospinal projections to segmental levels that actually control limb musculature.

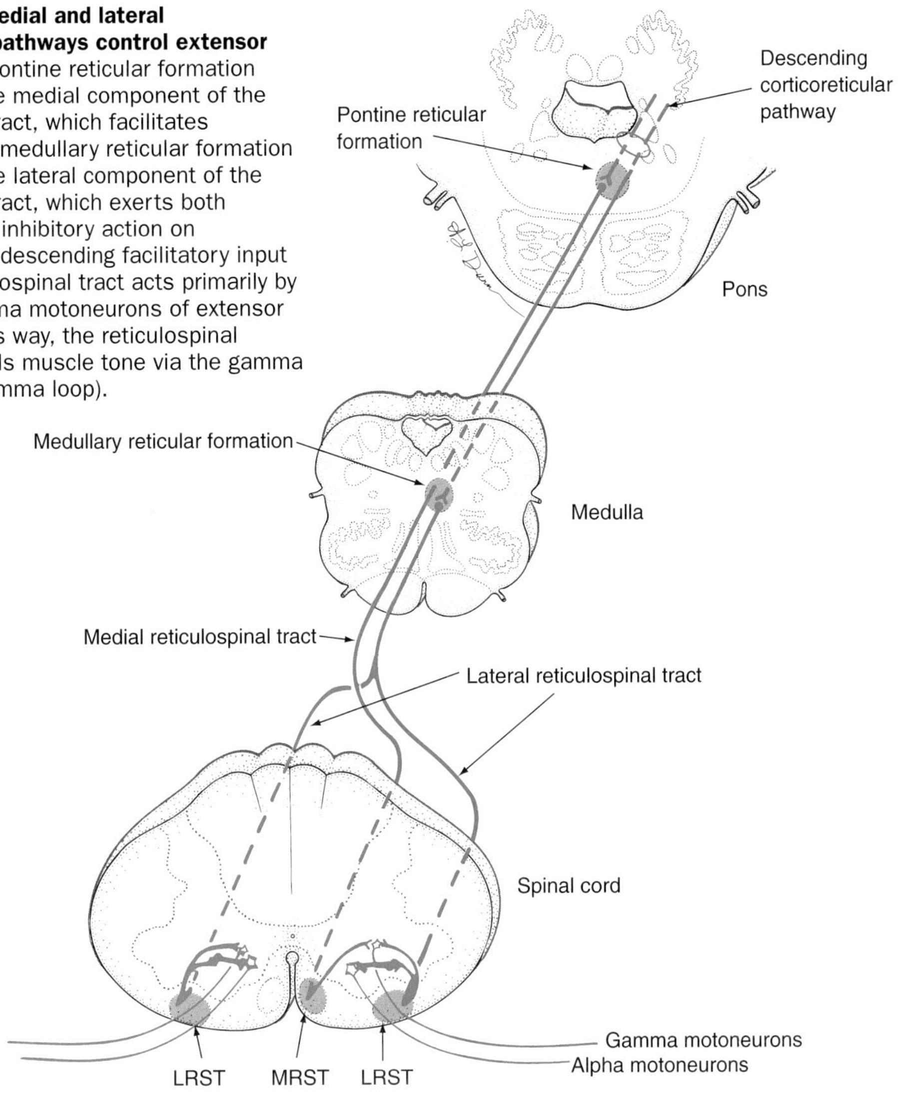

Figure 15.1. Medial and lateral reticulospinal pathways control extensor muscles. The pontine reticular formation gives rise to the medial component of the reticulospinal tract, which facilitates extensors. The medullary reticular formation gives rise to the lateral component of the reticulospinal tract, which exerts both facilitatory and inhibitory action on extensors. The descending facilitatory input from the reticulospinal tract acts primarily by activating gamma motoneurons of extensor muscles. In this way, the reticulospinal pathway controls muscle tone via the gamma system (the gamma loop).

You can demonstrate the involvement of the head and trunk in movements involving the limbs by analyzing a simple movement. Start from a standing position with your feet together and take a single step that ends with your feet together again. You will notice that the step begins with a forward movement of your head and body, which positions your weight ahead of your feet. One foot then moves forward. As the second foot is brought forward, your head and body tilt backward slightly so as to stop forward movement. If you carefully analyze what various parts of your body are actually doing, you will see that even a single step involves coordination of the entire body.

The Vestibulospinal Tract Activates Motoneurons Serving Extensor Muscles

The other medial brainstem pathway that is involved in postural regulation is the vestibulospinal tract. It acts in conjunction with the reticulospinal tract to activate extensor musculature. The vestibulospinal tract terminates on gamma motoneurons and also directly on alpha motoneurons. The direct innervation of alpha motoneurons is important because it means that the vestibulospinal pathway can operate both via the gamma loop and by directly controlling the discharge of alpha motoneurons. This is in contrast to the reticulospinal pathways, which operate primarily via the gamma loop. As discussed later, the direct activation of alpha motoneurons by the vestibulospinal tract is of clinical importance.

There are two components of the vestibulospinal tract that originate in different divisions of the *vestibular nucleus* (Figure 15.2). The lateral vestibulospinal tract originates in the lateral vestibular nucleus *(Deiters' nucleus)*. Neurons in the lateral vestibulospinal tract project ipsilaterally to facilitate extensor muscles (with

reciprocal inhibition of flexors being mediated by segmental circuitry). The medial vestibulospinal tract originates in the medial vestibular nucleus. Neurons in the medial vestibular nucleus project bilaterally to cervical and thoracic levels, where they terminate on motoneurons supplying muscles of the neck and trunk.

The Cells of Origin of the Vestibulospinal Pathway Receive Input From the Vestibular Labyrinth and the Cerebellum

The principal inputs to the neurons that give rise to the vestibulospinal tract are the fibers from the vestibular labyrinth and projections from the cerebellar cortex and the

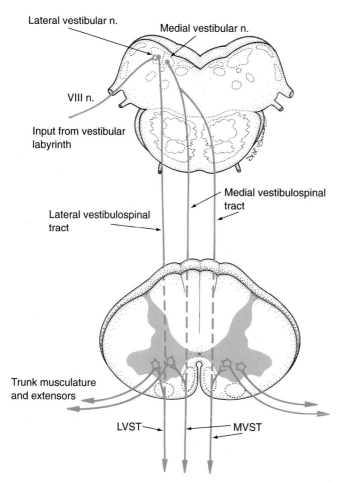

Figure 15.2. Vestibulospinal control of extensor muscles. There are two components of the vestibulospinal tract that originate in different divisions of the vestibular nucleus: the lateral vestibulospinal tract (LVST), which originates in the lateral vestibular nucleus, and the medial vestibular spinal tract (MVST), which originates in the medial vestibular nucleus. The LVST projects ipsilaterally to facilitate motoneurons supplying extensor muscles (with reciprocal inhibition of flexors being mediated by segmental circuitry). The MVST projects bilaterally to cervical and thoracic levels, where it terminates on motoneurons supplying muscles of the neck and trunk.

deep cerebellar nuclei. The labyrinthine input to vestibular neurons is considered in more detail in Chapter 25. Here we focus on the cerebellar projections.

The projections from the cerebellar cortex to the vestibular nucleus are inhibitory; thus, damage to the cerebellar cortex disinhibits the neurons of the vestibular nucleus. Vestibular neurons project to alpha motoneurons, where they form excitatory synapses. Thus, the increases in activity of vestibular neurons resulting from disinhibition lead to activation of the alpha motoneurons serving the extensor muscles.

Actually, the story is more complex because the vestibular nuclei also receive an excitatory projection from the deep cerebellar nuclei, which receive an inhibitory projection from the cerebellar cortex. Thus, damage to the cerebellar cortex eliminates a direct inhibitory projection to the vestibular nuclei and also disinhibits the excitatory pathway from the deep cerebellar nucleus. As we see later, the fact that the vestibulospinal tract controls alpha motoneurons is clinically important in the phenomenon of *alpha rigidity.*

LOCOMOTION DEPENDS ON SPINAL PATTERN GENERATORS THAT ARE REGULATED BY DESCENDING INPUTS

The central pattern generator in the spinal cord produces rhythmic stepping, but its activity is controlled by descending pathways and sensory inputs. Important evidence for this came from studies in experimental animals that demonstrated that stimulation of an area in the mesencephalon produced stepping, and that the rate of stepping depended on the intensity of stimulation. Thus, in cats, low-intensity stimulation produced walking, medium-intensity stimulation produced trotting, and high-intensity stimulation produced galloping. This area is now termed the *mesencephalic locomotor region.* In this way, the mesencephalic locomotor re-

gion is thought to operate as a command system for locomotion.

It is also obviously important to regulate locomotion so as to produce movement in particular directions. Thus, the activity of spinal pattern generators is regulated by descending input from the cortex. This feature of descending control falls within the framework of voluntary movement, to be discussed later.

Sensory Input Is Critical for Motor Function

During locomotion, it is necessary to adjust one's footfalls to the terrain and to compensate for irregularities such as slope and so forth. Such adjustments are possible because of afferent input from the limbs. You can demonstrate this to yourself by walking across rough terrain with your eyes closed. The importance of sensory, particular proprioceptive input, is obvious. For this reason, loss of sensory input severely impairs motor behavior. Indeed, animals will not use a limb that has been deprived of sensation. And a symptom of damage to proprioceptive pathways in humans is difficulty in walking with eyes closed or in the dark.

These considerations reinforce an important generalization; all aspects of motor behavior require *sensory–motor integration.* At the spinal level, sensory–motor integration is manifested by the afferent and efferent arms of spinal reflexes, and for the regulation of the spinal pattern generators. Sensory–motor integration is also important in regulating the medullary centers. Sensory information about the limb movements during locomotion plays a key role in regulating cerebellar output. This sensory information derives from Golgi tendon organs, muscle spindles, and joint afferents and is conveyed to the cerebellum via the spinocerebellar tract. In this way, the cerebellum is provided with what is called an *afferent copy* of the movement. In the case of the cortex, sensory information is important for planning how a given voluntary movement will be executed (i.e., one has to know the starting position of the body to accomplish a directed movement).

Respiration Is Mediated by Pattern Generators in the Lower Medulla That Project to Motoneurons in the Cervical Cord

An important rhythmic motor behavior is breathing. The respiratory pattern generator is located in the medullary reticular formation. For breathing to occur, this circuit must be connected with the motoneurons that control chest muscles. The relevant motor neurons are located at C2, C3, and C4. Spinal transections between the medulla and C4 lead to immediate death unless artificial respiration can be initiated within minutes.

Cortical Control of Voluntary Motor Functions

Voluntary Movement Depends on the Cerebral Cortex and Is Mediated by the Corticospinal Tract

Voluntary movement requires an interaction of the organism with its environment. It requires knowledge of the environment, knowledge about body position with respect to elements in the environment, and planning of the motor act, taking into account how best to achieve a given objective. Thus, voluntary movement requires the integration of information from different sources. This integration occurs in the cerebral cortex.

It is thought that different parts of the cortex are responsible for different aspects of the motor act. The *primary motor cortex* and other areas that give rise to descending axons of the corticospinal tract (CST) are responsible for controlling segmental circuitry. The *premotor cortex* and *supplementary motor areas* are thought to be involved in the planning of the motor strategy. The *posterior parietal cortex* is thought to be the site in which spatial information is integrated to permit the planning of the motor act with relationship to the environment. In the following consideration,

we begin with the primary motor cortex and then consider the supplementary areas.

The Primary Motor Cortex Is Organized Somatotopically

The primary motor cortex is somatotopically organized so that its different parts control the motor activities of different parts of the body. Muscles of the face, tongue, and mouth are controlled by neurons in the ventral portions of the *precentral gyrus*. The remainder of the body is represented in successively more dorsal portions of the gyrus, with the leg representation extending onto the medial wall of the hemisphere. The resulting body map for motor action is termed the *motor homunculus* (see Chapter 13).

The existence of a somatotopic organization was first inferred by observing the progression of motor seizures originating from a defined focus. Hughlings Jackson noted that focal motor seizures originated in a particular part of the body and progressed in a systematic fashion. Thus, a motor seizure that began in the finger progressed so as to involve the lower arm, then the upper arm, then the body. He correctly surmised that this progression reflects the propagation of seizure activity from a point source in the cortex into adjacent cortical regions. The progression of motor seizures is now termed a *Jacksonian march*.

The exact location of the primary motor cortex was defined by Charles Sherrington. Using stimulation techniques in monkeys, he demonstrated that motor responses were elicited at the lowest threshold by stimulation of the precentral gyrus. The neurosurgeon Wilder Penfield used the same sort of stimulation techniques to define the location of the primary motor cortex in human subjects during the course of neurosurgical procedures.

The two important features of the motor map are its somatotopic organization and the fact that the motor map is not proportional to the body area. Thus, portions of the body requiring fine motor control (the face and hands, for example) are represented by relatively larger portions of the primary motor cortex than areas in which motor control is less tightly controlled (for example, the trunk). In this way, the motor homunculus has roughly the same proportional organization as the somatosensory homunculus.

Although there is clear somatotopic organization of the primary motor cortex, recent studies have revealed that the nature of the map is complex. For example, it was initially thought that there was one-to-one correspondence between particular microdomains of the primary motor cortex and parts of the body. However, recent studies by a number of investigators have revealed that the map is a patchy mosaic in which there are multiple patches that control the movement of particular muscle groups, interspersed with other patches that control other muscle groups (a sort of *fractured somatotopy*). For example, detailed mapping studies in nonhumans have revealed that in the area of the primary motor cortex that controls the arm, the finger and arm areas are intermingled in a patchy mosaic.

The patchy representation in the primary motor cortex may be important for two reasons. First, the intermingling of neurons supplying different muscle groups of an individual limb may provide a substrate for coordinated activation (for example, coordination between muscle groups that collaborate during the execution of a movement). Second, the mosaic organization may help to preserve function when very small parts of the primary motor cortex are damaged or dysfunctional. If the primary motor cortex were organized so that it had a precise one-to-one correspondence between microdomains of the cortex and parts of the body, very small lesions of the cortex would paralyze individual muscles or sets of muscles. However, a distributed mosaic representation would enable the cortex to continue to exert at least some control, providing that some part of the mosaic remained.

The Output of the Primary Motor Cortex

Like other parts of the neocortex, the primary motor cortex has six layers, but the layers containing *projection neurons* are especially prominent (layers III and V). The neurons within layer V that give rise to some of the axons of the pyramidal tract (termed *Betz cells*) are among the largest neurons of the cerebral cortex. As in other cortical areas, *thalamic recipient neurons* are located in layer IV, but these are not nearly as numerous as in sensory cortices. Because of the lack of small neurons, the motor cortex has an appearance that is distinctly different from that of the sensory areas. For this reason, early neuroanatomists termed the motor cortical regions *agranular* to indicate their contrast to the sensory areas of the cortex that were termed *granular*. Some of this terminology may still be found in texts.

As in other cortical areas, neurons in different cortical layers are interconnected into functional columns. Thalamic inputs arrive in layer IV, where they terminate on stellate cells and to some extent on the dendrites of pyramidal neurons that pass through the layer; the stellate cells form intralaminar (vertical) connections with neurons in other layers, creating a substrate for intralaminar processing.

The output of the motor cortex is from pyramidal neurons. The descending pathways of the corticospinal and corticobulbar pathways originate from pyramidal neurons of layer V. It is important to recall, however, that the primary motor cortex is only one of the sources of the descending projections. For example, about one-third of the axons of the CST originate in the primary motor cortex; the remainder originate from pyramidal neurons in the premotor cortex, which lies anterior to the precentral gyrus and the somatosensory cortex in the postcentral gyrus.

Axons of the CST make direct connections with alpha motoneurons and interneurons. Corticospinal projections are excitatory on their target cells. The direct connections to motoneurons activate particular motor units, whereas the connections with interneurons are important for producing reciprocal inhibition. In this way, the CST produces direct excitation as well as projected inhibition and disinhibition.

Coupled with the fractured somatotopic organization of the motor cortex is the fact that single corticospinal axons often innervate different sets of motoneurons in a one-to-multiple pattern of connectivity. For example Figure 15.3 illustrates the distribution of the terminal arbors of an individual CST axon that innervates motoneurons supplying the arm. The important point illustrated by this figure is that the axon branches to innervate multiple motoneurons. One interpretation of these results is that individual CST axons actually control sets of muscles that would be activated together during coordinated movements involving multiple muscle groups. For example, single CST axons might link motoneurons that supply the shoulder, upper arm, and forearm musculature, all of which would be coactivated during a reaching movement. In any case, our understanding of the somatotopy of the motor cortex and the patterns of connectivity of CST axons is still evolving.

The output of neurons in the motor cortex is regulated based on the required force of the movement that is to be executed. The greater the force required to make a movement, the greater the increase in activity of the cortical neurons. This makes sense because greater force requires the activation of a greater number of motor units (and thus motoneurons). Different classes of cortical neurons encode different aspects of force. For example, one class of neurons encodes for the dynamic phase of the force, that is, the rate of force development. Their firing rate is highest as force develops. Another class of neurons encodes for the maintained level of force. A third class of neurons has mixed characteristics.

The Role of Sensory–Motor Integration in Cortical Function

As already noted, the output of the motor cortex depends critically on information from the somatosensory system. Neurophysiological recordings have revealed that neurons in the motor cortex receive sensory information from the portion of the body that they control. This information is thought to be conveyed by projections from the somatosensory cortex and by direct projections from the thalamus.

The Supplementary Motor Area and Premotor Cortex Play a Role in Planning Coordinated Motor Movements

The *supplementary motor area* and *premotor cortex* are areas just anterior to the primary

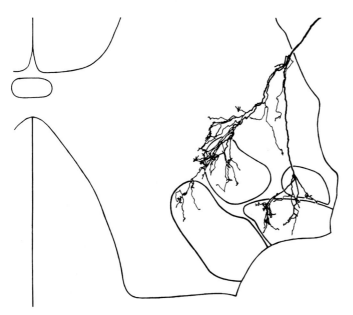

Figure 15.3. One-to-multiple innervation pattern of single CST axons. A reconstruction of the terminal arbor of a CST axon originating from a single neuron in the "hand area" of the motor cortex of a monkey. The neuron that gave rise to this axon was injected with a tracer so that the entire axon arbor was labeled. Terminals of this axon are distributed to motoneurons that supply the ulnar nerve (upper two encircled regions) and radial nerve (lower two encircled regions). The branches of this single axon project to at least four different motoneuron groups. From Shinoda Y, Yokota J, Futami T. *Neurosci Lett.* 1981;23:7–12.

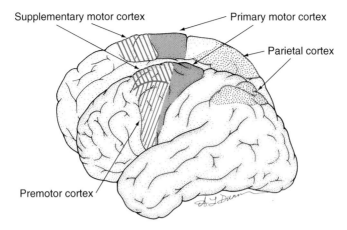

Figure 15.4. Adjuncts to the primary motor cortex: supplementary motor areas, premotor cortex, and posterior parietal cortex. The primary motor cortex receives input from several areas involved in higher-order motor control, including the supplementary motor area, premotor cortex, and posterior parietal cortex. The premotor and supplementary motor areas are somatotopically organized; therefore, local stimulation produces movements in defined portions of the body. The supplementary motor area plays a key role in planning complex movements involving distal musculature. The premotor area is thought to play a similar role for proximal musculature, especially in orientation of the body in preparation for a motor act. The posterior parietal cortex integrates information from different sensory modalities to guide motor behavior with respect to extrapersonal space.

motor cortex (Figure 15.4). The supplementary motor area lies on the dorsal part of the cortex, extending onto the medial wall; the premotor cortex lies just anterior to the primary motor cortex on the precentral gyrus. These areas were initially defined as motor, based on stimulation studies; local stimulation of either area produces movement. The supplementary motor area was defined by Penfield in his studies of humans, and the premotor cortex was defined by C. N. Woolsey in monkeys. Like the primary motor cortex, both areas are somatotopically organized; therefore, local stimulation produces movements in defined portions of the body.

The idea that the supplementary motor area and premotor cortex play a role in planning coordinated movements came initially from the stimulation studies. The movements elicited by stimulation of these areas were more complex and coordinated than those elicited by stimulation of the primary motor cortex. For example, whereas stimulation of the primary motor cortex activates particular muscle groups, stimulation of the supplementary motor area elicits complex movements such as opening or closing the hand, reorientation of the body, and so on.

The Supplementary Motor Area Plays a Key Role in Planning Complex Movements Involving Distal Musculature

Important evidence for the role of the supplementary motor area came from studies of cortical activity in human subjects during motor performance. Cortical activity can be assessed indirectly by measuring blood flow. This is because oxygen utilization increases during neuronal activity, and blood flow then increases to compensate for the increase in oxygen utilization. Blood flow is evaluated by injecting radiolabeled xenon into the bloodstream and assessing the distribution of the label by means of an array of detectors on the scalp.

When human subjects were asked to perform simple motor tasks (for example, holding the finger and thumb together), blood flow increased in the hand portions of both primary motor and sensory cortices, but not in supplementary motor areas. When the subjects were asked to perform more complex tasks (a sequence of movements of the digits) blood flow increased in primary and supplementary motor areas. The most fascinating result came when subjects were asked to mentally rehearse the complex movement without actually performing it. In this case, blood flow increased selectively in the supplementary motor cortex. These results suggested that the supplementary motor areas were responsible for the planning and programming of the complex movement.

The Premotor Area Primarily Regulates Proximal Musculature

The functional role of the premotor area is less well understood. It sends projections to the portion of the brainstem that controls medial descending systems (especially the reticulospinal systems) and to the portion of the spinal cord controlling proximal musculature. It receives input from the posterior parietal cortex. Based on these facts, H. J. M. Kuypers suggested that this part of the cortex is responsible for controlling the initial orientation of the body and limbs in preparation for a motor act. For example, to reach for an object, one must first orient the body toward the object. This orientation movement is thought to be mediated by the premotor area.

Damage to the premotor area leads to the release of certain reflexes. For example, the *grasp response* is greatly enhanced (in which grasping movements of the fingers are elicited by tactile stimulation of the palm).

The Posterior Parietal Cortex Integrates Information From Different Sensory Modalities to Guide Motor Behavior With Respect to Extrapersonal Space

The planning and execution of voluntary movements requires knowledge about the position of the body in space and the location of objects. Thus, planning of voluntary movements requires an integration of information from other sensory modalities, especially information from the visual system.

This integration is thought to occur primarily in the posterior parietal cortex. This region of the cortex receives important inputs from sensory association areas and projects to the motor cortex.

Cortical Control of Eye Movements

Eye movements are controlled by a special region of the cortex termed the *frontal eye field*. This area is well separated from the primary motor cortex. The control of eye movements will be discussed in more detail in Chapter 21.

Symptoms Resulting From Damage at Different Levels of the Motor System

Table 15.1 summarizes the symptoms that result from damage at different levels of the motor system. The symptoms are considered in more detail in what follows.

Damage to the Primary Motor Cortex or Its Projections Via the CST

Damage to the motor cortex or the CST produces a characteristic symptom complex. Complete destruction of corticospinal pathways on one side produces a symptom complex termed the *corticospinal* or *pyramidal tract syndrome*. Partial damage produces partial symptoms. Damage to the CST system occurs most commonly as a result of stroke, including the following types:

1. Middle cerebral artery infarcts that affect the lateral cortex including the portion of the motor cortex that serves all parts of the body except for the lower leg

2. Anterior cerebral artery infarcts that affect the medial wall of the cortex including the paracentral lobule, which controls the lower leg

3. Infarcts involving penetrating branches of the middle cerebral or posterior cerebral artery that damage the internal capsule

4. Infarcts involving penetrating vessels supplying the basis pedunculi, which contains the descending CST fibers

5. Infarcts involving medial penetrating vessels supplying the basal pons or medulla (medullary pyramid)

Table 15.1. Fixed postures resulting from damage at different levels of the motor system*

Tracts	Lesion site	Symptom	Syndrome
CST	Cortex Internal capsule Basis pedunculi Basal medulla (pyramid)	Paralysis with extensor hypertonus of the leg, flexor hypertonus of the arm	Pyramidal tract syndrome
CST and midbrain tegmentum	Midbrain above the level of the red nucleus	Coma or semicoma; paralysis with fixed extension of the leg, fixed flexion of the arm	Decorticate rigidity
CST + RuST	Midbrain at or below the level of the red nucleus, or pontine lesions that spare the pontine RF	Paralysis with fixed extension of the leg and arm	Decerebrate rigidity
CST + RuST + RST	Medulla	Paralysis with no fixed posture	

*CST indicates corticospinal tract; RuST, rubospinal tract; RST, reticulospinal tract; RF, reticular formation.

All these injuries have in common that they destroy part or all of the descending CST and may spare other descending motor pathways.

When the CST is injured above the level of the pyramidal decussation, the symptoms are contralateral to the lesion. When the lesions destroy the CST and spare other descending motor pathways, the following symptoms are present:

1. Loss of voluntary motor control *(paralysis)*, and varying degrees of muscle weakness *(paresis)*, depending on the size of the injury. This is the primary symptom.
2. *Extensor hypertonus* of the affected leg; *flexor hypertonus* of the affected arm. These are not seen immediately but develop over the first few days after the injury. The hypertonus leads to *fixed postures* in which the limbs are held in characteristic positions (extension of the affected leg and flexion of the affected arm).
3. A *Babinski sign*, the most characteristic component of which is dorsiflexion of the great toe.

Clinicians term this symptom complex the *corticospinal syndrome, upper motor neuron syndrome,* or *pyramidal tract syndrome.* An understanding of these symptoms, especially the

fixed postures, requires one to consider two questions: which pathways are damaged, and which pathways are spared?

Paralysis With Fixed Postures

Cortical injuries destroy the CST and eliminate the cortical input to the brainstem centers, but the brainstem centers themselves are spared. The same is true of injuries that affect the descending CST in the internal capsule, basis pedunculi, or medullary pyramid. The resulting loss of voluntary control over movement in the portion of the body supplied by the damaged fibers (paralysis) is easy to understand based on the functions of the CST.

Immediately after the injury, the affected limbs exhibit *flaccid paralysis*. But over the course of a few days, the affected limbs begin to exhibit hypertonus of certain muscle groups. The legs exhibit extensor hypertonus; the arms exhibit flexor hypertonus. As a result, the paralyzed limbs tend to be held in fixed positions (termed *fixed postures*). In particular, the affected leg is stiffly extended (*fixed extensor posture,* also termed *extensor rigidity*), whereas the affected arm, hand, wrist, and fingers are held in a fixed *flexed* position. The fixed postures are explained by the fact that brainstem pathways are spared and continue to exert their influences on spinal motoneurons.

The foundations of our understanding of the consequences of selective damage to descending motor pathways came from the work of Sir Charles Sherrington around the turn of the century. In a series of experimental studies in cats, Sherrington found that lesions at the level of the midbrain that interrupted corticospinal pathways but spared descending projections from the brainstem led to a condition in which animals maintained their limbs in a fixed and extended position. The experimental lesion that produced the fixed extensor posture was a complete transection at the level of the midbrain (termed *decerebration*). For this reason, the set of symptoms was termed *decerebrate rigidity*. The symptoms were characterized by a selective activation of the extensor muscles that are important for maintaining posture (extensor hypertonus).

Fixed Extensor Posture of the Leg Results From Persisting Activity in the Extensor Facilitatory Pathways That Mediate Posture and Balance

The fixed extension of the affected leg is explained by the survival and continued activity of medial brainstem pathways that regulate posture and balance at the same time that descending inputs that mediate voluntary control are lost. Because normal motor function depends on a balance of activity in the different descending pathways, when one component is destroyed and others are spared, the actions of the spared pathways are exaggerated and unmodulated (in this case, this consists of unopposed activation of extensor muscles).

Given that components of the reticulospinal tract both facilitate and inhibit extensors, why does elimination of input from the cortex cause extensor hypertonus? The reason is that although both the reticular extensor inhibitory and facilitatory areas receive descending input, the extensor facilitatory area remains tonically active after interruption of descending inputs, whereas the extensor inhibitory area is silenced. When descending inputs are lost, the balance of facilitation and inhibition of extensor tone swings in favor of facilitation.

The Medial Brainstem Pathways Control Muscles That Oppose Gravity (Antigravity Muscles); These Are Termed Physiological Extensors

Although the predominant feature of the fixed posture of the lower limb is extension, certain flexors are also activated. For example, in humans, the toes are flexed ventrally. The reason is that the medial brainstem pathways control all of the key muscle groups that are used to oppose gravity (antigravity muscles). Thus, in humans, the toes oppose gravity by flexion and so are held in a fixed flexed position as a result of the unapposed activity of the medial brainstem pathways.

Interestingly, in other animals that maintain posture using flexors (for example, opossums and other animals that hang from trees), decerebrate rigidity is reflected by hypertonus of the flexor muscles that are used to oppose gravity. For this reason, all muscles that are used to oppose gravity are termed *physiological extensors*.

Fixed Flexion of the Arm and Hand (Flexor Hypertonus) Results From Persisting Activity of the Rubrospinal Tract

The symptoms resulting from selective damage to the CST differ from those of decerebrate rigidity in that the affected arm is held in a fixed flexed position. The explanation for this fixed flexor posture of the arm is the preservation of the rubrospinal tract.

The rubrospinal tract is a lateral pathway that facilitates flexors. This pathway is especially important for fine motor control (for example, of the fingers). When the rubrospinal pathway is spared along with other brainstem pathways, it continues to exert a tonic facilitatory action on the flexors of the arm. As a result, the arm is held in a fixed flexed position. The critical role of the rubrospinal tract in the flexor hypertonus of the arm is demonstrated by the fact that the arms exhibit extensor hypertonus when the rubrospinal tract is damaged along with descending cortical pathways (for example, as a result of an injury at the level of the midbrain; see later discussion).

It is not well understood why facilitation of flexors (controlled by the rubrospinal tract) prevails in the arms while facilitation of extensors (reticulospinal tract) prevails in the legs. Some textbooks state that the rubrospinal pathway does not project beyond the cervical portions of the spinal cord. But experimental evidence suggests that this is not the case. However, it is true that in humans, there is relatively less fine motor control of the lower limb, and so perhaps the motoneuron pools supplying the leg are dominated by inputs from medial brainstem pathways.

The fact that damage to the CST leads to extensor hypertonus of the leg is of considerable clinical significance. It is for this reason that patients with extensive damage to the corticospinal pathway can often walk using a cane. Although they cannot directly control the muscles of the leg, they can use muscles of the trunk to bring the leg forward and then take advantage of the fixed extensor tone to bear weight.

The Special Case of Facial Musculature

In understanding the consequences of damage to the motor cortex or CST, it is important to recall the difference in the innervation of motoneurons controlling muscles of the face versus those controlling muscles in the body (specifically, that motoneurons supplying the upper face receive input from both sides of the cortex). Cortical lesions produce deficits in motor control of the contralateral limbs and trunk and lower face. However, the cortical input to the motoneurons that control the upper face (eyebrows and forehead) is bilateral. Thus, cortical injuries lead to a selective paralysis of the muscles of the lower face contralateral to the injury (termed *supranuclear palsy*).

The selective paralysis of the lower face following an injury to the upper motoneurons supplying the facial nucleus is in contrast to the consequences of direct injuries of facial motoneurons in the brainstem; like other lower motoneuron injuries, damage to the facial nucleus leads to flaccid paralysis of both the upper and lower face.

So, we can summarize as follows:

- Injuries involving lower motoneurons paralyze the entire half of the face.
- Injuries of upper motoneurons (cortex, internal capsule, basis pedunculi) lead to selective paralysis of the lower face.

For these reasons, one can quickly diagnose whether a facial paralysis results from damage to the CST (upper motoneuron injury), or the facial motoneurons or facial nerve (lower motoneuron injury). One asks the patient to raise his or her eyebrow or wrinkle his or her forehead. With a lower motoneuron injury, the upper face is paralyzed; with an upper motoneuron injury, motor function in the upper face is preserved.

It is interesting that although individuals with supranuclear palsy cannot voluntarily move the muscles of the lower face to produce a social smile, when they smile involuntarily in response to something humorous, the smile is bilateral. This suggests that the circuitry responsible for movements associated with emotional expression is separate from that responsible for voluntary facial movement.

Functional Reorganization Following Motor Cortex Lesions

Because of the somatotopic organization of the motor cortex, lesions that involve only part of the motor representation produce paralysis of regions of the body proportional to the site and size of the damage in the cortex. However, there is often partial recovery of function in the weeks and months after the injury. This recovery is thought to be due to a functional reorganization in which surviving regions of the motor system take over the functions of the areas of the cortex that are damaged.

For example, following a small cortical lesion (such as might occur with a focal stroke), motor functions normally mediated by one part of the motor cortex may be taken over by surviving cortical tissue surrounding the damaged region. Evidence for such a mechanism has come from experimental studies in monkeys, in which the organization of the motor cortex was defined using microstimulation techniques. In these studies, detailed maps of the somatotopic organization of the hand region of the motor cortex were obtained in individual monkeys before and several months after a small ischemic lesion had been created. After creation of the lesions, there were dramatic changes in the motor map in the area surrounding the infarct (Figure 15.5). Significantly, when monkeys received rehabilitation (in which there was specific retraining of hand use), the hand representation in the surviving parts of the cortex increased. However, without rehabilitation, the hand representation in the surviving parts of the cortex actually decreased. This reorganization of motor representation is an example of plasticity of cortical representation—a phenomenon that we encounter in other chapters. These important findings both indicate a potential mechanism for recovery of function following cortical injuries and demonstrate an important beneficial consequence of rehabilitation therapy.

Lesions at the Level of the Upper Midbrain

Lesions at the level of the upper midbrain that spare the red nucleus produce the

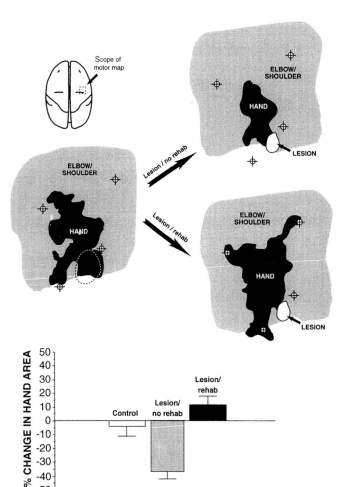

Figure 15.5. Remapping the motor cortex following partial lesions. Reorganization of motor maps has been documented using microstimulation techniques in primates. Motor maps were defined before and at various times after placing a small ischemic lesion. Following a small lesion that destroyed part of the hand representation, there was a gradual reorganization of the motor representation in surviving cortical tissue surrounding the lesion site. If monkeys received rehabilitation training in the use of the affected hand, the hand representation in the surviving parts of the cortex increased. In the absence of rehabilitation training, the hand representation in the surviving parts of the cortex decreased. The graph illustrates the average results from several different experiments. From Nudo RJ. Remodeling of motor representations after stroke: implications for recovery from brain damage. *Mol Psychiatry.* 1997;2(3):188–191.

symptom complex already described, along with symptoms that result from damage to lower motoneurons in cranial nerve nuclei or the nerve roots of the cranial nerves. Lesions at this level may also produce coma or disruptions of consciousness because of damage to the pathways from the ascending reticular-activating system (see Chapter 30). Comatose patients exhibit the fixed extension of the leg and fixed flexion of the arm. In this case, the symptom complex is termed *decorticate rigidity*.

Lesions at the Level of the Lower Midbrain or Pons

Lesions at the level of the midbrain or pons that interrupt descending rubrospinal pathways along with the CST but spare descending projections from the brainstem lead to a condition in which both arms and legs are maintained in a fixed and extended position. The symptoms are identical to the condition of decerebrate rigidity seen in experimental animals. Depending on the location of the lesion, there will also be damage to lower motor neurons in cranial nerve nuclei or the nerve roots of the cranial nerves.

Focal lesions of the lower midbrain or pons that lead to decerebrate rigidity also typically damage the ascending *reticular-activating system*. For this reason, decerebrate rigidity is often accompanied by coma or semicoma. However the fixed extensor posture may also be seen in individuals with widespread diffuse brain damage (for example, patients with cerebral palsy).

As is true of the extensor hypertonus seen with CST injuries, decerebrate rigidity also involves flexors. In particular, the toes are flexed ventrally, and the back is arched, with the head flexed dorsally. Again, the explanation is that the function of the medial brainstem pathways is to control antigravity muscles that are used to maintain balance (physiological extensors).

An important clinical manifestation of decerebrate rigidity in unconscious patients is the interaction with the *tonic neck reflex*. Decerebrate rigidity is expressed preferentially on the side of the body opposite to the way the head is facing. Thus, if a comatose patient's head is passively turned to face to the right, rigidity is expressed primarily on the left side and vice versa.

Lesions at the Level of the Medulla

Lesions at the level of the basal medulla (for example, a stroke involving the medial penetrating vessels from the basalar or vertebral arteries) damage the medullary pyramids and the nerve roots of the cranial nerves. If the medial brainstem pathways are spared, the result is a CST syndrome with fixed postures, along with symptoms resulting from lower motoneuron injury to the cranial nerves. If the lesion involves the dorsal medulla and damages the descending fibers of the medial reticulospinal tract from the pontine reticular formation (extensor facilitatory area), the result is an upper motoneuron paralysis of the body without fixed extensor posture (basically the same symptoms that result from a lesion of the spinal cord). But lesions involving the dorsal medulla also usually result in coma because of damage to the reticular formation, and most such lesions are fatal.

Gamma Rigidity

An important feature of extensor hypertonus is that it is eliminated in a particular spinal segment if the dorsal root input to that segment is severed. The reason is that facilitation of extensors by the reticulospinal pathway is via the gamma loop, which depends on activation of alpha motoneurons by Ia afferents. For this reason, extensor rigidity is also termed *gamma rigidity*.

The fact that decerebrate rigidity can be eliminated by dorsal root section suggested a surgical approach to the treatment of certain forms of rigidity. Some patients who suffer from rigidity as a result of cerebral palsy benefit from operations that selectively cut the dorsal roots serving particular segments *(selective dorsal rhizotomy)*.

16

The Basal Ganglia

The term *basal ganglia* was originally used to refer to all the nuclear groups buried beneath the cortical mantle (that is, cell groups that are organized in nuclei rather than being laminated; see Chapter 2). As it became clear that a subset of the nuclei played a special role in motor function, the definition was narrowed so that, in its modern usage, the term *basal ganglia* refers to five structures: the *caudate nucleus, putamen, globus pallidus, subthalamic nucleus,* and *substantia nigra.*

The role of the basal ganglia in motor functions was initially inferred from clinical observations: damage to components of the basal ganglia leads to movement disorders that are fundamentally different from those that occur following damage to the motor cortex or its descending projections. Specifically, rather than leading to paralysis or paresis, damage to the basal ganglia produces involuntary movements or an inability to initiate voluntary movement, which results in immobility.

Based on the qualitative differences in symptoms, the concept arose that there were two motor control systems, a *pyramidal system* regulated by the motor cortex and its projections via the pyramidal tract, and an *extrapyramidal system* regulated by the basal ganglia. However, as more information became available, it became clear that this concept was not correct. Indeed, it was found that the basal ganglia exert their effect by modulating the output of the cortex. Although the terms *pyramidal* and *extrapyramidal* are not very useful for understanding how the motor systems function, the terms are still used clinically. Thus, in the clinical setting, *pyramidal symptoms* refer to the symptoms resulting from damage to the direct descending motor pathways (paralysis or paresis); *extrapyramidal symptoms* refer to the symptoms that result from damage or disease involving the basal ganglia.

An important recent discovery has been that in addition to modulating the output of the motor portions of the cortex, the basal ganglia also play an important role in regulating nonmotor cortical areas. For this reason, damage or disease involving the basal ganglia produces important symptoms in addition to the movement disorders that first bring attention. In the following, the anatomical organization and functional properties of the basal ganglia are reviewed. We then consider the results from injury or disease affecting the basal ganglia.

RELATIONSHIP BETWEEN THE NUCLEI THAT COMPRISE THE BASAL GANGLIA

The relationship between the different nuclear groups comprising the basal ganglia can be understood most easily by recalling that some of the different structures are actually parts of the same nuclear group that are physically separated by the internal capsule. For example, the *caudate nucleus* and *putamen* possess identical cell types and neurotransmitter systems, similar interconnections, and similar physiological properties. Thus, they are considered two divisions of a single nucleus and are collectively termed the *caudate–putamen,* the *neostriatum,* or simply the *striatum.* The term *striatum* derives from the fact that there are collections of fiber bundles embedded within the nuclear masses, giving the region a striated appearance.

The overall form of the two divisions of the striatum can be understood by considering how the fibers of the internal capsule split the striatum. In the absence of the internal capsule, the striatum would be a single mass with a narrow tail-like extension that follows the curvature of the lateral ventricle into the temporal lobe. The fibers of the internal capsule grow through the mass in a broadly curved band (Figure 16.1). The portion of the striatum that lies nearest to the ventricle is the caudate nucleus. It has an overall form resembling a tadpole with its head located anteriorly and its body and tail following the curvature of the lateral ventricle into the temporal lobe. Thus, the terms *head, body,* and *tail* are used to describe the different portions of the caudate nucleus. The putamen is the portion of the striatum that lies lateral to the fibers of the internal

capsule surrounding the globus pallidus like a thick capsule.

The caudate nucleus and putamen merge just rostral to the anterior limb of the internal capsule. They are also connected by *cell bridges* that lie between the penetrating fibers of the internal capsule (Figure 16.1). Rostroventrally, the caudate and putamen merge with a closely related area termed the *ventral striatum.* The ventral striatum has cell types and neurotransmitter systems that are similar to those of the striatum; however, its connections are primarily with structures of the limbic system.

The spatial relationship between the caudate–putamen and the internal capsule can be appreciated by comparing coronal and horizontal sections that pass through the nuclei comprising the basal ganglia (Figures 16.2 and 16.3). In both types of sections, the putamen lies deep to the insula and is separated from the insula by the thin alternating bands of white matter and gray matter that represent the *external capsule,* the *claustrum,* and the *extreme capsule.* Students should be familiar with the relationships between the putamen, the internal capsule, and the caudate nucleus because these areas are of considerable functional importance and because they are a common site of vascular infarct and neurodegenerative diseases.

In coronal (frontal) sections, the caudate nucleus lies medial to the internal capsule. The body of the caudate nucleus forms the lateral wall of the lateral ventricle. Because the long tail of the caudate nucleus follows the ventricle into the temporal lobe, both the head and tail of the caudate nucleus can be visualized simultaneously in sections in which both the anterior and temporal horns of the lateral ventricle are visible (Figure 16.2).

In horizontal sections (Figure 16.3), the putamen and caudate are separated rostrally by the anterior limb of the internal capsule. The head of the caudate nucleus again forms the lateral wall of the ventricle. The posterior limb of the internal capsule separates the laterally located putamen and globus pallidus from the thalamus.

Figure 16.1. Separation of the caudate nucleus from the putamen by the fibers of the internal capsule.

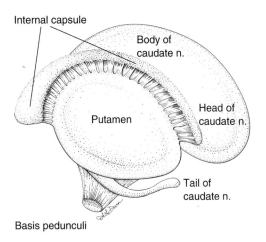

Internal capsule

Body of caudate n.

Head of caudate n.

Putamen

Tail of caudate n.

Basis pedunculi

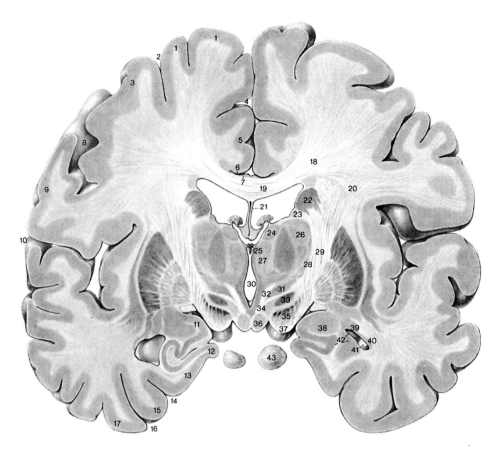

1. Superior frontal gyrus	16. Occipitotemporal sulcus	31. Zona incerta
2. Precentral sulcus	17. Inferior temporal gyrus	32. Mammillothalamic tract
3. Precentral gyrus	18. Corpus collosum, radiations	33. Subthalamic nucleus
4. Cingulate sulcus	19. Corpus collosum	34. Mammillary fasciculus
5. Cingulate gyrus	20. Corona radiata	35. Substantia nigra
6. Calloso-marginal sulcus	21. Septum pellucidum	36. Mammillary body
7. Induseum griseum	22. Caudate nucleus, body	37. Cerebral peduncle
8. Central sulcus	23. Stria terminalis	38. Amygdaloid body
9. Postcentral gyrus	24. Anterior thalamic nucleus	39. Stria terminalis
10. Lateral fissure	25. Choroid plexus, third ventricle	40. Caudate nucleus, tail
11. Uncinate gyrus	26. Ventral lateral nucleus	41. Lateral ventricle, inferior
12. Ambiens gyrus	27. Medial (thalamus) nucleus	horn
13. Parahippocampal gyrus	28. Reticular (thalamus) nucleus	42. Hippocampal foot
14. Collateral sulcus	29. Internal capsule, posterior limb	43. Pons
15. Lateral occipitotemporal gyrus	30. Third ventricle	

Figure 16.2. Coronal section at a midthalamic level. In coronal (frontal) sections, the caudate nucleus lies medial to the fibers of the internal capsule. The body of the caudate nucleus forms the lateral wall of the lateral ventricle. Because the long tail of the caudate nucleus follows the ventricle into the temporal lobe, both the head and tail of the caudate nucleus can be visualized simultaneously in sections in which both the anterior and temporal horns of the lateral ventricle are visible. From Nieuwenhuys R, Voogd J, van Huijzen C. *The Human Central Nervous System.* Berlin, Germany: Springer-Verlag; 1985.

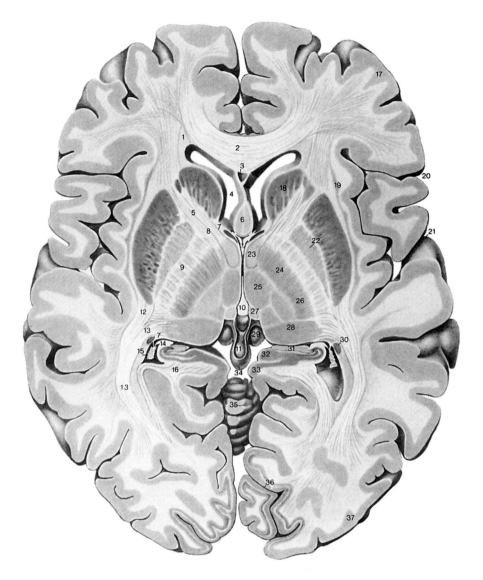

1. Superior occipitofrontal fasciculus
2. Genu of the corpus collosum
3. Septum pellucidum
4. Lateral ventricle, anterior horn
5. Internal capsule, anterior limb
6. Fornix
7. Stria terminalis
8. Internal capsule, genu
9. Internal capsule, posterior limb
10. Third ventricle
11. Pineal recess
12. Superior longitudinal fasciculus
13. Optic radiation
14. Hippocampal fimbria
15. Lateral ventricle, inferior horn
16. Corpus collosum, radiation
17. Frontal gyri
18. Caudate nucleus, head
19. Claustrum
20. Lateral sulcus, ascending branch
21. Lateral sulcus, posterior branch
22. Lentiform nucleus
23. Anterior nucleus (thalamus)
24. Ventral lateral nucleus (thalamus)
25. Medial nucleus (thalamus)
26. Lateral nucleus (thalamus)
27. Habenular nuclei
28. Pulvinar nuclei (thalamus)
29. Superior colliculus
30. Caudate nucleus, tail
31. Fasciola cinerea
32. Fasciolar gyrus
33. Medial temporal lobe
34. Pineal body
35. Cerebellar vermis
36. Calcarine sulcus
37. Occipital gyrus

Figure 16.3. Horizontal section at a midthalamic level. In horizontal sections, the putamen and caudate are separated rostrally by the anterior limb of the internal capsule. The head of the caudate nucleus again forms the lateral wall of the ventricle. The posterior limb of the internal capsule separates the laterally located putamen and globus pallidus from the thalamus. From Nieuwenhuys R, Voogd J, van Huijzen C. *The Human Central Nervous System*. Berlin, Germany: Springer-Verlag; 1985.

Types of Neurons in the Striatum

Within the striatum are two principal neuron types: (1) *spiny stellate cells,* accounting for 90% to 95% of the neurons in the striatum; and (2) *aspiny interneurons.* The spiny stellate cells are the principal recipient neurons of the striatum, that is, they are the cells upon which most of the incoming afferents terminate. As their name implies, they bear large numbers of dendritic spines. The spiny stellate cells also give rise to the principal outputs from the striatum to the globus pallidus.

An important feature of the basal ganglion is that many of the *projection neurons* (the neurons that project from one structure to another) use γ-aminobutyric acid (GABA) as their neurotransmitter and are thus inhibitory on their targets. This is true of the spiny stellate cells of the striatum.

The other principal cell type in the striatum is the aspiny interneuron. The axons of aspiny interneurons do not leave the striatum. Many of the aspiny interneurons use acetylcholine (ACh) as their neurotransmitter. Like other cholinergic neurons in the CNS these cells contain large amounts of acetylcholinesterase (AChE). For this reason, the striatum has one of the highest concentrations of AChE in the brain and stands out conspicuously in sections stained histochemically for AChE.

The Globus Pallidus Is Composed of Two Subdivisions

Just internal to the putamen lies the globus pallidus. In coronal sections, the globus pallidus is separated from the thalamus by the broad internal capsule (Figure 16.2). In horizontal sections, it lies within the V formed by the two limbs of the internal capsule. A thin lamina of myelinated fibers, termed the *lateral medullary lamina,* separates the putamen from the globus pallidus. The name *globus pallidus* derives from the fact that the nucleus appears very light (pallid) in unstained tissue because it contains large numbers of myelinated axons. In Weigert-stained sections, it appears dark for the same reason.

The globus pallidus is divided into external and internal divisions by another thin band of myelinated fibers termed the *medial medullary lamina.* The external portion is termed the *globus pallidus pars externa* (GPPEx), and the internal portion is the *globus pallidus pars interna* (GPPIn).

Other Terms You May Encounter

The internal division of the globus pallidus is also sometimes called the *entopeduncular nucleus,* especially in mammals other than primates. As described later, the external and internal divisions of the globus pallidus have different interconnections and function, but both contain neurons that use GABA as their neurotransmitter and thus are inhibitory on their targets.

The terms *lentiform nucleus* or *lenticular nucleus* are sometimes used to refer collectively to the putamen and globus pallidus because of their lenslike appearance. This is an archaic term based on gross morphology rather than function. Use of the term is becoming less common because it encompasses nuclei with different functional characteristics (the putamen and globus pallidus) and does not include the caudate nucleus, which is functionally related to the putamen. The term persists in part because names begat names—the penetrating arteries that supply the region are termed the *lenticulostriate arteries.*

The Substantia Nigra Is Made Up of Two Functionally Distinct Subdivisions

The substantia nigra is located ventral to the subthalamic nucleus and just mediodorsal to the massive fiber tract in the base of the cerebral peduncle (the *basis pedunculi,* see Figure 16.4). The substantia nigra is made up of two groups of cells with very different properties: the *substantia nigra pars compacta* (SNPC) and *substantia nigra pars reticulata* (SNPR).

The SNPC is made up of a collection of closely packed neurons that contain melanin pigment. It is from this cell group that the substantia nigra gets its name. The melanin

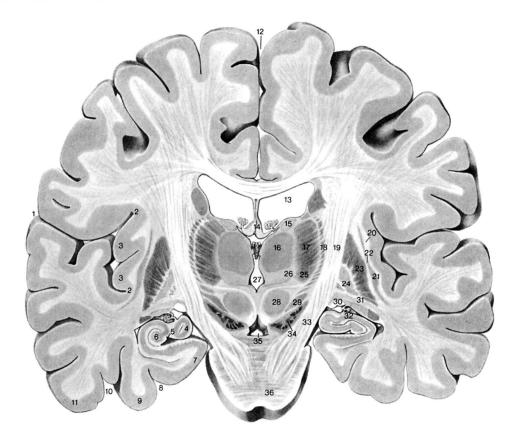

1. Lateral sulcus, posterior branch
2. Circular sulcus
3. Long gyrus of insula
4. Intralimbic gyrus
5. Hippocampal sulcus
6. Dentate gyrus
7. Parahippocampal gyrus
8. Collateral sulcus
9. Lateral occipitotemporal gyrus
10. Occipitotemporal sulcus
11. Inferior temporal gyrus
12. Longitudinal cerebral fissure
13. Central lateral ventricle
14. Fornix, body
15. Lateral dorsal nucleus (thalamus)
16. Medial nucleus (thalamus)
17. Ventral lateral nucleus (thalamus)
18. Reticular nucleus (thalamus)

19. Internal capsule, posterior limb
20. Extreme capsule
21. Claustrum
22. External capsule
23. Putamen
24. Globus pallidus
25. Ventral posterolateral nucleus (thalamus)
26. Centromedian nucleus (thalamus)
27. Third ventricle
28. Red nucleus
29. Subthalamic nucleus
30. Optic tract
31. Internal capsule
32. Lateral ventricular choroid plexus
33. Cerebral peduncle
34. Substantia nigra
35. Interpeduncular fossa
36. Pons

Figure 16.4. Coronal section at the level of the substantia nigra and cerebral peduncle. From Nieuwenhuys R, Voogd J, van Huijzen C. *The Human Central Nervous System.* Berlin, Germany: Springer-Verlag; 1985.

pigment is a metabolic by-product of dopamine, which is the neurotransmitter used by these neurons.

The more ventrally located SNPR is made up of neurons that are anatomically and physiologically similar to the GABAergic neurons in the internal division of the globus pallidus. Thus, like the caudate and putamen, the SNPR and the internal division of the globus pallidus (GPPIn) are part of the same nuclear group that becomes separated by fibers of the internal capsule. In what follows, the two nuclei are collectively called GPPIn/SNPR (say "gippin-snipper"). Understanding that the internal division of the globus pallidus and the SNPR are one and the same helps to simplify what is otherwise a very complex circuit.

The Subthalamic Nucleus

The subthalamic nucleus is located below the caudal portion of the thalamus and just dorsal to the substantia nigra. It is separated from the internal division of the globus pallidus by the internal capsule. The subthalamic nucleus is considered part of the basal ganglia because its connections are primarily with the globus pallidus (see later discussion).

CIRCUITRY OF THE BASAL GANGLIA

The interconnections between the components of the basal ganglia are very complex. Most of the possible permutations of connectivity exist (that is, almost every structure is connected in some way with every other). The following summarizes the most important of the interconnections in terms of basal ganglia function.

In considering the interconnections, it is helpful to recall the functional groupings already noted, particularly that the caudate nucleus and putamen consist of functionally the same nuclei (collection term is *striatum*), as do the GPPIn and SNPR. Taking this into

account, the consideration of interconnections is made more simple by using a schematic diagram that groups together the functionally related nuclei (Figure 16.5).

In considering the connections of the basal ganglia, it is useful to begin by summarizing the extrinsic inputs, then consider the interconnections between the components of the basal ganglion, and finally consider the outputs.

Extrinsic Inputs to the Striatum From the Cortex and Thalamus

The major extrinsic input to the basal ganglia is the *corticostriate projection,* which arises in essentially all parts of the cerebral cortex and terminates throughout the striatum in both the caudate nucleus and putamen (Figure 16.6). As discussed later in more detail, the corticostriate projections are glutamatergic and thus excitatory. The synapses of the corticostriate pathway terminate on the dendritic spines of the spiny stellate cells.

The corticostriate projection is topographically organized so that different parts of the cortex terminate in different parts of the caudate–putamen. The putamen receives

Figure 16.5. Schematic of the functional components of the basal ganglia.

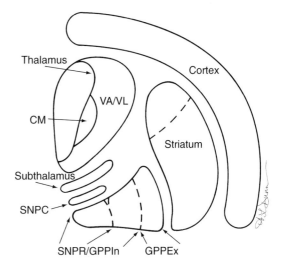

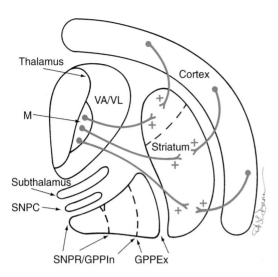

Figure 16.6. Inputs to the striatum from the cortex and centromedian nucleus of the thalamus.

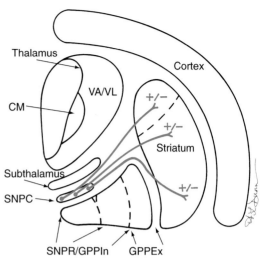

Figure 16.7. The nigrostriatal pathway.

input from areas of the cortex involved in motor functions. The caudate nucleus receives input from "association" areas and from the frontal eye fields and is thus involved in regulating complex cognitive functions as well as eye movements.

The other important extrinsic input to the striatum arises in the intralaminar nuclei of the thalamus, especially the *centromedian nucleus*. This projection also terminates on the spines of the spiny stellate cells. Again, this projection is topographically organized. The centromedian nucleus in turn receives input from the motor cortex.

Interconnections Between the Components of the Basal Ganglia

The Dopaminergic Nigrostriatal Pathway

Probably the best known of the pathways that interconnect components of the basal ganglia is the dopaminergic pathway from the SNPC to the striatum (Figure 16.7). The cells of origin of this pathway are the ones that are destroyed in Parkinson's disease. The dopaminergic axons from the substantia nigra form synapses on the spines of spiny neurons throughout the striatum and also

on the cholinergic interneurons (see later). The physiological effects of the dopamine released by the nigrostriatal pathway differs depending on the types of dopamine receptors that are expressed by the target neurons (D1 or D2, see later).

The Two-Component Striatopallidal Pathway

The principal outputs from the striatum are to the globus pallidus. But there are actually two functionally distinct pathways that originate from different populations of spiny stellate cells in the striatum. The spiny stellate cells giving rise to these two pathways make different connections, use a different mix of neurotransmitters, and express different neurotransmitter receptors. (For a summary, see Table 16.1.)

Connections

The two populations of spiny neurons in the striatum are the point of divergence of two functional circuits through the globus pallidus, which are termed the *direct* and *indirect striatopallidal circuits* (Figure 16.8).

1. The spiny neurons that give rise to the direct striatopallidal circuit project directly to the GPPIn/SNPR. These neurons use GABA as their neurotransmitter, along with a

Table 16.1. Characteristics of the spiny striatal neurons that give rise to direct and indirect striatopallidal circuits

Characteristic	Direct	Indirect
Projections	To GPPIn/SNPR	To GPPEx
Neurotransmitters	GABA + substance P and dynorphin	GABA + enkephalin
Receptors	D1	D2
Action of dopamine	Facilitates	Inhibits
Alternative names	Striatonigral	Striatopallidal

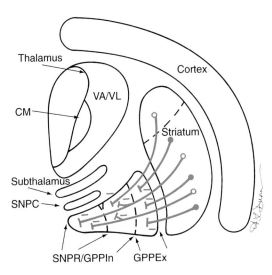

Figure 16.8. Striatopallidal pathways. One group of spiny stellate cells projects to the GPPEx (filled in). The other population projects to the GPPIn/SNPR (open). The former represents the cells of origin of the indirect striatopallidal circuit; the latter represents the cells of origin of the direct striatopallidal circuit.

particular mix of peptide neurotransmitters (see Figure 16.9 and later discussion).

2. The indirect striatopallidal circuit is so named because it provides an indirect polysynaptic pathway from the striatum to the GPPIn/SNPR. The circuit begins with the output from the GABA/enkephalin neurons, which project to the GPPEx. The circuit continues via the interconnections between the globus pallidus and the subthalamus (Figure 16.10). In particular, neurons in the GPPEx project to the subthalamic nucleus, which in turn projects to the GPPIn/SNPR. In this indirect circuit, all projections are inhibitory except for the projection from the subthalamic nucleus to the GPPIn/SNPR, which is excitatory (glutamatergic).

Experimental neuroscientists often use a somewhat different terminology to refer to the preceding circuits. In particular, the direct striatopallidal circuit is often called the striatonigral pathway, and the indirect striatopallidal circuit is referred to as simply the *striatopallidal pathway.* Unfortunately, this terminology engenders some confusion because there is another striatonigral pathway that projects to the neurons in the SNPC (see later). For this reason, the terms *direct* and *indirect striatopallidal pathways* are preferred.

Neurotransmitters

All spiny stellate cells use GABA as their neurotransmitter, but striatal neurons that project directly to the GPPIn/SNPR use GABA along with substance P or dynorphin; striatal neurons that project to the GPPEx use GABA along with enkephalin. It is the mix of cotransmitters that distinguishes the two groups of neurons with differing striatopallidal projections.

Neurotransmitter Receptors

The third important difference that distinguishes the two populations of spiny striatal cells is the type of receptors that the neurons express. As noted above, dopamine is the neurotransmitter used by the nigrostriatal pathway. There are two principal types of receptors for dopamine, termed D1 and D2, that exert different physiological actions (see later). The GABA/substance P-containing and dynorphin-

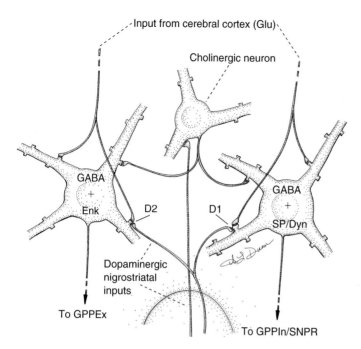

Figure 16.9. Different classes of spiny stellate neurons in the striatum. There are two classes of spiny stellate neurons within the striatum that differ in three ways: (1) the neurotransmitters that they use, (2) the neurotransmitter receptors that they express, and (3) their connections. Neurons that use GABA and enkaphalin express D2 dopamine receptors and project to the GPPEx. Neurons that use GABA and substance P or dynorphin express D1 receptors and project to the GPPIn. The physiological action of dopamine at D1 receptors is facilitatory, increasing activity in the direct striatopallidal circuit. The physiological action of dopamine at D2 receptors is inhibitory, decreasing activity in the indirect striatopallidal circuit.

containing neurons that are the output neurons for the direct circuit possess D1 receptors; the GABA/enkephalin-containing neurons that are the output neurons for the indirect circuit possess D2 receptors. The differences in the types of dopamine receptors explains the different physiological consequences of dopamine on the two neuron types.

Output of the Basal Ganglia

The output pathways of the basal ganglia are summarized in Figure 16.11 and Table 16.1. The most important output of the basal ganglia is from the GPPIn/SNPR to the ventral lateral (VL) and ventral anterior (VA) nuclei of the thalamus. The VA and VL nuclei in

Figure 16.10. Interconnections between the globus pallidus and the subthalamus.

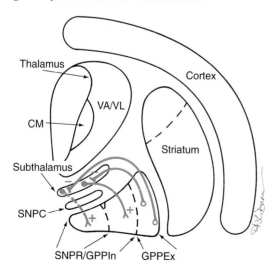

Figure 16.11. Output of the basal ganglia. The projections from the GPPIn/SNPR to the thalamus, and from the thalamus to the cortex.

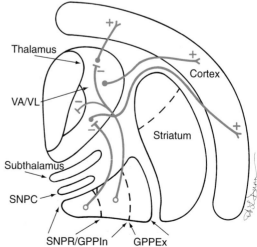

turn project to the cortex. It is through these projections that the basal ganglia regulate the activity of the motor cortex as well as other parts of the cortex. The SNPR also projects to the superior colliculus, where it plays a role in regulating eye movements.

Basal Ganglia Circuitry Operates Via Inhibition and Disinhibition

The direct striatopallidal circuit is made up of two sets of inhibitory neurons in series, which represent a substrate for disinhibition. Thus, circumstances that lead to increases in activity of the neurons, giving rise to the direct circuit, lead to inhibition of neurons in the GPPIn/SNPR and disinhibition of neurons in the VA/VL thalamus (Figure 16.12).

The indirect striatopallidal circuit is a polysynaptic pathway in which the actions of the neurotransmitters at each respective station are as follows: striatum to GPPEx is inhibitory; GPPEx to subthalamus is inhibitory; subthalamus to GPPIn/SNPR is excitatory; GPPIn/SNPR to thalamus is inhibitory.

Thus, circumstances that lead to increases in activity of striatal neurons lead to increased inhibition neurons in the GPPEx (Figure 16.13), which in turn disinhibit neurons in the subthalamus. The projections from the subthalamic nucleus back to the GPPIn/SNPR are excitatory; hence disinhibition of neurons in the subthalamus leads to excitation of neurons in the GPPIn/SNPR, which give rise to the inhibitory projections to the VA/VL thalamus. Thus the net effect of increasing activity over the indirect pathway is to inhibit neurons in the VA/VL thalamus.

Modulation of Striatal Output by the Nigrostriatal Pathway

The direct and indirect pathways are antagonistic in their effects on the thalamus. How

Figure 16.12. The physiology of the direct striatopallidal circuit. The direct striatopallidal circuit is made up of two sets of inhibitory neurons connected in series. This forms a substrate for disinhibition. Thus, circumstances that lead to increases in activity of the neurons giving rise to the direct circuit lead to inhibition of neurons in the GPPIn/SNPR and disinhibition of neurons in the VA/VL thalamus.

Figure 16.13. The physiology of the indirect striatopallidal circuit. The indirect striatopallidal circuit is made up of a chain of interconnected neurons in which the synaptic action at each station is as follows: inhibitory, inhibitory, excitatory, inhibitory. In this circuit, increases in activity in the spiny stellate cells lead to increased inhibition neurons in the GPPEx, which in turn disinhibits neurons in the subthalamus. Disinhibition of neurons in the subthalamic nucleus leads to increases in excitatory drive of neurons in the GPPIn/SNPR that give rise to the inhibitory projections to the VA/VL thalamus. The result is increased inhibition of neurons in the VA/VL thalamus.

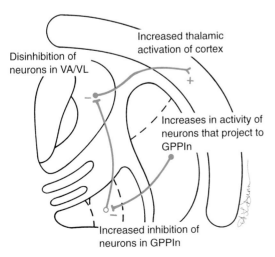

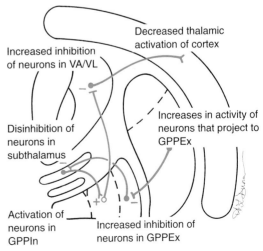

then are these opposing outputs modulated? The answer lies in the way the spiny stellate cells that express different types of dopamine receptor respond to the dopaminergic input from the substantia nigra.

The physiological actions of dopamine are not entirely straightforward because both D1 and D2 receptors are metabotrophic. Nevertheless, a reasonable summary is that the physiological action of dopamine at D1 receptors is facilitatory on the neurons that give rise to the direct striatopallidal circuit. In contrast, the physiological action of dopamine at D2 receptors is inhibitory on the neurons that give rise to the indirect striatopallidal circuit (Figure 16.9 and Table 16.1). In this way, dopaminergic input facilitates transmission along the direct route and inhibits transmission along the indirect route. Because of the opposing actions of the direct and indirect circuit already described, these two effects lead to a net disinhibition of neurons in the VA/VL thalamus.

This action of dopamine forms the basis of our understanding of the consequences of drugs that are selective D1 or D2 receptor agonists and antagonists. It also provides a simple explanation for the pathophysiology that underlies Parkinson's disease. When dopaminergic input is lost, the net effect is an increase in tonic inhibitory input to the VA/VL thalamus.

The cholinergic interneurons in the striatum form excitatory connections onto the GABAergic projection neurons. It is for this reason that cholinergic drugs also have an effect on modulating basal ganglion function (and have been useful to some extent for treating basal ganglion disorders including Parkinson's disease).

Independent Regulation of Different Parts of the Cortex

An important organizational principle is that all of the interconnections between the components of the basal ganglia are topographically organized. In this way, different parts of the circuitry can operate independently. The several different circuits that are recognized are characterized by topographically organized connections between different parts of the cortex and different portions of the striatum:

- The *motor circuit* involves interconnections between the areas of the cortex involved in motor functions (motor and sensory cortex, premotor cortex, and supplementary motor area) and the putamen.

- The *dorsolateral prefrontal circuit* involves interconnections between the dorsolateral prefrontal cortex and the head of the caudate nucleus.

- The *lateral orbitofrontal circuit* involves interconnections between the lateral orbitofrontal cortex and the body of the caudate nucleus.

- The *oculomotor circuit* involves interconnections between the frontal eye fields, the caudate nucleus, and the superior colliculus.

Each subdivision of the striatum then projects on through the globus pallidus, subthalamus, VA/VL thalamus, and so on in a topographically organized fashion in the manner described. The key concept to remember is that essentially every part of the cortex has important interconnections with the basal ganglia, and it is likely that the physiological role of these circuits is similar. Thus, the basal ganglia probably play an important role in regulating cortical function in general (and thus are important for the functions that the different parts of the cortex subserve).

The Patch and Matrix Organization of the Striatum

An important recent discovery is that the striatum itself is structurally heterogeneous. This concept arose from discoveries made independently by Ann Graybiel and Patricia Goldman-Rakic that projections to the striatum terminate in patches. The corticostriate projection terminates in large patches separated by areas that receive few or no corticostriate projections.

The areas between the patches receive input from limbic structures. The patchlike areas with few corticostriate projections are termed *striosomes;* the areas with dense corticostriate projections are termed the *matrix.* In addition to the differences in the pattern of termination of the corticostriate pathway, the striosomes and matrix are marked by different concentrations of neurotransmitters and enzyme systems, indicating a different functional organization.

Neurons within the matrix give rise to the GABAergic projections to the direct and indirect pathways already described. Thus, the matrix component corresponds to the circuit that is the key to regulating the output of the motor system through its connections with the thalamus.

Neurons within the *striosomes* give rise to a projection to the SNPC. Because the striosome compartment receives input from limbic structures, it is thought that this circuit provides a route by which input from limbic regions can regulate the dopaminergic pathway and thus the overall output of the basal ganglia.

Because of the Inhibitory Nature of Basal Ganglion Circuitry, Injury or Disease Leads to Release Phenomena Resulting in Positive Signs

Many of the signs and symptoms resulting from diseases of the basal ganglia can be understood as positive symptoms resulting from release from inhibitory control. It is easy to imagine how the removal of inhibition of a motor control area could result in positive signs such as involuntary movements. What is less intuitively obvious is that the same type of release process can also produce loss of function, specifically an inability to initiate motor acts *(akinesia).* The reason is that lesions can lead to disinhibition of neurons that are themselves inhibitory, leading to increases in net inhibition.

This is exactly what happens when Parkinson's disease destroys the dopaminergic neurons of the substantia nigra. There is a net disinhibition of neurons in the GPPIn/SNPR, leading to increases in tonic inhibition of the

VA/VL thalamus. This concept is the key to our understanding of the symptoms resulting from diseases of the basal ganglia.

DISEASES AFFECTING THE BASAL GANGLIA

The two best-known diseases that affect the basal ganglia are *Parkinson's disease* and *Huntington's disease.* Characteristic symptoms of diseases of the basal ganglia include *resting tremor* (as distinct from the *intention tremor* that occurs following cerebellar injury), *akinesia* (lack of movement), *bradykinesia* (slowness in the execution of movement), *athetosis* (slow writhing movements of the hands), *chorea* (sudden movements of the limbs and facial muscles), *ballismus* (sudden ballistic movements), and *dystonia* (persistent distorted postures or movements). All are positive symptoms that come about as a result of loss of inhibitory circuits (release phenomena).

Parkinson's Disease

As just noted, Parkinson's disease (also called *paralysis agitans*) is a neurodegenerative disease that affects dopaminergic neurons. The most important symptoms derive from the loss of the dopaminergic neurons of the substantia nigra, but other monoaminergic neurons are also affected. Why the disease affects some neurons selectively is not yet known.

Parkinson's disease typically strikes middle-aged or older individuals. The mean age of onset is 58. The disease progresses slowly over the course of many years. The symptoms include resting tremor with a frequency of 3 to 6 Hz, akinesia, bradykinesia, dystonia with impaired postural reflexes, and an increase in muscle tone (rigidity). The rigidity is different from that resulting from injury to the motor systems, however. When the limbs are passively manipulated, the resistance to movement releases periodically, like a ratchet or cogwheel. For this reason, the rigidity of Parkinson's disease is termed *cogwheel rigidity.*

The tremor, dystonia, and rigidity are recognized as positive symptoms due to release phenomena. The akinesia and bradykinesia are also positive symptoms that result from a disinhibition of the final output neurons of the basal ganglia, leading to increased net inhibition of neurons in the thalamus.

Parkinson's disease is treated by administering the drug L-dopa, which is the immediate precursor of dopamine. It is thought that the increased availability of the precursor allows surviving neurons to synthesize more dopamine than they would be able to without the added substrate. L-dopa therapy has unquestionably been one of the great success stories of modern medicine, in that it can, for a time, dramatically ameliorate the symptoms of the disease. However, L-dopa eventually loses its efficacy because of the continued loss of the dopaminergic neurons. When there are too few left to make use of the precursor, L-dopa treatment becomes ineffective.

A potentially important new strategy for the treatment of Parkinson's disease involves the transplantation of cells that release dopamine. For example, one approach involves transplantation of dopaminergic neurons obtained from fetal brains. These neurons differentiate and grow to form connections within the host brain. Animal experiments indicate that fetal transplants can be quite effective in helping to restore function. Other strategies involve the transplantation of cells that have been genetically engineered to synthesize and release dopamine. All of these treatments are highly experimental.

Another treatment strategy that has waxed and waned in popularity involves selective surgical ablation of portions of the globus pallidus or the subthalamus. The rationale for the former approach is that the symptoms of Parkinson's disease are positive signs resulting from a net increase in inhibitory output by the basal ganglia. Thus, destroying part of this inhibitory output would be expected to reduce the net inhibition of motor cortical output. Similarly, the rationale for removing parts of the subthala-

mus is that the disease results in disinhibition of these excitatory neurons, leading to an increased excitatory drive of the neurons of the GPPIn/SNPR.

These surgical strategies have met with varying degrees of success. Sometimes, symptom amelioration has been reported to be striking. Other times, the benefits are modest or nonexistent. Recent techniques involve the use of stereotaxic approaches, interoperative imaging, and interoperative physiological recording to precisely identify the regions to be destroyed. The efficacy of the surgical strategy must still be ascertained.

An important clue about the potential etiology of Parkinson's disease came from an unfortunate accident involving a designer drug. A small group of drug users in California, who had obtained and used a privately manufactured synthetic heroin derivative, developed the signs and symptoms of full-blown Parkinson's disease. It was found that the synthetic drug was contaminated by the toxin 1-methyl-4-phenyl-1,2,3,6-tetrahydropyridine, and that this agent had destroyed the dopaminergic neurons of the substantia nigra in the affected individuals. The discovery of this potent and selective toxin has both provided a useful animal model for the disease and suggested clues to the possible underlying pathophysiology.

Huntington's Disease

The other important disease of the basal ganglia is Huntington's disease (also called *Huntington's chorea*). Huntington's disease is a genetic disorder that results in the degeneration of the GABAergic spiny neurons in the striatum. Cholinergic neurons may also be involved, but to a variable extent. Symptom onset in Huntington's disease typically occurs in middle age. Huntington's disease is characterized by chorea, dementia, and eventual death after a prolonged period of virtually complete disability. There is no effective therapy, although certain drugs can be used to control the chorea during the early stages of the disease.

Huntington's disease is inherited as an autosomal dominant mutation, and there are several well-known pedigrees. In the United States, many of the individuals suffering from the disease are descendants of two individuals who migrated to Salem, Massachusetts, in 1630. It is thought that some of the women executed as witches were exhibiting symptoms of the disease. Another well-known pedigree is found in a Native American population that lives near Lake Maracaibo in Venezuela. This pedigree descends from a single woman who lived in the area in the early 1800s whose father was an English sailor. There are several thousand living descendants of the woman; about 100 of these descendants already have the disease, and there are about 1000 currently living children that have a 50% probability of developing the disease.

One of the most unfortunate aspects of the disease is its late onset. Individuals carrying the dominant mutation do not experience symptoms until after their childbearing years. Given the autosomal pattern of inheritance, there is a 50% probability that the child of an individual carrying the mutation will inherit the mutation. Genetic markers have now been identified through studies of the well-defined pedigrees, so that it is now possible to screen for the mutation early in life, before the childbearing years.

Other Disorders Involving the Basal Ganglia

The basal ganglia are often affected by vascular accidents. The caudate nucleus and putamen are supplied by penetrating branches of the internal carotid artery; the globus pallidus is supplied by the anterior choroidal artery. The penetrating branches of the internal carotid are one of the most common sites of stroke. The symptoms resulting from stroke are typically unilateral and depend on the specific structures that are affected.

Strokes can also damage the subthalamic nucleus. The result is a highly characteristic involuntary movement disorder called *ballism* in which there are "ballistic" movements of the limbs. The ballistic movements may be unilateral *(hemiballism)*.

Another important disorder involving the basal ganglia is *tardive dyskinesia*. Tardive dyskinesia occurs after prolonged treatment with any of a number of antipsychotic drugs (e.g., chlorpromazine, perphenazine, and haloperidol). These drugs are antagonists for dopamine, and the tardive dyskinesia is thought to result from drug-induced supersensitivity. The symptoms include involuntary movements, especially of the face and tongue.

17

Role of the Cerebellum in Motor Function

The cerebellum plays a key role in motor function by regulating the output of the neurons that give rise to descending motor pathways. The current thinking regarding the overall function of the cerebellum is that it operates as a *comparator*. It receives information about the output of motor centers (a measure of the intended movement), and sensory information about the actual results (from vestibular and ascending proprioceptive systems). Thus, it can compare the intended movement with the actual.

The input to the cerebellum from the motor centers in the cerebral cortex is termed *corollary discharge* (corollary because it is not along the direct output path from the motor cortex), or *internal feedback*. The sensory input about the results is termed *reafference*, or *external feedback*. It is thought that the cerebellum integrates the combined information and then projects to the nuclei in the brainstem that give rise to descending pathways, and also to the nuclei in the thalamus that in turn control the activity of the motor cortex. In this way, the output of motor centers is adjusted on a moment-by-moment basis to compensate for errors in the intended movement, allowing a smooth and accurate execution of the desired motor act.

This hypothesis of cerebellar function has been developed on the basis of four types of information: (1) the type of information that the cerebellum receives via its inputs, (2) the areas that receive input from the cerebellum, (3) the basic circuit organization of the cerebellum, (4) the physiology of the cerebellum, and (5) the symptoms that occur following injury to cerebellar circuitry. Each is discussed in the following sections.

MORPHOLOGICAL SUBDIVISIONS OF THE CEREBELLUM

The cerebellum is divided into gross morphological subdivisions and functional subdivisions. The former are based on obvious physical landmarks and divide the cerebellum into lobes that are separated by prominent fissures; the latter are based on connectivity and function. Unfortunately, the two are not in register (Figure 17.1). The functional subdivisions are the more important in terms of understanding cerebellar function. Thus, the *vermis,* along with the *intermediate hemisphere* represent one functional subdivision; the *lateral hemisphere* represents a second; and the *flocculonodular lobe* represents the third. The connections of these subdivisions are summarized in the following sections.

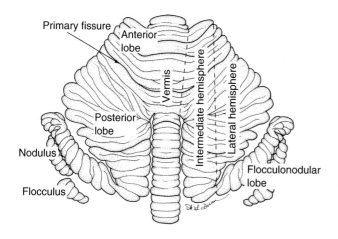

Primary fissure

Anterior lobe

Vermis

Intermediate hemisphere

Lateral hemisphere

Posterior lobe

Nodulus

Flocculonodular lobe

Flocculus

Figure 17.1. The subdivisions of the cerebellum. The cerebellum is divided into gross morphological subdivisions and functional subdivisions based on connectivity. The morphological subdivisions are illustrated on the left; the functional subdivisions on the right. The flocculonodular lobe receives input from the vestibular nucleus and sends a return projection to the vestibular nucleus. The vermis and intermediate hemisphere receive input from spinocerebellar pathways and project to the fastigial nucleus and interposed nuclei, respectively. The lateral hemisphere receives input from the pontocerebellar pathway and projects to the dentate nucleus.

Cellular Organization of the Cerebellum

The cerebellum is made up of a cortex and a core that contains a series of nuclear groups called the *deep cerebellar nuclei,* or simply cerebellar nuclei.

The cerebellar cortex contains five types of neurons: Purkinje cells, granule cells, Golgi cells, stellate cells, and basket cells. Purkinje cells give rise to the output pathways. The other cell types project intrinsically within the cerebellar cortex and control the discharge of the Purkinje cells. Of the intrinsically projecting cells, granule cells form excitatory synapses with Purkinje cells; Golgi cells, stellate cells, and basket cells are inhibitory interneurons.

The cerebellar cortex is organized in three layers that contain different cell types and cellular processes (Figures 17.2 and 17.3):

1. The *molecular layer* is a neuropil layer that contains the dendrites of the Purkinje cells and the axons terminating on these dendrites. The dendrites of Purkinje cells are unique in that the dendritic arbor is planar (very extensive in one dimension and essentially a flat sheet in the other dimension). In this respect, the dendritic arbor of Purkinje cells resembles the sea fan coral. Stellate cells and basket cells are also present in the molecular layer. These neurons are somewhat smaller than Purkinje cells and have a dendritic arbor that radiates outward from the cell body in a roughly spherical fashion.

2. The *Purkinje cell layer* contains the cell bodies of the Purkinje cells. The cell bodies are large and relatively widely spaced and the layer is one cell body thick. The extensive dendritic tree arises from the pole of the cell that is toward the molecular layer. A single axon arises from the opposite pole and projects toward the white matter, giving rise to recurrent collaterals en route.

3. The *granule cell layer* is a relatively thick layer of tightly packed small cell bodies. The granule cells have short stubby dendrites that radiate from the cell body. The axons of the granule cells project upward into the molecular layer (see later).

Below the cortex is the white matter, which contains the incoming and outgoing axons. Deep to the cerebellar white matter are three paired nuclei termed the *deep cerebellar nuclei.* The most medial is the *fastigial nucleus;* the next is the *interposed,* or *interpositus nucleus;* and the most lateral is the *dentate nucleus.* The interposed nucleus is subdivided into two subnuclei: the *globose* and *emboliform nuclei.* Neurons in these nuclei receive input from Purkinje cells in different subdivisions of the cerebellum.

Purkinje cells project to deep cerebellar nuclei in a topographically organized fashion (Figure 17.3). Purkinje cells in the vermis project to the fastigial nucleus; Purkinje cells in the intermediate hemisphere adjacent to the vermis project to the interposed nuclei; and Purkinje cells in the lateral hemisphere project to the dentate nucleus.

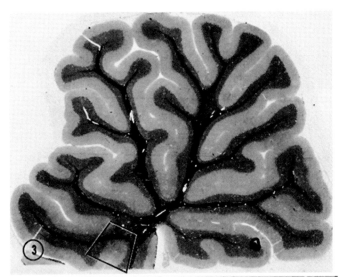

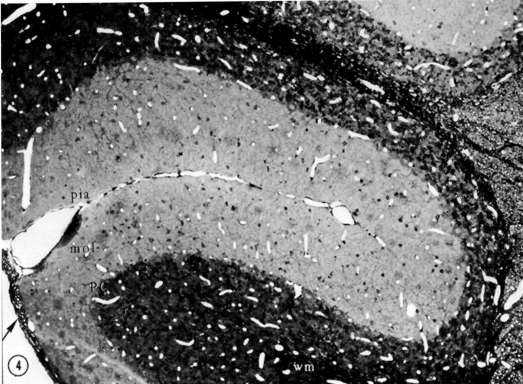

Figure 17.2. The cerebellar cortex has three layers. The upper panel illustrates a midsagittal section through the rat's cerebellum. The lobules of the cerebellum are evident. The lower panel illustrates a higher-magnification view of two lobules (area indicated by box in upper panel). The two lobules are folded against one another with their pial surfaces in apposition. mol indicates the molecular layer that contains the dendrites of the Purkinje cells, along with the innervating axons (especially parallel fibers) and the sparsely distributed cell bodies of stellate cells, and basket cells; PC, Purkinje cell layer; gr, granule cell layer; wm, white matter. From Palay SL, Chan-Palay V. *Cerebellar Cortex: Cytology and Organization.* Berlin, Germany: Springer-Verlag; 1974.

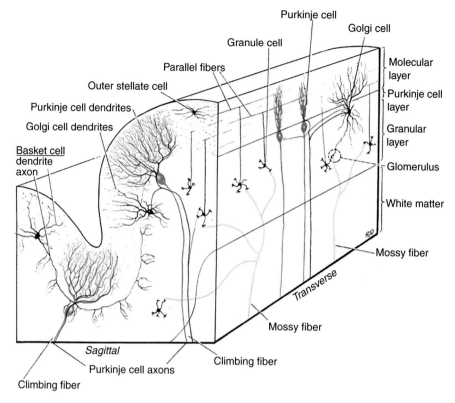

Figure 17.3. Cellular organization of the cerebellar cortex. The cerebellar cortex contains five types of neurons: Purkinje cells, granule cells, Golgi cells, stellate cells, and basket cells. Purkinje cells give rise to the output pathways. The two principal input pathways terminate either as climbing fibers or mossy fibers. Stellate cells and basket cells provide feedforward and feedback inhibition to the Purkinje cells. Golgi cells provide inhibitory input to the granule cells. Reproduced with permission from Carpenter MB. *Human Neuroanatomy.* Baltimore, Md: Williams & Wilkins; 1976.

The Purkinje cells in the phylogenetically oldest portion of the cerebellum (the flocculonodular lobe) project directly to the vestibular nucleus. Although the vestibular nucleus is not one of the deep cerebellar nuclei, it's connections with Purkinje cells in the flocculonodular lobe operate similarly to the connections between the different deep cerebellar nuclei and their respective subdivisions of the cerebellar cortex.

The generalizations summarized in Figure 17.4 are not absolute. For example, the vermis also receives projections from the vestibular nucleus, and some Purkinje cells within the vermis project back to the vestibular nucleus. Similarly, there is some input to the flocculonodular lobe from the pontine nuclei.

Neverthless, the generalizations define the predominant connections.

Purkinje Cells Receive Two Types of Excitatory Inputs

The principal innervation of Purkinje cell dendrites comes from granule cells. Axons from granule cells project into the molecular layer, bifurcate, and then project for some distance through the cerebellar cortex to innervate a row of Purkinje cells, much in the same way that a high-tension wire travels from tower to tower. Because the axons from granule cells run parallel to one another, they are called *parallel fibers.*

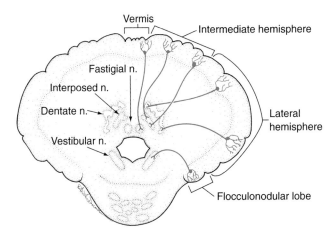

Figure 17.4. Projections from the cerebellar cortex to the deep cerebellar nuclei. Purkinje cells project to deep cerebellar nuclei in a topographically organized fashion. Purkinje cells in the vermis and intermediate hemisphere project to the fastigial and interposed nuclei, respectively; Purkinje cells in the lateral hemisphere project to the dentate nucleus; and Purkinje cells in the flocculonodular lobe project to the vestibular nucleus.

The other excitatory synaptic input to Purkinje cells is the *climbing fiber* input, which originates from the *inferior olivary nucleus.* Climbing fibers enter the cerebellum via the inferior cerebellar peduncle, contralateral to the nucleus from which they originate. They enter the cerebellar cortex, pass through the Purkinje cell layer, and then "climb" along the dendrites of an individual Purkinje cell. An important property of climbing fibers is that in the adult, each Purkinje cell is innervated by one and only one climbing fiber. However, an individual climbing fiber can innervate between 1 and 10 Purkinje cells.

Stellate cells and basket cells are innervated by parallel fibers but not climbing fibers. Stellate and basket cells use γ-aminobutyric acid as their neurotransmitter and form inhibitory synaptic connections with Purkinje cells. Thus, they operate in an analogous fashion to inhibitory interneurons in other brain regions. The axons from stellate cells terminate primarily on Purkinje cell dendrites; the axons from basket cells terminate on Purkinje cell somata. (For an example of the pattern of termination of a basket cell, see Chapter 4.)

Golgi cells are also inhibitory interneurons; however, the axons from the Golgi cells terminate on the dendrites of granule cells. The axon terminals of the Golgi cells enter the synaptic glomerulus, which is formed by a mossy fiber and a cluster of granule cell dendrites (see later); thus the inhibitory synapses are comingled with the excitatory mossy fiber terminals that carry information into the cerebellum.

Input and Output Pathways

The cerebellar input and output pathways enter and leave the cerebellum via the cerebellar peduncles. The inferior and middle cerebellar peduncles contain mostly input pathways; the superior cerebellar peduncle is for the most part an output pathway.

Input Pathways

Following are the cerebellar input pathways. Several have been mentioned in previous chapters.

- *Spinocerebellar pathways,* carrying proprioceptive information, enter the cerebellum via the inferior peduncles.
- *Vestibulocerebellar pathways* carry axons from the vestibular nucleus as well as direct projections from the vestibular labyrinth. These also arrive via the inferior cerebellar peduncles.
- *Pontocerebellar* projections carry information regarding motor output (corollary discharge); these arise from neurons in the pontine

nuclei, which in turn receive input from *corticopontine pathways*. The pontocerebellar projections arrive via the middle cerebellar peduncle.

- The *olivocerebellar pathway* originates from neurons in the inferior olivary nucleus and enters the cerebellulm via the inferior cerebellar peduncle.

Synaptic Connections Formed by Input Pathways to the Cerebellar Cortex

The input pathways to the cerebellar cortex form two types of synaptic connections.

1. Most inputs terminate as *mossy fibers*. Axons that form mossy fiber endings enter via the white matter and terminate in the granule cell layer where they form large club-type endings on the stubby dendrites of granule cells. The large terminals wrap around the dendrites of the granule cells, forming a complex termed a *synaptic glomerulus*. Quantitatively speaking, mossy fibers provide the main input to the cerebellar cortex.

2. Climbing fibers literally climb along the dendrites of individual Purkinje cells. Each Purkinje cell is innervated by one and only one climbing fiber, which forms multiple synaptic connections along the dendrite.

Other inputs to the cerebellar cortex originate from the *locus ceruleus* and *raphe nuclei*. The *cerulocerebellar pathway* is *noradrenergic;* the *raphecerebellar pathway* is *serotonergic*. These monoaminergic projections terminate in both the granular and molecular layers and modulate the firing of granule cells and Purkinje cells.

The basic circuit operation of the cerebellum is determined by the synaptic properties of its constituent neurons.

Activation of Granule Cells by Mossy Fibers

Mossy fibers form excitatory synaptic connections on the dendrites of cerebellar granule cells. Each mossy fiber innervates several granule cells, and each granule cell receives input from several mossy fibers. Thus there is both convergence and divergence. Activation of the mossy fiber initiates action potentials within granule cells. The action potentials are conveyed to the parallel fibers.

Activation of Purkinje Cells by Parallel Fibers

Parallel fibers form excitatory synaptic connections on the dendrites of Purkinje cells. It has been estimated that each individual Purkinje cell receives up to 200,000 parallel fiber synapses. Each one of the individual synaptic connections is relatively weak, so considerable summation must occur to lead to sufficient depolarization of the Purkinje cell to initiate an action potential.

Because of the parallel organization of parallel fibers, activation of a bundle of fibers produces a "beam" of activation of Purkinje cells, basket cells, and stellate cells. The axons of the stellate cells and basket cells project laterally with respect to this beam, producing a surround area of inhibition. This is reminiscent of the circuitry that produces *surround inhibition* in other systems. Golgi cells are also activated by parallel fibers, resulting in *feedback inhibition* of granule cells.

Activation of Purkinje Cells by Climbing Fibers

Climbing fibers also form excitatory synaptic connections on the dendrites of Purkinje cells. Because of the intimate relationships between climbing fibers and Purkinje cells, each action potential in a climing fiber powerfully activates the Purkinje cell, generating a characteristic *complex spike* made up of multiple action potentials of descending amplitude (that is, each spike in the train is smaller than the immediately preceding spike). Because Purkinje cells are contacted by only one climbing fiber, and because Purkinje cells are powerfully activated by the climbing fiber, the climbing fiber can be said to be a *command input* for the Purkinje cell.

How this elegant circuitry actually operates during motor behaviors is not known. It is likely that the circuit interactions will prove to be extremely complex. What is known is that in the absence of the cerebellum, motor functions are seriously disrupted in predictable ways, and this is the most important information for the physician.

Output Pathways

The output of the cerebellar cortex is via the axons of the Purkinje cells. Most of these projections terminate on neurons in the deep cerebellar nuclei. An important exception is that the Purkinje cells in the flocculonodular lobe project out of the cerebellum to the vestibular nuclei. Purkinje cells use γ-aminobutyric acid as their neurotransmitter, and so they exert an inhibitory influence on their targets.

The output of the entire cerebellum is largely via the neurons in the deep cerebellar nuclei. Except for the projections from the Purkinje cells in the flocculonodular lobe, the outputs from the cerebellum originate from neurons in the deep cerebellar nuclei. There are several component pathways: (1) the dentate nucleus and interposed nuclei send projections to the motor thalamus to control the activity of neurons supplying lateral motor pathways; (2) the dentate nucleus also projects to the red nucleus, where it terminates on the cells of origin of a pathway to the inferior olive; (3) the interposed nucleus also projects to the red nucleus, where it terminates on the cells of origin of the rubrospinal tract; and (4) the fastigial nucleus projects to the reticular formation and vestibular nuclei that control the activity of neurons of the medial motor pathways. The organization of these input and output pathways is described in more detail later.

In addition to the inhibitory projections from Purkinje cells, neurons in the deep cerebellar nuclei also receive collateral input from the systems that enter the cerebellar cortex. These collaterals are excitatory. Thus, for example, the systems that give rise to mossy fiber projections to the granule cells in different regions of the cerebellar cortex also give rise to a collateral projection to the deep nucleus, which is innervated by that portion of the cerebellar cortex. In this way, the deep cerebellar nuclei receive the direct excitatory input from the ascending systems, and then inhibitory input from the Purkinje cells.

Cells in the deep cerebellar nuclei also give rise to mossy fiber projections to the cerebellar cortex, so the connections between particular portions of the cerebellar cortex and the associated deep cerebellar nucleus are reciprocal.

Physiology of the Cerebellar Cortex

The cerebellum has been a favorite site for basic neurophysiological studies because of its elegantly simple organization. Hence, a good deal is known about its basic cellular physiology. Interestingly, though, it has been difficult to integrate the information about cellular functions into a unified hypothesis to describe exactly how cerebellar circuitry actually operates to mediate its functions. In this regard, the study of the cerebellum is an excellent example of the difference between clinical and scientific questions. On the one hand, the signs and symptoms that result from cerebellar damage are predictable, characteristic, and reasonably well understood. These *cerebellar signs and symptoms* make it possible for the clinician to accurately determine when injuries involve the cerebellum, what parts of the cerebellum are involved, and so on. On the other hand, our understanding about exactly how cerebellar circuitry operates to modulate motor function is fragmentary.

FUNCTIONAL SUBDIVISIONS OF THE CEREBELLUM

The cerebellum is divided into functional subdivisions based on the topographic organization of the principal cerebellar input, and the topography of corticonuclear projections (Table 17.1).

Following are the subdivisions:

1. The *vestibulocerebellum*
2. The *spinocerebellum*
3. The *cerebrocerebellum*

Table 17.1. Functional circuits of the cerebellum

Subdivision	Location	Principal input	Deep nucleus	Output	Function
Vestibulo-cerebellum	Flocculonodular lobe	Vestibular (direct from labyrinth, indirect from vestibular N.)	Lateral vestibular	Vestibular N.	Vestibulospinal pathways responsible for posture and balance
Spinocere-bellum (medial component)	Vermis	Spinocerebellar Vestibular	Fastigial	Vestibular N. Reticular Form.	Medial brainstem pathways Control of axial and proximal musculature during execution of movement
Spino-cerebellum (lateral component)	Intermediate lobe	Spinocerebellar	Interposed	Red Nucleus VA/VL thalamus	Lateral motor pathways (LCST rubrospinal tract). Control of distal musculature during execution of movement
Cerebro-cerebellum	Lateral hemisphere	Pontocerebellar	Dentate	VA/VL thalamus Red nucleus (parvocellular division)	Motor and premotor cortices Important for the planning, initiation, and timing of complex movements, especially skilled motor tasks

The Vestibulocerebellum Is Particularly Involved in Regulating Balance and Equilibrium

The vestibulocerebellum is the portion of the cerebellum that receives its primary input from the vestibular nuclei and sends a reciprocal projection back to the vestibular nuclei. The basic circuitry is illustrated in Figure 17.5. These interconnections are important for regulating the movements that help to maintain body equilibrium, and the eye movements that occur in response to vestibular signals (the vestibulo-ocular reflex).

The Spinocerebellum Is Particularly Involved in Adjusting Movements as They Are Being Performed Based on Proprioceptive Feedback

The spinocerebellum is the portion of the cerebellum that receives its principal input

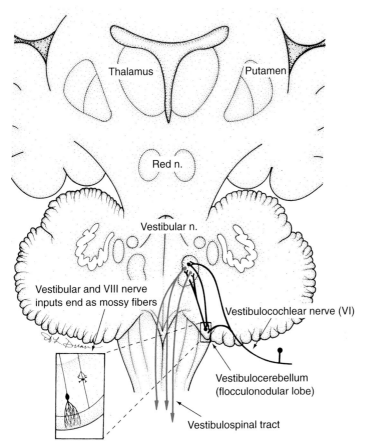

Thalamus

Putamen

Red n.

Vestibular n.

Vestibular and VIII nerve
inputs end as mossy fibers

Vestibulocochlear nerve (VI)

Vestibulocerebellum
(flocculonodular lobe)

Vestibulospinal tract

Figure 17.5. Circuitry of the vestibulocerebellum. The vestibulocerebellum is the portion of the cerebellum that receives its primary input from the vestibular nuclei, the semicircular canals, and the otolyth organs. These inputs enter the cerebellum via the inferior cerebellar peduncle. The vestibulocerebellum corresponds to the flocculonodular lobe. Purkinje cells project out of the cerebellum proper, back to the vestibular nucleus, which in turn gives rise to descending projections that control the extensor musculature involved in posture and balance (the lateral and medial vestibulospinal pathways).

from the *spinocerebellar tract*. It includes the centrally located vermis and the intermediate part of the hemisphere just lateral to the vermis (the intermediate lobe). Most of the spinocerebellar fibers that terminate in these regions enter the cerebellum via the posterior cerebellar peduncle (Figure 17.6). However, a small component (forming the *anterior spinocerebellar tract*) enters via the superior cerebellar peduncle. The spinocerebellar tract carries proprioceptive information relating to limb position and movement (that is, from stretch receptors and Golgi tendon organs, as well as from cutaneous receptors).

The spinocerebellar projection terminates in a somatotopically organized fashion. There are two body representations: one is in the anterior cerebellum, and a second is in the posterior lobe. The "homunculus" in the posterior lobe is oriented with the head region

facing rostrally; the map in the anterior lobe has the opposite orientation. However, the nature of the map is complex in that the same body parts are represented in multiple different locations. This pattern is termed *fractured somatotopy.*

Subcircuits in the Spinocerebellum Regulate Medial Versus Lateral Descending Motor Pathways

There are two subcircuits in the spinocerebellum that regulate medial versus lateral descending motor pathways.

The Vermis–Fastigial Nucleus–Brainstem Circuit

Purkinje cells in the vermis project to the medial-most of the deep cerebellar nuclei (the fastigial nucleus). The fastigial nucleus

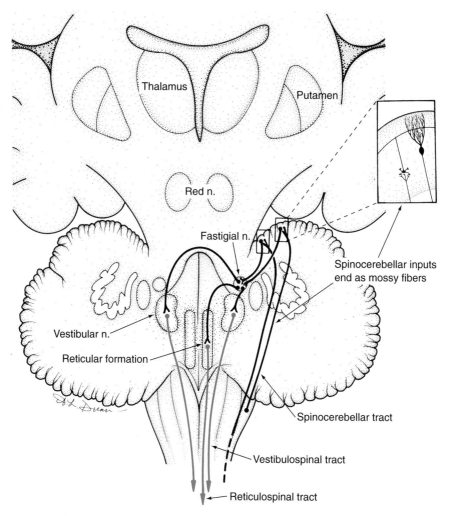

Thalamus

Putamen

Red n.

Fastigial n.

Spinocerebellar inputs
end as mossy fibers

Vestibular n.

Reticular formation

Spinocerebellar tract

Vestibulospinal tract

Reticulospinal tract

Figure 17.6. Circuitry of the spinocerebellum.
Vermis–fastigial nucleus–brainstem motor
pathways. The spinocerebellum is the portion of
the cerebellum that receives its principal input
from the spinocerebellar tract. It includes the
vermis and the intermediate lobe. One component
of this circuitry provides input to the vestibuli
nuclei and the portion of the brainstem reticular

formation that gives rise to descending motor
pathways that control proximal muscles and
extensors (the medial brainstem pathways). The
circuit involves a projection from the Purkinje cells
to the fastigial nucleus, which in turn projects to
the vestibular nuclei and the brainstem reticular
formation. The other component of this circuitry is
illustrated in Figure 17.7.

in turn projects to the portion of the reticular
formation that gives rise to the reticulospinal
pathways and to the lateral vestibular nucleus, which gives rise to the vestibulospinal
pathway (Figure 17.6). Thus, the vermis–
fastigial nucleus are particularly involved
with controlling the output of the medial
brainstem pathways that regulate the mus-

cles that are important for posture and balance.

The projections from the fastigial nucleus
leave the cerebellum via the superior cerebellar peduncle. The projections to the vestibular nucleus are bilateral. The projections to
the contralateral side loop over the superior
cerebellar peduncle as the *uncinate fasciculus*

and then enter the contralateral superior cerebellar peduncle. There is also a more minor projection from the fastigial nucleus to the ventrolateral nucleus of the thalamus.

The Intermediate Lobe–Interposed Nucleus–Red Nucleus/Thalamus Circuit

As shown in Figure 17.7, Purkinje cells in the intermediate lobe project to the interposed nucleus. The interposed nucleus projects in turn to the red nucleus via the superior cerebellar peduncle, and also to the ventral lateral nucleus of the thalamus. The pathways originating in the intermediate lobe of the spinocerebellum influence primarily the lateral components of the descending motor systems so as to regulate the corticospinal and rubrospinal projections. Thus, the lateral-most part of the spinocerebellum (the intermediate lobe) controls primarily the lateral component of the descend-

ing motor system, which regulates distal musculature.

The Double Cross

An important point on laterality is that the projections from the interposed nucleus decussate in the decussation of the superior cerebellar peduncle and thus affect corticospinal and rubrospinal pathways on the contralateral side. However, both the corticospinal pathway and the rubrospinal pathways decussate on their descent to the spinal cord. This is an example of the characteristic "double crossing" of cerebellomotor circuitry. For this reason, each side of the cerebellum regulates motor function on the ipsilateral side of the body.

The spinocerebellum plays an important role in regulating muscle tone via the descending projections of the medial brainstem pathways to gamma motoneurons. Damage

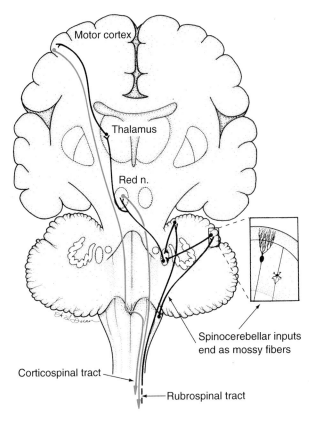

Figure 17.7. Circuitry of the spinocerebellum. The intermediate lobe–interposed nucleus–red nucleus/thalamus circuit. Purkinje cells in the intermediate lobe project to the interposed nucleus. The interposed nucleus projects to the red nucleus via the superior cerebellar peduncle, and also to the ventral lateral and ventral anterior nuclei of the contralateral thalamus. This circuit is involved in the control of lateral motor pathways.

Motor cortex

Thalamus

Red n.

Spinocerebellar inputs end as mossy fibers

Corticospinal tract

Rubrospinal tract

to the spinocerebellum or to the deep cerebellar nuclei leads to decreases in muscle tone (hypotonia).

Damage to the cerebellum can also contribute to hypertonia if it occurs in conjunction with injuries that produce decorticate (gamma) rigidity. The reason is the loss of the inhibitory pathway from the cerebellar cortex to the vestibular nucleus. The disinhibition of vestibular neurons that occurs as a result of injury to the cerebellar cortex leads to an uncontrolled facilitation of alpha motoneurons (alpha rigidity). An important point, however, is that alpha rigidity is seen only in conjunction with gamma rigidity; that is, injuries to the cerebellum alone do not produce rigidity and instead produce more complex motor symptoms (see later).

The Cerebrocerebellum Is Particularly Involved With the Preparation and Initiation of Complex Motor Sequences

The cerebrocerebellum (also termed the neocerebellum) corresponds to the lateral portion of the cerebellar hemisphere, which receives its principal mossy fiber input from the pontine nuclei (see Figure 17.8). The neurons in the pontine nuclei receive input from corticopontine fibers that originate in the ipsilateral cerebral cortex. The areas of the cortex that provide input to the pontine nuclei include the sensory and motor cortices as well as premotor and posterior parietal areas. Axons from neurons in the pontine nuclei cross the midline, ascend into the contralateral cerebellar cortex via the middle cerebellar peduncle, and terminate as mossy fibers.

Purkinje cells in the cerebrocerebellum project to the dentate nucleus. Thus, the lateral-most part of the cerebellar cortex projects to the lateral most of the deep nuclei. Neurons in the dentate nucleus send their axons into the superior cerebellar peduncle where they decussate and then project to the ventral lateral and ventral anterior thalamus. The thalamus in turn projects to the motor and premotor cortex. Via these pathways, one side of the cerebrocerebellum modulates the output of the motor and premotor cortices on the contralateral side. Again, the corticospinal projections decussate, and so, through double-crossing, one side of the cerebellum controls the motor cortex that supplies the ipsilateral side of the body.

Feedback Control Is Mediated Via a Projection From the Dentate Nucleus to the Red Nucleus to the Inferior Olive

In addition to its projection to the motor thalamus, the dentate nucleus also projects to the red nucleus. This dentatorubral projection terminates on neurons in the parvocellular division of the red nucleus (a portion of the red nucleus that does not give rise to the rubrospinal tract). Neurons in the parvocellular portion of the red nucleus project to the ipsilateral inferior olive. The inferior olive in turn gives rise to the climbing fiber projections that recross the midline to project back to Purkinje cells throughout the cerebellar cortex. The dentate nucleus–red nucleus–inferior olive pathway thus provides a route for feedback modulation of Purkinje cell activity in all of the different functional subdivisions of the cerebellum.

The lateral cerebellum–dentate nucleus–red nucleus–inferior olive–cerebellar cortex circuit is another example of double crossing. The first crossing takes place in the decussation of the superior cerebellar peduncle. The recrossing occurs in the projection from the inferior olivary nucleus to the contralateral cerebellar hemisphere (Figure 17.9).

The cerebrocerebellum is thought to be particularly important for complex movements involving multiple joints, tasks requiring fine dexterity, especially highly skilled tasks that are performed rapidly (playing the piano, typing). These types of tasks require coordination of all parts of the body and so require coordination between all of the different functional subdivisions of the cerebellum. One potential substrate for this coordination is the dentate nucleus–red nucleus–inferior olive–climbing fiber pathway.

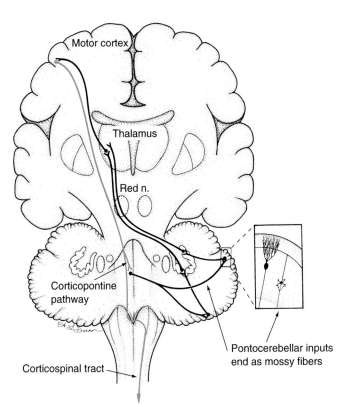

Figure 17.8. Circuitry of the cerebrocerebellum. The cerebrocerebellum receives its principal input from the cerebral cortex via the pontine nuclei. The neurons in the pontine nuclei receive input from corticopontine fibers that originate in the ipsilateral sensory and motor cortices, as well as premotor and posterior parietal areas. Axons from neurons in the pontine nuclei cross the midline and ascend into the contralateral cerebellar cortex via the middle cerebellar peduncle. The Purkinje cells in the cerebrocerebellum project to the dentate nucleus. The dentate nucleus projects to the ventral lateral and ventral anterior nuclei of the contralateral thalamus.

Physiological studies have suggested that the lateral cerebellum plays a role in triggering the final output of the primary motor cortex. Thus, whereas the spinocerebellum is thought to be involved primarily in modulating movement as it is occurring (that is, movement execution and feedback adjustment), the cerebrocerebellum is thought to be important for setting up the coordinated motor program that is to be played out. It is for this reason that lesions of the cerebellum are thought to lead to delays in initiating movements.

The Cerebellum and Motor Learning

Motor learning differs from other types of learning (for example, verbal learning) in two ways: (1) the learning requires extensive practice (repetitions of the sequence of movements); (2) when the skill is well de-

veloped, the motor sequences are played out without conscious effort. The unique features of motor learning are thought to be due to the types of processing that takes place in the cerebellum.

Different parts of the cerebellum play a role in different types of motor learning. In the vestibulocerebellum, motor learning is represented by adaptation of the vestibulo-ocular reflex. This reflex allows the eyes to remain focused on an object when the head is turned. When humans are fitted with goggles that reverse the left and right visual field, the vestibulo-ocular reflex is reversed. However, in a surprisingly short amount of time, humans can learn to reverse the reflex, allowing the eyes to remain fixed as the head moves.

Other parts of the cerebellum are thought to be involved in more complex motor learning involving distal musculature—playing a musical instrument, skilled typing, a physical

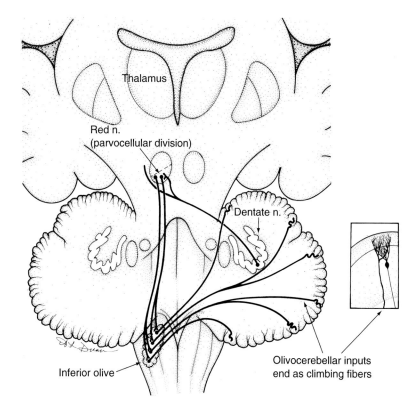

Thalamus

Red n.
(parvocellular division)

Dentate n.

Olivocerebellar inputs
end as climbing fibers

Inferior olive

Figure 17.9. The dentate nucleus–red nucleus–inferior olivary pathway. The dentate nucleus also sends a projection to the parvocellular portion of the red nucleus. This portion of the red nucleus does not give rise to the rubrospinal tract; instead these neurons project to the ipsilateral inferior olive. The inferior olive in turn gives rise to the climbing fiber projections back to Purkinje cells in the ipsilateral cerebellar cortex. The dentate nucleus–red nucleus–inferior olive pathway thus provides a route for feedback modulation of Purkinje cell activity throughout the different functional subdivisions of the cerebellum.

sport. These motor tasks are often complex temporal sequences that are executed without conscious effort (for example, when skilled pianists play the piano, they do not have to think of the sequence of notes to be played; instead, the motor program seems to run without conscious effort).

CLINICAL DISORDERS AFFECTING CEREBELLAR CIRCUITRY

Damage to cerebellar circuitry can result from direct physical injuries (trauma) and from interruption of the blood supply (stroke). In addition, the cerebellum can be affected by a number of other important disease processes and toxins.

Chronic alcoholic patients sometimes exhibit a selective degeneration of the anterior portions of the cerebellum (vermis and leg representation areas). As a result, these patients exhibit characteristic cerebellar symptoms involving the lower limbs, such as difficulty in walking.

An interesting and poorly understood cerebellar degeneration sometimes occurs as a result of metastatic carcinoma, especially ovarian carcinoma. Although there may be no direct invasion of the cerebellum by metastases, cerebellar Purkinje cells degenerate. This is thought to be due to an autoimmune response in which antibodies that are produced in response to the tumor recognize antigens on Purkinje cells.

The cerebellum is also affected by a number of hereditary disorders, collectively termed the *hereditary cerebellar ataxias*. The symptoms are similar to those produced by cerebellar injuries.

Cerebellar Signs and Symptoms

Clinicians have an extensive terminology to refer to the signs and symptoms that result from cerebellar injuries. Some of the terms that are used are defined briefly below.

- *Hypotonia* is decreased muscle tone.
- *Hypermetria* is a failure to terminate a movement at the appropriate time (also termed *lack of check, overshoot*).
- *Ataxia* is an unsteadiness and lack of coordination of movement. The symptom complex includes (1) delays in initiating and terminating movements, (2) errors in the range and force of movement *(dysmetria)*, (3) errors in the rate and regularity of movements *(dysdiadokokinesia)*.
- *Titubation* refers to staggering or stumbling while tring to walk.
- *Ataxic gait* is manifested by unsteady balance and a wide stance (the "drunken sailor's gait").
- *Decomposition of movement* refers to a disruption in the timing and coordination of motor sequences involving multiple joints. For example, rather than executing a smooth continuous motion, a movement will seem to be carried out in stages.
- *Action tremor* (also called *intention tremor*) refers to a tremor during the performance or especially at the end of a movement. Tremor resulting from cerebellar injury is not exhibited when the limb is at rest. This is different from the resting tremor of Parkinson's disease.

In terms of localization, these are the most important principles:

1. Unilateral lesions of the cerebellum produce ipsilateral symptoms. This is due to the double crossing of cerebellar circuits.
2. Because of the somatotopic organization of the spinocerebellar projection, midline lesions affect axial musculature, lateral lesions affect distal musculature.

Lesions Involving the Vestibulocerebellum

Damage to the flocculonodular lobe disrupts the ability to use vestibular information to coordinate movement, resulting in disturbances in balance and equilibrium as well as producing *nystagmus* because of the disruption of the vestibulo-ocular reflex.

Lesions Involving the Spinocerebellum

Midline lesions (vermis) affect trunk musculature, producing *titubation* and *gait ataxia*. Because the control of facial musculature is also affected, speech articulation is impaired *(dysarthria)*, leading to a slurring and slowing of speech. Patients may exhibit a speech rhythm in which successive syllables emerge slowly, termed *scanning speech*.

Lateral lesions (intermediate lobe) produce ataxia of limb movements and *action tremor* and lead to *decomposition of movements* involving the limbs.

Lesions Involving the Cerebrocerebellum

Lesions of the lateral part of the cerebellum produce three primary symptoms:

1. Delays in initiating and terminating movement
2. Action tremor and terminal tremor after the completion of a movement
3. Decomposition of movement

The delay in initiating movement is thought to reflect the fact that the cerebrocerebellum is important for setting up the coordinated motor program before it begins. Terminal tremor is explained by a disruption in timing. Normally, rapid motor movements are terminated at the correct time by activation of antagonist muscles. With damage to the lateral cerebellum, this activation is delayed, leading to overshoot *(hypermetria,* or *lack of check*). In compensating for the overshoot, the same thing occurs, so there is overcorrection, leading to an oscillation of the limb. The same sort of oscillation (tremor) occurs when the limb is passively displaced.

Decomposition of movement involving multiple joints is reflected by changes in the speed and regularity of movements (termed *diadochokinesia*). Temporal coordination is assessed clinically by asking the patient to pat his or her hand, alternating between the back and the palm of the hand. The timing and regularity of the tapping is grossly abnormal in patients with cerebellar injuries (a sign termed *dysdiadochokinesia*).

Patients with cerebellar injuries may describe their symptoms in terms of a *loss of automaticity*. That is, that prior to the injury, they did not have to think about their motor performance. Following the injury, their motor performance requires conscious control over each of the fractional movements comprising a motor sequence. It is primarily for this reason that rapid, skilled movements are no longer possible.

An interesting aspect of cerebellar injuries is that there may be considerable recovery, especially of more simple motor skills. However, deficits in complex, learned motor functions are often permanent.

5

The Visual System

The visual system is one of the "special" senses. The components that are involved in conscious perception of visual information include the following:

1. Photoreceptors in the neural retina in the eye

2. Neuronal circuitry in the retina that carries out the initial processing of visual information

3. Cranial nerve II (the optic nerve), which carries information from the eye to the brain

4. Relay nuclei in the thalamus (specifically, the lateral geniculate nucleus)

5. The visual cortex

Parallel pathways convey visual information from the eye to other stations including the superior colliculus, pretectal area, and hypothalamus. These parallel pathways carry visual information to coordinate conjugate eye movements, mediate optic reflexes, and regulate the centers responsible for circadian behaviors (see Figure V.1).

We divide our consideration of the visual system into four major sections: Chapter 18 deals with the photoreceptors and the neuronal circuitry in the eye; Chapter 19 considers central visual pathways; Chapter 20 considers visual neurophysiology and the basis of visual perception; and Chapter 21 considers the visual inputs to the centers that coordinate eye movements, mediate optic reflexes, and regulate circadian behaviors.

There are several key concepts for physicians:

1. The optical components of the eye function like a camera, producing a *visuotopic map* of the *visual field* upon the surface of the retina. In this way, different parts of the retina "represent" particular parts of the visual field. (The visual field is the total visual image that is seen by the two eyes.)

2. One side of the brain receives information from the contralateral visual field. The total visual field can be divided into right and left halves. Within each eye, the parts of the retina that represent the half of the visual field on the same (ipsilateral)

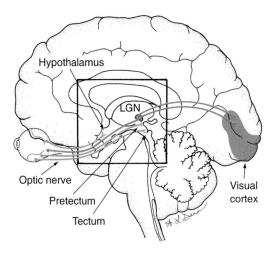

Figure V.1. Overview of visual system organization. The overall organization of the projections from the eye to the brain projected on a schematic illustration of the brain in midsagittal section. Retinal axons project from the eye through the optic nerve to terminate in the following regions: (1) the lateral geniculate nucleus (LGN), which is the thalamic relay nucleus for projections to the primary visual cortex (striate cortex). This is termed the *retinogeniculostriate pathway.* (2) The superior colliculus in the tectum (termed the *retinotectal pathway*). (3) The pretectal area. (4) The pregeniculate nucleus (not shown). (5) The suprachiasmatic nucleus of the hypothalamus (termed the *retinohypothalamic pathway*). Modified from Kandel ER, Schwartz JH, Jessell TM. *Principles of Neural Science* 3rd ed. New York, NY: Elsevier; 1991.

sis of the site of lesions that affect the central visual pathways.

3. Convergence of visual information from the two eyes first occurs at the level of the visual cortex.

4. There are several distinct "parallel pathways" or "channels" that convey information regarding different features of the visual environment.

5. Conscious perception of visual sensation depends on inputs to the cerebral cortex. Thus, interruption of the visual pathways anywhere along their course from the retina to the cortex leads to a loss of visual perception. Nevertheless, subcortical regions mediate important visual functions, and these survive after cortical injuries.

The *retinogeniculostriate* pathway is specialized for *form vision,* which is the ability to perceive different configurations and spatial relationships between objects in the visual environment. This type of vision allows us to identify objects in the environment. The *retinotectal pathway* is important for *visually guided behaviors,* especially orientation toward visual stimuli. The projections to the *pretectal area* provide the sensory limb for ocular reflexes (accommodation and pupillary light reflex). The projections to the *pregeniculate nucleus* are important for controlling conjugate eye movement. The function of these pathways is discussed in Chapter 21. The projections to the *suprachiasmatic nucleus* (the retinohypothalamic pathway) play an important role in regulating *circadian rhythms* (from the Latin *circa* and *dies,* meaning "around the day").

side project to the contralateral cerebral cortex. The parts of the retina that represent the contralateral half of the visual field project ipsilaterally. The decussation occurs at the optic chiasm. The partial crossing at the level of the optic chiasm provides the basis for differential diagno-

18

Visual System I

The Retina

Visual transduction and a fair amount of the initial neural processing of visual information occurs with the local circuitry that is present in the *neural retina* within the eye. Most of the visual disorders with which physicians must contend involve the eye or the retina rather than the central visual pathways.

But the eye also provides a "window" to the brain. It is possible for the physician to look into the eye with an opthalmoscope and actually see the retina, which is in fact part of the central nervous system (CNS). These simple examinations can provide important diagnostic information about disorders affecting the CNS. This is true because the neural component of the eye (the retina) derives from and remains part of the CNS, and the circulation of the eye originates from central vasculature. In addition, many important disorders of the CNS are expressed as disorders of vision, disordered eye movements, or pupillary asymmetry. For this reason, there is a fair amount of information that physicians need to know about the eye.

We begin by identifying the physical structures of the eye that allow light to be focused on the retina. Students should be familiar with the names and functions of the different parts of the eye and the regional specializations of the retina that are important for vision. We will then describe the cellular organization of the retina, focusing especially

on the different types of photoreceptors and their interconnections. Students should be familiar with the names of the different cell types and should be able to describe the interconnections.

STRUCTURE OF THE EYE

The key structural features of the eyeball are indicated in Figure 18.1. The white outer rind of the eyeball is the *sclera*. The anterior portion of the eyeball is the clear *cornea*. The junction between the cornea and sclera is termed the *limbus*. Behind the cornea is the pigmented *iris,* which confers the signature eye color. The whole in the center of the iris is the *pupil*. Behind the iris is the *lens,* which is held in place laterally by the suspensory ligaments, which in turn attach the lens to the *ciliary body*. The *ciliary muscles* then attach the ciliary body to the interior surface of the sclera.

The iris divides the space between the cornea and the lens into the anterior chamber and posterior chambers, which are filled with a clear fluid termed the *aqueous humor*. The aqueous humor is secreted by the ciliary body and is reabsorbed into the venous drainage of the sclera via the *canal of Schlemm*.

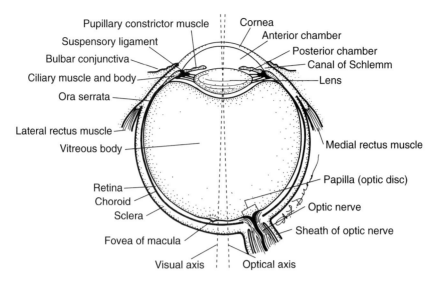

Figure 18.1. Cross section through the eye illustrating the key landmarks.

The aqueous humor is important for hydrating and nourishing the lens and cornea. Normally, the rate of production is balanced with the rate of reabsorption, so that an intraocular pressure of 15 to 20 mm Hg is maintained. If the balance is disrupted, intraocular pressure increases. This is what occurs in *glaucoma*. Eventually, the increased pressure resulting from glaucoma can restrict blood flow to the retina, causing blindness. For this reason, it is important to detect and treat glaucoma early in the course of the disease.

The large space behind the lens is filled with a viscous substance termed the *vitreous humor,* or *vitreous body*. The inner surface of the rind of the eyeball is composed of an inner *neural retina,* or just the *retina* and a *choroid layer* that lies between the retina and the sclera.

Students should be familiar with the terms used to refer to different parts of the eye and the functions of the different parts, as summarized in Figure 18.1 and Table 18.1.

FUNCTIONAL COMPONENTS OF THE EYE

The elements of the eye that are directly involved in vision can be divided into two components, according to their function: (1) an *optical system* for focusing light on the retina and adjusting the amount of light passing into the eye, and (2) a *receptive system* for transducing light into neural activity. The focusing system comprises the cornea, lens, and suspensory ligaments. The amount of light that is transmitted is regulated by the iris, which is a diaphragm that opens and closes depending on the amount of ambient light. The receptive system is the neural retina, which is made up of photoreceptors (rods and cones) and several types of neurons.

The Optical System

Operation of the Optical System

The light that enters the eye is focused on the retina by an optical system that *refracts* (bends) the light, altering its path. The first refraction occurs at the curved surface of the cornea because the cornea has a different refractive index than air. The light then passes through the anterior chamber, pupil, and posterior chamber, to a second refractive structure, the lens.

Focusing the Visual Image

The importance of the lens is that its refractive index can be changed by adjusting

Table 18.1. Structures comprising the eye

Structure	Description/function
Central retinal artery and vein	Vascular supply and drainage
Choroid	Vascularized nutritive layer between the sclera and the retina
Ciliary body	Ring of tissue composed of ciliary epithelium (for production of aqueous humor) and ciliary muscle
Ciliary muscle	Alters lens shape to focus light rays on the retina
Cornea	The clear external membrane through which light enters; provides initial refraction
Iris	Diaphragm regulating amount of light admitted
Pupil	Opening at center of iris
Lens	Focuses light on retina
Macula lutea	"Yellow body": the area of highest cone density and the area of central vision
Fovea	Central portion of the macula lueta with highest receptor density
Optic disc	Circular depression where optic nerve fibers collect to pass out of the eyeball.
Optic nerve	Axons from retinal ganglion cells to CNS
Retinal Pigment	Single layer of pigmented epithelial cells behind retina
Epithelium (RPE)	Neural retina; provides trophic support to photoreceptors
Anterior chamber	Fluid-filled space between iris and cornea
Posterior chamber	Fluid-filled space between iris and lens
Posterior ciliary artery	Blood supply for choroid
Sclera	Tough outer covering of eyeball
Suspensory ligaments	Also called zonule fibers; attach lens to ciliary body

lens shape so as to focus the visual image on the retina. The normal shape of the lens would be spherical were it not for the fact that suspensory ligaments attach to the capsule's equator and draw the lens into a flattened biconvex shape. The suspensory ligaments stretch between the lens and the ciliary body. The ciliary body has bands of muscle fibers (ciliary muscles) running like an annulus around the lens, attaching between the *choroid* and the *scleral roll*.

Ciliary muscles operate like a sphincter. When the ciliary muscles are relaxed, the aperture formed by the annulus is widest. In this configuration, the tension in the suspensory ligaments is maximal and the ligaments pull the lens into a flattened shape. When flattened, lens power is minimal, and the eye is focused for distant objects. When the ciliary muscles contract, the aperture decreases in diameter, and the tension in the *zonule fibers* is reduced, allowing the lens to assume a more spherical shape. In this configuration, lens power is maximal, and the eye is accomodated (focused for near objects).

Light rays are, for practical purposes, parallel if their source is 6 m or more from the eye. In a normal *(emmetropic)* eye, the shape of the lens with ciliary muscles relaxed is such that parallel rays focus on the retina. Light rays from objects closer than 6 m are divergent rather than parallel; for these rays

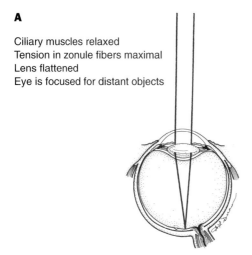

A

Ciliary muscles relaxed
Tension in zonule fibers maximal
Lens flattened
Eye is focused for distant objects

B

Ciliary muscles contracted
Tension in the zonule fibers reduced
Lens assumes more spherical shape
Eye focused for near objects

Figure 18.2. The operation of the lens. The refractive index of the lens can be changed by adjusting lens shape so as to focus the visual image on the retina. Suspensory ligaments attach to the equator of the lens and extend to the ciliary body. The ciliary body has bands of muscle fibers *(ciliary muscles)* running like an annulus around the lens, attaching between the choroid and the scleral roll. These ligaments draw the lens into a flattened biconvex shape. Ciliary muscles operate like a sphincter. When the ciliary muscles are relaxed, the aperture formed by the sphincter is widest and the suspensory ligaments draw the lens into a flattened biconvex shape. When flattened, lens power is minimal, and the eye is focused for distant objects (**A**). When the ciliary muscles contract, the aperture decreases in diameter, and the tension in the zonule fibers is reduced, allowing the lens to assume a more spherical shape. In this configuration, lens power is maximal, and the eye is accommodated (focused for near objects) (**B**). As the lens becomes inelastic with age, it is less capable of undergoing the shape change that underlies accommodation.

to be focused on the retina, lens shape must be altered (Figure 18.2).

Optical Defects

There are three types of optical defects:

1. Refractive errors cause light to be focused in a plane that is either in front of or behind the retina. Refractive errors result from a mismatch between corneal curvature and eyeball length (Figure 18.3). Eyes with optical defects are termed *ametropic.*

    (a) *Hyperopia* (farsightedness) occurs when the eyeball is too short with respect to the refraction provided by the cornea and lens. In such an eye, the principal focus occurs in a position behind the retina.

    (b) *Myopia* (nearsightedness) occurs when the eyeball is too long with respect to the refraction provided by the cornea and lens, so the principal focus occurs in front of the retina (Figure 18.3).

2. Abnormalities in the curvature of the refracting surface (the cornea).

    (a) *Spherical aberrations* refer to a condition in which the lens curvature is not precisely spherical. For example, in a condition called *keratoconus,* the cornea thins at its margin and protrudes from the center, such that the cornea assumes a conelike shape. This condition is treated by corneal transplants.

    (b) *Astigmatism* is a condition in which the curvature of the cornea, and thus the refractive index, varies at different locations across the corneal surface. As a result, the visual image is brought into focus over a distance between two focal planes rather than within a single plane (Figure 18.4). Astigmatism can be corrected with external lenses or by contact lenses that replace the defective corneal refractive surface.

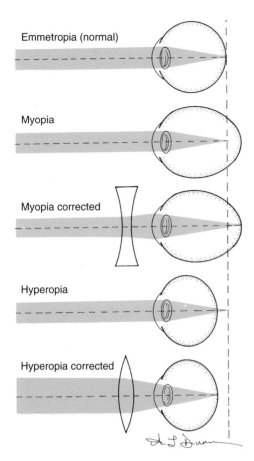

Emmetropia (normal)

Myopia

Myopia corrected

Hyperopia

Hyperopia corrected

Figure 18.3. Optical defects. In a normal (emmetropic) eye, light focuses in the plane of the retina. Refractive errors in eyes with optical defects cause light to be focused in a plane that is either in front of or behind the retina. Myopia (nearsightedness) occurs when the eyeball is too long with respect to the refraction provided by the cornea and lens, so the principal focus occurs in front of the retina. Hyperopia (farsightedness) occurs when the eyeball is too short with respect to the refraction provided by the cornea and lens. In such an eye, the principal focus occurs in a position behind the retina. After Kingsley RE. *Concise Text of Neuroscience.* Baltimore, Md: Williams & Wilkins; 1996.

3. Abnormalities in accommodation, called *presbyopia,* result from a loss of lens elasticity. Accommodation for near-viewing requires that the lens assume a spherical shape when it is not under tension from the suspensory ligaments. As the lens loses elasticity, this transition is prevented. The result is a loss in ability to accommodate for near vision. Presbyopia occurs with aging. The rate of loss

of elasticity increases substantially at about 45 years of age, and by about 60 years of age, the lens has very little elasticity. Presbyopia can be corrected by bifocals, which provide a different refraction for near versus far objects.

Mapping the Visual Image Onto the Retina

The optic system operates much like a camera; the visual image is projected onto the retina such that each point in the retina maps a point in the visual field. This is termed a *visuotopic map.* Thus, damage to a part of the retina leads to a loss of visual perception in the corresponding portion of the visual field in that eye. For this reason, it is important for the physician to understand the nature of the map.

Understanding the topography of the map is straightforward. One has only to imagine a straight line extending from any point in the visual environment, through the center of the iris, to the retina (see Figure 18.1). Such a straight line points to the site on the retina that maps the point in visual space from which the line originated.

The visual field is the portion of the environment that we see at any one time. We refer to positions in the visual field with reference to the *vertical* and *horizontal meridians* (the perpendicular lines that divide right and left, and upper and lower portions of the visual field). One can think of these lines as an imaginary crosshair, with the center of the crosshair being the focal point. When referring to the entire visual field that is seen when both eyes are open, we refer to right and left halves, upper and lower halves, and quadrants (right upper quadrant, left upper quadrant, and so on).

The terminology is different when speaking of the visual field of an individual eye; in this case, the portion of the visual field that appears near the nose is the *nasal* portion of the visual field; the lateral portion of the field is termed the *temporal* portion of the field (Figure 18.5).

The visual field of each eye extends from about 90° laterally to about 60° medially

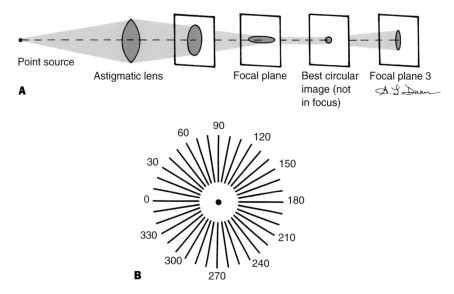

Figure 18.4. Astigmatism. Astigmatism is a condition in which the curvature of the cornea varies at different locations across the corneal surface. As a result, a point is brought into focus over a distance between two focal planes rather than within a single plane (**A**). The condition can be demonstrated using a set of radially oriented lines called a *Lancaster–Regan chart* (**B**). All lines appear in focus to an observer with normal vision. For individuals with astigmatism, the lines at one orientation appear in focus, whereas perpendicular lines will appear out of focus. Reproduced with permission from Kingsley RE. *Concise Text of Neuroscience.* Baltimore, Md: Williams & Wilkins; 1996.

(with respect to the point of focus). The central portion of the visual field that can be seen simultaneously by both eyes is the *binocular segment.* The lateral-most 30° that can be seen by only the ipsilateral eye is the *monocular segment.*

Lesions at various locations in the visual pathways produce highly characteristic visual field defects (that is, blind spots in particular portions of the visual field of one or both eyes). For this reason, it is important for physicians to be able to map visual fields. Routine mapping can be done without special equipment. Patients are positioned about 1 m from a dark screen and asked to focus on a point in the center of the screen. Each eye is then tested individually (ask the patient to cover one eye). During testing, the physician moves a small, white stimulus object across the screen. The patient is asked to report when the stimulus appears and disappears. Compliance can be determined by accurate reporting of the disappearance of the stimu-

lus as it enters the *blind spot* (see Figure 18.9). Such routine exams are often sufficient to deduce the site of the lesion. More precise determinations of visual field defects can be obtained using a perimeter apparatus, as illustrated in Figure 18.6.

The nomenclature for positions on the retina is based on anatomical location (Figure 18.6). The *nasal hemiretina* is the half that is closest to the nose; the *temporal hemiretina* is the lateral half that is near the temple. The retina is also divided into dorsal and ventral halves. In this way, locations in the retina are referenced with regard to quadrants: *upper temporal, lower temporal, upper nasal,* and *lower nasal.*

Remember, the terms *nasal* and *temporal* are used to refer both to parts of the visual field and to parts of the retina, and the two terminologies do not correspond. Thus, light from the temporal visual field of a given eye strikes the nasal half of the retina, and so forth (see Figure 18.5).

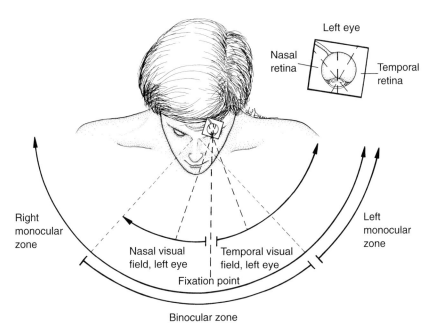

Figure 18.5. Nomenclature for locations in the visual field and retina. The nomenclature for positions in the visual field is based on the vertical and horizontal meridians (the perpendicular lines that divide right and left, and upper and lower portions of the visual field). When speaking of the visual field of an individual eye, the portion of the visual field that appears near the nose is the nasal portion of the visual field; the lateral portion of the field is termed the temporal portion of the field.

The nomenclature for positions in the retina is based on anatomical location. The nasal hemiretina is the half that is closest to the nose; the temporal hemiretina is the lateral half that is nearest the temple. The retina is also divided into dorsal and ventral halves. Locations in the retina are referenced with regard to quadrants: upper temporal, lower temporal, upper nasal, and lower nasal.

The Receptive System

The retina is the receptive system of the eye. It is made up of the photoreceptors and several different types of neurons that represent the initial substrate for processing visual information.

The retina is inside out in terms of light. We refer to layers of the retina with respect to the spherical eyeball. Thus *inner* means toward the center of the sphere; *outer* means toward the surface. In the retina, the photoreceptor layer is actually the outer layer (that is, the layer immediately apposed to the retinal pigment epithelium [RPE]). As a result, light must pass through all of the other neural layers before reaching the photoreceptors (Figures 18.7 and 18.8).

The retina has a total of 10 layers composed of the cell bodies and processes of photoreceptors and neurons, and *limiting membranes* formed by the processes of specialized astroglial cells termed *Müller cells*. Beginning at the outer edge, the layers are as follows:

1. The epithelial cells of the RPE contain pigments that reduce light scattering.

2. The *photoreceptor layer* contains the *outer segments* of the photoreceptors.

3. The outer limiting membrane is formed by processes of Müller cells (specialized astrocytes).

4. The *outer nuclear layer* contains the cell bodies of the photoreceptors (the rods and cones).

5. The *outer plexiform layer* contains the synaptic connections between

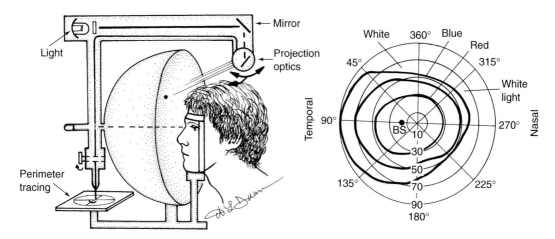

Figure 18.6. Mapping the visual field. Detailed maps of visual fields are obtained using a perimeter apparatus. In this apparatus, a small light stimulus is projected onto various locations on the inner surface of the hemisphere of the apparatus. The light can be of different sizes, luminances, and colors. The subject is asked to report when he or she sees the light. The dimensions of the visual field can be marked precisely on a card onto which the stimulus is projected via mirrors in the projection optics. Modified from Grusser O-J, Grusser-Cornehls U. The sense of sight. In: Schmidt RF, Thews G, eds. Biederman-Thorson MA, trans. *Human Physiology*. Berlin, Germany: Springer-Verlag; 1983:237–276.

photoreceptors and the neurons in the inner nuclear layer (that is, it is a neuropil layer).

6. The *inner nuclear layer* contains the cell bodies of biploar cells, horizontal cells, and amacrine cells. These neurons interconnect to process activity generated at the receptor level.

7. The *inner plexiform layer* contains the synaptic connections between the neurons of the inner nuclear layer and the *retinal ganglion cells*.

8. The *ganglion cell layer* contains the cell bodies of the ganglion cells.

9. The *nerve fiber layer* contains the axons of the ganglion cells.

10. The inner limiting membrane.

The Neural Retina Depends on the Retinal Pigment Epithelium

There is an important functional relationship between the cells of the RPE and the photoreceptors. The apposition between the neural retina and the RPE is fragile. The two are firmly attached only at the *ora serrata* and the *optic disk*. Trauma to the eye or certain disease processes can result in detachment of the retina from the RPE *(retinal detachment)*. This is of immense importance because, when separated from the RPE, the photoreceptors begin to degenerate. This implies that the RPE provides trophic support for the photoreceptors. Fortunately, it is now possible to

Figure 18.7. Histology of the retina. A, The photomicrograph illustrates a cross section through the human retina close to the fovea. The processes in the fiber layer are from the photoreceptors. These travel at an angle because the inner plexiform layer is displaced laterally in the region of the fovea to minimize the absorption of the light reaching the cones in the fovea (see **C**). **B** illustrates a scanning electron micrograph of the photoreceptors in the retina of a rhesus monkey (courtesy M. Miyoshi). **C** is a drawing of a section through the fovea, illustrating the displacement of the inner layers of the retina, which reduces the dispersion and absorption of light reaching the cone-rich fovea. From Heimer L. *The Human Brain and Spinal Cord*. New York, NY: Springer-Verlag; 1995. **A** is modified from Boycott RB, Dowling LE. *Phil Trans R Soc Lond (Biol)*. 1969;255:109–184. **C** is from *Lehrbuch der Histologie und der Mikroskopischen Anatomie des Menschen mit Einschluss der mikroskopischen Technik* 23. Auflage, 1933.

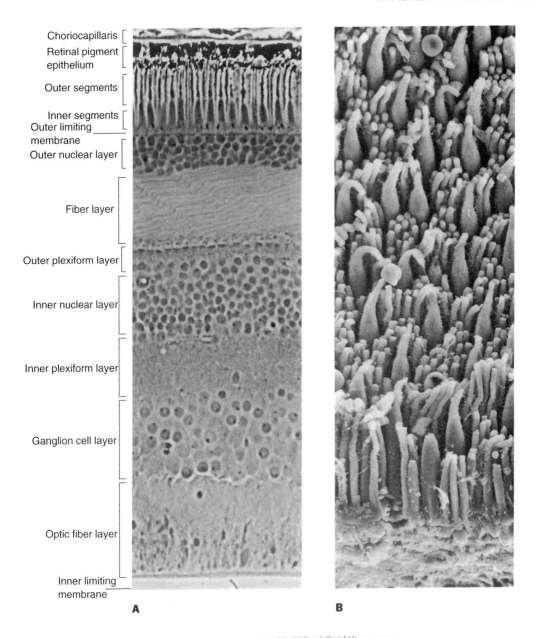

Choriocapillaris

Retinal pigment epithelium

Outer segments

Inner segments
Outer limiting membrane

Outer nuclear layer

Fiber layer

Outer plexiform layer

Inner nuclear layer

Inner plexiform layer

Ganglion cell layer

Optic fiber layer

Inner limiting membrane

A

B

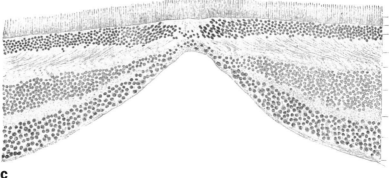

C

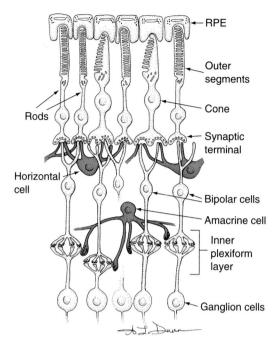

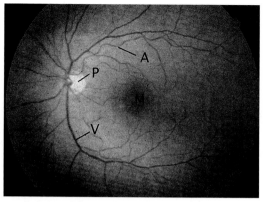

Figure 18.9. Regional specializations of the retina. The photomicrograph is a view of the fundus of the eye (left retina) as seen through the opthalmoscope. M indicates macula lutea with central foveal depression; P, optic papilla, also called optic disk; A, branches of the central artery of the retina; V, branches of the central veins of the retina. Photomicrograph courtesy of S. A. Newman, University of Virginia.

Figure 18.8. Cellular organization of the retina. The principal cell types of the retina and their synaptic interconnections. The outer segments of rods and cones contain stacks of membrane in which the photopigments are located. The outer segments are connected with the cell body of the receptor cell by a ciliary bridge. The inner segments of the receptors contain the nucleus and most of the cytoplasm. At the end of the cell opposite to the outer segment is the receptor terminal, which forms synaptic connections with bipolar and horizontal neurons.

- *Macula lutea.* Most of the cones in the retina are concentrated within the *macula lutea* (Latin for "yellow spot"). The color comes from the cone pigments.

- *Fovea.* The center of the visual field is the *fovea* (Latin for "pit"). It is a depressed area in the center of the macula lutea that appears as a dark spot. It contains only cone photoreceptors and is the area of highest visual acuity.

- *Optic disk and optic papilla.* The optic disk is the area in which the optic nerve fibers converge to leave the eyeball. The area appears pale pink. Because the area contains no photoreceptors, it is a blind spot. You can demonstrate the blind spot to yourself using Figure 18.10. The *optic papilla* is a circular ridge around a central depression termed the *physiological cup.* Together, the optic papilla and the physiological cup comprise the optic disk.

reattach the retina and the RPE surgically and so prevent the retinal degeneration that would otherwise occur. For these reasons, retinal detachment calls for rapid surgical intervention.

Regional Variations in the Retina Provide Important Clinical Landmarks

There are important structural features of the retina that are important for vision and for clinical diagnosis. These features can be directly visualized by looking at the back of the eye *(fundus)* with an opthalmoscope (Figure 18.9). These features include the following:

An important feature of the eye is that the retina receives its blood supply via the *opthalmic artery,* which enters the eye along with the nerve fibers of the optic nerve; venous drainage is via the same route. The vessels radiate outward from the optic disk into the retina. Interruption of this blood supply leads to the death of the neural elements within the retina, resulting in blindness.

Figure 18.10. The blind spot. The blind spot in the left eye can be mapped by covering the right eye and fixating on the white circle on the right. Hold the book about 12 inches away and move it back and forth slowly. The circle with the cross on the left will disappear when it lies directly in the blind spot. To illustrate how the brain fills in for missing information, close the left eye and fixate on the circle with the cross on the left. When the white circle in the black square lies within the blind spot, the square appears totally black. From Kingsley RE. *Concise Text of Neuroscience.* Baltimore, Md: Williams & Wilkins; 1996.

Papilledema or "*choked disc*" is an important clinical sign; the condition occurs when there is increased intracranial pressure, usually as a consequence of disruption of cerebrospinal fluid flow, tumors, or injury. Because of the increased intracranial pressure, the venous drainage of the eye is compromised, leading to a dilation of the veins; the optic disk is pushed forward and the disk itself appears white rather than pink. Prolonged papilledema can result in optic nerve damage.

Specialized Circuits of the Retina

The key to understanding the retina is to understand that it is made up of different classes of receptors and different types of intraretinal circuitry, all of which are specialized to process different features of visual stimuli. In this regard, the retina is similar to the skin: different receptor types are specialized to respond to different submodalities. Similarly, the information from the different receptor types is conveyed through the retina and to the brain via separate "channels," or parallel lines. This is what is meant by parallel processing. The brain then puts the information from the different channels together into an integrated perception.

The Photoreceptors

There are two main classes of receptor that differ in structure and function: *rods* and *cones* (Figure 18.8 and Table 18.2). Rods are sensitive to low levels of light and are thus responsible for vision under low illumination (*scotopic vision,* more commonly called *night vision).* Rods are insensitive to differences in spectrum (color). They contain the visual pigment *rhodopsin.*

Table 18.2. Characteristics of retinal photoreceptors

Receptor type	Distribution within retina	Sensitivity	Photopigment
Rods	Highest density in parafovea; absent from central fovea	Low light	Rhodopsin
Cones	Highest density in fovea	Moderate light	Cone opsin

Cones mediate daylight vision *(photopic vision)*. There are three types of cones that contain one of three variants of the visual pigment *cone opsin*. The three types of cone opsin absorb light maximally in the red, green, and blue ranges of the visual spectrum. Color perception depends on the mix of activation of the three types of receptor (more on this in the section on color vision).

Both rods and cones have specialized receptive elements (the *outer segments*) that contain stacks of membrane in which the photopigments are located. The *inner segments* of the receptors contain the nucleus and most of the cytoplasm. At the end of the cell opposite to the outer segment is the receptor terminal, which forms synaptic connections with bipolar and horizontal neurons.

Rods and cones are differentially distributed across the retina. Cones are present at maximal density in the central fovea (the area of highest visual acuity) and decrease in density in portions of the retina that represent the peripheral visual field. Rods are most dense in the parafoveal region, are present at moderate density in the peripheral retina, and are absent from the central fovea, where the cones predominate (Figure 18.11). Because the highest density of rods is found in the parafoveal region, the portion of the visual field that is represented in the parafovea (about 20° from the center of the field) is the most sensitive under low-light conditions. You have probably experienced this fact yourself if you have tried to look at a dim star in the night sky. Often, a very dim star that you can't see when you look directly at it can be seen by looking a few degrees to the right or left.

Circuitry of the Retina

The basic circuitry of the retina is illustrated in Figures 18.12 and 18.13. Photoreceptors form synapses on two types of neurons: *bipolar neurons* and *horizontal neurons*. Bipolar neurons form synapses with ganglion cells and amacrine cells. The relay from photoreceptor to bipolar neuron to ganglion cells is the throughput pathway. Horizontal

neurons project laterally within the external plexiform layer to synapse on nearby photoreceptors, forming the substrate for lateral interactions (specifically, lateral inhibition). Amacrine cells are the substrate for lateral interactions within the inner plexiform layer.

The First Synaptic Relay: Connections of Photoreceptors With Bipolar and Horizontal Neurons

The Throughput Pathway

The throughput pathway is made up of three separate lines of communication that are based on different types of bipolar neurons (Figure 18.12). The bipolar neurons are differentiated on the basis of the nature of the photoreceptor input they receive (rods versus cones) and the nature of the synaptic contact that is formed between the photoreceptor and the bipolar neuron. These differences in morphology correspond to differences in the way the bipolar neurons respond to the neurotransmitter that is released by the photoreceptor.

Photoreceptor Inputs

Rods and cones form synapses on separate populations of bipolar neurons termed *rod bipolar neurons* and *cone bipolar neurons*. These separate populations are the first relays in the separate channels of communication that convey information from rods and cones to ganglion cells (Figure 18.12).

Synaptic Connections

Within the classes of cone bipolar neurons are two subclasses of bipolar neurons that are distinguished by the types of synaptic connections they receive from photoreceptors.

One class of cone bipolar neuron can be distinguished by the highly characteristic *invaginating synapse* that is formed on these neurons by the photoreceptors (also called a *ribbon synapse*). An invaginating synapse is actually a *synaptic triad* in which a dendritic process from a bipolar neuron is sandwiched between two dendrites from horizontal neu-

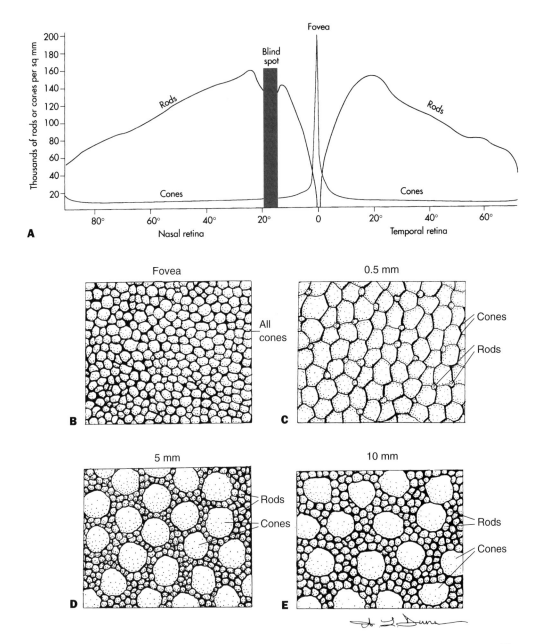

Figure 18.11. Distribution of rods and cones across the retina. A, The graph illustrates the packing densities of rods and cones across the human retina. Counts were made in a horizontal band that extends through the fovea and the blind spot. The drawings in **B–E** illustrate the density of cones and rods at different locations in the retina of a pigtail macaque monkey. **B,** The fovea contains small tightly packed cones and no rods. **C,** Just outside the fovea (0.5 mm), the size of cones increases, and the density of cones decreases (number of cones per unit area). A few rods are present. **D,** At 5 mm, cones are still larger and are intermixed with rods. **E,** Cone density decreases further at 10 mm. **A** is from Nolte J. *The Human Brain.* St. Louis, Mo: Mosby Year Book Inc; 1988. Modified from Osterberg GA. *Acta Opthalmol.* 1935;(suppl 6). **B–E** are modified from Packer O, Hendrickson AE, Curcio CA. Developmental redistribution of photoreceptors across the *Macaca nemestrina* (pigtail macaque) retina. *J Comp Neurol.* 1990;293:473–493. Reprinted by permission of Wiley-Liss, Inc., a subsidiary of John Wiley & Sons, Inc.

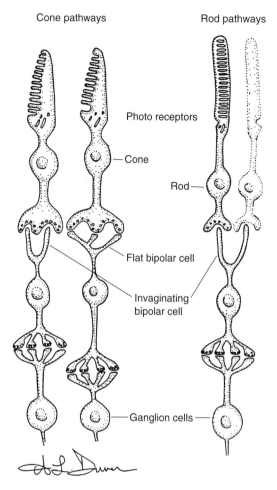

Cone pathways   Rod pathways

Photo receptors

Cone

Rod

Flat bipolar cell

Invaginating
bipolar cell

Ganglion cells

Figure 18.12. Throughput pathways of the retina. There are three throughput pathways, based on three types of bipolar neurons. Cones and rods synapse with separate populations of bipolar neurons termed cone bipolar neurons and rod bipolar neurons. Within the class of cone bipolar neurons are two subclasses that form invaginating synapses or flat synapses with the respective photoreceptor types. Rods form only the invaginating types of synapses. Invaginating versus flat bipolar neurons respond in opposite ways to the neurotransmitter released by the photoreceptors (see Chapter 20).

photoreceptor. The dendrites of these neurons do not participate in synaptic triads, and the synapse is noninvaginating; as a result, the synapse is termed a *flat synapse*. Bipolar neurons that receive flat synapses are termed *flat bipolar neurons.*

Individual photoreceptors form both invaginating and flat synapses with bipolar neurons; however, the converse is not true. Individual bipolar neurons receive only one or the other type of contact from photoreceptors. That is, invaginating bipolar neurons do not participate in flat synapses, and flat bipolar neurons do not participate in invaginating synapses. This fact is important because, as we see in Chapter 20, the two types of bipolar neuron respond in opposite ways to the neurotransmitter released by the photoreceptor. The differences in their response are due to differences in the type of neurotransmitter receptor that the two types of bipolar neuron express. In the Table 18.3, we summarize both anatomical and physiological characteristics of the two types of bipolar neuron. The physiological differences are discussed in more detail in Chapter 20.

Lateral Interactions: Interconnections Between Photoreceptors and Horizontal Neurons

Horizontal neurons are the other neurons that are directly innervated by photoreceptors. Their dendrites make up the other component of the synaptic triad. The horizontal neurons then send their axons laterally within the outer plexiform layer to form synapses on photoreceptors in surrounding areas of the retina. Horizontal cells thus form the substrate for lateral interactions between receptors.

The Second Synaptic Relay: Connections of Bipolar Neurons With Ganglion Cells and Amacrine Cells

Bipolar cells send their axons from the outer layer of the retina to the inner, where they form synapses on ganglion cells and amacrine cells. Connections with ganglion cells form the throughput pathway (Figure 18.12); connec-

rons; the entire triad then invaginates (protrudes into) the presynaptic "terminal" of the photoreceptor (Figure 18.12). Bipolar neurons that send dendrites into this synaptic triad are termed *invaginating bipolar neurons.*

The other class of cone bipolar neuron receives a different type of synapse from the

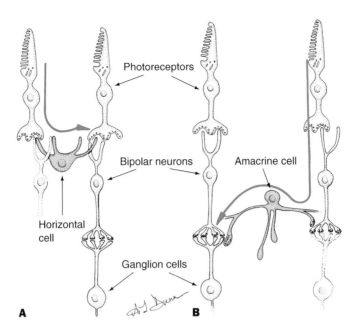

Figure 18.13. Pathways that mediate lateral interactions within the retina. A, Photoreceptors form synapses onto the dendrites of horizontal neurons. The horizontal neurons then send their axons laterally within the outer plexiform layer to form synapses on photoreceptors in surrounding areas of the retina. Horizontal cells thus form the substrate for lateral interactions between receptors. **B,** Amacrine cells receive synaptic input from the axons of bipolar neurons; they then send their axons laterally within the inner plexiform layer to synapse on ganglion cells in surrounding regions. These lateral interconnections form the basis for lateral inhibition. Thus ganglion cells receive direct input via the throughput pathway, which can be modulated by the lateral interactions mediated by the amacrine cells.

Table 18.3. Classes of bipolar neurons

Class and subclass	Synapse type	Response to neurotransmitter	Response to light
Rod, invaginating	Synaptic triad	Hyperpolarization	Depolarization
Rod, flat	Single	Depolarization	Hyperpolarization
Cone, invaginating	Synaptic triad	Hyperpolarization	Depolarization
Cone, flat	Single	Depolarization	Hyperpolarization

tions with amacrine cells form the substrate for lateral interactions (Figure 18.13).

The Throughput Pathway

Different classes of ganglion cell are innervated by the two types of bipolar neurons, and this innervation determines the response properties of the ganglion cells. One class of ganglion cell is innervated by invaginating bipolar neurons; another class is innervated by flat bipolar neurons, and a third class is innervated by both types of bipolar neurons.

It is interesting that the segregation is maintained even at the level of the synaptic connections within the internal plexiform layer. Ganglion cells that receive input from invaginating bipolar neurons ramify their dendrites in a sublamina of the internal plexiform layer proximal to the ganglion cell layer; ganglion cells that receive input from flat bipolar neurons ramify their dendrites within a sublamina of the internal plexiform layer that is more distal to the ganglion cells. Some ganglion cells receive input from both invaginating and flat bipolar neurons and ramify their dendrites in both sublaminae. In this way, the different classes of ganglion cell can be identified on the basis of the distribution of their dendritic arbor.

Lateral Interactions: Amacrine Cells

Amacrine cells are innervated by the axons of bipolar neurons; they send their axons laterally within the inner plexiform layer to innervate ganglion cells. These connections form the basis for lateral interactions (again, lateral inhibition). Thus, ganglion cells receive direct input via the throughput pathway, and the input can be modulated by the lateral interactions mediated by the amacrine cells.

Ganglion Cells Give Rise to the Axons That Leave the Retina Via the Optic Nerve

Ganglion cells are the final stage of processing within the retina. These neurons give rise to the axons that form the optic nerve and tract. Probably the most important single facts about the organization of the visual system pertain to the patterns of projection of ganglion cell axons and their patterns of termination in the central nervous system. This is the subject of Chapter 19.

What a Patient Sees When There Is Damage to the Retina

What one sees when a small area of the retina is damaged can best be understood by considering the blind spot. When you map your blind spot using Figure 18.9, you will discover that objects simply disappear when they are represented in the blind spot, or images that extend through the blind spot appear to be filled in. You do not perceive a blank space. This is also true when there are small lesions of the retina that produce focal blindness in part of the visual field (termed *scotomas*). Thus, patients may be unaware of small scotomas and may be quite surprised to learn that they are blind in a part of their visual field.

With retinal detachment, the image appears out of focus in the part of the visual field in which the retina is detached. The reason is that the retina floats off the RPE at the back of the eyeball, such that the visual image is not focused on the photoreceptors. Symptom onset is often sudden, especially when associated with physical trauma. A common cause of retinal detachment is a blow to the eyeball, which sets up a shock wave through the fluid-filled chamber of the eye. Even a seemingly innocuous blow that does no other damage (for example, a basketball or volleyball striking the face) can produce retinal detachment. Given the rapid degeneration of photoreceptors following retinal detachment, blurred vision after a blow to the eye calls for immediate opthalmoscopic evaluation of the integrity of the retina, and detachment calls for immediate surgical intervention.

Clinical Syndromes Can Reveal Scientific Principles: An Anecdote

Retinal detachment provides a striking clinical experiment that can demonstrate the way in which the visual image is represented in the visual system. As already described, the visual environment is projected onto the retinal surface so that each point in the retina maps a particular part of the visual world. When the retina is detached and then reattached surgically, the precise point-to-point mapping may be slightly distorted. A brief anecdote illustrates this.

At the age of 15, the author's son experienced retinal detachment due to trauma. The precise cause was never determined, but the probable cause was a basketball striking his face some months prior to symptom onset. The detachment apparently began in the peripheral retina and then, at some point after the initial injury, there was a sudden detachment that involved the macula. With the macular involvement, visual acuity in the affected eye went from 20/20 to 20/200. (He couldn't see the big E on the eye chart!)

The retina was successfully reattached by cryo surgery. Nevertheless, although visual acuity recovered to near normal after the surgery, the patient described his vision as "distorted, as if looking into a carnival mirror." This was a result of a slight misalignment of the retina during reattachment.

Over time, the patient ceased to perceive a distorted visual image. The probable reason is that there was a reorganization of central visual circuitry that compensated for the slight distortion of the peripheral receptor sheet. This process of reorganization is discussed in further detail in Chapter 20.

19

Visual System II

Central Pathways That Mediate Visual Perception

This chapter considers the overall organization of the central visual pathways. Students should know the names of each component of the central visual pathways and be able to identify each component in brain dissections, histological sections, or drawings. We then define how the information from the two eyes is mapped within the central visual pathways and define the retinotopic organization of the pathways. Students should be familiar with how the fibers of the optic nerve decussate at the optic chiasm, and how different parts of the visual field are represented in the optic tract, lateral geniculate nucleus, and cortex. This information is the key to understanding the visual field deficits that result from lesions at particular locations along the central visual pathways.

OVERALL ORGANIZATION OF CENTRAL VISUAL PATHWAYS

Axons from retinal ganglion cells project to the brain via cranial nerve II. The nerves from the two eyes converge at the *optic chiasm,* and the axons are then routed into one or the other *optic tracts.* By convention, the prechiasmatic portion of cranial nerve II is called the *optic nerve,* despite the fact that it is a central tract in terms of its cellular organization; the postchiasmatic portion is called the *optic tract.* Both portions are central tracts in every respect because the axons in both nerve and tract are myelinated by oligodendrocytes.

From the optic chiasm, the retinal ganglion cell axons project to several locations: (1) Most axons project to the lateral geniculate nucleus (LGN), which is the thalamic relay nucleus for projections to the primary visual cortex (striate cortex). This is termed the *retinogeniculostriate pathway.* (2) *Extrageniculostriate pathways* carry information from the retina to several locations, including the superior colliculus (the *retinotectal pathway*), pretectal area, pregeniculate nucleus, accessory optic nucleus, and suprachiasmatic nucleus of the hypothalamus (Figure 19.1 and Figure V.1).

The geniculostriate pathway is specialized for *form vision,* which is the ability to perceive different configurations and spatial relationships between objects in the visual environ-

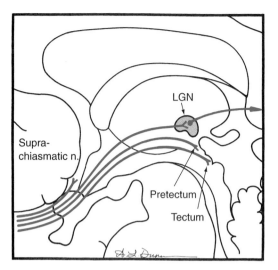

Figure 19.1. Central projections of the optic nerve. An enlarged view of the regions in which the optic nerve terminates, as seen in a schematic drawing of a midsagittal section (the boxed-in area in Figure V.1). Retinal axons project from the eye through the optic nerve to terminate in (1) the LGN, which is the thalamic relay nucleus for projections to the the primary visual cortex (striate cortex); this is termed the *retinogeniculostriate pathway;* (2) the superior colliculus in the tectum (the retinotectal pathway); (3) the pretectal area; (4) the pregeniculate nucleus (not shown); and (5) the suprachiasmatic nucleus of the hypothalamus (termed the *retinohypothalamic pathway*). Modified from Kandel ER, Schwartz JH, Jessell TM. *Principles of Neural Science.* 3rd ed. New York, NY: Elsevier; 1991.

ment. This type of vision allows us to identify objects in the environment. The retino-tectal pathway is important for *visually guided behaviors,* especially orientation toward visual stimuli. The projections to the pretectal area provide the sensory limb for ocular reflexes (accomodation and pupillary light reflex). The projections to the pregeniculate nucleus are important for controlling conjugate eye movement. The function of these pathways is discussed in Chapter 21. The projections to the suprachiasmatic nucleus play an important role in regulating *circadian rhythms* (from the Latin *circa* and *dies,* "around the day"). These too are considered further in Chapter 21.

Pathways That Mediate Form Vision: The Retinogeniculostriate System

As in other sensory systems, conscious perception of visual information depends on the cortex, in particular, the portion of the cortex that receives information from the LGN of the thalamus. Thus, lesions that interrupt the pathways from the eye to the brain lead to an inability to perceive visual information (blindness). For this reason, it is very important for physicians to understand the topography of the central visual pathways because this information provides the basis for diagnosing the site of lesions that affect the visual system.

Central visual pathways are organized so that information from the two eyes converges in the visual cortex. A fundamental organizational principle of the visual system is that visual images from the two eyes are brought together in the cortex, so that inputs from homonymous parts of the visual field are mapped in overlapping fashion (homonymous means the same point in the visual field). The overlapping maps from the two eyes form the basis for binocular vision. In order for the maps to be constructed, axons from the different parts of the retina in two eyes must be routed appropriately.

The first step in this routing occurs at the optic chiasm (Figure 19.2). At the optic chiasm, axons from the nasal half of each retina cross the midline and join the contralateral optic tract; axons from the temporal half of the retina join the ipsilateral optic tract. In this way, the axons from portions of the retina that map homonymous portions of the visual field are brought together in the optic tract and project centrally to their site of termination.

The second step in the routing occurs in the LGN. The LGN comprises six cellular layers, numbered I through VI, with I being the most ventral, and VI the most dorsolateral. Layers I and II contain large neurons and are thus termed *magnocellular layers.* Layers III through VI contain small- and medium-sized neurons and are termed *parvocellular layers.*

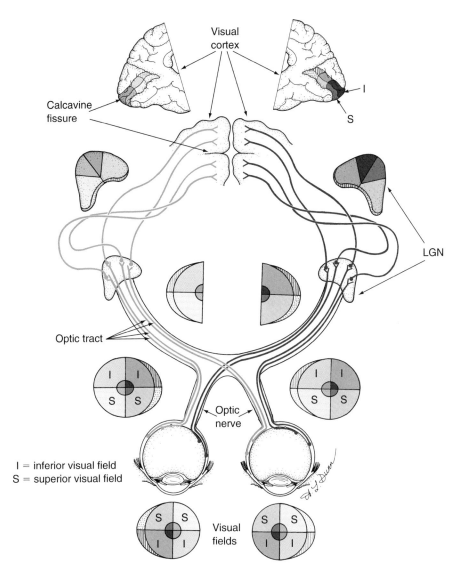

Figure 19.2. Routing of retinofugal projections in the optic chiasm. Axons from the two eyes merge at the optic chiasm and are sorted according to their site of origin within the retina. Axons from the nasal half of each retina cross the midline and join the contralateral optic tract; axons from the temporal half of the retina join the ipsilateral optic tract.

Axons from the two eyes terminate in different layers. Thus, axons from the contralateral eye terminate in layers I, IV, and VI, whereas axons from the ipsilateral eye terminate in layers II, III, and V. There is no convergence of retinal fibers from the two eyes at the level of the LGN. That is, neurons in each lamina receive input from one eye or the other.

Retinogeniculate axons terminate in a highly ordered topographic map within the different layers (termed a *retinotopic map*). The projections are arranged such that individual points in the visual field are mapped on corresponding portions of each lamina; hence, the retinotopographic maps from each eye are in register (Figure 19.3).

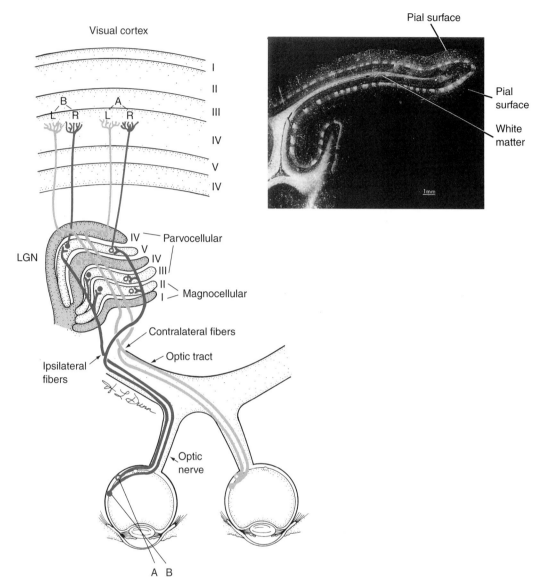

Figure 19.3. Retinotopic mapping in LGN and cortex. Axons from the two eyes terminate in different layers in the LGN. Axons from the contralateral eye terminate in layers I, IV, and VI, whereas axons from the ipsilateral eye terminate in layers II, III, and V. Individual points in the visual field are mapped on corresponding portions of each layer, such that the retinotopographic maps from each eye are in register. For example, the bottom of the figure illustrates the mapping of two points in visual space (A and B) on the retina. Each eye maps these same points when the eyes are in register, and the two images are then represented sequentially at the level of the LGN (in the different laminae) and the cortex (by the arborizations of the axons from the LGN neurons). The labels A and B in the cortex indicate the representation of points A and B in visual space. The arbors carrying information from each eye terminate in ocular dominance stripes (labeled L and R for the axons carrying information from the left and right eye, respectively). This figure schematically illustrates only the terminations of the axons from neurons in the parvocellular laminae. A parallel pathway exists from the magnocellular laminae. After Kingsley RE. *Concise Text of Neuroscience.* Baltimore, Md: Williams & Wilkins; 1996.

The inset illustrates the pattern of termination of geniculocortical axons representing one eye in the primate visual cortex. To selectively label the

Whereas the topographic map is a continuous representation of the visual field, the representation is not proportional with respect to the physical dimensions of the field. The macula has a higher density of receptors than the periphery of the retina and has an expanded area of representation in the LGN.

Neurons in the LGN give rise to axons that project to the primary visual cortex via the *geniculostriate projections* (also called the *geniculocalcarine tract*). These axons loop around the lateral ventricle through the temporal and parietal cortex (termed *Meyer's loop*), continuing as the *optic radiations,* to terminate in the area surrounding the *calcarine fissure* (Figure 19.4).

The geniculocalcarine tract maintains a topographic organization, so the upper medially located portion of the tract contains fibers for the lower field of vision, whereas the lower laterally located portion of the tract carries fibers for the upper field of vision. Damage to the temporal lobe often interrupts the fibers traveling in Meyer's loop, leading to visual deficits in the upper field of vision.

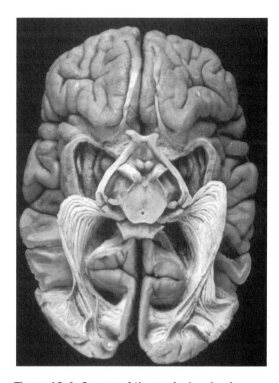

Figure 19.4. Course of the geniculocalcarine tract through the temporal lobe. Meyers loop. The photomicrograph illustrates a brain in which the optic radiations through the temporal lobe have been revealed by carefully dissecting away the overlying cortex. From Gluhbegovic N, Williams TH. *The Human Brain, a Photographic Guide.* Hagerstown, Md: Harper-Row; 1980.

Visuotopic Mapping in the Primary Visual Cortex

The *primary visual cortex* (Brodmann's area 17) is the portion of the cortex that surrounds the calcarine fissure on the medial surface of the occipital lobe. It has the same basic cellular organization as other sensory cortices. Thalamic projections from the LGN terminate primarily in layer IV. Because of the dense thalamic input, layer IV stands out as a prominent stripe in unstained sections (termed the *stripe of Gennari*). For this reason, early anatomists named the primary visual cortex the *striate cortex.*

The visual cortex is the first point along the central visual pathway where individual neurons receive inputs from both eyes. This

Figure 19.3. *(Continued)*
terminals representing one eye, radiolabeled amino acids were injected into one eye. These amino acids were incorporated into proteins, transported to the terminals of retinal ganglion cells in the LGN, and transported transneuronally to LGN neurons and then on to the cortex. The distribution of the terminals was revealed by taking histological sections and coating them with a photographic emulsion. Over time, the emulsion is exposed by the radioactive emissions. Labeling is then revealed by the presence of silver grains, which appear white in the illustration because they reflect light. From Hubel DH, Wiesel TN. Brain mechanisms of vision. *Sci Am.* 1979;241:150–162.

binocular convergence is essential for stereoscopic vision. The projections from the different layers in the LGN terminate in a retinotopic fashion within the striate cortex so that the maps from the two eyes are in register. The lower half of the visual field is mapped upon the *cuneate gyrus,* dorsal to the calcarine fissure; the upper half of the visual field is mapped on the *lingual gyrus,* ventral to the calcarine fissure. Within each region, fibers conveying information from different parts of the retina terminate in a visuotopic map. In this way, different parts of the visual cortex represent different parts of the visual field (Figure 19.5).

Geniculostriate Projections Terminate in Discrete Patches Related to the Two Eyes

An important aspect of the projection from the LGN to the cortex is the pattern of termination of fibers from the two eyes. The terminal fields of LGN axons carrying information from each eye are organized in alternating stripes. These stripes were first revealed through experimental studies of ocular projections in primates (Figures 19.3 and 19.6). Cortical neurons located within the center of the stripes receive input primarily from one eye; neurons at the borders between the stripes receive inputs from both eyes. In this way, the stripes define domains in which inputs from one or the other eye dominate *(ocular dominance domains).* As we see later, the visual cortex is organized in functional columns, so the ocular dominance domains actually form the basis for *ocular dominance columns* within the cortex (see Chapter 20).

Stereoscopic Vision Requires Proper Alignment of the Eyes

Stereoscopic vision requires that the visual fields of the two eyes are in register; this in turn depends on proper alignment of the eyes. You can demonstrate this to yourself by focusing on a distant object and gently pressing on the side of one eye; you will see a double image. Thus, an important compo-

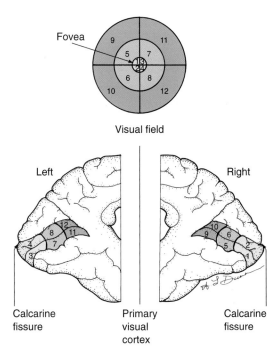

Figure 19.5. The visuotopic map in the primary visual cortex. The lower half of the visual field is mapped upon the cuneate gyrus, dorsal to the calcarine fissure; the upper half of the visual field is mapped on the lingual gyrus, ventral to the calcarine fissure. Within each region, fibers conveying information from different parts of the retina terminate in a visuotopic map (the numbers refer to locations in the visual field). A large portion of the cortex is devoted to macular vision, with progressively less being devoted to the representation of areas of the peripheral retina. Redrawn from Kandel ER, Schwartz JH, Jessell TM. *Principles of Neural Science.* 3rd ed. New York, NY: Elsevier; 1991.

nent of stereoscopic vision is the control of eye position and movement, which is mediated by specialized circuitry involving the cranial nerve nuclei that control the extraocular muscles (see Chapter 21).

Visual Association Cortex

Surrounding the primary visual cortex are the visual association areas (Brodmann's areas 18 and 19, Figure 19.7). These areas receive input from the primary visual cortex. In addition, visual association areas receive

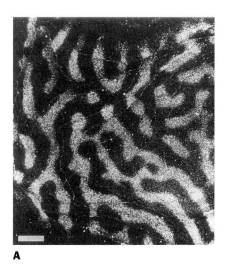

A

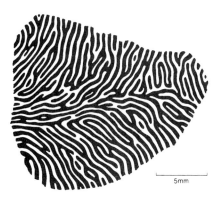

B

Figure 19.6. Geniculocortical projections terminate in ocular dominance stripes. A, The stripelike terminations of geniculocortical projections were revealed using the autoradiographic tract-tracing technique described in Figure 19.3. In this case, however, the cortex was sectioned in such a way that the plane of section went through layer 4. Labeled projections appear as stripes; the unlabeled bands between the labeled stripes represent the site of termination of geniculocortical projections

representing the noninjected eye. From Levay S, Wiesel TN, Hubel DH. The development of ocular dominance in normal and visually deprived monkeys. *J. Comp Neurol.* 1980;191:1–51. **B,** Reconstruction of the pattern of termination of the axons from one eye defined on the basis of a silver stain that reveals the boundaries between columns. From Hubel DH and Freeman DC. Projection into the visual field of ocular dominance columns in macaque monkey. *Brain Res.* 1977;122:336–343.

thalamic input from the pulvinar nucleus, which in turn receives information from the superior colliculus.

LESIONS INVOLVING THE RETINOGENICULOSTRIATE PATHWAY

Because of the topographic organization of the central visual pathways, lesions at particular locations lead to predictable *visual field deficits* (deficits affecting part of the visual field). Conversely, these deficits are highly diagnostic for lesions at particular sites. For this reason, the information summarized in Figure 19.8 is very important for physicians and should be committed to memory.

Students should be familiar with the terms used to define visual field defects. Blindness in a part of the visual field is termed an *anopsia.* Anopsia in the same part of the visual field in each eye is termed *homonymous anopsia.* Anopsia involving different parts of the

visual field in the two eyes is termed *heteronymous anopsia.*

Injuries to one eye or to a single optic nerve disrupt the visual input from one eye. The portion of the visual field normally seen only by the damaged eye is lost; the remaining visual field is seen only by the undamaged eye. Complete loss of vision in one eye is termed *monocular blindness.*

Lesions at the level of the optic chiasm can selectively affect the crossing fibers, leading to a loss of vision in the temporal portion of the visual field from both eyes *(tunnel vision).* The field deficit is termed *bitemporal hemianopsia*—a heteronymous visual field defect. Bitemporal hemianopsia is a common symptom of pituitary tumors that expand upward to press upon the optic chiasm.

Damage to any part of the geniculostriate system central to the optic chiasm leads to homonymous visual field deficits. For example, unilateral lesions of the optic tract lead to *homonymous hemianopsia* (a loss of vision in the entire contralateral visual hemifield).

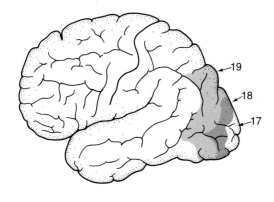

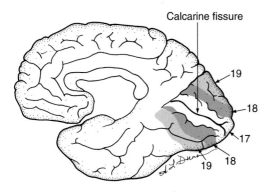

Calcarine fissure

Figure 19.7. Primary and secondary sites of cortical processing of visual information. Lateral (upper drawing) and midsagittal (lower drawing) views of the cortex. The primary visual cortex (area 17) lies on the banks of the calcarine fissure and extends onto the lateral surface of the cortex. The visual association areas (areas 18 and 19) surround area 17.

Damage to the optic radiations produces a similar deficit, depending on the extent of the lesion. Lesions in the temporal lobe can selectively damage the more laterally located fibers of Meyer's loop, leading to visual field defects in the upper quadrant of the contralateral visual hemifield (termed an *upper quadrantopsia*). Damage to the visual cortex of one hemisphere also leads to homonymous visual field defects in the contralateral visual field. The location of the defect depends on the site of injury. Thus, complete unilateral destruction of the visual cortex leads to the same sort of homonymous hemianopsia as destruction of the optic tract.

Damage to the cuneate gyrus produces a *lower quadrantopsia;* damage to the lingual gyrus produces an *upper quadrantopsia.*

In the case of the homonymous anopsias due to lesions involving the optic radiations or the visual cortex, macular vision is often spared *(macular sparing).* The reason macular vision is relatively spared is not entirely known. One possibility is that the macular representation is so large that lesions rarely if ever affect all of the representation. In the case of vascular lesions affecting the cortex, an explanation for macular sparing is collateral circulation. Finally, it is hypothesized that some retinal ganglion cells in the central macula actually project bilaterally to the LGN. Thus, macular sparing is one of those important clinical facts for which we do not have a completely satisfactory scientific explanation.

DEVELOPMENT AND PLASTICITY OF VISUAL CORTICAL ORGANIZATION

Experimental studies have revealed that the basic synaptic organization of the visual cortex is highly modifiable. These studies have revealed principles of sensory system organization and development that are turning out to apply to other systems as well. For example, we now know that the development of the highly characteristic stripelike pattern of termination of thalamocortical projections depends on binocular visual experience during a critical period of development. Moreover, there are mechanisms that permit a substantial reorganization of visuotopic representation in the cortex after injuries. These fundamental discoveries have directly impacted on clinical practice, even for family physicians.

Appropriate development of ocular dominance domains in the visual cortex depends on binocular experience early in life. The discovery that the organization of thalamocortical projections depended on visual experience came from studies in primates by D. Hubel, T. Weisel, and their colleagues. They found that if young monkeys were de-

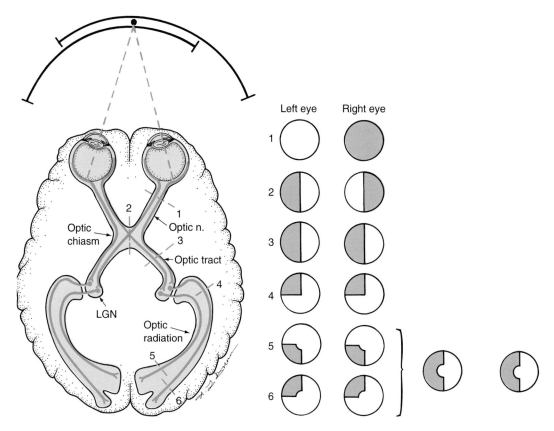

Figure 19.8. Visual field defects resulting from lesions of central visual pathways. Lesions at particular locations lead to predictable visual field deficits. The drawing illustrates the sites of lesions and the visual field deficits that result. Black indicates an area of visual loss.

prived of visual experience in one eye (by suturing the eyelid closed), the cortical representation of the deprived eye was reduced; at the same time, the cortical representation of the nondeprived eye expanded. This can be directly visualized by the same tract-tracing techniques used to define the normal stripelike pattern of termination of thalamocortical fibers in layer IV (see Figure 19.6). These changes in cortical representation were associated with significant impairment of vision in the deprived eye.

The changes in thalamocortical projections were seen only if the deprivation occurred in young animals during the time that the projections were still developing. These results led to the concept that there was a critical period during which balanced input from both eyes was important for establishing normal thalamocortical topography. Moreover, once changes were induced by monocular deprivation early in life, they were relatively permanent. Even if the deprived eye was opened later in life, thalamocortical termination patterns and the response properties of cortical neurons remained abnormal.

These findings have modified how patients with eye injuries are managed. It was once standard practice to have patients (including children) wear eye patches for extended periods of time after an injury to the eye. This is effectively monocular deprivation. Because monocular deprivation can have profound effects on visual pathways, it is now considered prudent to reduce to a minimum the time spent wearing a patch.

The visuotopic map can be modified in response to injury. In the same way that cortical representation in the somatosensory system can be modified following peripheral nerve injury, cortical representation in the visual system can be modified following injuries to the retina. Indeed, the nature of the change is very similar. Following focal lesions of the retina, the part of the primary visual cortex that represents the damaged zone loses its retinal input. The areas that no longer receive visual input can be mapped by recording functional activity in response to visual stimulation (Figure 19.9). When such studies are carried out soon after the placement of a focal lesion in the retina, a silent area can be defined that is unresponsive to visual stimuli. Over a period of months, the area that normally represents the damaged portion of the retina becomes responsive to surviving regions of the retina. In this way, the area in the cortex that is unresponsive shrinks, and the visual representation is remapped.

Exactly how the reorganization occurs is not yet known. Experimental evidence suggests that horizontal pathways that interconnect nearby areas within the visual cortex become more powerful. In any case, the reorganization seems to occur in such a way that residual vision is remapped onto the areas of the primary visual cortex that are made available. It remains to be seen whether the reorganization in the visuotopic representation plays a role in optimizing function after retinal injuries.

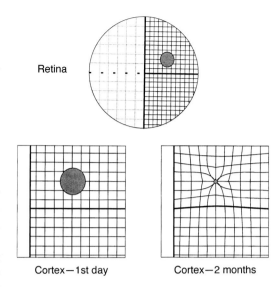

Figure 19.9. Reorganization of the visuotopic map in the cortex after focal injuries to the retina. In this study, small lesions were placed in the retina of a primate using a laser. The area of the lesion is illustrated schematically in the upper diagram. When the visual cortex was mapped immediately after the creation of the lesion, using physiological recording techniques, the area of the cortex that normally received input from the damaged area in the retina was unresponsive. For example, in the primate, a lesion subtending about 5° of the visual field in the retina that is centered about 4° peripheral to the fovea silences an area of cortex about 10 mm in diameter. When the response properties of neurons in the cortex were reevaluated two months after the injury, there was a reorganization of the retinotopic map, such that the area of cortex that had lost its normal input became responsive to nearby sites in the retina. This reorganization produced an expanded representation of the portion of the retina immediately surrounding the lesion. From Gilbert CD. Horizontal integration and cortical dynamics. *Neuron.* 1992;9:1–20.

KEY POINTS

1. The visual system is constructed so that homonymous parts of the visual field map together. Binocular vision depends on proper eye alignment.

2. In the geniculostriate pathway, the first point of convergence of inputs from the two eyes is the cortex.

3. The retinogeniculostriate pathway subserves form vision. The retinotectal pathways are (a) the projection from the retina to the superior colliculus, which mediates visually guided behavior; and (b) the projection to the pretectum, which subserves the pupillary light reflex.

4. The projections to the accessory optic nucleus and pregeniculate nucleus are the first limb of the pathway to the vestibular portion of the cerebellum via the mossy fibers. These pathways are involved in vestibuloocular reflexes.

5. The retinohypothalamic pathway to the suprachiasmatic nucleus regulates circadian rhythmicity.

20

Visual System III

Visual Neurophysiology and Visual Perception

In considering how the visual system processes information, we encounter again the themes of *parallel* and *hierarchical processing*. One manifestation of parallel processing is the existence of separate "labeled lines" originating from rods and cones. But as we see later, there are other forms of parallel processing that involve the two types of bipolar neurons (which form both invaginating and flat synapses). In terms of hierarchical processing, the overall story is reminiscent of the somatosensory system. At each level, cells respond best to more and more complicated stimulus parameters. As a result, neurons at progressively higher levels of the neural processing hierarchy "represent" particular stimulus configurations.

In what follows, we first consider how light is transduced into neural activity in the retina and then consider the manifestations of parallel and hierarchical processing at each level. Students should be able to define the patterns of activity that are produced in retinal ganglion cells in response to visual stimuli and explain the response properties of neurons at the different hierarchical levels of the central visual pathways.

THE PHYSIOLOGY OF THE RETINA

The processing of information in the retina is unusual in that photoreceptors and all other neurons in the retina, except for ganglion cells, are nonspiking. Intercellular communication in the retina is mediated by regulation of the membrane potential in a graded fashion.

The process of neurotransmitter release in these nonspiking neurons is regulated in a different way than in spiking neurons. In neurons that generate action potentials, neurotransmitter release is minimal or nonexistent under resting conditions and occurs in an explosive fashion in response to the action potential. In contrast, in nonspiking cells such as the photoreceptors and interneurons of the retina, neurotransmitter release varies in a graded fashion as a function of the membrane potential. Hyperpolarization leads to decreases in ongoing release; depolarization leads to increases in release.

The level of the membrane potential can vary over a relatively wide range because the neurons do not give rise to action potentials; as a result, release can also vary over a wide range.

The Receptor Event

Transduction of light *(phototransduction)* occurs in the outer segments of the photoreceptors. In the dark, there is a resting Na^+ current that causes a resting depolarization of the receptors. This resting Na^+ current (termed the dark current) is produced be-cause Na^+ channels are maintained in the open state as a result of very high levels of cGMP in the cytoplasm of the outer segment. When light bleaches the photopigments (rhodopsin or cone opsin), the activated photopigment stimulates a G protein (transducin, in the case of rods), which in turn activates a phosphodiesterase that breaks down the cGMP by converting it to 5'-GMP. As a result of the decrease in cGMP concentration, the Na^+ channels close, reducing the Na^+ current, and the membrane potential becomes more hyperpolarized (Figure 20.1).

Figure 20.1. Phototransduction.
Phototransduction occurs in the outer segments of the photoreceptors. In the dark, there is a resting Na^+ current that causes a resting depolarization of the receptors. This resting Na^+ current (termed the dark current) is produced because Na^+ channels are maintained in the open state as a result of very high levels of cGMP in the outer segment. When light bleaches the photopigments (rhodopsin or cone opsin), the activated photopigment stimulates a G protein (transducin, in the case of rods), which in turn activates a phosphodiesterase that breaks down the cGMP by converting it to 5'-GMP. As a result of the decrease in cGMP concentration, the Na^+ channels close, reducing the Na^+ current, and the membrane potential becomes more hyperpolarized. The receptor potential that is produced by light striking the photoreceptor is thus a hyperpolarizing potential.

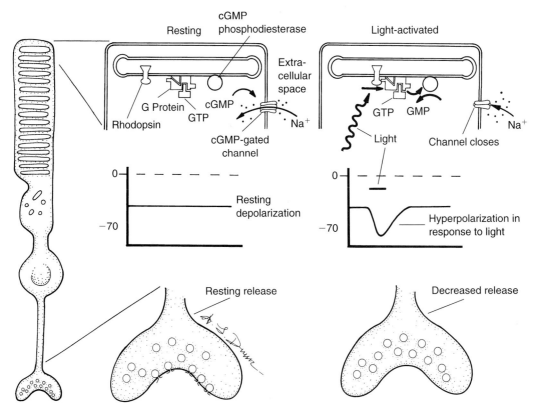

The receptor potential that is produced by light striking the photoreceptor is thus a hyperpolarizing potential.

Because of the resting Na^+ current, photoreceptors are depolarized in the resting state, so there is a continuous release of neurotransmitter from their synaptic terminals. The hyperpolarization of the photoreceptor that occurs in response to light reduces this tonic neurotransmitter output. Thus, the intercellular signaling event in response to light activation of a photoreceptor is a decrease in neurotransmitter release at the photoreceptor terminals.

Synaptic Interactions Between Photoreceptors and Bipolar Neurons

Like the photoreceptors, bipolar neurons are nonspiking. They tonically release neurotransmitter, and the amount of release varies as a function of their membrane potential. Thus, depolarization increases release, and hyperpolarization decreases release.

All photoreceptors hyperpolarize in response to light. Bipolar neurons, on the other hand, exhibit different responses depending on their type. In particular, the bipolar neurons that are contacted by cone photoreceptors can exhibit either depolarizing or hyperpolarizing responses to light activation of the photoreceptor. In contrast, in at least some mammals, the bipolar neurons contacted by rods exhibit only depolarizing responses to light activation of the photoreceptor. We begin by describing the cone pathways because these are the prototype for retinal processing.

Cones Contact "On" and "Off" Bipolar Neurons

Cones contact both invaginating and flat bipolar neurons (see Chapter 18). These two morphological types respond in opposite ways to the neurotransmitter released by cones. Invaginating bipolar neurons exhibit depolarization in response to light activation of the photoreceptor (what is termed an *"on" response*). Flat bipolar neurons exhibit hyperpolarization (an *"off" response*). The differences are summarized in Table 20.1. In this way, the two types of bipolar neurons represent the first stage of two retinal circuits that lead to either increases or decreases in activity of retinal ganglion cells (Figure 20.2).

Invaginating bipolar neurons are hyperpolarized by the neurotransmitter released by photoreceptors. The decrease in release of this hyperpolarizing neurotransmitter from the photoreceptor terminal in response to light leads to depolarization of invaginating bipolar neurons. This is basically a process of disinhibition.

Flat bipolar neurons are depolarized by the neurotransmitter released by photoreceptors. The decrease in release of this depolarizing neurotransmitter from the photoreceptor terminal in response to light leads to hyperpolarization of flat bipolar neurons. This is basically a process of disfacilitation.

The changes in membrane potential of the bipolar cells are sustained rather than transient. That is, if a set of photoreceptors are illuminated and the illumination is maintained, the changes in membrane potential of the bipolar neuron is also sustained. As we see later, this is in contrast to the response characteristics of amacrine cells.

Table 20.1. Classes of bipolar neurons

Class and subclass	Synapse type	Response to neurotransmitter	Response to light
Invaginating	Synaptic triad	Hyperpolarization	Depolarization
Flat	Single	Depolarization	Hyperpolarization

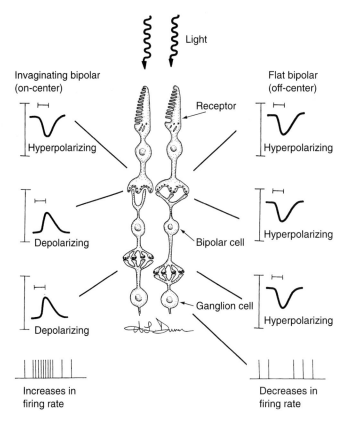

Figure 20.2. Responses of bipolar neurons to the neurotransmitter released by photoreceptors. Invaginating and flat bipolar cells respond in opposite ways to the neurotransmitter released by photoreceptors.

Invaginating bipolar neurons are hyperpolarized by the neurotransmitter released by photoreceptors. The decrease in release of this hyperpolarizing neurotransmitter from the photoreceptor terminal in response to light leads to a depolarization of invaginating bipolar neurons. The depolarization results in an increased release of neurotransmitter at the terminal of the bipolar neuron, which depolarizes the ganglion cell, increasing its rate of discharge. Consequently, the invaginating bipolar neuron forms an on-center circuit.

Flat bipolar neurons are depolarized by the neurotransmitter released by photoreceptors. The decrease in release of this depolarizing neurotransmitter from the photoreceptor terminal in response to light leads to a hyperpolarization of flat bipolar neurons. The hyperpolarization results in a decreased release of neurotransmitter at the terminal of the bipolar neuron, which hyperpolarizes the ganglion cell, decreasing its rate of discharge. Consequently, the invaginating bipolar neuron forms an off-center circuit.

The figure also illustrates the important fact that individual photoreceptors innervate both invaginating and flat bipolar neurons; hence, the difference in responses between on-center and off-center circuits is determined by the bipolar neuron (specifically, which receptors it expresses and, thus, how it responds to the neurotransmitter released by the photoreceptor).

The bipolar neurons contacted by rods exhibit only depolarizing responses to light activation of the photoreceptor (on responses). Moreover, many of the rod bipolar neurons contact amacrine cells rather than directly synapsing on ganglion cells. The nature of the responses generated by amacrine cells are discussed later.

Transmission Between Bipolar Neurons and Ganglion Cells

The bipolar neurons themselves release a neurotransmitter that depolarizes the ganglion cells. This is true of both invaginating and flat bipolar neurons. Consequently, when bipolar neurons depolarize, their release of neurotransmitter is increased, causing depolarization of ganglion cells. Conversely, when bipolar neurons hyperpolarize, their release of neurotransmitter decreases, causing hyperpolarization of the ganglion cells. In this way, ganglion cells that are contacted by invaginating bipolar neurons are depolarized when light strikes the photoreceptors (an *on-center circuit*, Figure 20.2). Ganglion cells that are contacted by flat bipolar neurons are disfacilitated and thus hyperpolarized (an *off-center circuit*). Ganglion cells are typical spiking neurons, so depolarization and hyperpolarization lead to increases and decreases in the frequency of action potentials in the ganglion cell's axon (Figure 20.2).

Generation of the Receptive Field of Ganglion Cells

Each ganglion cell receives input from a small number of bipolar neurons that in turn are contacted by a group of photoreceptors within a small patch of the retina. Thus, each ganglion cell "represents" a small part of the visual field. This is the *receptive field* of that ganglion cell, which is defined as the area on the retina from which the activity of the neuron can be influenced (Figure 20.3).

The actual physical size of the area that is represented by a single ganglion cell varies across the retina. Receptive fields are smallest where receptor density is highest. Stated in terms of receptive fields, the receptive fields of ganglion cells are smallest in the fovea (where receptor density is higher) and largest in the peripheral retina.

The response properties of the ganglion cells are determined by the summed input they receive from bipolar neurons (that is, whether the cell is on-center or off-center). The other important feature of the receptive field (the *center–surround* organization) is determined by the action of horizontal neurons. More complex response characteristics are determined through the action of amacrine cells.

Ganglion Cells Have Antagonistic Center–Surround Receptive Fields

An important feature of the receptive fields of retinal ganglion cells is that they have a center–surround organization. One type of response occurs when light activates the center of the field, and an opposite (antagonistic) response occurs when light activates the surround. An on-center receptive field will have an off-surround, whereas an off-center receptive field will have an on-surround.

These responses can be demonstrated by recording from ganglion cells while stimulating the retina with spots of light. When a small spot of light strikes the center of a particular ganglion cell's receptive field, action potential frequency either increases or decreases depending on whether the ganglion cell is on-center or off-center. When the same spot of light strikes nearby regions of the retina that are part of the inhibitory surround, the opposite changes in activity are produced in the given ganglion cell.

The antagonistic surround receptive field is produced through the action of *horizontal cells*. Horizontal neurons receive synaptic input from photoreceptors and then project to other photoreceptors in surrounding portions of the retina (Figure 20.4). Like the bipolar neurons, horizontal neurons are nonspiking; thus, their membrane potential is controlled directly by the neurotransmitter released by the photoreceptors.

Horizontal neurons are depolarized by the neurotransmitter released by photoreceptors. Thus, light activation of photoreceptors (which decreases their resting release of neurotransmitter) leads to hyperpolarization of horizontal neurons. The neurotransmitter released by horizontal neurons hyperpolarizes their photoreceptor targets. When the release of this neurotransmitter is decreased, the target photoreceptors become more depolarized.

To appreciate the significance of receptor depolarization by horizontal neurons, one has to recall that photoreceptors are hyperpolarized by light. Thus, the depolarization by horizontal cells produces a response opposite to that of light. This response is then communicated to the ganglion cells that

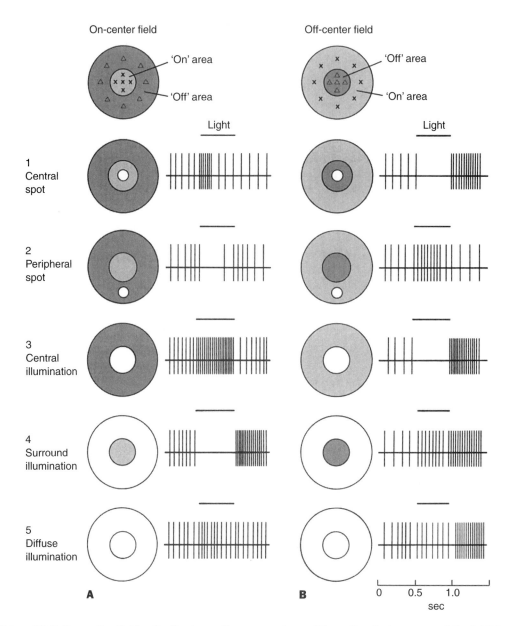

Figure 20.3. Receptive fields of retinal ganglion cells. The receptive fields of retinal ganglion cells are circular and have a center–surround organization. One type of response occurs when light activates the center of the field, and an opposite (antagonistic) response occurs when light activates the surround. An on-center receptive field will have an off-surround, while an off-center receptive field will have an on-surround. These responses can be demonstrated by recording from ganglion cells while stimulating the retina with spots of light (the white areas in the diagram represent the spot of light). When a small spot of light strikes the center of a particular ganglion cell's receptive field, action potential frequency either increases or decreases, depending on whether the ganglion cell is on-center or off-center. When the same spot of light strikes nearby regions of the retina that are part of the inhibitory surround, the opposite changes in activity are produced in the given ganglion cell. On-center cells (panel **A**) are maximally activated when the entire central part of the receptive field is illuminated (3) and respond less well when only a portion of the center is illuminated (1). On-center cells are inhibited when any part of the surround is illuminated (2) and are maximally inhibited when the entire surround is illuminated (4). Immediately after a period of inhibition, the cells may exhibit a brief period of more rapid discharge (termed *postinhibitory rebound*). Diffuse illumination (5) causes minimal change in discharge. The opposite relationships apply to off-center cells (panel **B**). From Kandel ER, Schwartz JH, Jessell TM. *Principles of Neural Science.* 3rd ed. New York, NY: Elsevier; 1991.

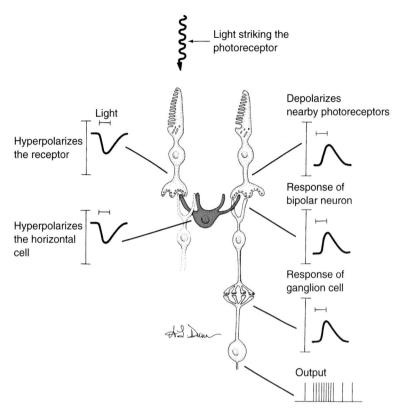

Light striking the photoreceptor

Light

Hyperpolarizes the receptor

Depolarizes nearby photoreceptors

Response of bipolar neuron

Hyperpolarizes the horizontal cell

Response of ganglion cell

Output

Figure 20.4. Generation of the antagonistic surround by horizontal neurons. The antagonistic surround portion of a ganglion cell's receptive field is produced through the action of horizontal cells. Horizontal neurons receive synaptic input from photoreceptors and then project to other photoreceptors in surrounding portions of the retina. Horizontal neurons are depolarized by the neurotransmitter released by photoreceptors. Thus, light striking a photoreceptor (which decreases its resting release of neurotransmitter) leads to hyperpolarization of horizontal neurons. The neurotransmitter released by horizontal neurons hyperpolarizes the photoreceptor targets. When the release of this neurotransmitter is decreased, the target photoreceptors become more depolarized. The circuit illustrated is one involving a flat bipolar cell (off-center). Consequently, light striking the surround produces the opposite response (increased activity of the ganglion cell).

are supplied by those photoreceptors. This is the substrate for the antagonistic surround.

The on and off portions of a particular ganglion cell's receptive field each contribute (antagonistically) to the overall activity of the ganglion cell. Thus, in the case of an on-center receptive field, when a light stimulus is centered in the on portion of the field, and does not extend into the off portion, the ganglion cell is maximally active. If light covers the off portion of the receptive field and does not extend into the on portion, the ganglion cell is maximally inhibited. When both on and off portions of the receptive field are simultaneously illuminated, the ganglion cell's activity will depend on the relative contribution of on and off influences. When these influences are equal, there will be no change in the activity of the neuron. It is for this reason that ganglion cells respond poorly if at all to overall changes in illumination, but respond well to light–dark boundaries. This last is probably the most important point to remember. Pattern vision detects light–

dark boundaries that represent the edges of objects in the visual environment; these edges define the physical form of the object.

The Role of Amacrine Cells

The other principal type of interneuron in the retina is the *amacrine cell*. Amacrine cells receive synaptic connections from bipolar and horizontal neurons and, in turn, form synaptic connections upon ganglion cells. These too are nonspiking neurons. However, unlike the bipolar and horizontal cells, which give sustained responses with given levels of illumination, amacrine cells respond transiently to either the onset or offset of light. The amacrine cells then produce responses to both light onset and offset in a class of ganglion cell that is separate from the sustained type of responses already described. The transient responses produced by amacrine cells are thought to be important for motion and direction detection.

Response Properties of Ganglion Cells

From our discussion, one would predict the existence of populations of ganglion cells that exhibit different response properties:

1. On-center, sustained
2. Off-center, sustained
3. On-center, transient
4. Off-center transient
5. On transient–off transient

In fact, ganglion cells with these response properties are seen. Neurophysiologists who first defined the different classes of cells designated them by letters: X, Y, and W. Importantly, it turns out that ganglion cells with different response properties project to different locations in the CNS and are responsible for different aspects of visual function.

Ganglion cells that exhibit sustained (tonic) responses to visual stimuli project to the lateral geniculate nucleus. These are termed *X cells*. These neurons have a rela-

tively small receptive field and are thought to be responsible for high-resolution analysis of the visual image.

Ganglion cells that exhibit transient (phasic) responses to visual stimuli are termed *Y cells* or *W cells*, depending on their exact response properties. Y cells have larger receptive fields than X cells and project to both the lateral geniculate nucleus and the superior colliculus. Y cells are thought to be important for analyzing the gross form of the visual image. W cells project to the superior colliculus and are thought to be responsible for the information that controls head and eye movement. In keeping with the general distribution of central projections, about 80% of the retinal ganglion cells are X cells, 10% are Y cells, and 10% are W cells.

Electrical Potentials That Are Important for Clinical Diagnosis

There are two forms of electrical activity generated in the retina that are of diagnostic importance. These can be evaluated by the electroretinogram and the electro-oculogram.

The electroretinogram. The depolarizing and hyperpolarizing responses of receptors and interneurons lead to electrical potentials that can be recorded by electrodes placed on the cornea. The most prominent responses are elicited by light flashes, which lead to synchronous activation of receptors throughout the retina. This test is useful for diagnosing retinal pathology.

The electro-oculogram. There is also a steady direct current potential between the front and back of the eye, with the back being more negative than the front. This is probably due to the ongoing activity of the photoreceptors and interneurons. This potential can be recorded by placing electrodes below the medial and lateral corners of the eye. When the eye is turned, the electrode nearest to the front of the eye will be positive relative to the other electrode. This test is also useful as an early indicator of retinal disease.

PHYSIOLOGY OF CENTRAL VISUAL PATHWAYS

The parallel processing that begins in the retina continues in the central visual structures. Retinal ganglion cells that have different response properties project to different central targets: X and Y cells project to the lateral geniculate nucleus (LGN); W cells project to the superior colliculus. Within the retinogeniculate pathway, X and Y inputs terminate in different laminae so that segregation is maintained (see later). The different parallel pathways then project onto the cerebral cortex, where they terminate in different cortical laminae.

There is also hierarchical processing (also called *serial processing*). At each synaptic station, the response properties of target neurons are determined by the combination of inputs. In this way, the receptive fields of higher-order neurons become more and more complex. This is the basis of *feature abstraction*—a process through which higher-order neurons respond selectively to stimuli that meet certain abstract criteria (a corner of a particular orientation at a particular location in the visual field). The principle can best be illustrated by considering how the receptive fields of individual neurons at each station along the central visual pathway are constructed from the combination of receptive fields of neurons at lower levels.

Processing of Visual Information at the Level of the LGN

Neurons in the LGN have receptive fields that are very similar to the receptive fields of retinal ganglion cells. The receptive fields have a simple center–surround antagonistic organization and are monocular because there is no convergence of input from the two eyes at the level of the LGN.

The separate channels for different types of information that began in the retina are maintained in the LGN. Thus, the response properties of neurons in the *parvocellular* and *magnocellular* layers differ based on the input

these layers receive from X and Y retinal ganglion cells.

Processing of Visual Information at the Level of the Visual Cortex

The projections from the different cell types in the LGN terminate in different sublaminae within layer IVc of the primary visual cortex (area 17 or V1). The projections from relay cells in the parvocellular laminae terminate in the lower part of layer IVc (layer IVc-beta). The projections from relay cells in the magnocellular laminae terminate above the terminal field of the X cells in layer IVc-alpha, as well as in layer VI. These are termed the X-parvocellular and Y-magnocellular pathways.

The X-parvocellular pathway is thought to be especially important for the finer aspects of form vision, for color vision, and for the interocular disparity cues that underlie binocular vision. The Y-magnocellular pathway is thought to be especially important for a coarser-grained analysis of spatial relationships, motion, and again interocular disparity cues that are important for binocular vision. These functions depend on the unique response properties of visual cortical neurons.

The Receptive Field Characteristics of Neurons That Mediate Form Vision

Many of the neurons in layer IVc have concentric receptive fields (on- or off-center) similar to the receptive fields of LGN neurons. In other layers, the receptive fields are more complex. The changes in receptive field characteristics reflect serial processing.

Unlike neurons in the LGN, most of the neurons in V1 do not have antagonistic center–surround receptive fields. Instead, cortical neurons respond best to edges (light–dark boundaries). Three principal cell types are recognized based on their response properties—*simple, complex,* and *hypercomplex*. It is not critical that students memorize the features that distinguish the different cell types.

However, a discussion of the response properties helps to illustrate the transformations that occur in the primary visual cortex.

Simple cells have elongated receptive fields with distinct antagonistic on and off zones (Figure 20.5A). The fields also have a characteristic axis of orientation. The most effective stimulus for such a field is illumination in the on portion, and dark in the off portion, that is, a light–dark boundary (edge) aligned with the on–off boundary of the receptive field. A light–dark boundary with a different orientation is ineffective. Thus, each simple cell is specialized to detect edges of a particular orientation located in a particular portion of the receptive field. There are other cells that receive input from the same retinotopic location and respond to different orientations. Thus, every site in the visual field has cells that recognize each axis of orientation.

Complex cells typically have larger receptive fields. Again, these cells respond best to light–dark boundaries with a particular orientation. However, for these cells, the position within the receptive field is less critical because there are no definite on and off zones. Instead, a light–dark boundary of the preferred orientation elicits a response when it appears in any portion of the receptive field (Figure 20.5B).

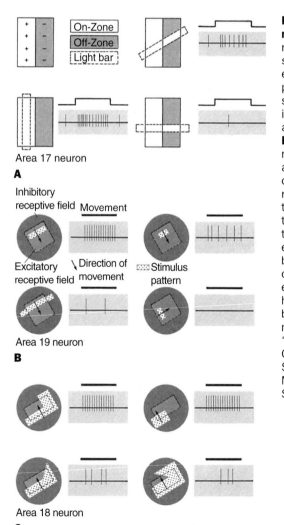

Area 17 neuron

A

Inhibitory receptive field   Movement

Excitatory receptive field   Direction of movement   Stimulus pattern

Area 19 neuron

B

Area 18 neuron

C

Figure 20.5. Response characteristics of neurons in the visual cortex. A illustrates the response characteristics of a neuron with a simple receptive field. Typically, the neuron has elongated on and off zones that are aligned in parallel. Maximal activation occurs when a light stimulus fills the on zone and does not extend into the off zone. Hence, such a cell "represents" a light–dark boundary of a particular orientation. **B** illustrates the response characteristics of a neuron with a complex receptive field. Maximal activation occurs when a stimulus of a particular orientation and defined length moves across the receptive field. A stimulus that does not extend to the boundary of the on region is less effective than one that extends to the boundary. A stimulus that extends beyond the on region is less effective than a stimulus that extends just to the boundary. A stimulus moving in the direction orthogonal to the preferred direction of movement elicits no response. **C** illustrates a neuron with a hypercomplex receptive field. This cell responds best when there are two contrast boundaries that meet at right angles (that is, this cell "represents" a corner). From Grusser O-J, Grusser-Cornehls U. The sense of sight. In: Schmidt RF, Thews G, eds. Biederman-Thorson MA, trans. *Human Physiology*. Berlin, Germany: Springer-Verlag; 1983:237–276.

Hypercomplex cells have receptive fields that respond best to even more complex stimulus configurations. For example, whereas simple and complex cells respond best to light–dark boundaries (edges), hypercomplex cells respond best to configurations in which the edge has a stopping point (that is, there is a corner, Figure 20.5C). In this way, activity of neurons with particular response properties represents (that is, signals the presence of) a particular stimulus configuration in the visual field (a line, a corner, etc.).

The receptive fields of cortical neurons are built from the receptive fields of neurons at lower levels. The receptive fields of simple cells are determined by the receptive fields of the thalamic inputs to the cortical cells. That is, simple cells are thought to receive input from neighboring LGN cells that lie along an orientation axis in the retina; this collection of retinal ganglion cells would be strongly activated by an edge of the appropriate orientation in the visual field. Complex cells are thought to receive input from several simple cells that respond to a stimulus of the same axis of orientation, but represent slightly different locations in the visual field, and so forth.

Orientation Columns

An interesting and important feature of visual cortical organization is that neurons with particular orientation selectivities are organized in vertical columns. Throughout a column, all simple and complex cells have receptive fields with a similar orientation selectivity. Moreover, there is an orderly shift in the preferred axis of orientation as one moves from one column to the next and across the cortical surface. Thus, every 30 to 100 μm, the orientation preference shifts about 10°. In this way, there is a continuous map of orientation preferences across the cortical surface.

Binocularity in the Visual Cortex

The visual cortex is the first station at which inputs from the two eyes converge on single cells. Most neurons in area 17 respond to both eyes, although they often respond best to one eye or the other (see later). The degree to which a neuron is dominated by one or the other eye is termed *ocular dominance*.

When neurons respond to both eyes, the receptive field for each eye is identical. The receptive field maps the same retinotopic location in each eye, and the orientation selectivity is the same. Thus, if a cell responds best to a horizontally oriented stimulus when activated by one eye, the preferred stimulus will be identical for the opposite eye.

Ocular Dominance Columns

As already noted, geniculocortical axons that carry information from the two eyes terminate in alternating patches in layer IV of the visual cortex. This is the anatomical basis for *ocular dominance domains*. Neurons located in the center of a "patch" of terminals carrying information from one eye will be dominated by inputs from that eye. Neurons located at the boundary between patches of terminals carrying information from the two eyes will receive more or less equivalent input from each eye.

Neurons at various depths within the cortex tend to have the same response properties because of strong interlaminar connections. Inputs to layer IV define the response characteristics of the thalamic recipient neurons, and then interlaminar projections of these neurons define the response characteristics of neurons in other layers. In this way, the alternating patches of terminals from different laminae in the LGN produce alternating columns of neurons with similar ocular dominance characteristics (*ocular dominance columns*).

Cortical Modules, or Hypercolumns

Orientation columns and ocular dominance columns are mapped orthogonally across the cortical surface. In this way, each sector of the cortex that represents a particular location in the retinotopic map is composed of subregions within which there are ocular dominance columns carrying inputs for the two eyes, and a complete set of orientation columns. Each *cortical module* (about 1 mm² in cross-sectional area) has an entire set of columns for analyzing lines of all orientations and inputs from both eyes. Each such module thus maps a particular area of the visual field. These cortical modules are also termed *hypercolumns*. The hypercolumns are repeated for each portion of the visual field, so the entire visual cortex has a repetitive structure reminiscent of a crystal.

Color Vision

Color vision is mediated by the three different types of cones that express different cone opsins that absorb light maximally in the red, green, and blue ranges of the visual spectrum. The cones that express the respective opsins are termed red, green, and blue cones. But cone opsins have broadly overlapping absorption spectra and so individual cones do not "encode" colors. Instead, the visual system perceives color by combining the responses produced in different types of cones. This combinatorial process begins in the retina and continues at subsequent stages along the central visual pathways.

Our understanding of color vision is based on the *opponent process theory* set forth by E. Hering. Hering proposed that vision depends on neural "channels" in which there are mutually antagonistic opponents (white–black, red–green, and blue–yellow). The channels respond in one direction to one opponent (that is, on or off), and in the opposite direction to the other. The black–white channel corresponds to the on-center and off-center circuits already described.

The opponent process that mediates color vision works through the same basic circuits. For example, red-sensitive cones are the input stage for an on-center circuit supplying a particular ganglion cell; green-sensitive cones are the input stage for an off-center circuit supplying the same ganglion cell (Figure 20.6). The receptive fields of color oppo-

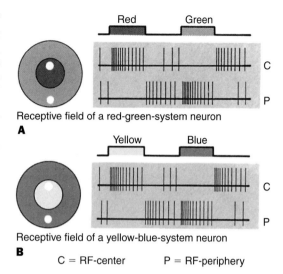

Receptive field of a red-green-system neuron
A

Receptive field of a yellow-blue-system neuron
B

C = RF-center   P = RF-periphery

Figure 20.6. Color perception is mediated by retinal ganglion cells that combine input from opponent pairs. A and **B** illustrate, respectively, the receptive field organization of two color opponent retinal ganglion cells. The receptive field illustrated in **A** is red on-center and green off-surround. The receptive field in **B** is yellow on-center and blue off-surround. C indicates receptive field center; P, receptive field surround. From Grusser O-J, Grusser-Cornehls U. The sense of sight. In: Schmidt RF, Thews G, eds. Biederman-Thorson MA, trans. *Human Physiology.* Berlin, Germany: Springer-Verlag; 1983:237–276.

nent cells have a center–surround organization, except that the center is one of the opponent colors and the surround is the other of the pair. Ganglion cells on which opponent channels converge are termed *color opponent cells.* Ganglion cells then integrate the inputs in different ways, depending on the types of ganglion cells, and then convey that information to the CNS.

Color Blindness

Individuals who are color blind are missing one or more of the cone opsins. The genetic condition in which only two cone opsins are expressed is termed *dichromatopsia.* If only one cone opsin is present, the condition is termed *monochromatopsia.*

Higher-Order Processing

The processing of the visual image discussed so far occurs within the primary visual cortex. There is further processing and feature extraction within *visual association areas* that receive input from the primary visual cortex. There are a number of such association areas located in the occipital, parietal, and temporal lobes. Neurons within these areas respond to visual stimuli, but the stimulus configurations to which neurons exhibit optimal responses are more and more abstract.

It is thought that different visual association areas receive input from the X-parvocellular and Y-magnocellular circuits of V1. In particular, the X-parvocellular system is thought to be the primary source of visual input to the parietal association areas. The Y-magnocellular system is thought to be the primary source of visual input to the association areas in the temporal lobe.

Injuries involving the visual association areas do not produce blindness but do lead to an inability to use visual information in a normal way. For example, injuries to the parietal lobe lead to a phenomenon in which visual stimuli in the contralateral hemifield are ignored *(visual neglect)*. It is as if the patient does not see the stimuli even though visual function is demonstrably intact. Circumscribed injuries in other cortical locations can lead to an inability to recognize or name objects based on visual cues *(visual agnosia)*.

KEY POINTS

1. The receptor potential is produced by turning off a resting dark current. The receptor potential is hyperpolarizing.

2. Light activation of the photoreceptor leads to a decrease in resting neurotransmitter release.

3. Processing in the retina is by nonspiking neurons. Transmitter release varies as a function of membrane potential.

4. Transmission in the retina involves disinhibition and disfacilitation.

5. Ganglion cells have receptive fields with an antagonistic center–surround organization.

6. Different types of ganglion cells convey different types of information to the CNS. The ganglion cells represent the starting point for different labeled lines conveying different types of information. Processing is also hierarchical in that at each station in the geniculocortical system, neurons are specialized to respond to more and more complex stimulus configurations.

21

Visual System IV

Visual Afferent Pathways That Control Eye Movement and Ocular Reflexes

The previous chapters focus on the retinogeniculostriate system, which mediates form and color vision. Retinal projections to other sites mediate other visually guided behaviors and ocular reflexes. The retinal projections to areas other than the lateral geniculate nucleus (LGN) are termed *extrageniculostriate pathways*, and include (1) the *retinotectal pathway*, (2) the *retinopretectal pathway*, (3) the retinal projections to the *pregeniculate nucleus*, (4) the *retinohypothalamic pathway*, and (5) the retinal projections to the *accessory optic nucleus*. The organization of these pathways in comparison to the geniculostriate pathway is schematically illustrated in Figure 21.1, and the organization of the different extrageniculate pathways is summarized in Table 21.1.

In what follows, we briefly summarize the connections and functions mediated by each pathway, focusing especially on the pathways that control eye movements and ocular reflexes. Students should be familiar with the names of the different pathways and the basic circuitry of each and should be familiar with the functions that are mediated by the different pathways. Students should

be especially familiar with the circuitry that controls eye movements, because disorders of eye movements provide important clues about the location of lesions.

THE RETINOTECTAL PATHWAY PROVIDES THE VISUAL INFORMATION NECESSARY TO DIRECT GAZE TO OBJECTS IN THE VISUAL ENVIRONMENT

The retinotectal pathway branches off from the optic track fibers that enter the LGN. The pathway then enters the superior colliculus via the *brachium of the superior colliculus*. The projection is organized in a *visuotopic map* in much the same way as are the retinal projections to the lateral geniculate nucleus. The projection comes from the portion of the retina that maps the contralateral half of the visual field (see Figure 21.1B). In this visuotopic map, the upper quadrants of the visual field are represented medially in the superior col-

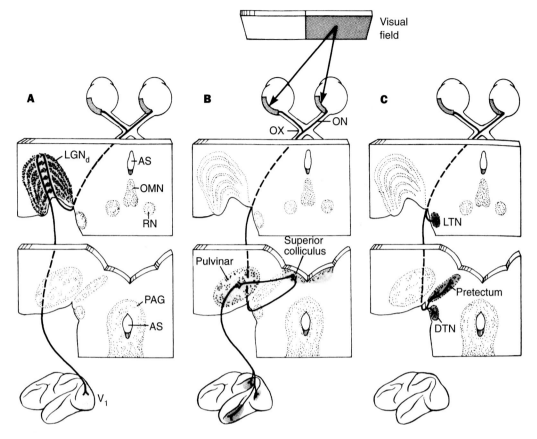

Figure 21.1. Extrageniculostriate pathways. A is a simplified schematic of the retinogeniculostriate pathway; **B** and **C** illustrate various extrageniculostriate pathways including (1) the retinotectal pathway, (2) the retinopretectal pathway, (3) the retinal projections to the pregeniculate nucleus, and (4) the retinal projections to the accessory optic nucleus. Not shown is the retinohypothalamic pathway. ON indicates optic nerve; OX, optic chiasm; LGNd, dorsal lateral geniculate nucleus; AS, aqueduct of Sylvius; OMN, oculomotor nucleus; RN, red nucleus; PAG, periaqueductal gray; LTN, lateral terminal nucleus; DTN, dorsal terminal nucleus. The LTN and DTN are subdivisions of the pregeniculate nucleus. Reproduced with permission from Fuchs AF. Excitable cells and neurophysiology. In: Patton HD, Fuchs AF, Hille B, Scher AM, Steiner R, eds. *Textbook of Physiology 1*. Philadelphia, Pa: WB Saunders; 1989.

liculus; lower quadrants are represented laterally. These fibers terminate in the superficial layers of the superior colliculus (specifically, the *stratum opticum*).

The ganglion cells that give rise to the retinotectal pathway are the cells that exhibit transient (phasic) responses to visual stimuli *(W cells)*. Hence, the pathway is specialized to convey information regarding moving stimuli, which is important for controlling eye and head movements so as to direct gaze to objects in the visual environ-

ment. The pathway also provides an alternate route for information flow from the retina to the cerebral cortex. This is relayed from the superior colliculus to the *pulvinar nucleus* and then from the pulvinar nucleus to areas in the parietal and temporal lobes (Figure 21.1B). This pathway is thought to be important for conveying information regarding moving visual stimuli to the cerebral cortex.

Other sensory systems project to the deeper layers of the superior colliculus (Fig-

Table 21.1. Extragenicuolostriate pathways

Pathway	Target	Converging inputs	Outputs of target	Function
1. Retinotectal	Superior colliculus	Somatosensory Auditory Visual cortex Frontal eye field	Pontine RF Cervical spinal cord Vestibular nucleus Pulvinar nucleus of thalamus	Control of head and eye movement; visual tracking; Visual input to secondary visual areas of the cortex
2. Retinopretectal	Pretectal nucleus		Edinger–Westphal nucleus	Pupillary light reflex
3. Retinohypothalamic	Suprachias-matic nucleus		Hypothalamus	Circadian rhythms
4. Retinoaccessory optic	Accessory optic nucleus		Inferior olivary nucleus	Control of head and eye movement; visual tracking

ure 21.2). There are inputs from the somatosensory system that terminate in somatotopic maps, as well as inputs from the inferior colliculus that convey information about the location of sounds in space. In the somatotopic map, a relatively large proportion of the representation is for the upper body, neck, and head. These pathways provide multisensory information that allows the superior colliculus to integrate input from the different sources to coordinate eye and head movements that permit orientation to, and tracking of, visual stimuli.

The superior colliculus also receives important projections from the cortex on the ipsilateral side (the *corticotectal pathway*). One component arises from the visual cortex. This projection terminates in a retinotopic fashion across the superficial layers of the superior colliculus. Both the visual cortex and the superior colliculus receive input from the contralateral visual field, and so the visuotopic map that is created by the corticotectal pathway corresponds with the visuotopic map of the retinotectal pathway. Another important component derives from the *frontal eye fields* in the cerebral cortex. The frontal eye fields, located just rostral to the motor cortex, operate as a cortical motor control area for conjugate eye movement. Activation of the frontal eye field in one hemisphere causes the eyes to deviate toward the side of the stimulation (more on this later).

Neurons in the deep layers of the superior colliculus project to areas that are involved in the control of eye and head movement, including the so-called gaze centers in the reticular formation (see later), the vestibular nuclei, and the cervical spinal cord (via the tectospinal tract) (Figure 21.2). Lesions of the superior colliculus disrupt the ability to orient to visual stimuli in the contralateral visual field. This is consistent with the fact that each side of the superior colliculus receives input from the contralateral visual field.

THE CONTROL OF EYE MOVEMENTS

The mechanisms of motor control of eye movements are somewhat different from the mechanisms that control other voluntary motor functions. In the first place, voluntary eye movements are *conjugate* (the two eyes

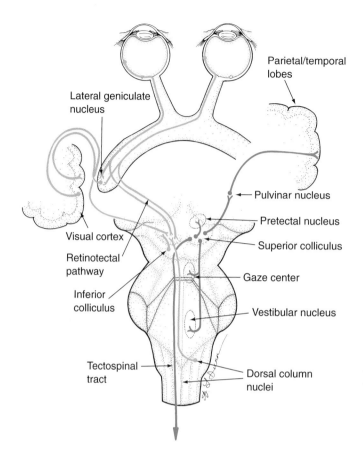

Figure 21.2. The superior colliculus receives multimodal sensory input so as to control eye and head movements to direct gaze. Schematic illustration of the inputs to the superior colliculus (left side) and the principal outputs (right side). The retinotectal pathway terminates in a retinotopic fashion across the superficial layers of the superior colliculus. Projections carrying auditory information from the inferior colliculus and somatosensory information from the dorsal column nuclei terminate in the deep layers. Neurons in the deep layers of the superior colliculus then project to areas that are involved in the control of eye and head movement, including the so-called gaze centers in the reticular formation, the vestibular nuclei, and the cervical spinal cord (via the tectospinal tract). Collicular neurons also project to the pulvinar nucleus of the thalamus, which then projects to areas in the parietal and temporal lobes.

move together). This is essential for binocular vision (where the same visual image is seen by both eyes, see Chapter 20). Also, voluntary eye movements are controlled by gaze control centers located in the brainstem, which coordinate eye movements through important interconnections between the nuclei containing upper motoneurons that innervate the *extrinsic eye muscles*. The projections from the superior colliculus terminate in these gaze centers rather than directly on the motoneuron pools that supply the extrinsic muscles of the eye. In what follows, we begin by considering the muscles that control eye movement and then consider the central circuitry.

The Eye Muscles

Extrinsic eye muscles are the muscles that control the movements of the eyes within the orbit. Extrinsic eye muscles exist in antagonistic pairs: (1) the *medial and lateral rectus muscles,* (2) the *superior and inferior rectus muscles,* and (3) the *superior and inferior oblique muscles*. The direction of pull of each antagonistic muscle pair is summarized in Table 21.2 and Figure 21.3.

The medial rectus and lateral rectus muscles rotate the eye medially and laterally along the vertical axis. The lateral rectus *abducts* the eye (turns the eye away from the midline); the medial rectus *adducts* the eye (brings the eye toward the midline).

The other muscle pairs exert forces along oblique axes. Their anatomical organization is illustrated in Figure 21.3, and the operation of the different muscle pairs is illustrated in Figure 21.4 and summarized in Table 21.1.

When gaze is straight ahead, contraction of the superior rectus muscle causes elevation, medial rotation, and adduction of the eye. Contraction of the inferior rectus mus-

Table 21.2. Control of extrinsic eye muscles

Muscle	Innervation	Function (when gaze is straight ahead)
Lateral rectus	Abducens (VI)	Abduction
Medial rectus	Oculomotor (III)	Adduction
Superior rectus	Oculomotor (III)	Elevation, medial rotation, adduction
Inferior rectus	Oculomotor (III)	Depression, lateral rotation, adduction
Superior oblique	Trochlear (IV)	Depression, medial rotation, abduction
Inferior oblique	Oculomotor (III)	Elevation, lateral rotation, abduction

cle causes depression, lateral rotation, and adduction. Contraction of the superior oblique muscle causes depression, medial rotation, and abduction. Contraction of the inferior oblique muscle causes elevation, lateral rotation, and abduction. See Figure 21.3 for details.

If the gaze is shifted about 23° to one side, the superior and inferior rectus muscles are aligned such that they simply elevate and depress the eye on the side to which the gaze is shifted (Figure 21.3D, right eye). The superior and inferior oblique muscles are aligned such that they simply elevate and depress the eye opposite to the side to which the gaze is shifted (Figure 21.3D, left eye).

Physicians can test the operation of the different pairs of muscles by evaluating elevation and depression of the eyes during lateral gaze. Thus, when the eyes are directed to the right, elevation and depression tests the operation of the superior and inferior rectus muscles on the right, and the superior and inferior oblique muscles on the left. The opposite is true when the eyes are directed to the left.

The lower motor neurons that control extrinsic eye muscles are located in three cranial nerve nuclei: (1) the *oculomotor,* (2) the *trochlear,* and (3) the *abducens.* The axons of the motoneurons join the oculomotor (III), trochlear (IV), and abducens (VI) nerves, respectively. The location of the cranial nerve nuclei that control eye movement is illustrated in Figure 21.5.

The oculomotor nerve also carries fibers that innervate the *levator palpebrae muscle,* which lifts the eyeball, and parasympathetic axons from the Edinger–Westphal nucleus (part of the oculomotor complex), which control the pupil and modify lens shape (more on this later).

The distribution of innervation to the extrinsic eye muscles is summarized in Figure 21.4 and Table 21.1. The abducens nerve innervates the lateral rectus muscles; the trochlear nerve innervates the superior oblique; and the oculomotor nerve innervates the other muscles.

The mode of control of extrinsic eye muscles by motoneurons is essentially identical to that of other somatic motoneurons. Muscle tension is regulated as a function of the rate of action potentials over the motor axons. The balance of tension in the opposing muscles determines the position of the eye. During eye movements, the speed of movement is directly proportional to the discharge frequency over the motor axons. During the rapid movement of eyes from one point to another, there is a pulse of activity that rapidly shifts the eye. Smooth tracking (following a moving object) is accomplished by slow changes in action potential frequency. These two types of eye movements are discussed in more detail later.

There are some differences in the control of the motoneurons, however. First, as we discuss later, the motoneurons are not directly innervated by corticobulbar fibers. Hence, voluntary eye movements are mediated indirectly. Second, extrinsic muscles of

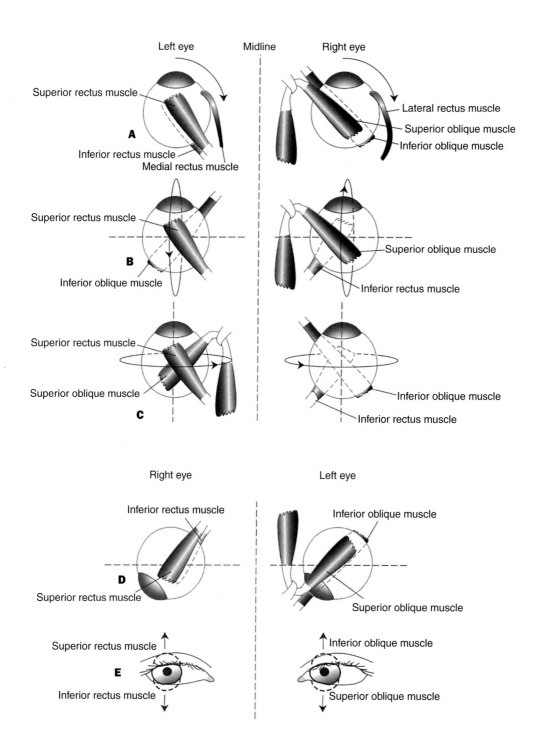

Figure 21.3. Anatomical organization of extrinsic eye muscles. Extrinsic eye muscles exist in antagonistic pairs: (1) the medial and lateral rectus muscles, (2) the superior and inferior rectus muscles, and (3) the superior and inferior oblique muscles. The medial rectus and lateral rectus muscles rotate the eye medially and laterally (**A**). The other muscle pairs exert forces along oblique axes. If the gaze is shifted about 23° to one side, the superior and inferior rectus muscles are aligned so that they simply elevate and depress the eye on the side to which the gaze is shifted (**D,** right eye). The superior and inferior oblique muscles are aligned so that they simply elevate and depress the eye opposite to the side to which the gaze is shifted (**D,** left eye). Reproduced with permission from Kingsley RE. *Concise Text of Neuroscience.* Baltimore, Md: Williams & Wilkins; 1996.

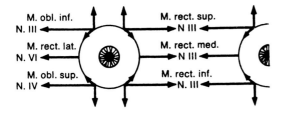

Figure 21.4. A summary of the action of extrinsic eye muscles. Note that the arrows refer to the directions that the eyes move in response to activation of the different muscles. From Brodal A. *Neurological Anatomy*. New York, NY: Oxford University Press; 1981.

the eye do not experience changes in loading; hence, they do not respond to muscle stretch (although muscle spindles are present). Third, there is no recurrent inhibition of the oculomotor neurons; however, there are inhibitory interneurons in the nuclei that allow descending inputs to operate through projected inhibition (see Chapter 14). Thus,

the inputs to the cranial nerve nuclei can control the motoneurons through both excitation and inhibition.

Eye Movements Are Coordianted by Visuomotor Centers in the Brainstem

The coordination of eye movements during *gaze* (that is, the voluntary direction of the eyes to an object in the visual environment) requires that different muscles be activated in the two eyes. For example, to look at an object on your right, your right eye must turn away from the nose (lateral rectus), and the left eye must turn toward the nose (medial rectus). This is called *conjugate eye movement.*

There are two types of conjugate eye movements, termed *smooth pursuit* (also called *tracking movements*) and *saccades*. Smooth pursuit movements are exemplified by the movements that occur as one is tracking a

Figure 21.5. Cranial nerve nuclei contain lower motoneurons that control eye movement. Lower motoneurons that control extraocular muscles are located in three nuclei in the brainstem: (1) the oculomotor, (2) the trochlear, and (3) the abducens. The axons of the motoneurons join the oculomotor (III), trochlear (IV), and abducens (VI) nerves, respectively.

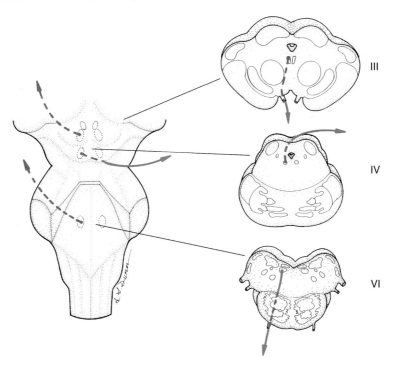

moving object. Saccades are the movements one makes to move the eyes from one focal point in the visual environment to another. You can demonstrate the two types of movement to yourself. Move your finger back and forth in front of your eyes and track the movements smoothly (you may want to be sure no one is watching you when you try this). However, it is not possible to slowly move your eyes so as to scan across your visual environment. Instead, your eyes will jump from point to point, bringing one point and then another into focus. These rapid movements are termed *saccades.*

Another type of eye movement is important for aligning the eyes to focus on near or far objects. The two eyes *converge* to look at a nearby object and *diverge* to look at a distant object; these are termed *vergent movements.* Again, you can demonstrate these movements to yourself by moving your finger back and forth in front of your nose. Because vergent movements cause the eyes to move in different directions, they are called *disconjugate movements.*

The three types of eye movements mediate different functions: (1) smooth pursuit allows one to follow moving stimuli, (2) saccades allow one to quickly point the eyes to objects of interest in the visual world, and (3) vergent movements allow the eyes to focus on objects at different distances from the viewer.

These different voluntary eye movements are coordinated by visuomotor centers *(gaze centers)* in the brainstem. To a large extent, these movements are controlled involuntarily, although deciding *what* to look at is voluntary and is thus under cortical control (more on this later).

Brainstem Circuits That Control Gaze

The key gaze centers are located in the *pontine paramedian reticular formation* and in the *rostral interstitial nucleus of the medial longitudinal fasciculus* in the mesencephalic reticular formation.

The pontine paramedian reticular formation is a collection of neurons in the area between the abducens and trochlear nuclei,

just lateral to the medial longitudinal fasciculus—the fiber tract that carries fibers from the vestibular nuclei. This group of neurons is especially important for controlling *horizontal gaze.* The neurons in the nucleus project directly to motoneurons in the cranial nerve nuclei that move the eye in the horizontal plane. Unilateral lesions of the pons that involve this area lead to an inability to voluntarily direct gaze toward the side where the lesion is located.

The rostral interstitial nucleus of the medial longitudinal fasciculus is located near the midline, just rostral to the nucleus of cranial nerve III. It and the nearby nucleus of Darkschewitsch receive direct projections from the medial and superior vestibular nuclei. The neurons in the nucleus project directly to motoneurons in the cranial nerve nuclei that move the eye in the vertical plane. Obviously most eye movements have both horizontal and vertical components. Hence, the horizontal and vertical gaze centers work in conjunction with one another.

Inputs to the gaze centers provide the sensory information that is the key to coordinating eye movements to track objects in the visual environment and compensate for changes in the position of the head in space. Information pertaining to objects in the visual environment is relayed from the superior colliculus; information relating to the position of the head in space comes from the vestibular nuclei.

The pathway from the vestibular nuclei originates from the medial and superior vestibular nuclei (see Chapter 25) and travels via the medial longitudinal fasciculus—a paired tract on each side of the midline in the dorsal brainstem, just beneath the fourth ventricle (Figure 21.4). The medial vestibular nucleus also projects to the cervical spinal cord, where it terminates on motoneurons supplying the neck. In this way, the medial vestibular nucleus influences motoneurons that control both eye and head movements.

The Vestibular-Ocular Reflex

The previous discussion focuses on the control of gaze, that is, the voluntary control of eye movements to direct the eyes to objects of interest in the visual world. Often, however, it is necessary to maintain focus while the head is moving (for example, while one walks or runs). Maintaining stability of focus when the eyes are located in an unstable platform (the head) is accomplished via the vestibular ocular reflex.

The vestibular-ocular reflex (VOR) is mediated by a three-neuron reflex arc. The afferent limb is provided by the vestibular afferents from the ear, which terminate in the brainstem vestibular nuclei. The brainstem vestibular nuclei project directly to the motoneurons of the third, fourth, and sixth cranial nerve nuclei via the medial longitudinal fasciculus (see Chapter 25). The efferent limb is the projection from the cranial nerve nuclei to the extrinsic ocular muscles. The organization of the circuitry responsible for the vestibular-ocular reflex is discussed further in the chapter on the vestibular system (Chapter 25).

Cortical Control of Eye Movements

Cortical control of eye movements (that is, directing the eyes to a particular point in the visual environment) is mediated by a special motor region called the *frontal eye field*. The frontal eye field was identified experimentally by the fact that stimulation of the region caused convergent movement of the eyes to the side opposite the stimulation. The frontal eye fields control eye movement through projections to the superior colliculus. These corticotectal projections terminate in the intermediate layers of the superior colliculus. The neurons in the superior colliculus then project to the gaze centers in the pontine and mesencephalic reticular formation.

The Optokinetic Reflex

The fixation of the eyes on an object that moves across the visual field is mediated by the brainstem circuitry already described, in conjunction with the cerebral cortex (which decides to attend to the moving object). Once having decided to attend to a moving object, the eyes follow without conscious effort. Hence, this tracking has the characteristics of a reflex and is termed the *fixation reflex*.

When there is continuous movement (for example, in a moving vehicle) or when a moving object passes out of the field of view, the eyes rapidly move to focus on a new point by means of a saccade. The smooth tracking–saccade sequence is termed *optokinetic nystagmus*. By convention, we refer to the direction of nystagmus based on the saccade. For example, if you are looking out the right-hand window of a car, the visual environment is moving to the right. The resulting optokinetic nystagmus is termed *left optokinetic nystagmus*.

The decision to attend to a moving stimulus appears to be mediated by the parietal lobe. When there is unilateral injury to the parietal lobe, patients are unable to smoothly track objects moving toward the side of the lesion. This can be assessed by eliciting the optokinetic reflex by moving some patterned stimulus one way or the other across the patient's visual field.

Retinal Projections to the Pregeniculate Nucleus and Nucleus of the Optic Tract Provide a Route for Visual Input to the Portions of the Cerebellum That Coordinate Eye and Head Movements

The *pregeniculate nucleus,* also called the *ventral lateral geniculate nucleus,* also receives a direct projection from the eye. This nucleus sends a projection to the pontine nuclei, which in turn projects to the vestibular portion of the cerebellum via mossy fibers (see Chapter 17).

The *accessory optic nucleus* also receives a projection from the retina. This nucleus projects to the inferior olivary nucleus, which in turn projects to the vestibular portion of the cerebellar cortex via mossy fibers. The accessory optic nucleus and the pregeniculate nucleus operate in concert to coordinate eye–head movement in conjunction with the vestibular system and cerebellum.

Retinal Projections to the Suprachiasmatic Nucleus Provide Information Regarding the

Light/Dark Cycle, Which Entrains Circadian Rhythms

The *suprachiasmatic nucleus* is innervated by a small component of optic tract axons that leave the tract near the level of the optic chiasm. These projections provide a direct retinal input to the hypothalamus. This nucleus conveys information pertaining to the light–dark cycle, which is critical for entraining *circadian behaviors* (those with a 24-hour cycle). The pathway also plays a role in light-induced gonadotrophic effects in the estrous cycle.

Retinal Projections to the Pretectum Represent the Afferent Limb of the Circuit That Mediates the Pupillary Light Reflex

The *retinopretectal pathway* is made up of axons that arrive via the brachium of the superior colliculus but diverge from the main retinocollicular pathway to enter the pretectal nuclei (Figure 21.6). These retinopretectal connections are the afferent limb of the *pupillary light reflex*, which mediates pupillary constriction in response to light. The pretectum then projects to the Edinger–Westphal nucleus—a division of the nucleus of cranial nerve III. The efferent limb is mediated by the *preganglionic parasympathetic projections* from the Edinger–Westphal nucleus to the ciliary ganglia in the eye. Axons from the ciliary ganglia then innervate the constrictor muscle of the iris.

Illumination of one eye leads to a constriction of the pupil bilaterally. The bilateral response is termed a *consensual response*. The consensual response occurs for two reasons: the retinopretectal pathway projects bilaterally, and the pretectal nuclei project bilaterally to the Edinger–Westphal nucleus (Figure 21.6). Because of the bilateral representation, pupillary inequality *(anisocoria)* usually results from lesions involving the efferent projections from the oculomotor complex, or lesions involving the posterior commissure.

The pupil is also controlled by postganglionic sympathetic fibers, which cause dilation when activated. These originate from the superior cervical ganglion (Figure 21.6).

The pupil also constricts during *accommodation,* which is the process through which the shape of the lens is adjusted to focus on near objects. Lens shape is controlled via preganglionic parasympathetic neurons that project from the Edinger–Westphal nucleus to the ciliary ganglia, which in turn receive their input from the pretectal nuclei (Figure 21.7). Postganglionic neurons in the ciliary ganglia then control the ciliary muscles. There must be cortical (voluntary) control, and the afferent limb for this control must come via the retinogeniculostriate pathways that mediate form vision, but the exact circuitry that mediates this is not entirely clear.

Because the pupil constricts during accommodation, one should not test the pupillary light reflex using a small light held near the eye. The reason is that patients tend to focus on the light, so pupillary constriction due to accommodation may be confused with constriction in response to light. In some disease processes (syphilis, for example), the pupil is unresponsive to light but still responds during accommodation. This is termed the *Argyll–Robertson pupil.* The exact cause of the dissociation is not clear.

Symptoms Resulting From Injury to the Circuitry That Controls Eye Movement

The symptoms that result from selective injuries to the cranial nerves themselves are easy to predict on the basis of the muscle groups that the nerves control. Selective injury to a cranial nerve on one side leads to paralysis of the muscles supplied by that nerve on the side ipsilateral to the lesion. This is true even of the trochlear nerve. Although the nerve decussates, most injuries to the nerve itself occur after the decussation.

Injury to the oculomotor nerve leads to (1) paralysis of the medial rectus muscle; the eye then deviates laterally because the action of the lateral rectus muscle is unapposed; (2) paralysis of the levator palpebrae, leading to eyelid droop *(ptosis);* (3) pupillary dilation *(mydriasis)* due to loss of parasympathetic innervation; (4) loss of pupillary light reflexes

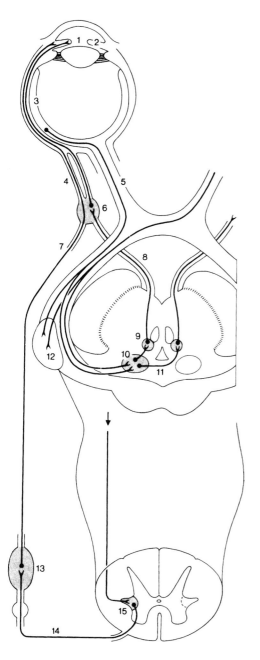

Figure 21.6. Retinal projections to the pretectum are the afferent limb for the pupillary light reflex. The retinopretectal pathway is made up of axons that diverge from the main retinocollicular pathway to enter the pretectal nuclei. Neurons in the pretectum then project bilaterally to the Edinger–Westphal nucleus, which gives rise to the parasympathetic fibers of the oculomotor nerve that mediate the pupillary light reflex. Illumination of one eye leads to a constriction of the pupil bilaterally (termed the *consensual response*) because the retinopretectal pathway projects bilaterally, and the pretectal nuclei project bilaterally to the Edinger–Westphal nucleus. 1 indicates pupil; 2, iris; 3, retina; 4, ciliary nerve; 5, optic nerve; 6, ciliary ganglion; 7, projections from superior cervical ganglion; 8, oculomotor nerve; 9, Edinger–Westphal nucleus; 10, pretectal region; 11, posterior commissure; 12, lateral geniculate nucleus; 13, superior cervical ganglion; 14, sympathetic preganglionic fibers; 15, preganglionic motoneurons in the intermediolateral column. From Nieuwenhuys R, Voogd H, van Huijzen C. *The Human Central Nervous System.* Berlin, Germany: Springer-Verlag; 1988.

(both direct and consensual) due to the loss of the efferent limb of the reflex; and (5) impaired accommodation. Patients may also experience double vision *(diplopia)* with *horizontal displacement* of the images (images are side by side) due to the misalignment of the eyes resulting from lateral deviation of the affected eye.

Injury to the trochlear nerve leads to paralysis of the superior oblique muscle, which leads to an impairment of downward gaze. Patients may also experience diplopia with images displaced vertically.

Injury to the abducens nerve leads to paralysis of the lateral rectus muscle. The affected eye tends to deviate medially because

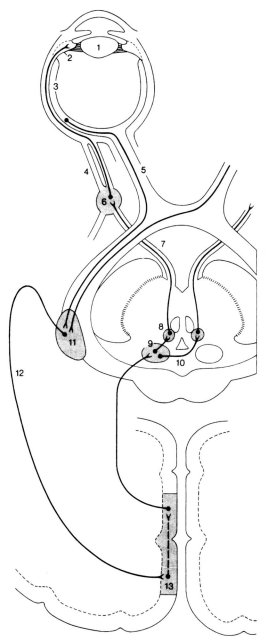

Figure 21.7. Neural pathways that mediate accommodation. Accommodation is the process through which the shape of the lens is adjusted to focus on near or far objects. Lens shape is controlled via preganglionic parasympathetic neurons that project from the Edinger–Westphal nucleus to the ciliary ganglia. Postganglionic neurons in the ciliary ganglia then control the ciliary muscles. The pupil also constricts during accommodation, which is mediated by the retinogeniculostriate circuitry. 1 indicates lens; 2, ciliary body; 3, retina; 4, ciliary nerve; 5, optic nerve; 6, ciliary ganglion; 7, oculomotor nerve; 8, Edinger–Westphal nucleus; 9, pretectal region; 10, posterior commissure; 11, lateral geniculate nucleus; 12, optic radiation; 13, striate cortex. From Nieuwenhuys R, Voogd H, van Huijzen C. *The Human Central Nervous System.* Berlin, Germany: Springer-Verlag; 1988.

the action of the medial rectus is unapposed. Patients may experience diplopia with horizontal displacement.

Injury to the internal circuitry within the brainstem that controls eye movements also disrupts eye movements. However, the impairments of eye movements that occur as a result of brainstem lesions are part of symp-tom complexes *(syndromes)* that also involve other sensory and motor symptoms.

Lesions involving the frontal eye fields cause transient deficits in *contralateral gaze.* That is, individuals cannot move the eyes to focus on an object in the contralateral visual field. The eyes are not paralyzed and can move laterally under the control of the vestibular-ocular reflex. Paresis of contralateral gaze following frontal cortical lesions usually resolves fairly quickly. Lesions involving the frontal eye fields also disrupt the optokinetic reflex. In this case, it is the fast component of the movement (the saccade) that is disrupted.

It is of considerable clinical significance that the corticobulbar fibers that mediate voluntary eye movement do not terminate directly on the lower motoneurons in the respective cranial nerve nuclei. This fact explains why damage to the descending corticobulbar pathways does not paralyze the eyes.

6

The Auditory System

The auditory system is another of the "special" senses. The components of the auditory system that are involved in the conscious perception of sound include the following (Figure VI.1):

1. The external ear, auditory meatus, and tympanic membrane (eardrum)

2. The mechanical transduction system in the middle ear (the ossicles)

3. The cochlea including the first-order neurons (cochlear ganglion cells)

4. The auditory component of cranial nerve VIII, which carries information from the ear to the brain

5. Auditory brainstem nuclei

6. The inferior colliculus

7. The medial geniculate nucleus of the thalamus

8. The auditory cortex

We divide our consideration of the auditory system into two major sections: the first deals with the ear and the receptive apparatus in the cochlea; the second considers central auditory pathways, auditory neurophysiology, and the basis of auditory perception.

There are several key concepts for physicians:

1. *Frequencies are mapped as a place code,* so that particular parts of the cochlea are maximally responsive to particular frequencies of sound.

2. There is a topographic mapping of frequencies at each successive stage of the auditory system. This is termed tonotopic mapping.

3. Convergence of information from the two ears first occurs at the level of the third-order auditory brainstem nuclei.

4. *There are several distinct "channels" of information flow.* Different subdivisions of the cochlear nucleus and different third- and

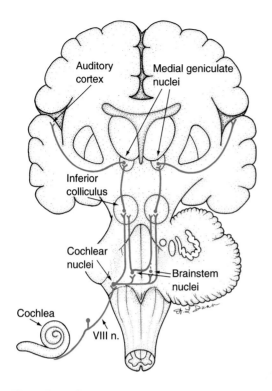

Figure VI.1. Overview of auditory system organization: Schematic illustration of the organization of the auditory pathways. The auditory receptors are found in the cochlea in the inner ear. The receptors activate the distal endings (functionally, the dendrites) of cochlear ganglion cells in the spiral ganglion. The axons from the ganglion cells (first-order axons) travel to the brain via cranial nerve VIII and enter the brain at the cerebellopontine junction, where they terminate on second-order neurons in the cochlear nuclei. These second-order neurons give rise to a complicated set of ascending projections that relay through the inferior colliculus in the midbrain and the medial geniculate nucleus of the thalamus, eventually reaching the auditory cortex in the temporal lobe.

higher-order auditory brainstem nuclei are responsible for frequency discrimination and sound localization.

5. *Conscious perception of auditory sensation depends on inputs to the cerebral cortex.* In principle, interruption of the auditory pathways anywhere along their course from the ear to the cortex would lead to a loss of auditory perception. However, the pathways pass through the core of the brainstem, which also contains areas that are critical for consciousness and the maintenance of life. This, together with the fact that information from the two ears converges at the level of the auditory brainstem nuclei, means that central lesions rarely produce hearing disorders that are important for lesion localization.

22

Auditory System I

The Ear and the Auditory Receptive Apparatus in the Cochlea

The auditory system allows animals to detect different frequencies of sounds (pitches) and to localize sounds in space. These two functions underlie a wide variety of behaviors. Sounds allow animals to recognize and evade predators, recognize and capture prey, and communicate with conspecifics. Indeed, the perception of sound underlies what is certainly one of the most important functions for humans: language. For this reason, disorders of auditory function interfere with the most important of human functions, the ability to communicate. Communicative disorders are among the most common disorders affecting the nervous system and can be among the most debilitating if they occur before the development of language.

In this and the following chapters, we consider how the nervous system detects and processes sound. We begin with a brief consideration of the nature of sound and the units of measurement that define sound. Then we consider the mechanical and neural machinery that transduce sound and relay this information to the central nervous system.

THE NATURE OF SOUND

Sound is generated by mechanical vibrations that generate pressure waves in the air (con-densation and rarifaction of molecules) or in some other medium (for example, water). These pressure waves travel through the medium in a way that depends on the physical characteristics of that medium. At standard conditions of temperature and pressure, sound travels through air at a rate of 33 cm/ms, which translates to 1058 ft/s. The pressure waves generated by sound set up vibrations in the tympanic membrane of the ear, which in turn are transmitted through the middle ear to the cochlea. Thus, sound perception is essentially a special form of vibration sensitivity.

There are three important aspects of sound: (1) its point of origin (location of the sound source in the environment), (2) its spectral content (that is, the frequencies represented in the sound), and (3) the intensity of the sound at each frequency.

The *location* of a sound source is defined with respect to the listener and involves a determination of the azimuth (location in the horizontal plane), elevation (location in the vertical plane), and distance. The location of a sound source is thus expressed in degrees azimuth and elevation, with 0° azimuth being a point directly in front of the listener, and 0° elevation being in the same horizontal plane as the listener.

Sound *frequency* is defined on the basis of the *period* of the sinusoidal waves comprising a sound. The period of the wave is the time *(T)* it takes to complete one cycle. Sound frequency is the inverse of the period (frequency = 1/*T*). The units of measurement of frequency are cycles per second (cps), also called *hertz* (Hz). Humans are sensitive to sounds ranging from 20 to 20,000 Hz.

Pure tones are sounds at a single frequency. Noise, by definition, contains multiple frequencies. Sounds in speech are a combination of lower-frequency vowels and high-frequency consonants, and much of the information content in speech is carried by the higher-frequency consonants. It is for this reason that common forms of hearing loss, which affect high frequencies above the "speech range," interfere with the ability to understand speech (more on this later).

Sound *intensity* is defined on the basis of relative force. The measurement is relative because it is with respect to the threshold for auditory perception. Human auditory threshold at 1000 Hz is 0.0002 dyne/cm^2. Sound intensity is referenced to this threshold using *decibels* (dB) sound pressure level (SPL) as the unit of measurement, where

$$N \text{ (in decibels)} = 20 \log^{10}(a_{observed}/a_{reference})$$

Thus, a sound that is 10 times higher in pressure than a reference sound is 20 dB greater relative to that reference; a sound that is 100 times higher in pressure than a reference sound is 40 dB greater than that reference, and so forth. Humans can distinguish differences in sound intensities that are between 0 and about 120 dB(SPL). This is a dynamic range of 1 to 1 million. A sound that is a few decibels louder than 120 dB is perceived as painful. Table 22.1 lists the approximate decibel level of some representative sounds.

MECHANICAL TRANSDUCTION OF SOUND

Sound waves enter through the *external auditory meatus*, causing the *tympanic membrane*

(eardrum) to vibrate. The vibrations are conducted from the tympanic membrane to the *oval window of the cochlea* via the *ossicles,* the three small bones of the middle ear, which include the *malleus* (hammer), which is attached to the tympanic membrane, the *incus* (anvil), and the *stapes* (stirrup), which is pressed against the membrane covering the oval window of the cochlea (Figure 22.1).

The transduction of sound to neural activity occurs in the *cochlea,* also known as the *auditory labyrinth.* The cochlea is part of the *vestibulo-cochlear labyrinth*—a complex set of cavities in the petrous portion of the temporal bone. The cavities in the bone are called the *bony labyrinth.* Within these is a set of epithelial membranes that form a *membranous labyrinth* that contains the sensory neuroepithelium.

The components of the vestibulo-cochlear labyrinth are the *cochlea,* the *utricle,* the *saccule,* and three *semicircular canals.* All but the cochlea are components of the vestibular system and are considered in that chapter.

The cochlea itself, illustrated in Figure 22.2, is essentially a spiral tunnel through the temporal bone, which contains the specialized receptor apparatus. The overall appearance of the cochlea is rather like a snail's shell that coils 2.5 times. Like the shell of a snail, the tunnel narrows from a large end where the oval and round windows are (the base) to a narrow end (the apex). The terms *basal* and *apical* refer to positions along the cochlear spiral.

The tunnel of the cochlea is divided into two more or less equal longitudinal chambers *(scalae)* by the *basilar membrane.* The two chambers are termed the *scala tympani* and *scala vestibuli.* Another membrane (*Reissner's membrane*) attaches near the midpoint of the basilar membrane and extends to the wall of the tunnel, forming a third, small, triangular compartment (the *scala media,* also known as the *cochlear duct*). The floor of this triangular compartment is formed by the *organ of Corti,* a highly specialized *neuroepithelium* that contains the auditory receptor cells (the cochlear *hair cells*).

The scala vestibuli and scala tympani are filled with a fluid termed the *perilymphatic*

Table 22.1. Approximate sound pressure level (SPL) of some representative sounds

	dBSPL	
(Painful)	140	Jet engine nearby
Jet engine (takeoff power) at 100 ft	130	Shotgun or high-powered rifle
Boom box	120	Rock concert
Jack hammer	110	Chainsaw
Arcade game parlor	100	Radio headset
Motorcycle	90	Lawnmower (5 ft)
City traffic noise	80	Hair dryer
Dishwasher	70	Vacuum cleaner
Inside moving car (windows up)	60	Normal conversation
Quiet office	50	
Refrigerator compressor	40	Quiet living room
Broadcasting studio	30	Whisper
Rustling leaves	20	
Normal breathing	10	

fluid, or simply *perilymph.* The two chambers are connected via a small opening at the apex of the cochlea, termed the *helicotrema,* so the fluid composition of the two chambers is identical. The fluid within the scala media (termed the *endolymphatic fluid,* or simply *endolymph*) has an ionic composition that is very different from that of the perilymph, in that it has very high levels of potassium. This is important for the physiological operation of the cochlear hair cells (see later). The ionic differences between the perilymph and the endolymph are maintained by the ionic pumping activity of cells of the *stria vascularis* within the scala media.

The vibration of the stapes that is produced by sounds puts oscillating pressure on the membrane covering of the oval window and, hence, sets up vibratory waves in the fluid in the scala vestibuli of the cochlea. These waves are transmitted throughout the cochlea, ultimately causing oscillations of the membrane of the round window of the scala tympani. These fluid oscillations cause the basilar membrane to vibrate, which in turn causes small deflections of the *stereocilia* of the cochlear hair cells. These processes are described in more detail later. In the meantime, become familiar with the structures illustrated in Figures 22.1 and 22.2 and defined in Table 22.2.

PHYSIOLOGICAL FUNCTIONS OF THE EXTERNAL EAR

The external ear directs sound waves toward the eardrum. The structure of the external ear also alters the acoustic spectrum of sound reaching the eardrum. You can demonstrate this to yourself by cupping your hand over your ear and concentrating on the nature of the sound that you hear. With your hand cupped behind your ear (as you might do if you were having trouble hearing) you will notice that you hear high frequencies somewhat better. This is in fact why people with hearing loss use this strategy (the author speaks from experience; more on this later). A considerable amount of word identification in spoken language is provided by the high-frequency components ("s," "t," and "st").

The modification of spectral frequency produced by the external ear depends on the frequency of the sound and its relative location. Thus, the intensities of different frequencies are different in the two ears, depending on the spatial location of the sound. This information helps in sound localization.

PHYSIOLOGICAL FUNCTIONS OF THE MIDDLE EAR

The tympanic membrane together with the ossicles in the middle ear transform the large-amplitude, low-force, airborne waves into small-amplitude, high-force vibrations of the membrane at the oval window, which then generate waves in the fluid-filled cochlea. The middle ear *force transformation*

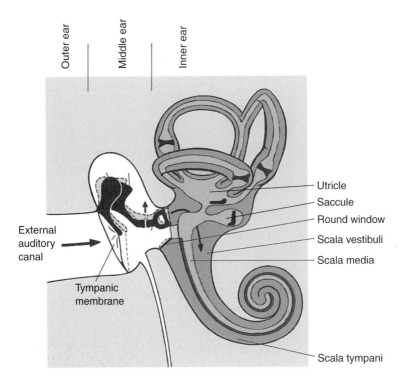

Figure 22.1. The anatomy of the ear. The external ear includes the pinna, the external auditory meatus, and the tympanic membrane. The middle ear contains the three ossicles (the malleus, incus, and stapes). The internal ear is composed of the cochlea and vestibular apparatus. The cochlea is divided into three, parallel, fluid-filled compartments: the scala vestibuli; scala tympani; and scala media, or cochlear duct, which contains the basilar membrane. The scala vestibuli and scala tympani are connected via a small opening at the apex of the cochlea termed the helicotrema. The fluid within the scala vestibuli and scala tympani is termed the perilymphatic fluid, or simply perilymph. The fluid within the scala media is termed the endolymphatic fluid, or simply endolymph. The ionic composition of the endolymph is maintained by the ionic pumping activity of cells of the stria vascularis. From Heimer L. *The Human Brain and Spinal Cord.* New York, NY: Springer-Verlag; 1995. Modified from Klinke R. Physiology of the sense of equilibrium, hearing, and speech. In: Schmidt RF, Thews G, eds. Biederman-Thorson MA, trans. *Human Physiology.* Berlin, Germany: Springer-Verlag; 1983:277–305.

system is necessary because of the *impedence mismatch* between air and the aqueous environment of the inner ear.

The force transformation by the middle ear results from two factors: (1) the ratio between the areas of the tympanic membrane and the footplate of the stapes (about 20:1), and (2) the lever ratio of the ossicular chain. The large-amplitude, low-force movements of the tympanic membrane are concentrated, leading to smaller, more forceful movements of the stapes. The mechanical transformer action of the middle ear produces a 70- to 100-fold gain in force.

Some physiological processing also occurs in the middle ear as a result of the operation of two muscles that attach to the ossicles (the *tensor tympani* and *stapedius muscles*). The tensor tympani attaches to the malleus, whereas the stapedius attaches to the stapes. The tensor tympani is innervated by the trigeminal nerve, and the stapedius is innervated by the facial nerve. Contraction of these muscles damps vibration of the ossicles, attenuating sound transmission. These muscles are responsible for the *acoustic middle-ear reflex,* during which sound transmission is reduced in response to loud noises. This reflex is impor-

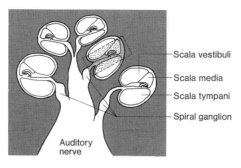

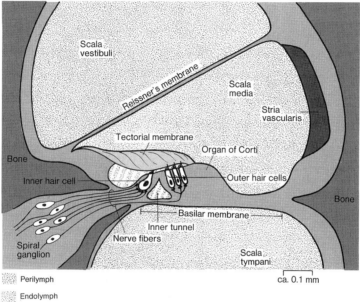

Figure 22.2. Anatomy of the cochlea. Cross sections through the coiled cochlea. The upper drawing illustrates several turns of the cochlear coil and the relative position of the spiral ganglion and auditory nerve. The lower drawing is a higher-magnification view of one turn of the coil. From Klinke R. Physiology of the sense of equilibrium, hearing, and speech. In: Schmidt RF, Thews G, eds. Biederman-Thorson MA, trans. *Human Physiology*. Berlin, Germany: Springer-Verlag; 1983:277–305.

tant for (1) preventing damage due to overstimulation, (2) increasing the dynamic range of hearing, and (3) reducing masking of high frequencies by low-frequency sounds. The latter two functions are important for improving speech discrimination, especially in noisy environments.

The "Place Code" for Frequency Discrimination

The stapes produces vibration of the membrane of the oval window, leading to vibra-tory waves in the fluid environment of the cochlea. The pressure transients are dissipated at the round window (Figure 22.3A). The vibratory waves in turn produce a vibration of the basilar membrane. Because of its physical properties, the maximal vibration of the basilar membrane occurs at specific locations, depending on the sound frequency (Figure 22.3B–D).

The physical property that determines basilar membrane function is the membrane's width. The membrane is narrowest at the end of the cochlea nearest the round and oval windows and becomes progres-

Table 22.2. Important structures of the ear

Structure	Description/function
External auditory meatus	Ear canal
Tympanic membrane	Ear drum
Ossicles	The small bones of the middle ear; malleus, incus, and stapes
Oval window	Contacted by stapes
Round window	Allows equalization of pressure transients
Organ of Corti	Neuroepithelium that contains auditory receptors (hair cells)
Scala vestibuli	Contains perilymph
Scala tympani	Contains perilymph
Helicotrema	Portal between the scala tympani and scala vestibuli
Scala media (cochlear duct)	Contains endolymph
Perilymph	Fluid within the scala tympani and scala vestibuli
Endolymph	Fluid within the scala media (high K^+)
Stria vascularis	Produces the endolymph
Basilar membrane	Divides scala tympani from scala vestibuli

sively wider toward the apex. The differences in width result in differences in the tautness of the membrane per unit area. The more taut, narrow portion of the basilar membrane resonates at high frequencies, whereas the less taut, wide portion in the apex resonates at low frequencies. In this sense, the basilar membrane is similar to the strings of a piano. When struck, the long, thick strings resonate at low frequencies (lower tones), and the short, thin strings resonate at high frequencies.

The actual spatial pattern of vibration is in the form of a *traveling wave*. This was discovered by George von Bekesy, who examined basilar membrane movements in cadavers. Using a stroboscope (which flashes a light at the same frequency as a tone that is played), von Bekesy found that the basilar membrane vibrated at the frequency of the tone. The deflections began at the oval window and traveled in a wavelike fashion down the cochlea toward the apex, hence the term *traveling wave*. The shape of the traveling wave, the distance that it travels, and its position of maximal deflection were all dependent on the sound frequency. High frequencies produced narrow waves that were maximal near the oval window and were quickly dampened in more apical regions (Figure 22.3). Low frequencies produced waves that were much wider, traveled further toward the apex, and produced maximal deflections apically. Thus, the basilar membrane operates as a mechanical acoustic spectrum analyzer, transforming the spectrum of mechanical displacement of the oval window into a spatial array of basilar membrane deflections. This place code, based on physical properties, is sharpened as a result of active dampening of basilar membrane movements in the areas bordering the area of maximal displacement (see later).

The original studies of von Bekesy were carried out using very intense stimuli (130 to 140 dB). Recently, techniques have been developed that make it possible to measure the tiny movements of the basilar membrane that occur in response to threshold stimuli in living animals. Using these techniques, one can "map" the position on the basilar membrane that is maximally responsive to particular frequencies. The way that this is done is to measure the movements produced at a particular location across a range of frequencies and define the minimal intensity that produces movement at that location (that is, the *threshold intensity*). One then plots the threshold intensities across a range of frequencies. The result is what is called a *tuning curve*, in which there is a V-shaped function with the base of the V at the optimal fre-

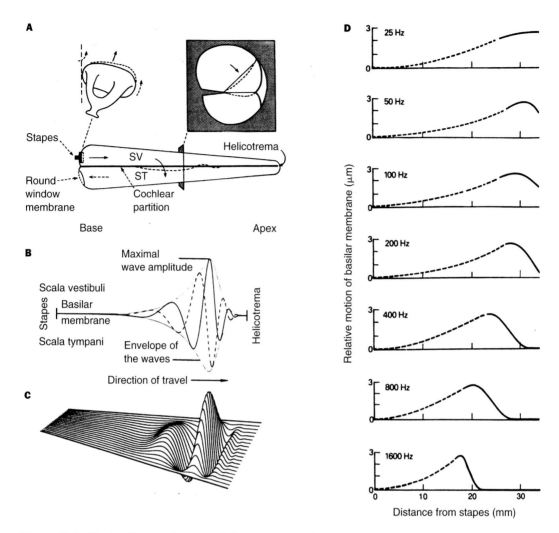

Figure 22.3. The basilar membrane and the physical representation of the "place code." A, The vibration of the stapes produces vibratory waves in the fluid environment of the cochlea, drawn here in an uncoiled form for ease of illustration. The pressure transients produced by inward movement of the stapes footplate are dissipated by compensatory outward movements of the round window membrane. The vibratory waves produce a vibration of the basilar membrane. The spatial pattern of vibration is in the form of a traveling wave. **B** is a diagram of the wave coutour at different times, generated by a fixed frequency. **C** is a three-dimensional representation of the wave. **D** illustrates plots of data from von Bekesy's experiments, which demonstrated that the position of maximal deflection is dependent on the sound frequency. High frequencies produce narrow waves that are maximal near the oval window. Low frequencies produce waves that travel further toward the apex and produce maximal deflections apically. **A** is from Dobie RA, Rubel EW. The auditory system: acoustics, psychoacoustics, and the periphery. In: Patton HD, Fuchs AF, Hille B, Scher AM, Steiner R, eds. *Textbook of Physiology, 1: Excitable Cells and Neurophysiology.* Philadelphia, Pa: WB Saunders; 1989. **B** and **C** are from Klinke R. Physiology of the sense of equilibrium, hearing, and speech. In: Schmidt RF, Thews G, eds. Biederman-Thorson MA, trans. *Human Physiology.* Berlin, Germany: Springer-Verlag; 1987:277–305. **D** is from Kandel ER, Schwartz JH, Jessell TM. *Principles of Neural Science.* 3rd ed. New York, NY: Elsevier; 1991. **A** and **D** are adapted from von Bekesy G. In: Wever EG, ed. and trans. *Experiments in Hearing.* New York, NY: McGraw-Hill; 1960.

quency (see Figure 22.3). Similar plots are used to map the frequencies to which neurons respond best. These types of studies have revealed a steeper gradient in response (that is, a narrower range of frequencies to which the membrane responds) than suggested by von Bekesy's experiments. This, and the fact that the sharp tuning curves are seen only in living tissue, indicates that there is an active mechanism that sharpens the place code. As we see later, this active process involves the outer hair cells.

CELLULAR ORGANIZATION OF THE AUDITORY NEUROEPITHELIUM

Transduction of vibrational energy into neural activity occurs in the organ of Corti, which is a highly specialized neuroepithelial sheet (Figures 22.2, 22.4, and 22.5). The sheet is made up of sensory receptor cells, termed *hair cells*, and various types of *supporting cells* (Figure 22.4). It is not necessary for students to memorize the names of the different types of supporting cells, but it is important to understand the cytoarchitectural organization of the organ of Corti, as illustrated in Figure 22.4 (that is, the arrangement of different types of hair cells into rows).

Hair cells are so called because of the prominent modified microvilli, called *stereocilia*, that extent from the apical surfaces of the hair cells to become embedded in the overlying *tectorial membrane*. The opposite pole of the hair cell functions as a presynaptic terminal that releases neurotransmitter onto the distal process of the primary afferent nerve (Figures 22.4 and 22.6).

There are two types of hair cells, inner and outer. *Inner* and *outer* refer to the relative proximity to the *modiolus*. Inner hair cells are found in a single row on one side of the row of inner pillar cells, and outer hair cells are found in three rows on the opposite side of the inner pillar cells (Figure 22.5). The two types have highly characteristic stereocilia; stereocilia on inner hair cells are arranged in a linear array, whereas the stereocilia on outer hair cells are arranged in a V-shaped configuration (Fig. 22.5). As we see later, the two types of hair cells also have different patterns of innervation and mediate different functions (Table 22.3).

The stereocilia of hair cells are anchored to the gelatinous material of the tectorial membrane. As a result, when the basilar membrane vibrates in response to sound, the hair cells are deflected.

The morphology of the stereocilia varies along the length of the organ of Corti. The stereocilia at the basal end of the cochlea (high frequency) are very short, whereas the stereocilia at the apex (low frequency) are long. These physical differences correspond to the frequencies of vibration that the stereocilia experience.

Innervation of Hair Cells

There are both afferent and efferent projections to hair cells, but the innervation patterns are different for the two types of hair cells. These differences in innervation pattern are related to the functions that the two types of hair cells subserve.

Afferent innervation of hair cells comes from auditory ganglion cells (also known as *spiral ganglion cells*) that are found within the *spiral canal* of the modiolus (also called *Rosenthal's canal*). The distal process of the ganglion cell innervates hair cells; their central process projects centrally in the eighth nerve.

Inner hair cells are considered to be the principal receptors because they receive most of the afferent innervation. The basal pole of the hair cell forms a synaptic contact with the afferent ending (the hair cell is presynaptic; the afferent ending is postsynaptic).

The afferents to inner hair cells arise from *type I ganglion cells;* these are typical bipolar neurons with a peripheral axon that innervates the hair cell, and a central axon that enters the eighth nerve. Both the peripheral and central processes are heavily myelinated. Each individual type I ganglion cell projects to only one inner hair cell, but inner hair cells receive projections from several ganglion cells. Inner hair cells do not receive direct efferent innervation, although efferent

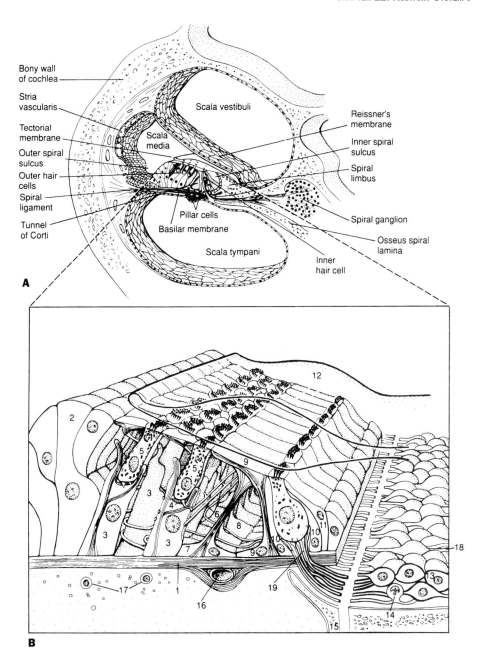

Bony wall
of cochlea

Stria
vascularis

Tectorial
membrane

Outer spiral
sulcus

Outer hair
cells

Spiral
ligament

Tunnel
of Corti

Scala vestibuli

Scala
media

Pillar cells

Basilar membrane

Scala tympani

Reissner's
membrane

Inner spiral
sulcus

Spiral
limbus

Spiral ganglion

Osseus spiral
lamina

Inner
hair cell

A

B

Figure 22.4. The organ of Corti. Illustration of the cellular architecture of the mammalian organ of Corti. Students should be familiar with the terms in italics: 1, *basilar membrane;* 2, Hensen's cells; 3, outer phalangeal cells (also called Deiter's cells); 4, *endings of spiral afferent fibers on outer hair cells;* 5, *outer hair cells;* 6, outer spiral fibers; 7, outer pillar cells; 8, tunnel of Corti; 9, inner pillar cells; 10, inner phalangeal cells; 11, border cells; 12, *tectorial membrane;* 13, *type I spiral ganglion cell* (innervates inner hair cells); 14, *type II spiral ganglion cells* (innervate outer hair cells); 15, bony spiral lamina; 16, blood vessel; 17, cells of the tympanic lamina; 18, *axons of spiral ganglion cells* (auditory nerve fibers); 19, radial fiber. From Kandel ER, Schwartz JH, Jessell TM. *Principles of Neural Science.* 3rd ed. New York, NY: Elsevier; 1991. **A** is adapted from Bloom W, Fawcett DW. *A Textbook of Histology.* 10th ed. Philadelphia, Pa: WB Saunders Co; 1975. **B** is adapted from Junqueira LC, Carneiro J, Contopoulos AN. *Basic Histology.* 2nd ed. Los Altos, Calif: Lange Medical Publications; 1977.

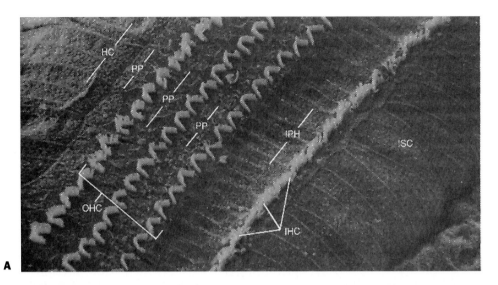

Figure 22.5. Cochlear hair cells. Scanning electron micrograph of cochlear hair cells after removal of the tectorial membrane. The stereocilia on outer hair cells (OHC) are arranged in a V-shaped array; those on inner hair cells (IHC) are in a linear array. At the apical pole of the hair cell is a specialized structure termed the *cuticular plate* (CP), from which the stereocilia emerge. **B** is a higher-magnification view of the stereocilia. From Kandel ER, Schwartz JH, Jessell TM. *Principles of Neural Science.* 3rd ed. New York, NY: Elsevier; 1991.

Table 22.3. Inner and outer hair cells

Type	No. of rows	Cilia	Afferents	Efferents	Myelin	Function
Inner	1	Linear array	Type I (one-to-one)	Few*	Yes	Receptor
Outer	3	V-shaped array	Type II (one-to-many)	Many	No	Effector

*The efferents synapse on the peripheral process of the type I ganglion cell, not directly on the hair cell.

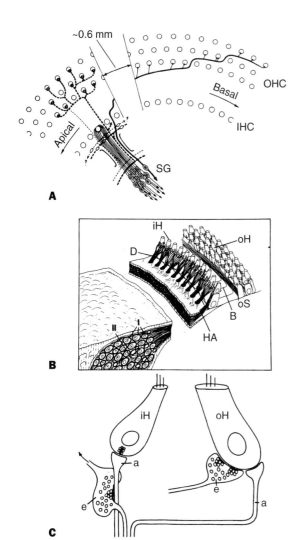

Figure 22.6. Afferent and efferent innervation of hair cells. A illustrates the innervation patterns for inner and outer hair cells. Thick lines represent type I afferents; thin lines represent type II. Dashed lines indicate efferent innervation. Inner hair cells receive most of the afferent innervation from type I ganglion cells. Each individual type I ganglion cell projects to only one inner hair cell, but inner hair cells receive projections from several ganglion cells. Inner hair cells do not receive direct efferent innervation, although efferent axons do innervate the terminal process of the ganglion cell (**C**). Outer hair cells receive the efferent innervation and also receive a projection from type II ganglion cells. **B** illustrates a three-dimensional view of the organization of the basilar membrane, the inner hair cells (iH) and outer hair cells (oH), and the type I and II ganglion cells in the spiral ganglion. HA indicates the habenula perforata, where the afferent axons collect to penetrate the bone of the modiolus. **C** illustrates the synaptic organization of afferent and efferent projections to inner and outer hair cells. Note that inner hair cells are not contacted directly by efferents; instead, the efferents terminate on the peripheral process of the afferent. From Kingsley RE. *Concise Text of Neuroscience.* Baltimore, Md: Williams & Wilkins; 1996. **A** is adapted from Spoendlin H. The afferent innervation of the cochlea. In: Naunton RF, Fernandez C. *Evoked Electrical Activity in the Auditory Nervous System.* New York, NY: Academic Press; 1978:21–41.

axons do innervate the terminal process of the ganglion cell (see Figure 22.6).

The afferent projections that exist to inner hair cells arise from type II ganglion cells in the spiral ganglion. These cells apparently do not give rise to a centrally projecting axon. The peripheral axon is unmyelinated and branches extensively so that an individual ganglion cell innervates many outer hair cells (see Figure 22.6).

Role of Outer Hair Cells in Sharpening the Place Code

As already noted, experiments using living tissue have revealed that the wave of maxi- mal displacement is actually much sharper than can be accounted for by the physical properties of the basilar membrane. Subsequent experiments demonstrated that the resiliency of the basilar membrane is directly modulated as a result of active changes in the shape of outer hair cells. When isolated outer hair cells are stimulated with a sound of a particular frequency, the hair cells contract or lengthen depending on where along the cochlea the hair cells originate (Figure 22.7).

By contracting hair cells in areas of the cochlea on each side of an area of maximal displacement, the hair cells increase the stiffness of the basilar membrane. In this way, basilar membrane oscillations on either side

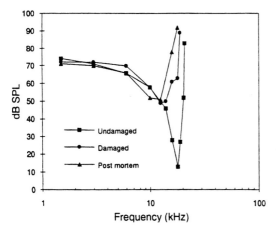

Figure 22.7. The response of the basilar membrane is tuned through an active process that involves outer hair cells. The graphs illustrate basilar membrane frequency-threshold tuning curves in experimental animals. Normal animals exhibit sharp tuning curves. Damage to the outer hair cells disrupts the tuning. When basilar membrane responses are measured postmortem, only crude tuning curves are present, reflecting the mechanical properties of the basilar membrane. Reproduced with permission from Yates GK, Johnstone BM, Patuzzi RB, Robertson D. Mechanical preprocessing in the mammalian cochlea. *Trends Neurosci.* 1992;15:57–61.

of a zone of maximal displacement are dampened, sharpening the focus of displacement.

Interestingly, the mechanical response of outer hair cells is highly selective to the frequency of the stimulation. In particular, individual isolated hair cells have the ability to distinguish different frequencies of sound so that the cell's mechanical response is "tuned" to particular frequencies. It is thought that this tuning is somehow determined by the length of the stereocilia. In particular, it is thought that long stereocilia tune the cell to respond selectively to low frequencies; short stereocilia tune the cell to respond to high frequencies. Exactly how this is accomplished is not known.

The important role of the outer hair cells, and the fact that tuning is an active process, has been documented experimentally. For example, the sharp tuning curves seen in normal animals are disrupted when the outer hair cells are damaged (Figure 22.7).

Most types of sensorineural hearing loss are due to the loss of outer hair cells. Outer hair cells are especially sensitive to noise-induced damage and the damage that occurs as a result of certain toxins. Indeed, the loss of outer hair cells is the most common cause of hearing impairment (more on this later). This fact explains two of the key features of hearing loss: (1) difficulties in auditory discrimination (for example, speech perception) are greatest in noisy environments, and (2) simple amplification (for example, through hearing aids) cannot completely restore function. The reason, in both cases, is that the hearing loss is a result of the loss of basilar membrane tuning, which in turn disrupts the precise tuning of ganglion cell responses to sound. Hence, it becomes difficult to screen out one sound in order to detect another. It is a bit like what happens when there is a loss of lens elasticity in the eye; the visual image is blurred because of an inability to detect the distinct boundaries that define an object in the visual world. In the same way, a loss of outer hair cells leads to a "blurring" of sound.

The mechanical movements of the outer hair cells actually produce minute sounds (termed *otoacoustic emissions*) that can be measured in patients by sensitive equipment. Assessments of otoacoustic emissions provide a direct objective measure of cochlear function, which can be used even in infants. This technique has provided a valuable means to detect hearing impairment early in life so that appropriate measures can be taken to prevent the disruption of normal language acquisition during the critical period for language development.

Efferent Feedback Regulates Cochlear Output

An important feature of sensory processing in the auditory system is feedback regulation. At the level of the middle ear, feedback is reflected in the operation of the middle-ear muscles (stapedius and tensor tympani).

There is also efferent control of receptor output, which is mediated by the efferent synapses onto outer hair cells and the peripheral processes (dendrites) of spiral ganglion cells that are contacted by inner hair cells. These synapses arise from axons that project from the brainstem to the cochlea via the *olivocochlear bundle*. The precise role that these synapses play in auditory processing is not yet known. Gross stimulation of the pathway during noise presentation dampens out the evoked activity of ganglion cell axons, but exactly how the system works during normal auditory processing is unclear. It is thought that the pathway functions to enhance signal to noise. As we see later, feedback also plays an important role in the central pathways of the auditory system.

23

Auditory System II
Central Auditory Pathways

The CNS circuitry that mediates auditory perception is very complex and is not completely understood. As in other sensory systems, there are *parallel circuits* that convey information about different features of the auditory environment (the axons that represent particular frequencies and the circuitry that is involved in sound location, for example), and a *hierarchical organization* so that information is conveyed from one synaptic station to the next in a sequence. Also, there are a large number of feedback pathways, so information flow along the central auditory circuits is gated by descending inputs. For the most part, these feedback pathways are not well understood and are mentioned only in passing.

It is important for physicians to understand the basics of this circuitry. There are, however, only a few situations in which damage to the central circuitry of the auditory system has diagnostic significance. There are two reasons. First, the auditory pathways of the brainstem lie near other important structures. In most cases, damage to nonauditory structures provides better localizing signs. Second, there is convergence of information from the two ears at the level of third-order relay neurons in the brainstem, and extensive crossing connections at each level of the auditory system. For this reason, except for lesions affecting the eighth nerve or cochlear nuclei, there are no CNS lesions that completely disrupt frequency discrimination. Uni-

lateral lesions involving certain structures (the inferior colliculus, for example) can disrupt the ability to localize sound on the side toward the lesion, but these types of highly specific lesions are not common.

DIFFERENT CENTRAL PATHWAYS ARE SPECIALIZED FOR FREQUENCY DISCRIMINATION VERSUS SOUND LOCALIZATION

As we have seen, *frequency discrimination* begins in the cochlea and depends primarily on *labeled lines.* That is, different hair cells are activated by different frequencies, and the hair cells "representing" different frequencies innervate different populations of ganglion cells. The axons from the different populations of ganglion cells then convey this information to the CNS via the eighth nerve and terminate in a topographically specific fashion on the neurons in the cochlear nuclei of the brainstem. This topographically specific organization on the basis of tones is termed a *tonotopic organization.*

All of the information required for frequency discrimination is available from a single ear. Hence, frequency discrimination is a *monaural* function. In contrast, *sound localization* can be accomplished only because the sounds coming to the two ears differ slightly as a function of the location of the sound source in the environment. Hence, sound localization is a *binaural function*. As we see later, different components of the central circuitry are specialized to mediate one or the other of these two key functions.

To facilitate the presentation, it is useful to consider the *brainstem circuits* first and then consider how information is relayed from the inferior colliculus to the thalamus and on to the auditory cortex (what we call the *forebrain circuits*).

Auditory Circuits of the Brainstem

The general distribution of the auditory pathways in the brainstem is illustrated in Figure 23.1. Students should be familiar with

Figure 23.1. Overview of central auditory pathways. This is a dorsal view of the brainstem, on which the principal auditory pathways are schematically illustrated. Students should become familiar with the numbered structures: 1, planum temporale—the dorsal surface of the temporal lobe, which forms the floor of the lateral ventricle; 2, temporal gyrus of Heschel; 3, acoustic radiations; 4, medial geniculate nucleus 5, brachium of the inferior colliculus; 6, commissure of the inferior colliculus; 7, inferior colliculus; 8, lateral lemniscus; 9, nucleus of the lateral lemniscus; 10, lateral superior olive; 11, medial superior olive; 12, medial nucleus of the trapezoid body (MNTB); 13, trapezoid body; 14, dorsal acoustic stria; 15, ventral cochlear nucleus 16, dorsal cochlear nucleus 17, inferior cerebellar peduncle; 18, cranial nerve VIII. From Nieuwenhuys R, Voogd J, van Huijzen C. *The Human Central Nervous System*. Berlin, Germany: Springer-Verlag; 1988.

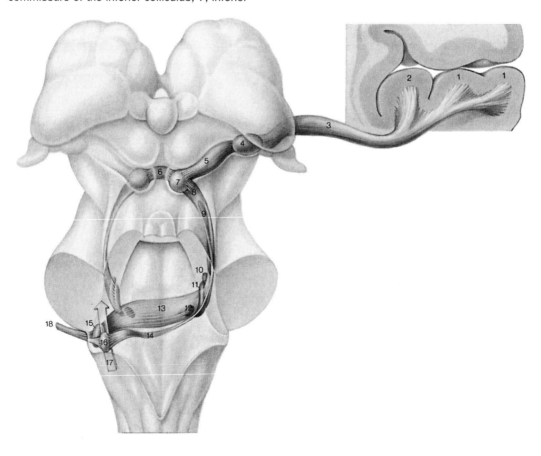

the general location of the cochlear nuclei and the nuclei containing third-order neurons and the principal ascending pathways (especially the lateral lemniscus) and should be able to recognize these in cross sections through the brain, such as those illustrated in Figure 23.2.

Figure 23.2. The pattern of termination of eighth nerve afferents in the cochlear nuclear complex. A, The axons of cochlear ganglion cells (the primary afferents, or first-order neurons of the auditory system) enter the brainstem via cranial nerve VIII. The fibers enter the cochlear nuclear complex and bifurcate, giving rise to anterior and posterior branches. The anterior branch terminates in the anteroventral cochlear nucleus (AVCN); the posterior branch bifurcates again and then terminates in the posteroventral cochlear nucleus (PVCN) and the dorsal cochlear nucleus (DCN). Together, these nuclei contain the neurons that are the second-order neurons of the auditory system. In each nucleus, the pattern of termination of eighth nerve axons is

Second-Order Nuclei of the Auditory System

The axons of cochlear ganglion cells (the *primary afferents,* or *first-order neurons* of the auditory system) enter the brainstem via the eighth nerve at the cerebellopontine recess. The fibers enter the *cochlear nuclear complex*

tonopically organized so that the fibers terminate in isofrequency bands. **B** schematically illustrates the pattern of termination in the AVCN. From Dobie, R.A., Rubel, E.W. "The auditory system: Central Auditory Pathways." In: Patton, H.D., Fuchs, A.F., Hillie, B., Scher, A.M., Steiner, R. eds. "Textbook of Physiology, 1: Excitable Cells and Neurophysiology" Philadelphia, PA: WB Saunders; 1989. **C** illustrates a transverse section through the medulla of a 1-month-old infant, which passes through the cochlear nuclei on the right-hand side. HF indicates high-frequency region; LF, low-frequency region; G, spiral ganglion. Reproduced with permission from Carpenter MB. *Human Neuroanatomy.* Baltimore, Md: Williams & Wilkins; 1976.

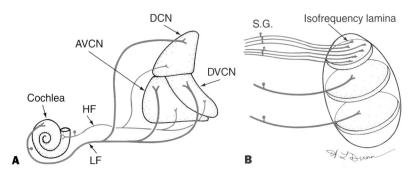

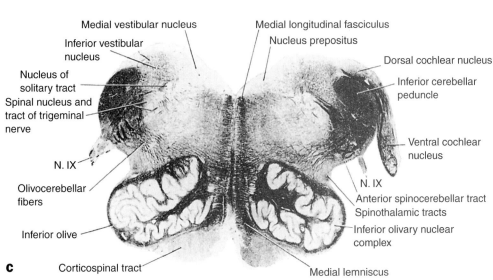

and immediately bifurcate, giving rise to anterior and posterior branches (Figure 23.2). The anterior branch terminates in the *anteroventral cochlear nucleus* (AVCN). The posterior branch bifurcates again and then terminates in the *posteroventral cochlear nucleus* (PVCN) and the *dorsal cochlear nucleus* (DCN). Together, these nuclei contain the neurons that are the *second-order neurons* of the auditory system. In each nucleus, the pattern of termination of eighth nerve axons is tonotopically organized (Figure 23.2), and a tonotopic organization is maintained at all of the subsequent stations along the central pathways (more on this later).

The eighth nerve afferents synapse on a number of different cell types in the cochlear nuclear complex, and the morphology of the synapses varies depending on the type of cell that is contacted. Some of the synapses are large calyxlike endings that almost completely surround the cell body of the second-order neuron (termed *calyces of Held* because of their shape and the fact that they were

first described by a neuroanatomist named Held; also called *endbulbs of Held*). Others are smaller boutons that are similar to the synapses in other locations. The different types of second-order neuron give rise to second-order projections to different locations (Figure 23.3) and these different pathways are involved in different aspects of central auditory processing (more on this later).

All of the different types of synapses are excitatory and very powerful (that is, they are *detonator synapses*). When the eighth nerve afferent fires, a large excitatory postsynaptic potential is produced in the postsynaptic cell, so there is a very high probability that an action potential will be generated. Hence, the second-order neurons function as *relay neurons* that convey information to higher levels with high fidelity.

In considering the projections from the different nuclei of the cochlear nuclear complex, it is useful to consider separately the circuitry that allows a comparison of inputs from the two ears (binaural circuits) versus

Figure 23.3. The pattern of innervation of second-order neurons in the cochlear nuclear complex. Schematic illustration of the types of contacts made by eighth nerve afferents on the different cell types in the cochlear nuclear complex. As illustrated, the different types of second-order neuron give rise to second-order projections to different locations that are involved in different aspects of central auditory processing. Reproduced with permission from Brodal A. *Neurological Anatomy.* New York, NY: Oxford University Press; 1981.

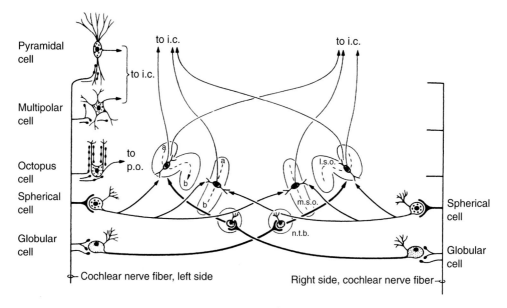

the circuitry that conveys information from only one ear (monaural circuits). As already noted, sound localization depends on comparing the sounds reaching the two ears. Hence, the division into monaural and binaural circuits corresponds to a relative specialization of the circuits for one or the other of the two key functions of sound localization and frequency discrimination. It is important to emphasize, however, that such divisions are an oversimplification. First, information pertaining to frequency is in fact conveyed by all of the pathways. Second, all of the brainstem pathways converge onto the inferior colliculus, which is a way station along the route of information flow from the ear to the auditory portions of the cerebral cortex. Hence, the breakdown into monaural and binaural circuits applies only to the first few synaptic stations in the brainstem.

Monaural Brainstem Circuits

The circuits that are thought to be especially important for conveying information regarding frequency to the thalamus and then the cortex begin in the DCN and PVCN (Figure 23.4).

Neurons in the DCN give rise to second-order axons that project across the midline in the *dorsal acoustic stria* (see Figure 23.4). The axons then collect in the lateral brainstem and ascend in the *lateral lemniscus.* These fibers terminate in a tonotopically organized fashion in the inferior colliculus on the contralateral side of the brain. A component of these fibers terminates in the *nuclei of the lateral lemniscus* (NLL), which in turn give rise to projections to the inferior colliculus and also to the medial geniculate nucleus (MGN) of the thalamus (the latter being part of the forebrain circuits considered later). In this way, there is both a direct projection from the DCN to the inferior colliculus and an indirect projection via the NLL.

As an aside, it should be noted that there are two divisions of the NLL: dorsal and ventral. Although the two subdivisions have somewhat different connections, it is not critical for physicians to know these details.

Hence, we lump the two subdivisions together as the NLL.

The circuit that begins in the PVCN is a bit more complex. The second-order neurons give rise to axons that collect in the *intermediate acoustic stria.* One component of these fibers crosses the midline, joins the lateral lemniscus on the contralateral side, and ascends to the NLL and the inferior colliculus. En route, the axons give rise to collaterals that terminate in the *superior olivary complex.* Another component of the fibers ascends ipsilaterally to terminate in the NLL. Again, the neurons in the NLL project onto the inferior colliculus and the MGN of the thalamus.

Binaural Brainstem Circuits

The binaural circuitry that is especially important for sound localization begins in the *anteroventral cochlear nucleus* (AVCN). Eighth nerve fibers form excitatory synapses on large spherical cells in the AVCN, and these neurons in turn give rise to the second-order projections to higher levels of the auditory system (Figure 23.5).

There are two components of the projection from the AVCN that project to different parts of the *superior olivary nuclear complex:* one component projects bilaterally to the *medial superior olive* (MSO); the other component projects to the *lateral superior olive* (LSO) (Figure 23.5). These two components can be differentiated according to the frequencies that they convey. The projections involving the MSO carry low-frequency information; the projections to the LSO carry high-frequency information.

The projections from the AVCN to the MSO leave the AVCN ventromedially and bifurcate. One branch courses ventromedially to terminate in the ipsilateral MSO. The other branch descends into the ventral part of the brainstem and crosses the midline in a large fiber bundle termed the *trapezoid body,* to terminate in the contralateral MSO. The projections to ipsilateral and contralateral sides are branches of individual axons, and the two branches terminate at frequency-specific locations in each MSO.

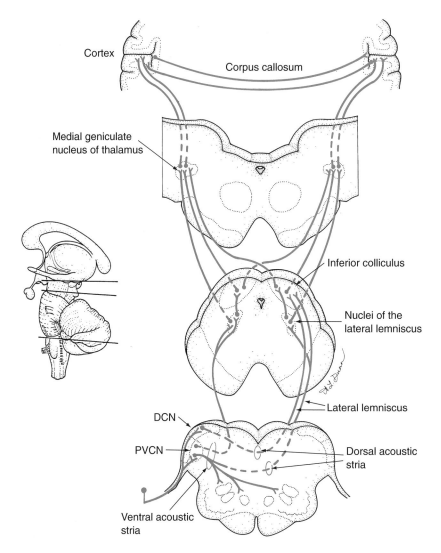

Figure 23.4. Monaural brainstem circuits. The circuits that are thought to be especially important for conveying information regarding frequency to the thalamus and then the cortex begin in the DCN and PVCN. Neurons in the DCN give rise to second-order axons that project across the midline in the dorsal acoustic stria and then ascend in the lateral lemniscus. These fibers terminate in a tonotopically organized fashion in the inferior colliculus on the contralateral side of the brain. A component of these fibers terminates in the nuclei of the lateral lemniscus (NLL), which in turn gives rise to projections to the inferior colliculus and also to the medial geniculate nucleus of the thalamus. Second-order neurons in the PVCN give rise to axons that collect in the intermediate acoustic stria. One component of these fibers crosses the midline, joins the lateral lemniscus on the contralateral side, and ascends to the NLL and the inferior colliculus. En route, the axons give rise to collaterals that terminate in the superior olivary complex. Another component of the fibers ascends ipsilaterally to terminate in the NLL, and the neurons in the NLL project onto the inferior colliculus and the medial geniculate nucleus of the thalamus. After Rubel EW, Dobie RA. The auditory system: central auditory pathways. In: Patton HD, Fuchs AF, Hille B, Scher AM, Steiner R. *Textbook of Physiology.* Philadelphia, Pa: WB Saunders; 1989.

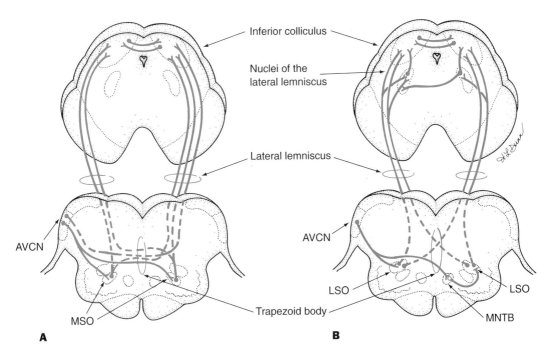

Figure 23.5. Binaural brainstem circuits. The binaural circuitry that is especially important for sound localization begins in the AVCN. Neurons in the AVCN then project to different parts of the superior olivary nuclear complex: one component projects bilaterally to the medial superior olive (MSO); the other component projects to the lateral superior olive. The projections to the MSO carry low-frequency information; the projections to the LSO carry high-frequency information. The MSO and LSO then project on to the inferior colliculus. After Rubel EW, Dobie RA. The auditory system: central auditory pathways. In: Patton HD, Fuchs AF, Hille B, Scher AM, Steiner R. *Textbook of Physiology*. Philadelphia, Pa: WB Saunders Co; 1989.

The projections from the AVCN to the LSO take a similar course, with one branch projecting ipsilaterally and another crossing the midline in the trapezoid body to terminate on the contralateral side. The ipsilateral component projects directly to the LSO, where it forms excitatory synaptic connections on the neurons in the LSO. The contralateral component terminates in the *medial nucleus of the trapezoid body* (MNTB); the MNTB in turn projects to the LSO. Hence, the projections from AVCN to the ipsilateral LSO are direct; the projections to the contralateral LSO are disynaptic and relayed through the MNTB.

As discussed further in Chapter 24, the projections from the AVCN are excitatory. However, the projections from the MNTB to the LSO are inhibitory. As a result of the interposition of the MNTB, the excitation produced by a single set of cochlear nucleus neurons is transformed into inhibition on the contralateral side. This circuitry forms the basis for detecting differences in sound intensity in the two ears.

The MSO and LSO components converge in the inferior colliculus. The neurons in both MSO and LSO that convey low- and high-frequency components, respectively, project to the inferior colliculus and the NLL via the lateral lemniscus. The two sets of projections terminate in a tonotopic fashion across the inferior colliculus.

The inferior colliculus gives rise to projections to (1) the MGN of the thalamus (the pathway conveying auditory information to the forebrain, which is discussed in greater detail later); (2) the superior colliculus and pretectal area; (3) descending projections

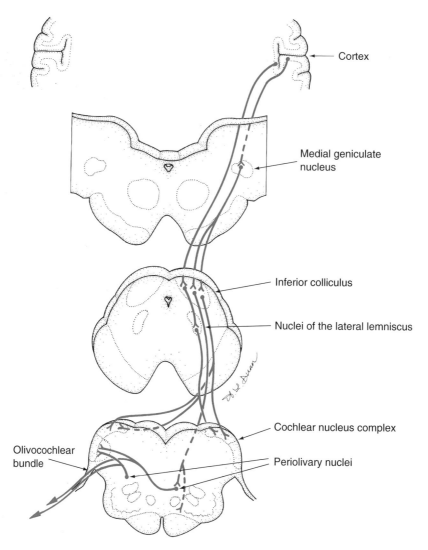

Cortex

Medial geniculate nucleus

Inferior colliculus

Nuclei of the lateral lemniscus

Cochlear nucleus complex

Olivocochlear bundle

Periolivary nuclei

Figure 23.6. Descending projections of the auditory system. Efferent projections of the olivocochlear bundle, which innervate the primary afferent endings, originate from the periolivary nuclei of the brainstem. The periolivary nuclei receive collateral projections from second-order neurons in the cochlear nuclear complex. These projections are primarily crossed, and the pathway is termed *the crossed olivocochlear tract*. There are also descending projections from the inferior colliculus to the NLL and to the cochlear nuclei, and from the NLL to the cochlear nuclei. These descending connections are thought to play a role in feedback modulation (gating) of transmission through the lower centers.

that play a role in gating information flow through lower relays (see later); and (4) interconnections between the inferior colliculi on the two sides of the brain that travel via the *commissure of the inferior colliculus.*

The projections to the superior colliculus terminate in the deep layers beneath the layers that receive input from the eye. Neurons in the deep layers of the superior colliculus also receive visual input and, hence, respond to both visual and auditory stimuli. Indeed, the neurons are arranged topographically such that the neurons that represent a particular location in visual space also respond preferentially to sounds originating from that location. In this way, the superior colliculus integrates auditory and visual information to provide a congruent map of the spatial environment. It is thought that this information plays an especially important role in coordinating the head and eye movements that are necessary to direct gaze to sites from which pertinent visual, auditory, or combined stimuli originate.

Descending Projections of the Auditory System

Each subdivision of the auditory system has important descending components in addition to the ascending ones (Figure 23.6). The efferent projections of the olivocochlear bundle, which innervate the primary afferent endings near their point of contact with cochlear hair cells, have already been mentioned. These projections are thought to originate from the *periolivary nuclei* of the brainstem, which in turn receive collateral projections from second-order neurons in the cochlear nuclear complex. These projections are primarily crossed, and the pathway is termed the *crossed olivocochlear tract.* This pathway is thought to be important for efferent regulation of cochlear output.

There are also descending projections from the inferior colliculus to the NLL and to the cochlear nuclei, and from the NLL to the cochlear nuclei. These descending connections are thought to play a role in feedback modulation (gating) of transmission through the lower centers.

Auditory Circuits of the Forebrain

Except for a very small component of fibers from the NLL, most of the auditory information that reaches the thalamus and then the cortex is relayed through the inferior colliculus. The projections from the inferior colliculus travel rostrally via the *brachium of the inferior colliculus* and terminate in a tonotopically organized fashion in the MGN of the thalamus—the primary auditory relay nucleus.

The inferior colliculus can thus be considered to be an obligatory link in the auditory pathways to the cortex.

The Auditory Cortex

The thalamocortical projections from the MGN extend beneath the pulvinar nucleus

Figure 23.7. Location of the primary auditory cortex in humans. A, Frontal section through the cerebral cortex. **B,** A view of the dorsal surface of the temporal lobe. Reproduced with permission from Kingsley RE. *Concise Text of Neuroscience.* Baltimore, Md: Williams & Wilkins; 1996.

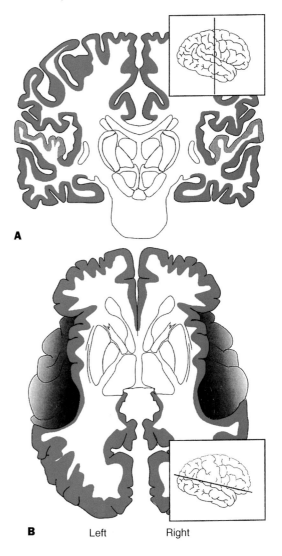

A

B   Left   Right

of the thalamus and then into the posterior limb of the internal capsule. From there, the axons extend under the putamen, into the cortical radiations, and then into the primary auditory cortex (Brodmann's area 41), which is located on the superior surface of the superior temporal gyrus *(temporal gyrus of Heschl)*. This is the portion of the temporal lobe that forms the floor of the lateral (Sylvian) fissure (Figure 23.7).

Surrounding the primary auditory cortex are a series of secondary auditory fields that may also receive direct projections from the MGN as well as projections from the primary auditory cortex. Especially important in humans are the appurtenant "association areas" such as *Broca's area* and *Wernicke's area,* which are important for the perception of speech. These receive projections from the primary and secondary auditory cortices. These areas are discussed in greater detail in Chapter 31.

There are transcallosal projections that interconnect the auditory regions of the two cerebral hemispheres, and interconnections between the primary auditory cortex and auditory association areas in the parietal cortex. The transcallosal projections terminate in patches, so some parts of the auditory cortex receive input from the contralateral hemisphere, whereas other parts do not. Also, there are descending projections to the MGN of the thalamus that are thought to gate information flow through the thalamic relay.

24

Auditory System III

Auditory Neurophysiology and Auditory Perception

In considering how the auditory system processes information, we again encounter the themes of parallel and hierarchical processing. Parallel processing is manifested by the frequency-specific pattern of afferent innervation of hair cells by spiral ganglion cells so that different frequencies are conveyed to the CNS along separate "labeled lines." The axons from the different populations of ganglion cells then terminate in a topographically specific fashion on the neurons in the cochlear nuclei of the brainstem. This topographically specific organization on the basis of tones is termed a *tonotopic organization*. Within the CNS, different components of the circuitry are specialized for frequency discrimination versus *sound localization*.

Hierarchical processing is reflected in the same way as in other sensory systems. At each synaptic station along the pathway, information is brought together and recombined; as a result, the neurons at each successive level exhibit more and more complex response characteristics. In this way, neurons at progressively higher levels of the neural processing hierarchy "represent" particular stimulus configurations. In what follows, we first consider how sound is transduced into neural activity and then consider the manifestations of parallel and hierarchical processing.

MECHANOELECTRICAL TRANSDUCTION BY COCHLEAR HAIR CELLS

Vibration of the basilar membrane produces a shearing force that deflects the stereocilia of the hair cells (Figure 24.1). The actual amount of the deflection is remarkably small. For example, if a stereocilium were scaled up to be the size of the Empire State building, the deflection produced by sound would be a few inches (about the amount that the Empire State building sways in a strong wind).

The deflection of the cilia leads to an opening of cation channels located on the tip of the stereocilium. The tips of the stereocilia are exposed to the endolymph in the scala media (Figure 24.2). Hence, the opening of the channels leads to a large influx of K^+, causing a depolarization of the hair cell. This depolarization is the *receptor potential*. When the hair cell becomes depolarized, the rate of neurotransmitter release increases (Figures 24.1 and 24.2).

The release of neurotransmitter by the hair cell is graded, rather than all or none. That is, the rate of release is determined on a moment-by-moment basis by the membrane

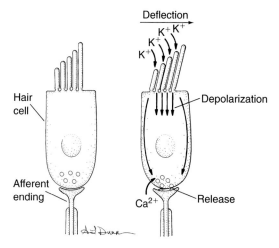

Figure 24.1. Transduction of vibration to neural activity. The displacement of the basilar membrane as a result of the traveling wave leads to deflection of the cilia of the hair cells; this in turn causes the cation channels located on the tips of the cilia to open, producing a large influx of K^+, which depolarizes the hair cells (*receptor potential* of the cochlea). Depolarization leads to an increased release of neurotransmitter from the synaptic specialization at the base of the hair cell. The neurotransmitter depolarizes the afferent ending (the generator potential) and, when the level of depolarization is sufficient, action potentials are generated.

potential of the hair cell. The neurotransmitter released by the hair cell then produces a depolarization of the afferent ending (the *generator potential).* Again, the generator potential is a graded response, the amplitude of which depends on the amount of neurotransmitter that is released by the hair cell. When the level of depolarization is sufficient, action potentials are produced in the myelinated portion of the axons of spiral ganglion cells (Figures 24.1 and 24.2).

NEURAL CODING

Neural Coding by Ganglion Cells

The activity of spiral ganglion cells is controlled directly by excitatory input from the hair cells. Physiological recordings from indi-

vidual axons of spiral ganglion cells reveal that most are spontaneously active, but this spontaneous activity is abolished when hair cells are damaged. The absence of spontaneous activity means that the activity of spiral ganglion cells is a direct function of input from the hair cells.

Two Aspects of Sound Are Encoded by Ganglion Cells

Frequency coding is based primarily on the *place code* of the basilar membrane. Because the deflection of the basilar membrane is maximal at a particular location, depending on sound frequency (see earlier), hair cell activation is also limited to a portion of the cochlea. As already noted, the connectivity between inner hair cells (the principal receptors) and type I ganglion cells is one to one. It is thus not surprising that ganglion cells respond best to a particular frequency (the frequency represented by the hair cells that provide their input). In this way, the place code of the cochlea is transformed into a *tonotopic pattern* of activation of eighth nerve axons. As we see later, this *tonotopic organization* is maintained throughout the auditory system.

Individual ganglion cells respond at lowest threshold to sounds at a particular frequency, termed the *best frequency,* or *characteristic frequency.* The response properties of ganglion cells are defined quantitatively by constructing *tuning curves,* as already described for the basilar membrane (Figure 24.3). Tuning curves are plots of response threshold across a range of frequencies. The resulting plot is V-shaped, with the base of the V pointing toward the characteristic frequency. For example, the ganglion cell that is illustrated in Figure 24.3 responds at lowest threshold (about 15 dB) to a sound at 2 Hz. The cell will respond to other frequencies, but at a much higher threshold.

Information regarding stimulus frequency is encoded by *phase-locking* of action potentials to the frequency of the sound. Each de-

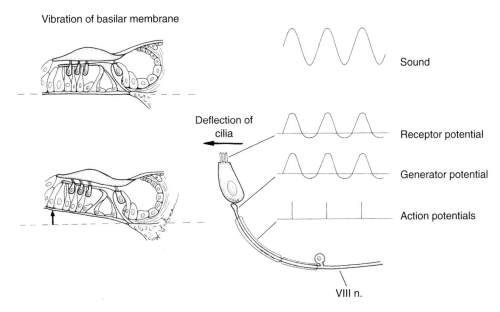

Vibration of basilar membrane

Deflection of cilia

Sound

Receptor potential

Generator potential

Action potentials

VIII n.

Figure 24.2. Sensory transduction in the cochlea. The basilar membrane vibrates in resonance with the sound, producing depolarizing generator potentials in the hair cell. These in turn cause an increased release of the excitatory neurotransmitter onto the afferent ending (the generator potential). When the level of depolarization is sufficient, action potentials are produced in the myelinated portion of the ganglion cell axon, and these are conveyed to the brain over cranial nerve VIII. The drawing on the left is modified from Kandel ER, Schwartz JH, Jessel TM. *Principles of Neural Science.* 3rd ed. New York, NY: Elsevier; 1991.

flection of the basilar membrane produced by a frequency cycle leads to a receptor potential, generator potential, and if the depolarization is sufficient, an action potential (Figure 24.2). The receptor and generator potentials are phase-locked to the sound; that is, each cycle (period) of the sound produces a depolarization–repolarization cycle. Individual ganglion cell axons do not fire with each cycle; but collectively, some axons representing a single frequency will fire for each cycle. In this way, the *population response* encodes the sound frequency. This is termed *volley coding* (see Figure 24.4).

Intensity coding depends on (1) modulation of the rate of discharge of individual fibers and (2) the number of fibers activated by a given stimulus.

Rate coding is based on the fact that the release of neurotransmitter by the hair cell occurs in a graded fashion. The magnitude of the receptor potential varies as a function of the intensity of the sound; the amount of neuro-transmitter released by the hair cell varies as a function of the receptor potential; the generator potential within the afferent ending varies as a function of the amount of neurotransmitter released; and the number of action potentials generated in the myelinated axon then vary as a function of the generator potential. In this way, cochlear nerve fibers display a monotonic increase in firing rate, up to a plateau, as the intensity of the sound is increased.

At the same time, higher-intensity sound also leads to activation of fibers with a broader range of characteristic frequencies, so that there is an increase in the number of active fibers as intensity increases. For example, the ganglion cell that is illustrated in Figure 24.3 responds at lowest threshold to a sound at 2 Hz but will also respond to a 55-dB sound at 1 Hz or 2.3 Hz. Of course, the sound at 55 dB and 2.3 Hz will also activate a set of ganglion cells whose characteristic frequency is 2.3 Hz. Thus, higher-intensity sounds acti-

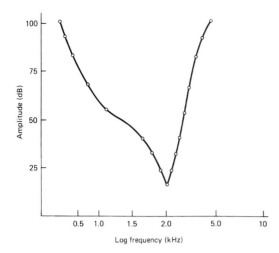

Figure 24.3. Tuning curves for auditory nerve fibers. The response properties of ganglion cells can be defined quantitatively by tuning curves. For this purpose, one plots the response threshold across a range of frequencies. The ganglion cell that is illustrated responds at lowest threshold (about 15 dB) to a sound at 2 Hz. The cell responds to other frequencies, but at much higher threshold. From Kandel ER, Schwartz JH, Jessell TM. *Principles of Neural Science.* 3rd ed. New York, NY: Elsevier; 1991.

vate ganglion cells across a broader range of frequencies and, hence, a greater total number of ganglion cells.

Neural Coding in the Cochlear Nuclei of the Brainstem

As noted in Chapter 23, eighth nerve afferents synapse on a number of different cell types in the cochlear nuclear complex, and the morphology of the synapses vary depending on the type of cell that is contacted. All of the different types of synapses formed by eighth nerve fibers are excitatory and very powerful (that is, they are *detonator synapses*). When the eighth nerve afferent fires, a large *excitatory postsynaptic potential* is produced in the postsynaptic cell, so that there is a very high probability that an action potential will be generated. Hence, the second-order neurons function as *relay neurons* that convey information to higher levels with high fidelity.

There are, however, important differences in the response characteristics of the different neuron types (Figure 24.5). These different types of neuronal responses thus "represent" different features of the sound. As noted

Figure 24.4. Phase-locking and volley coding. Each deflection of the basilar membrane produced by a frequency cycle produces a depolarization–repolarization cycle in the hair cell and then in the afferent ending. Hence, the action potentials generated in the afferent fibers are

phase-locked to the frequency of the sound. Individual ganglion cell axons do not always fire with each cycle; but some of the population representing a single frequency will fire at each cycle, producing a population response at the sound frequency. This is termed *volley coding.*

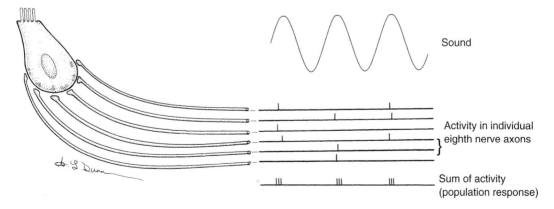

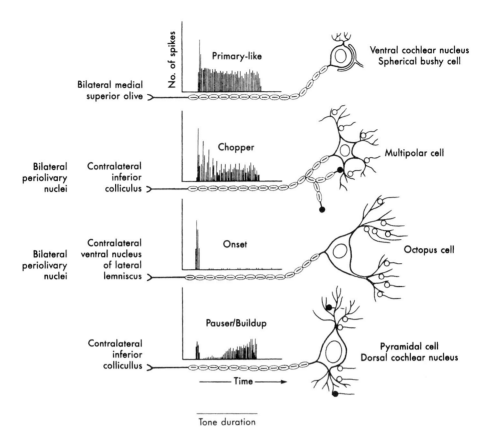

Figure 24.5. Differences in response characteristics of different neuron types in the cochlear nuclei. The response characteristics of different neuron types in the cochlear nuclear complex to presentation of a tone. The graphs are illustrations of poststimulus time histograms. These are constructed by repeatedly presenting a tone and then summing the number of action potentials (spikes) that occur during particular time bins. The higher the point, the greater the number of spikes in that time bin. These types of plots are used to determine the firing characteristics of neurons to a particular stimulus.

In some neurons, the pattern of action potentials that is generated is similar to the pattern of the primary afferent. Other neuron types exhibit different responses. For example, some respond to sound onset; others exhibit an increasing response over time. This information is then conveyed to different second-order structures. Reproduced with permission from Haines DE. *Fundamental Neuroscience.* Edinburgh, Scotland: Churchill Livingstone; 1997.

earlier, the different cell types give rise to different components of the ascending pathways. In this way, information regarding different aspects of sound are conveyed along separate CNS pathways.

In keeping with the fact that eighth nerve fibers terminate in a tonotopically organized fashion in each nucleus, individual neurons respond selectively to sounds at a particular frequency. The tuning curves for second-order neurons are similar to the tuning curves for spiral ganglion cells.

Neural Coding in Third-Order Auditory Relay Nuclei

Sound localization is mediated via the "binaural circuits," which originate from the *anteroventral cochlear nucleus* (AVCN) (see Chapter

23). These binaural circuits are further subdivided into two subdivisions that are specialized for the localization of low-frequency versus high-frequency sounds.

Localization of Low-Frequency Sounds: The Timing Pathway

The localization of low-frequency sounds is based on intra-aural phase differences that result from the fact that sounds originating from different spatial locations reach the two ears at different times. The slightly different timing means that the phase of the sound is slightly shifted in one ear versus the other (Figure 24.6A). The information pertaining to this phase shift is detected by the binaural pathways from the AVCN to the medial superior olive (MSO). It is for this reason that this circuit is called the *timing pathway.*

The AVCN gives rise to symmetrical projections to the MSO on the two sides of the brain. Each axon branches just after leaving the nucleus, and one branch projects to the ipsilateral MSO, the other branch projects to the contralateral MSO. Neurons in the MSO possess bipolar dendritic trees that radiate in each direction from the cell body (Figure 24.7). The lateral dendritic arbor is innervated by axons from the ipsilateral AVCN, whereas the medial arbor is innervated by the contralateral AVCN. In this way, individual neurons in the MSO receive complementary input from each ear. These neurons are thus ideally suited to detect differences in the phase relationships of sound from the two ears (termed *intra-aural phase differences*). In this way, neurons in the MSO respond selectively to sounds at a characteristic frequency that reaches the two ears with a particular phase delay.

Sounds higher than 5000 Hz cannot be localized on the basis of phase relationships because the period of the sound is too short. In particular, the differences in timing resulting from the longer conduction distance to the ear opposite the sound source can lead to a phase shift of more than one cycle. As a result, a sound originating from a position near 0° with respect to a listener would have the same phase shift as a sound originating from a lateral position. Hence, the timing differences necessary for localizating a high-frequency sound are confounded. For this reason, additional information is required to localize high-frequency sounds, and this is provided by the differences in the intensity of sound reaching the two ears.

Localization of High-Frequency Sounds: The Intensity Pathway

The localization of high-frequency sounds is based on the fact that there are differences in sound intensity in the two ears when the sound originates from the side. When sounds originate from positions lateral to the midline, the ear on the side opposite to the sound source is in the *acoustic shadow* produced by the head (Figure 24.6B). This shadowing attenuates the sound reaching the ear on the side opposite to the sound source.

The differences in sound intensity reaching the two ears are detected and magnified by the projections from the AVCN to the lateral superior olive (LSO)—hence the name, the *intensity pathway*. A sound originating from a lateral position produces asymmetrical activity in the cochlear nucleus. Then, the more active cochlear nucleus neurons ipsilateral to the sound source produce greater activation of the ipsilateral LSO.

At the same time, the crossed projections from the more active cochlear nucleus produce a net inhibition of the contralateral LSO. This is mediated by the projection from the AVCN to the medial nucleus of the trapezoid body (MNTB) (which is excitatory) and from the MNTB to the LSO (which is inhibitory). Hence, activation of the crossed projection results in projected inhibition in the LSO. The net result of these interactions is that neurons in the LSO ipsilateral to the sound source are excited, whereas neurons contralateral to the sound source are inhibited (Figure 24.8). Stated another way, neurons in the LSO respond selectively to sounds at a characteristic frequency that are loudest in the ipsilateral ear.

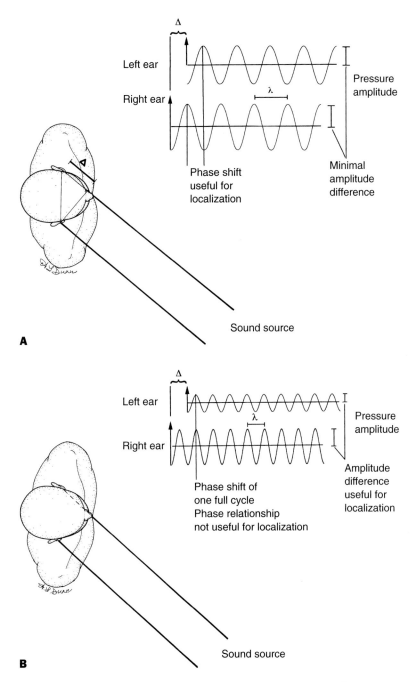

Figure 24.6. Localization of sound based on differences in the sounds reaching the two ears.
A, The localization of low-frequency sounds is based on phase differences that result from the fact that sounds originating from different spatial locations reach the two ears at different times. The slightly different timing means that the phase of the sound is slightly shifted in one ear versus the other. **B,** The localization of high-frequency sounds is based on the fact that there are differences in sound intensity in the two ears when the sound originates from one side. When sounds originate from positions lateral to the midline, the ear on the side opposite the sound source is in the acoustic shadow produced by the head. This shadowing attenuates the sound reaching the ear on the side opposite the sound source.

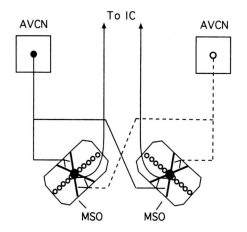

Figure 24.7. The timing pathway. The AVCN gives rise to symmetrical projections to the MSO on the two sides of the brain. Each axon branches just after leaving the nucleus, and one branch projects to the ipsilateral MSO, the other branch projects to the contralateral MSO. Neurons in the MSO possess bipolar dendritic trees that radiate in each direction from the cell body. The lateral dendritic arbor is innervated by axons from the ipsilateral AVCN, whereas the medial arbor is innervated by the contralateral AVCN. Neurons in the MSO then project to the inferior colliculus (see Figure 23.5).

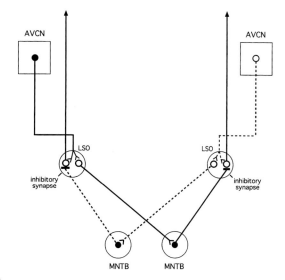

Figure 24.8. The intensity pathway. The differences in sound intensity reaching the two ears are detected and magnified by the intensity pathway. A sound originating from a lateral position produces asymmetrical activity in the cochlear nucleus. Then, the more active cochlear nucleus neurons ipsilateral to the sound source produce greater activation of the ipsilateral LSO. The LSO in turn projects to the MNTB. The neurons in the MNTB are inhibitory and project to the LSO. In this way, the crossed projections from the more active cochlear nucleus produce a net inhibition of the contralateral LSO. The net result is that neurons in the LSO ipsilateral to the sound source are excited, whereas neurons contralateral to the sound source are inhibited. Neurons in the LSO then project to the inferior colliculus (see Figure 23.5).

Intensity differences between the two ears can be used to localize high-frequency sounds but are less useful for low-frequency sounds because low-frequency sounds are not attenuated to the same extent by the head. For this reason, the AVCN-MTB-LSO–inferior colliculus pathway is particularly important for high-frequency sound localization.

Neural Coding in the Inferior Colliculus

As already noted, the various auditory brainstem nuclei project to the inferior colliculus, where they terminate in a tonotopically organized fashion. Hence, it is not surprising that electrophysiological experiments reveal a frequency selectivity in the responses of neurons in the inferior colliculus. In the monaural circuit, second-order neurons in the cochlear nuclei project directly to the inferior colliculus. In the binaural circuit, the second-order neurons project to third-order

neurons in the MSO and LSO, and the superior olivary neurons then project to the inferior colliculus. Thus, the responses of neurons in the inferior colliculus are a reflection of an integration of these different inputs.

Although collicular neurons do respond selectively to sounds at a particular frequency, the optimal stimulus for neurons is more complicated than in the lower levels of the auditory pathway. In particular, the responses of neurons in the inferior colliculus vary as a function of differences in the sounds presented to the two ears (that is, differences in frequency phase and the intensity of the sounds reaching the two ears).

Information pertaining to differences in the sounds reaching the two ears is integrated to allow the animal to localize sound in space. Exactly how this is mediated is not clear. It is interesting, however, that there is a topographic organization of spatial representation, at least in experimental animals. This has been documented through studies of animals that use sound to localize objects in the spatial environment (for example, owls, Figure 24.9). Thus, neurons in the inferior colliculus respond selectively to sounds at a characteristic frequency that originates from a particular point in the environment.

Although each side of the inferior colliculus clearly has to integrate information from each of the two ears in order to localize sounds, one side of the inferior colliculus

seems to "represent" auditory space on the contralateral side. This is demonstrated in the physiological experiments schematized in Figure 24.7 and by the fact that unilateral lesions involving the inferior colliculus lead to an inability to localize sounds originating on the side contralateral to the lesion.

Sound localization in the vertical plane is accomplished in the same way as sound localization in the horizontal plane. To use the same mechanisms, animals can simply tilt their heads. Information on head position is then integrated with information on timing differences and intensity differences. Again, it is thought that this integration is accomplished in the inferior colliculus.

Figure 24.9. Topographic mapping of auditory space in the inferior colliculus of the barn owl. In this experiment, recordings were made of neuronal responses to sounds presented at different locations in the spatial environment. The coordinates of "auditory space" are depicted as a globe surrounding the owl. The drawings indicate the derived maps of auditory space in the inferior colliculus as seen in transverse, horizontal, and sagittal planes. Reproduced with permission from Knudsen EI, Konishi M. *Science*. 1978;200: 795–797.

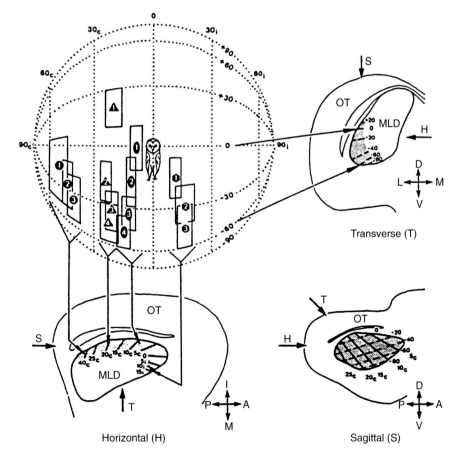

Neural Coding in the Medial Geniculate Nucleus

The medial geniculate nucleus is usually divided into three subdivisions (dorsal, ventral, and medial), each of which has a different cellular organization. The most thoroughly studied is the ventral subdivision, which contains neurons that respond to sound in ways that are generally similar to the responses exhibited by neurons in the brainstem nuclei. There is a precise tonotopic map, so the neurons exhibit excitatory responses to particular frequencies of sound. The rate of discharge increases monotonically as sound intensity increases. Hence, these neurons faithfully encode sound frequency and intensity.

Neurons in the dorsal and medial subdivisions have more complicated response characteristics. Although the neurons respond preferentially to certain frequencies, the tonotopic maps are complex; indeed, there appear to be multiple maps. Also, the rate/intensity functions (the relationship between sound intensity and neuronal discharge) are often nonmonotonic in that some neurons are maximally excited by sounds at a particular intensity and are suppressed by more intense sounds. What is important for physicians is that the nucleus serves as a typical thalamic sensory relay nucleus for the auditory information reaching the primary auditory cortex.

THE REPRESENTATION OF SOUND IN THE CEREBRAL CORTEX

Studies in experimental animals indicate that thalamocortical projections terminate in a tonotopically organized fashion across the primary auditory cortex. In primates, the frequency map is such that low frequencies are represented rostrally, and high frequencies represented caudally along the dorsal surface of the superior temporal gyrus. The tonotopic map is actually in the form of a series of stripes that contain neurons that respond to particular frequencies (termed *isofrequency bands*). Functional imaging studies suggest that in humans, high frequencies are represented deep within the lateral fissure, whereas lower frequencies are represented toward the top of the superior temporal gyrus.

Like other cortical areas, the auditory cortex is organized into columns. There are bands of neurons that respond to similar frequencies (the isofrequency bands) and orthogonal bands in which cells respond preferentially to one ear (monaural neurons) or both ears (binaural neurons). The bands of binaural neurons are further segregated into bands in which activation from each summates to produce an enhanced response (summation columns), and bands in which binaural activation is less effective than activation of one ear alone (suppression columns). Combined physiological and anatomical studies indicate that summation columns correspond to the zones that receive callosal (commissural) connections, whereas suppression columns correspond to the zones that receive inputs from other cortical areas on in the same hemisphere *(associational connections)*.

The orthogonal bands of cells with different response characteristics lead to columns of cells *(cortical modules)* that "represent" particular combinatorial stimulus configurations. For example, a particular column might contain neurons that respond best to a particular frequency and exhibit summation when both ears are activated, whereas an adjacent column would contain neurons that respond to the same frequency but exhibit suppression.

HIERARCHICAL PROCESSING: A RECAPITULATION

In the preceding descriptions of the response characteristics of neurons at different stations along the auditory pathway, the nature of hierarchical processing becomes evident. Specifically, hierarchical processing is reflected by an increasing degree of specialization in the types of stimuli to which a neuron responds best. In this way, neurons at progressively higher levels "represent" more and more complex stimulus configurations (Table 24.1).

CLINICAL ASSESSMENT

Assessing Hearing Loss

Damage to any part of the auditory system, from the eardrum to the auditory cortex, can produce hearing loss. Clinicians distinguish between *conductive hearing loss,* which results from a problem with the mechanical transmission of sound to the cochlea, and *sensorineural hearing loss,* which results from damage to any of the neural elements of the auditory system.

Conductive and sensorineural hearing loss can be distinguished by simple tests of audition that can be carried out in the physician's office. More-sophisticated follow-up testing can then be performed if called for.

Assessing Auditory Function

The differential diagnosis of conductive versus sensorineural hearing loss is made by comparing how the patient detects airborne sounds (termed *air conduction*) versus sounds that travel through bone (termed *bone conduction*). Airborne sounds can be transmitted to the cochlea only via the mechanical transduction system that includes the eardrum and ossicular chain. However, when tonal vibrations are induced in the skull, these activate the basilar membrane directly, bypassing the mechanical transduction system.

Simple Tests That Can be Done During a Routine Office Visit

The simplest test to distinguish air conduction from bone conduction is called *Rhinne's test.* A tuning fork is struck, and the stem is held against the patient's mastoid process. As the tuning fork's vibrations decrease, the patient is asked to indicate when he or she can no longer hear the tone. As soon as the tone can no longer be heard by bone conduction,

Table 24.1. Hierarchical processing

Neuron types	Response characteristics
Ganglion cells	Respond selectively to sounds at a characteristic frequency
Neurons in the cochlear nuclei	Respond selectively to sounds at a characteristic frequency and exhibit different response characteristics depending on the neuron type
Neurons in the MSO	Respond selectively to sounds at a characteristic frequency that reach the two ears with a particular phase delay
Neurons in the LSO	Respond selectively to sounds at a characteristic frequency that are loudest in the ipsilateral ear
Neurons in the inferior colliculus	Respond selectively to sounds at a characteristic frequency that originate from a particular point in space
Neurons in the medial geniculate nucleus	Respond selectively to sounds at a characteristic frequency and have complex rate/intensity response curves
Neurons in the auditory cortex	"Represent" particular combinatorial stimulus configurations

the tuning fork is held in the air near the ear. Normal patients will continue to hear the tone via air conduction for about 15 seconds because the threshold for air conduction is lower than the threshold for bone conduction. This is repeated for each ear. An inability to hear the tone via air conduction indicates conductive hearing loss.

Another simple test (termed *Weber's test*) is used to assess whether there is sensorineural hearing loss that is greater in one ear than in the other. In this case, the tuning fork is struck and the stem is held against the forehead at the midline. The patient is then asked to localize the sound (that is, whether the sound seems be centered or seems to come from one side or the other). If the sound seems to be centered, the two cochleas are equally sensitive. If there is sensorineural hearing loss that affects one ear more than the other, then the sound will seem to originate from the side opposite the more seriously affected ear (sound is said to localize to the side of the good ear).

If there is conductive hearing loss that affects one ear more than the other (which would have been detected by the Rhinne's test), the sound will seem to localize to the side of the bad ear. The reason is that in the ear with normal hearing, there is interference between air-conducted and bone-conducted sounds, and this does not occur in an ear in which sensitivity to air conduction is lost. Of course Weber's test is useful only for detecting hearing loss that affects one ear more than the other.

The Rhinne's and Weber's tests provide the physician with a rough indication of the nature of a patient's hearing loss. If hearing loss is documented, more-sophisticated tests are required, and these are usually carried out by audiologists. The two most common tests are *pure-tone audiometry* and the *auditory brainstem response*.

Pure-Tone Audiometry

For pure-tone audiometry, the patient is fitted with earphones. Pure tones are then presented across the frequency spectrum (from 250 Hz to 8 kHz), and the patient is asked to simply indicate when he or she hears a tone. Each ear is tested separately. By varying the intensity, the hearing threshold for each frequency is defined. The threshold is the minimum sound pressure level (in decibels) at which a patient can detect a tone. Then, one calculates the difference between the patient's thresholds at different frequencies, and the threshold of a normative group. For individuals with normal hearing, there is no difference between their threshold values and the values of the normative population (a difference of 0 to 20 dB is considered within the normal range). If there is hearing loss, the threshold for sound detection will be higher, and so the intensity of the sound will have to be greater to allow for detection. One then plots the difference (in decibels) between the patient's threshold and that of the normal population across a range of frequencies (Figure 24.10). Hearing loss is reflected in these plots (termed *audiograms*) by a downward deviation of the points. The thresholds for both bone conduction and air conduction are measured separately for each ear.

Conductive hearing loss is reflected by a difference in the threshold for air-conducted versus bone-conducted sounds. Sensorineural hearing loss is reflected by an increase in thresholds for both air-conducted and bone-conducted sounds. Pure-tone audiometry also allows an assessment of hearing function at different frequencies. In general, conductive hearing loss is reflected by increased thresholds across the frequency range. Sensorineural hearing loss resulting from damage to the cochlea is usually reflected by selective hearing loss at particular frequencies (most often in the high-frequency range). Sensorineural hearing loss resulting from damage to cranial nerve VIII or from damage to central pathways may be reflected by increased thresholds across the frequency range.

The patient illustrated in Figure 24.10A illustrates a hearing loss in the left ear in the 4- to 6-kHz frequency range. This is a highly characteristic pattern (termed a *4- to 6-kHz notch*) that is commonly seen in patients with *noise-induced hearing loss*. The selective hear-

Subject audiogram 1985

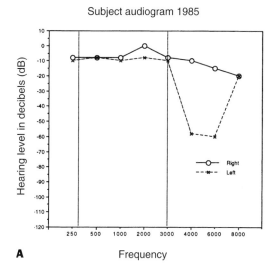

A   Frequency

Subject audiogram 1997

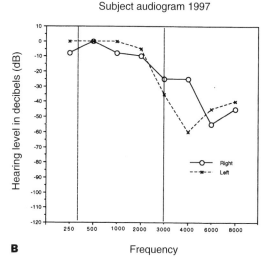

B   Frequency

Figure 24.10. Audiograms illustrating sensorineural hearing loss. The plots illustrate the results of pure-tone audiometry conducted on the author at age 35 (**A**) and 48 (**B**). The audiogram in **A** illustrates a selective hearing loss in the left ear in the 4- to 6-kHz frequency range, with normal hearing at 8 kHz. This is a highly characteristic pattern (termed a 4- to 6-kHz notch) that is commonly seen in patients with noise-induced hearing loss. The audiogram in **B** illustrates high-frequency hearing loss in both ears, with no sparing at 8 kHz. This is an example of the hearing loss that occurs with aging (termed *presbycusis*). Bummer!

ing loss in the left ear is commonly seen in right-handed individuals who have discharged high-powered firearms without adequate ear protection. The ear closest to the muzzle (which was exposed to the highest-intensity sound during the discharge) is the one that is most affected. The patient in this case is unfortunately the author of this text. The audiogram was taken when the author was 35 years old and is a "textbook" example of this form of noise-induced hearing loss.

The audiogram in Figure 24.10B is also from the author, but taken 13 years after the audiogram in Figure 24.10A. The difference is that, rather than exhibiting a selective hearing loss in the left ear, both ears are now affected. Moreover, whereas there was sparing at the high-frequency range (8 kHz) in the earlier audiogram, there is now a loss at all frequencies above 4 kHz. This is unfortunately an excellent example of another of the most common forms of hearing loss—the loss that occurs with aging (termed *presbycusis*).

Both noise-induced hearing loss and presbycusis have the same biological basis: outer hair cells are damaged. Intense noise damages the hair cells directly (Figure 24.11). The initial reflection of the injury is that the stereocilia are damaged. If the injury is severe enough, the hair cell itself dies. It may be that noise-induced injury makes hair cells more susceptible to the pathological processes that cause hair cell loss with aging.

Audiograms can reveal whether hearing loss is conductive or sensorineural and can also provide hints about whether the injury involves the cochlea (that is, whether there is a selective loss at particular frequencies). If there are no other symptoms, it is generally safe to conclude that the lesion involves the cochlea. However, if there are other symptoms, or if there is sensorineural hearing loss across the frequency range, a CNS lesion must be suspected.

The Auditory Brainstem Response

Pure-tone audiometry can only be carried out in compliant individuals who can indicate that they hear the tone. Audiometry

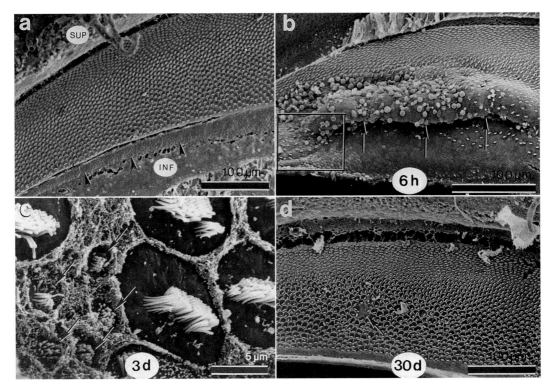

Figure 24.11. Noise-induced injury to outer hair cells. A is a scanning electron micrograph of normal hair cells in the chicken cochlea. **B** illustrates degenerating hair cells in the chicken cochlea 6 hours after exposure to an intense sound. **C** is a high-magnification view of the damaged region of the cochlea 3 days after noise-induced injury. **D** illustrates the repopulation of the cochlea by regenerated hair cells 30 days after injury. From Girod DA, Duckert LG, Rubel EW. Possible precursors of regenerated hair cells in the avian cochlea following acoustic trauma. *Hearing Res.* 1989;42:175–194.

cannot be carried out in infants or in any individual who, for whatever reason, cannot or will not cooperate with the tester. Auditory function can, however, be assessed by using electrophysiological techniques to measure the responses that are produced at each station along the auditory pathway. For this purpose, external recording electrodes are placed on the skull (in much the same way as for an electroencephalogram). Brief sounds are then presented (clicks), and the resulting responses are recorded. The responses are quite small, and computer averaging is required to detect the *auditory brainstem response*, which is also called the *brainstem auditory evoked response.*

The auditory brainstem response is a complex waveform in which peaks at different latencies represent the evoked activity at the different synaptic relays along the pathway (Figure 24.12). Peaks and troughs in the response during the first 10 ms reflect (1) the compound action potential in eighth nerve axons, (2) the synaptic responses in the cochlear nuclear complex, (3) the synaptic responses in the superior olivary nuclei, (4) the compound action potentials in the axons of the lateral lemniscus, and (5) the synaptic responses in the inferior colliculus. Abnormalities in the individual peaks indicate a failure of transmission at a particular location.

Tympanometry

One other important clinical test is *tympanometry,* which can detect conditions that in-

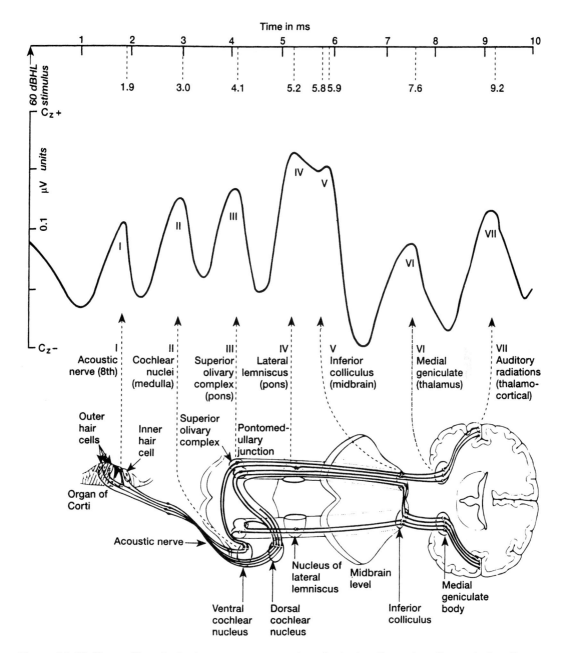

Figure 24.12. The auditory brainstem response. A typical waveform generated by clicks of 60 dB presented at a rate of 10 per second. Each component of the wave reflects neuronal activity at particular locations along the central auditory pathway. Reproduced with permission from Stockard JJ, Stockard JE, Sharbrough FW. *Mayo Clinic Proc.* 1977;52:761.

terfere with the vibration of the tympanic membrane. Abnormalities in the middle ear (for example, the buildup of fluid in the condition of otitus media) impair hearing by interfering with mechanical transduction. If fluid is present in the middle ear, the vibra- tion of the tympanic membrane is damp- ened. This can be detected by bouncing sound waves off the tympanic membrane and analyzing the echo. This is a useful rou- tine diagnostic tool to detect otitis media in children.

CAUSES OF HEARING DISORDERS

There are a large number of pathological processes that can produce hearing loss—far too many to consider here. In what follows, the most common causes of hearing loss are briefly considered.

Causes of Conductive Hearing Loss

Conductive hearing loss can result from occlusion of the auditory canal, damage to the eardrum, or disruption of the operation of the ossicular chain. The most common cause of occlusion is *cerumen impaction* (ear wax), which is usually easy to remove. The function of the eardrum can also become impaired as a result of perforation or infection. Small perforations usually heal spontaneously. More extensive damage can be treated surgically with tissue grafts.

The operation of the ossicular chain can be disrupted by an accumulation of fluid in the middle ear (a condition termed otitis media), disarticulation of the ossicular chain, or otosclerosis (abnormal bone formation, which usually occurs around the oval, resulting in impaired movement of the stapes).

Otitis media is most common in children and is usually due to blockage of the eustachian tube. Acute episodes due to infections (termed generally *middle-ear infections*) cause pain and are easily identified. The infections are treated with antibiotics and generally resolve quickly.

There are also cases in which there is chronic otitis media without infection. This condition is often not recognized because it does not cause pain. Instead the primary symptom is a loss of hearing, but this may not be reported by the child or the parents. Given the serious consequences of hearing loss on normal language development, early detection and treatment is a high priority. However the condition is easily diagnosed with tympanometry. Treatment involves the temporary placement of a drainage tube through the tympanic membrane.

Otosclerosis is a common cause of adult-onset hearing loss. The condition is determined by a single autosomal dominant gene with variable penetrance. Treatment involves removal of the stapes and insertion of a teflon prosthesis.

Causes of Sensorineural Hearing Loss

Sensorineural hearing loss is further subdivided into disorders affecting the cochlea (sensory), problems in the transmission of information along the auditory pathways (neural), and problems with the processing of information (cortical).

Disorders Affecting the Cochlea

The most common form of sensorineural hearing loss is *presbycusis*—the loss of hearing with advancing age. Presbycusis results from the loss of outer hair cells in the first turn of the cochlea (the high-frequency region). At this time, there is no treatment.

Intense sound is also a common cause of hearing loss and, again, results from the loss of outer hair cells. Exposure to loud sounds (for example, a rock concert) causes a temporary decrease in auditory acuity termed a *temporary threshold shift*, sometimes with *tinnitus* (ringing in the ears). These symptoms disappear within a few hours. Prolonged exposure causes permanent damage. Sudden-onset sounds are especially damaging (firing a high-powered rifle, a jackhammer), probably because the sounds cause damage before the protective middle ear reflex can be triggered. Fortunately, all forms of noise-induced hearing loss can be prevented by wearing ear protection.

One important and preventable cause of cochlear damage is *ototoxic agents*. Aminoglycoside antibiotics are quite toxic to hair cells. These drugs become concentrated in the endolymph if the drugs are given over a period of days. The problem is especially serious in patients with kidney disorders because the drugs are normally cleared through the kidney. The drugs can be used, however, if given over brief periods. Other agents that can be ototoxic in high doses are quinine and aspirin.

Certain diseases also disrupt cochlear function. For example, in *Meniere's disease* there is an increase in endolymphatic fluid, which leads to cochlear and vestibular damage. The auditory symptoms are tinnitus and hearing loss, especially at low frequencies. However the vestibular symptoms are far more debilitating and are the focus of treatment. Infection with rubella during pregnancy can disrupt cochlear development in the fetus, causing deafness.

Genetic Disorders That Cause Sensorineural Hearing Loss

There are a number of genetic disorders that cause sensorineural hearing loss. Indeed, it is estimated that up to 60% of the cases of hearing loss have a hereditary component. Some of these disorders cause congenital deafness. Other disorders lead to progressive loss of hearing over time.

Some of these genetic disorders are *syndromic,* in that the auditory symptoms are part of a symptom complex. Others are *nonsyndromic* and affect the auditory system selectively. Studies of different human pedigrees have recently led to the identification of the genes involved in these different forms of deafness, and parallel studies in experimental animals have revealed the mechanisms that underlie the deafness.

An example of a syndromic disorder for which the gene has been identified is Usher's syndrome, which causes deafness and blindness. Usher's syndrome is caused by a mutation in a gene encoding one form of myosin termed *myosin VIIA*. Myosin, along with actin, is an important component of the cytoskeleton in motile cells. Hence, it is thought that the mutation interferes with the cellular machinery that is responsible for the sound-induced motility of cochlear hair cells.

Genes causing several types of nonsyndromic hearing loss have also been identified (Figure 24.13). There are two examples of mutations of the myosin VIIA gene that cause hearing loss but no other symptoms. These involve different types of mutations

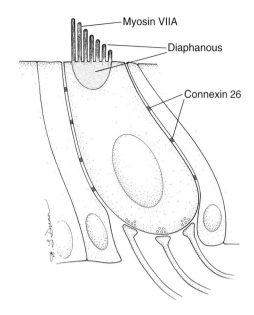

Figure 24.13. Mutations that cause the degeneration of cochlear hair cells and hence cause hearing loss. Three genes have been identified that, when mutated, cause deafness. These are myosin VIIA, connexin 26, and diaphanous. Myosin VIIA and diaphanous encode cytoskeletal proteins that are important for the formation and maintenance of stereocilia. The connexin 26 product is one of the components of gap junctions, which play a role in clearing excess K^+ from cochlear hair cells and thus preventing prolonged depolarization.

than the one that causes Usher's syndrome. At least some of the mutations of myosin VIIA appear to cause hearing loss as a result of disrupting the normal structure of the stereocilia. For example, in mice carrying a mutation in myosin VIIA, the characteristic organization of the stereocilia is severely disrupted. Rather than being configured in the characteristic V shape, the stereocilia are clumped.

Recent findings have revealed that a mutation in a gene encoding one form of *connexin* (connexin 26) also causes nonsyndromic deafness. Connexin is one of the structural components of the gap junctions that are normally present between hair cells and other cells. It is thought that the gap junctions serve as a channel through which the excess K^+ that floods the cell during activation is removed. Mutations of the channel protein interfere with K^+ clearance, leading to prolonged depolarization and, hence, cell death through an "excitotoxic" process.

Finally, mutations in a third gene (the human equivalent of a *Drosophila* gene called *diaphanous*) also are responsible for hair cell degeneration and hearing loss. *Diaphanous* encodes a protein that apparently serves as a temporary scaffold for actin as actin-containing structures are constructed during early development. The principal cytoskeletal constituents of the stereocilia of hair cells are actin filaments, and so mutations in diaphanous are thought to interfere with the construction of the stereocilia.

The chromosomal locations of about 30 genes that are involved in nonsyndromic deafness have now been identified. Identifying these genes is now a very high priority. In the first place, it may be possible to develop genetic screens that would allow physicians to identify newborns who will later lose their hearing. These individuals could be trained to use sign language in conjunction with spoken language during the period before hearing loss was complete. Moreover, knowing the molecular genetic basis of hearing loss could lead to the development of therapeutic interventions to prevent the hair cell degeneration that would otherwise occur.

Therapies for Sensorineural Hearing Loss

Some patients with profound hearing loss due to cochlear dysfunction can benefit from the implantation of computer-driven stimulating devices *(cochlear implants)* that directly stimulate the fibers of the eighth nerve. The restoration of hearing in patients with cochlear implants is sometimes remarkable, in that individuals who previously had profound hearing loss can even converse on the phone—a remarkable accomplishment considering that there are no attendant visual cues.

Considerable efforts are also being directed toward developing ways to induce hair cell regeneration to replace the hair cells lost to injury or disease. The stimulus for this work came from discoveries made independently by B. M. Ryals and E. W. Rubel and by J. T. Corwin and D. A. Cotanche that there

was remarkable regeneration of cochlear hair cells in birds after acoustic trauma or toxin-induced hair cell degeneration (Figure 24.11). Indeed, the regenerative process led to almost complete repopulation of the cochlea with hair cells. Unfortunately, however, it has not yet been possible to induce significant amounts of cochlear hair cell regeneration in mammals.

Disorders Affecting Transmission Along Auditory Pathways

Lesions central to the cochlea are termed *retrocochlear lesions*. Damage to cranial nerve VIII affects hearing and also vestibular function. One potential cause is a schwannoma of the eighth nerve, termed an *acoustic neuroma*. This term is something of a misnomer, however, because the tumor usually arises from the vestibular portion of the nerve and is a schwannoma, not a "neuroma." The tumors are also sometimes termed *cerebellopontine angle* tumors because they arise in that region.

Usually, the principal symptom of an acoustic neuroma is unilateral deafness across the range of frequencies, and loss of the *stapedius reflex* (because the tumor also usually affects the facial nerve). Loss of the stapedius reflex is important diagnostically, because it will not occur as a result of deafness due to damage of the cochlea. These tumors are benign and are removed surgically, although it is usually not possible to preserve hearing.

Lesions Involving Central Auditory Pathways

Because of the heavy callosal connections at most levels of the auditory pathway, central lesions usually do not cause hearing deficits. This is also true of unilateral lesions involving the auditory cortex. Nevertheless, although unilateral cortical lesions do not disrupt frequency discrimination or the ability to localize sound, they may lead to a disruption of language perception. The reason is that language function is represented unilaterally, usually on the left side. These functions will be considered in Chapter 31.

The Vestibular System

The vestibular system senses the static position of the head with respect to gravity and the movement of the head in space. This information is the key to adjusting posture (especially for controlling the antigravity muscles) to maintain balance, and for controlling eye movements to compensate for movement of the head. This control is to a large extent mediated unconsciously, and the most important components of the vestibular circuitry are located in the brainstem. An overview of the system is presented in Figure VII.1. The key components of the system are (1) the mechanoelectrical transduction system in the vestibular labyrinth, (2) the vestibular ganglion cells, (3) the vestibular component of cranial nerve VIII, (4) the vestibular nuclei in the brainstem, and (5) the projections from the vestibular nuclei to the motoneurons in the spinal cord and the cranial nerve nuclei that control eye movement.

There are several key concepts for physicians:

1. There are two different structures in the vestibular labyrinth that detect different types of positional information. Receptors in the macula of the utricle and saccule are responsible for detecting static head position and linear acceleration. Receptors in the semicircular canals are responsible for detecting angular (rotational) movements.

2. Axons in the vestibular division of the eighth nerve project to vestibular nuclei in the brainstem and also project directly to the "vestibulocerebellum" (the flocculonodular lobe).

3. Neurons in the different brainstem vestibular nuclei project directly to the cranial

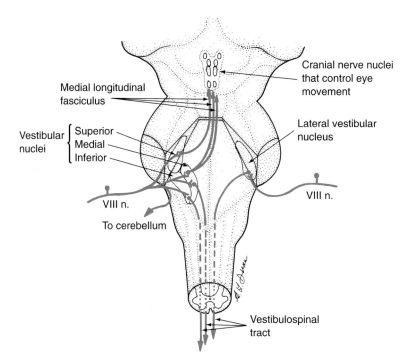

Figure VII.1. Overview of vestibular system organization. The vestibular receptors are found in the vestibular labyrinth in the inner ear. The receptors activate the distal endings (functionally, the dendrites) of vestibular ganglion cells in Scarpa's ganglion. The axons from the ganglion cells (first-order axons) travel to the brain via cranial nerve VIII and enter the brain at the cerebellopontine junction, where they terminate on second-order neurons in four vestibular nuclei in the dorsal medulla. One component of eighth nerve axons also projects directly to the flocculonodular lobe of the cerebellum. The two principal outputs of the vestibular nuclei are to the spinal cord via the vestibulospinal tracts and to the cranial nerve nuclei that contain the motoneurons that control eye movement. The superior, medial, and inferior vestibular nuclei are shown on the left. The lateral vestibular nucleus is shown on the right.

nerve nuclei that contain the motoneurons that control extraocular muscles, and also give rise to the descending vestibulospinal tract.

4. The central circuitry of the vestibular system controls eye and head movement and the antigravity muscles responsible for maintaining balance.

25

The Vestibular System

The vestibular system detects the static position of the head with respect to gravity, and two types of head movements: linear acceleration in a particular direction and angular (rotational) acceleration. This information is the key to maintaining balance and for controlling eye movements to compensate for movement of the head. The receptors that are responsible for detecting head position and the various types of movement are located in the vestibular labyrinth in the middle ear. Information from the receptors is conveyed to the brain via the vestibular component of the eighth nerve. In what follows, we first consider the mechanoelectrical transduction system in the vestibular labyrinth and then the central pathways that mediate vestibular function.

The Vestibular Labyrinth

The vestibular component of the *vestibulocochlear labyrinth* has the same basic cytological organization as the cochlear labyrinth described in Chapter 22. There is a series of bony compartments; each is lined with connective tissue and filled with *perilymph* (like the scalae vestibuli and tympani of the cochlea). Within each bony compartment is a membranous compartment that contains *endolymph* (like the scala media of the cochlea). Indeed, all of the compartments in the vestibulocochlear labyrinth that contain

perilymph interconnect with one another, and the same is true of the compartments that contain endolymph (Figure 25.1).

The endolymph-filled compartments of the vestibular labyrinth consist of a series of chambers with a complicated architecture; these chambers are termed the *saccule*, the *utricle*, and the *three semicircular canals* (Figure 25.1). The saccule is an ovoid compartment that is essentially an enlargement of the duct that interconnects the scala media of the cochlea with the utricle of the vestibular labyrinth. The saccule is located in the much larger, perilymph-filled chamber termed the *vestibule*. The utricle is a large chamber from which the three semicircular canals arise.

The structure of the different components of the vestibular receptor apparatus is remarkably specialized for the functions that the components mediate. Indeed it is most useful to explain the structure in terms of function.

The Utricle and Saccule Are Specialized to Detect Static Head Position and Linear Acceleration of the Head

The utricle and saccule contain a specialized sensory neuroepithelium termed the *macula*. Each macula is made up of a neuroepithelial sheet that contains *vestibular hair cells*, which

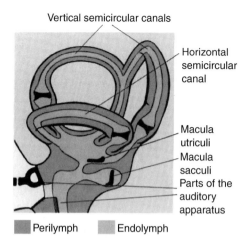

Vertical semicircular canals

Horizontal semicircular canal

Macula utriculi

Macula sacculi

Parts of the auditory apparatus

Perilymph Endolymph

Figure 25.1. The vestibular labyrinth. The vestibular component of the vestibulocochlear labyrinth is made up of a series of bony compartments (the bony labyrinth). Each is lined with connective tissue and filled with perilymph. Within each is a membranous labyrinth that contains endolymph. All of the endolymph-containing compartments interconnect with one another, and the same is true of the compartments that contain endolymph. From Klinke R. Physiology of the sense of equilibrium, hearing, and speech. In: Schmidt RF, Thews G, eds. Biederman-Thorson MA, trans. *Human Physiology.* Berlin, Germany: Springer-Verlag; 1983:277–305.

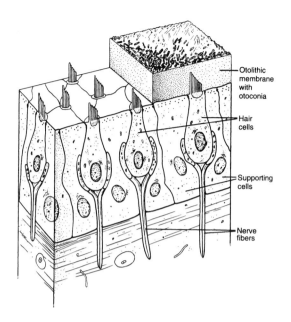

Otolithic membrane with otoconia

Hair cells

Supporting cells

Nerve fibers

Figure 25.2. The stereocilia of vestibular hair cells define a morphological axis of polarity. Vestibular hair cells have a collection of stereocilia extending from their surface into the endolymph and also have a single kinocilium. The stereocilia vary systematically in length, with the largest being adjacent to the kinocilium and the smallest at the opposite side of the cell. The alignment of the row of stereocilia defines the hair cell's morphological axis of polarity. The cilia of the macular hair cells extend into the endolymph and embed themselves in a gelatinous matrix termed the *otolithic membrane.* From Kandel ER, Schwartz JH, Jessell TM. *Principles of Neural Science.* 3rd ed. New York, NY: Elsevier; 1991; adapted from Lurato S. *Submicroscopic Structure of the Inner Ear.* Oxford, Mass: Pergamon Press; 1967.

are for the most part similar to cochlear hair cells, along with various supporting cells.

Like cochlear hair cells, the vestibular hair cells have a collection of *stereocilia* extending from their surface into the endolymph. Vestibular hair cells differ from cochlear hair cells in that the vestibular hair cells also have a single *kinocilium.* The stereocilia vary systematically in length, with the largest being adjacent to the kinocilium and the smallest at the opposite side of the cell (Figure 25.2). The row of cilia give the hair cell a morphological *axis of polarity.* The tips of the cilia contain mechanosensitive K^+ channels that are modulated by deflecting the stereocilia along this axis of polarity (more on this later).

The cilia of the macular hair cells extend into the endolymph and embed themselves in a gelatinous matrix termed the *otolithic membrane* (Figure 25.2). The overall organi-

zation is reminiscent of the stereocilia of cochlear hair cells and their relationship with the tectorial membrane. Embedded in the otolithic membrane are masses of calcium carbonate crystals that increase the inertial mass of the membrane. The individual crystals are called *otoconia,* and the masses of otoconia are called *otoliths.*

The two maculae in the saccule and utricle are oriented perpendicular to one another (Figure 25.3). When standing, the macula in the utricle is oriented in the horizontal plane, whereas the macula of the sac-

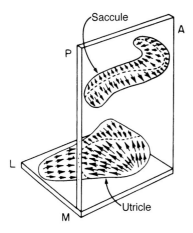

Figure 25.3. The orientation of the saccula and macula and the representation of direction of movement on the macular surface. Schematic illustration of the relative orientation of the maculae in the saccule and utricle from the vestibular apparatus on the left-hand side of the brain. Arrows indicate the direction of hair cell polarization on each side of the striola (dotted line). The arrows point from the short to the tall stereocilia of individual hair cells (and thus in the direction in which a movement causes depolarization, see Figure 25.7). A indicates anterior; P, posterior; M, medial; L, lateral. From Fuchs AF. The vestibular system. In: Patton HD, Fuchs AF, Hille B, Scher AM, Steiner R, eds. *Textbook of Physiology, 1: Excitable Cells and Neurophysiology.* Philadelphia, Pa: WB Saunders; 1989.

cule is oriented in the sagittal plane. In the upright position, gravity deflects the otolythic membrane of the vertically oriented saccule. This deflects the stereocilia of the hair cells, modulating the mechanosensitive K^+ channels at the tips of the stereocilia. When the head is tilted in various directions, the otolythic membrane of the utricle is also deflected. By combining information deriving from the two receptive sheets, the static position of the head with respect to gravity can be determined.

The hair cells of the maculae are also activated as a result of linear acceleration of the head. Because of the inertial mass of the otolithic membrane, it lags behind when the head (and the macula) are accelerated. The inertial lag of the otolithic membrane deflects the stereocilia of the hair cells,

modulating the mechanosensitive K^+ channels at the tips of the stereocilia.

Because the sheet is oriented in a particular plane, the structure is most sensitive to movements in that plane. To detect movements in a variety of planes, information from the two maculae are integrated in the central nervous system.

Directions of Movement Are Represented in Each Macula by a "Place Code"

An important feature of vestibular hair cells is that the modulation of membrane current flow occurs only when the stereocilia are deflected in a particular direction (along the morphological axis defined by the row of hair cells). Thus, stereocilia must be properly oriented with respect to the direction of movement. To detect movements in a variety of different directions, the hair cells are oriented in a remarkably precise matrix. Within each macula the hair cells are oriented with respect to a line called the *striola*. The stereocilia on one side of the striola are oriented with the short stereocilia toward the striola; those on the opposite side are also oriented with their short stereocilia toward the striola, but because the two sets lie on opposite sides of the line, the direction of polarization of the stereocilia is opposite on the two sides. The striola then curves approximately 90° across the surface of the macula so that each orientation is represented somewhere along the macular surface (Figure 25.3). In this way, different locations along the macula are specialized to detect a particular direction of movement. By way of analogy with the cochlea, there is a place map along each macula for particular directions of movement, so that directions of movement are represented by a "place code."

The Semicircular Canals

The three semicircular canals are specialized to detect angular (rotational) acceleration. They are oriented in three orthogonal planes (Figure 25.4). When upright, the horizontal canal is oriented at about 25° with respect to

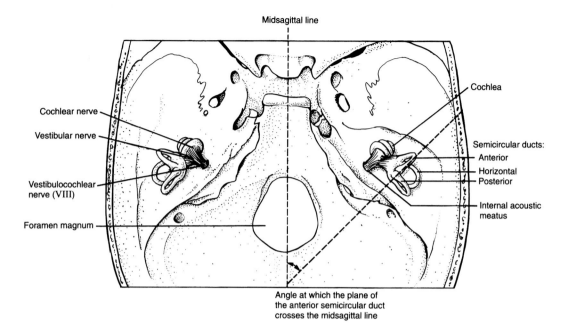

Angle at which the plane of
the anterior semicircular duct
crosses the midsagittal line

Figure 25.4. The semicircular canal. The orientation of the three semicircular canals within the skull. When upright, the horizontal canal is oriented at about 25° with respect to the horizontal plane, the anterior canal is inclined about 41° with respect to the sagittal plane (the vertical midline), and the posterior canal is oriented at about 56° with respect to the anterior canal. With this configuration, the posterior canal on one side of the head is in approximately the same plane as the anterior canal on the opposite side. From Kandel ER, Schwartz JH, Jessell TM. *Principles of Neural Science.* 3rd ed. New York, NY: Elsevier; 1991.

the horizontal plane, the anterior canal is inclined about 41° with respect to the sagittal plane (the vertical midline), and the posterior canal is oriented at about 56° with respect to the anterior canal. With this configuration, the posterior canal on one side of the head is in approximately the same plane as the anterior canal on the opposite side.

The three semicircular canals are filled with endolymph; turning the head leads to inertial movement of the endolymph within the semicircular canals (which are oriented in the plane of the movement). The inertial movement of the endolymph within the semicircular canals occurs when head rotation is initiated (the inertia of the endolymph causes it to lag behind) and when the rotation ends (the inertia of the endolymph causes it to continue to move within the canal). Hence, the semicircular canals are specialized to detect the beginning and the end of a rotational

movement, that is, *rotational acceleration* and *rotational deceleration*.

At one end of each semicircular canal is an enlarged compartment termed the *ampulla,* which contains another specialized sensory neuroepithelium termed the *ampullary crest* (Figure 25.5). Hair cells are located in the ampullary crest and extend their stereocilia into the endolymph. Again, the tips of the stereocilia are embedded in a gelatinous mass, this time called the *cupula.* The hair cell stereocilia and cupula thus form a structure something like a swinging gate across the ampulla. Inertial movement of the endolymph within the canal deflects the cupula with its embedded hair cell stereocilia, modulating the K^+ channels in the stereocilia.

The morphological axis of polarity of the hair cells in the ampulla is such that the tall stereocilia are oriented toward the utricle. Hence, inertial movement of the endolymph

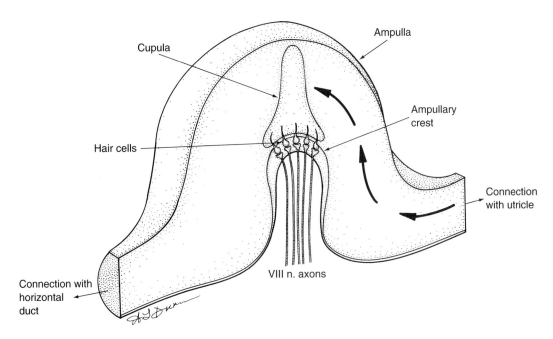

Cupula

Ampulla

Ampullary
crest

Hair cells

Connection
with utricle

Connection with
horizontal
duct

VIII n. axons

Figure 25.5. The ampulla. The drawing illustrates the organization of the ampulla and the receptor apparatus of the ampullary crest. Hair cells in the ampullary crest extend their stereocilia into the overlying cupula. The cupula is displaced when there is inertial movement of the endolymph within the semicircular canal as a result of head movement. The hair cells release their neurotransmitter onto the nerve endings of eighth nerve axons in the *ampullary nerve.*

that deflects the cupula toward the utricle depolarizes the hair cells and causes an increase in the firing rate of eighth nerve axons; a movement that deflects the cupula away from the utricle depolarizes the hair cells and causes a decrease in the firing rate of eighth nerve axons.

The cell bodies of the first-order ganglion cells of the vestibular system lie in the *vestibular ganglion* (also called *Scarpa's ganglion*), which lies near the internal auditory meatus. The ganglion cells give rise to distal branches that are contacted by the hair cells, and central branches that form the vestibular division of the eighth nerve. The two types of hair cell (type I and II) have somewhat different synaptic connections with the afferent ending, but the physiological significance of the differences in synaptology are not clear.

The hair cells are also contacted by efferent fibers. The efferent connections arise from a small group of neurons that lie just lateral to the abducens nucleus in the medulla. The physiological significance of the efferent pathway is not known.

CENTRAL VESTIBULAR PATHWAYS

The axons that make up the vestibular division of the eighth nerve terminate in four distinct *vestibular nuclei* that lie in the dorsal part of the medulla just beneath the fourth ventricle (see Figures 25.6 and 25.7). One component of eighth nerve axons also projects directly to the flocculonodular lobe of the cerebellum (the so-called vestibulocerebellum, see Chapter 17). The two principal outputs of the vestibular nuclei are to the spinal cord via the vestibulospinal tracts and to the cranial nerve nuclei that contain the motoneurons that control eye movement (Figures 25.6 and 25.7).

The vestibular nuclei are termed the *superior, inferior, medial,* and *lateral nuclei* (Figure

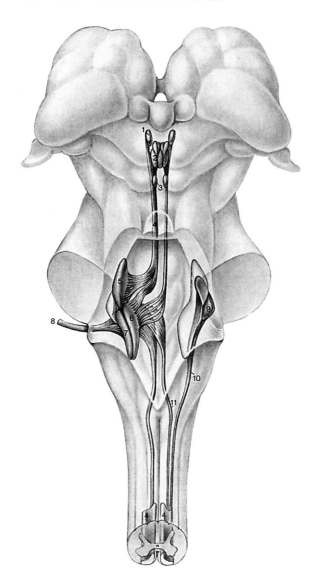

Figure 25.6. The vestibular nuclei of the brainstem. Dorsal view of the brainstem. The location of the vestibular nuclei is indicated. Also indicated are the cranial nerve nuclei that contain motoneurons that control eye movement, and the descending vestibulospinal tracts. 1, Interstitial nucleus; 2, oculomotor nucleus; 3, trochlear nucleus; 4, medial longitudinal fasciculus; 5, superior vestibular nucleus; 6, medial vestibular nucleus; 7, inferior vestibular nucleus; 8, eighth nerve; 9, lateral vestibular nucleus; 10, lateral vestibulospinal tract; 11, medial vestibulospinal tract. From Nieuwenhuys R, Voogd J, van Huijzen C, *The Human Central Nervous System*. Berlin, Germany: Springer-Verlag; 1988.

25.7). Each also has a name associated with the neuroanatomist who first described it. However, the only one that is commonly used is *Deiter's nucleus*, which refers to the lateral vestibular nucleus. In addition to the input from the eighth nerve, these nuclei also receive input from proprioceptors, especially the proprioceptors in the neck, and from areas that receive visual input.

As we see later, some of the output pathways from the vestibular nuclei are excitatory, others are inhibitory. This allows the vestibular projections to exert both positive and negative control on their targets.

The Lateral Vestibular Nucleus Is Especially Important for Posture and Balance

The *lateral vestibular nucleus* receives input from the utricle and semicircular canals. The neurons in the nucleus give rise to descending projections to the spinal cord that travel via the *lateral vestibulospinal tract*. The organi-

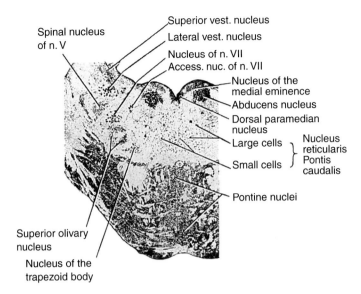

Spinal nucleus
of n. V

Superior vest. nucleus

Lateral vest. nucleus

Nucleus of n. VII

Access. nuc. of n. VII

Nucleus of the
medial eminence

Abducens nucleus

Dorsal paramedian
nucleus

Large cells ⎤ Nucleus
reticularis
Small cells ⎦ Pontis
caudalis

Pontine nuclei

Superior olivary
nucleus

Nucleus of the
trapezoid body

Figure 25.7. The vestibular nuclei of the brainstem in cross section. Photograph of a cresyl violet–stained section through the pons and midbrain tegmentum at the level of the vestibular nuclei that illustrates the location of the vestibular nuclei. The neurons in the vestibular nuclei, especially those in the lateral vestibular nucleus, are among the largest in the brainstem. Reproduced with permission from Carpenter MB. *Core Text of Neuroanatomy.* Baltimore, Md: Williams & Wilkins; 1991.

zation and operation of this pathway is described in the chapters pertaining to the motor system. Activation of the second-order neurons facilitates motoneurons that supply antigravity muscles. In this, the pathway operates primarily through gamma motoneurons, although some vestibulospinal fibers also innervate alpha motoneurons.

The lateral vestibular nuclei receive an important projection from the cerebellar vermis. This is an inhibitory pathway. When the cerebellum is damaged, the neurons in the vestibular nucleus are released from inhibition, causing hyperactivity of the descending vestibulospinal projections to antigravity muscles. If descending motor pathways are intact, this hyperactivity can be counterbalanced by descending excitation. However, when both the descending inputs from the cerebellum and the descending motor pathways are interrupted, the unbalanced hyperactivity of the vestibulospinal pathway causes *extensor rigidity* (discussed further in the chapters on the motor system). This can occur as a consequence of an occlusive stroke involving the anterior–inferior cerebellar artery, which

supplies the brainstem and the anterior cerebellum.

The Medial and Superior Vestibular Nuclei Are Especially Important for Vestibulo-ocular Reflexes

The *medial vestibular nucleus* and the *superior vestibular nucleus* receive input primarily from the semicircular canals. Second-order neurons in the medial vestibular nucleus give rise to axons that project to the cervical spinal cord via the *medial vestibulospinal tract.* These terminate on the motoneurons supplying the muscles of the neck. This circuitry is important for coordinating head movements that stabilize head position during movements of the body.

Other second-order neurons in the two nuclei play a role in coordinating eye movements to compensate for movements of the head (specifically, the *vestibulo-ocular reflexes*). The axons of the second-order neurons join a tract called the *medial longitudinal fasciculus,* which is located on each side of the midline just beneath the periventricular gray matter. These axons from the vestibular nuclei terminate in the cranial nerve nuclei that control eye movements.

The *inferior vestibular nucleus* receives input from all of the different parts of the peripheral vestibular apparatus (saccule, utricle, and the three semicircular canals). The neurons in the nucleus also receive input from the cerebellar vermis. Axons of the second-order neurons in the nucleus join the vestibulospinal tract and also project to the reticular formation. There is no single functional role that can be ascribed to these connections other than the integration of information from the interconnected structures (vestibular labyrinth and cerebellum).

The Vestibulocerebellar Pathways

There are important interconnections between the vestibular system and the cerebellum. One component of eighth nerve axons projects directly to the flocculonodular lobe, where the axons terminate as mossy fibers. There is also an indirect pathway via the lateral vestibular nuclei. These axons travel via the inferior cerebellar peduncle. It is because of the important interconnections between the vestibular system and the cerebellum that damage to the cerebellum produces the "vestibular" symptoms of *nystagmus* and sometimes *vertigo.*

Conscious Perception of Information From the Vestibular System

Although the vestibular system operates primarily at a subcortical level, information regarding head position and movement in space does reach consciousness. Information is conveyed to the cortex via projections from the vestibular nuclei to the thalamus and then on to the cerebral cortex.

There are also projections from the vestibular nuclei to the hypothalamus. These pathways seem to affect the autonomic control centers in the hypothalamus. The physiological role of these connections in normal situations is not clear. However they are thought to be responsible for some of the symptoms of vestibular dysfunction (nausea, sweating, and increases in heart rate) that are also consciously perceived.

MECHANOELECTRICAL TRANSDUCTION BY VESTIBULAR HAIR CELLS

An important difference between cochlear and vestibular hair cells is the frequency range of the stimuli to which the two respond. Auditory hair cells are sensitive to displacements that occur at frequencies ranging from 20 Hz to 20 kHz (the frequencies of sound). In contrast, vestibular hair cells must be sensitive to static displacements (induced by gravity) and the slow displacements that are produced by accelerations of the head.

At rest, there is a leakage current through the mechanosensitive K^+ channels in the stereocilia such that the hair cells are somewhat depolarized. This allows hair cells to respond differentially to deflections of the stereocilia along the principal axis of movement. Deflection toward the tall stereocilia opens the K^+ channels in the stereocilia causing, additional depolarization; deflection toward the short stereocilia closes the K^+ channels causing, hyperpolarization (see Figure 25.8).

Because the hair cells are somewhat depolarized at rest, there is a resting release of neurotransmitter onto the afferent ending. The depolarization caused by deflecting the hair cell toward the tall stereocilia increases the rate of neurotransmitter release; the hyperpolarization produced by deflecting the hair cell toward the short stereocilia decreases the rate of release. Both the depolarization and hyperpolarization can be considered *receptor potentials.*

The afferent ending is depolarized by the neurotransmitter that is released by the hair cell. Thus, increases and decreases in the rate of neurotransmitter release cause graded depolarization and hyperpolarization, respectively, of the afferent ending (*generator potentials).* These in turn are integrated so as to

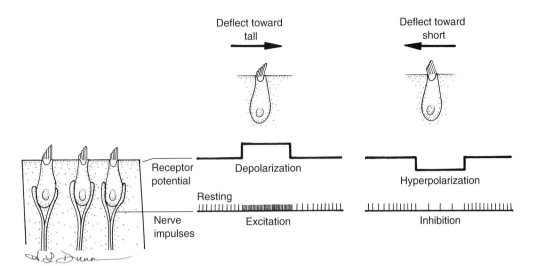

Figure 25.8. Hair cells respond differentially to deflections of the stereocilia along the principal axis of movement. At rest, there is a leakage current through the mechanosensitive K^+ channels in the stereocilia such that the hair cells are somewhat depolarized. Deflection of the stereocilia toward the tall stereocilia opens the K^+ channels in the stereocilia, causing additional depolarization, which in turn increases the rate of firing in the afferent axons. Deflection toward the short stereocilia closes the K^+ channels, causing hyperpolarization, which decreases the rate of firing in the afferent axons. After Flock A. Structure of the macula utriculi with special reference to directional interplay of sensory responses as revealed by morphological polarization. *J Cell Biol.* 1964;22:413–431.

increase or decrease the rate of action potentials in the myelinated portion of the afferent nerve. Because of the resting release of neurotransmitter, there is a relatively high rate of *resting activity* in eighth nerve axons, which increases and decreases depending on the input from the hair cells.

Integration of Signals From the Different Components of the Vestibular Apparatus

As we have already seen, the semicircular canals on the two sides of the head are organized in complementary pairs that are oriented in different axes of movement: Horizontal–horizontal, anterior (L)–posterior (R), anterior (R)–posterior (L).

The complementary pairs are, in a functional sense, mirror images of one another. Consider for example the two horizontal canals. Rotation of the head in one direction (left, for example) causes inertial movements in the endolymph of both canals.

When turning to the left, the head rotates in a counterclockwise direction (when looking down upon the head); hence, inertia causes the endolymph to move in a clockwise direction within both left and right horizontal canals (Figure 25.9).

However, the direction of inertial movement with respect to the *morphological axis* of the canals on the two sides is exactly opposite. On the left side, inertial flow of endolymph deflects the stereocilia toward the tall bundle, causing depolarization; on the right side, the stereocilia are deflected toward the short bundle, causing hyperpolarization. In this way, the rate of action potentials over eighth nerve axons would increase on the left side and decrease on the right side. Following is the rule for the horizontal canals: there is excitation of eighth nerve axons on the side toward which the head is turned.

Similar "push–pull" relationships exist with respect to the physiological responses that are generated by rotations in the planes

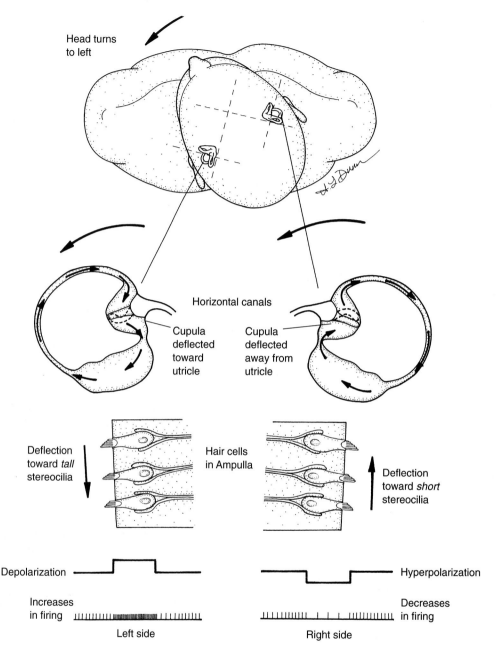

Figure 25.9. Rotation of the head induces counterrotational movement of endolymph within the semicircular canals. When the head is turned to the left, there is inertial counter movement of endolymph within the semicircular canals. With a rotation to the left, the movement of the endolymph in the horizontal canals is clockwise. On the left side, inertial flow of endolymph deflects the stereocilia toward the tall bundle causing depolarization; on the right side, the stereocilia are deflected toward the short bundle, causing hyperpolarization. In this way, rate of action potentials over eighth nerve axons would increase on the left side and decrease on the right side. There is excitation of eighth nerve axons on the side toward which the head is turned.

represented by the other pairs of canals. This is readily apparent if you take a bit of time to see how the complementary pairs of canals on the two sides line up in three-dimensional space and then consider the inertial flows that are generated within the pairs of channels by rotation in different planes.

It is important to recall that the different canals are not perfectly oriented to the cardinal axes of movement. The cardinal axes are nodding the head forward and back (pitch), turning the head to the right and left (yaw), and tilting the head to the left and right (roll). All these movements, and everything in between, activate more than one set of semicircular canals. This information is then conveyed to the brain, where it is integrated to determine the actual direction of movement.

Integration of Signals From the Maculae

The same sort of positive and negative signaling occurs in the case of the maculae of the saccule and utricle. As already noted, in each macula the hair cells are aligned along the striola so that the stereocilia on the two sides of the dividing line are oriented at 180°. Hence, movement in any given direction causes depolarization of hair cells on one side of the striola and hyperpolarization of the hair cells that are oriented in the opposite direction.

Transient Versus Long-Lasting Rotational Stimuli: Why One Gets "Dizzy" in a Revolving Chair

Turning the head in normal activities produces a transient signal due to inertial movements of the endolymph. If rotation continues, however, inertia is overcome and the endolymph moves at the same rate as the canal in which it is contained. In this situation, there is no deflection of the stereocilia of the receptors. With constant rotation (as in a revolving chair), receptor activation and, hence, the rate of neural discharge are as they would be at rest. Sudden cessation of rotation then triggers inertial movement of the endolymph within the semicircular

canal, and receptor activation. The deflection of the cupula in this situation occurs slowly; it takes 10 to 30 seconds for the cupula to return to its resting position. During this time, there is receptor activation that triggers nystagmus (more on this later), and during this period, one feels as if one is still spinning. It is interesting that children find this sensation pleasurable (or at least interesting), whereas adults find it distinctly unpleasant!

The Vestibulo-ocular Reflex

The vestibulo-ocular reflexes (VORs) keep the eyes focused on a point in space during movements of the head. That is, the circuit stabilizes gaze during head movements. The afferent limb of the reflex is provided by the first-order projections to the vestibular nuclei of the brainstem. The second-order neurons in these nuclei project to the cranial nerve nuclei that control the extrinsic eye muscles. The efferent limb of the reflex is the projection of the motoneurons to the extrinsic eye muscles.

Some of the circuitry that underlies the VOR reflex is illustrated in Figure 25.10. As already noted, some of the projections from the vestibular nuclei are excitatory (indicated as open symbols); others exert a net inhibitory effect (filled symbols). Students need not memorize the details of this circuitry. What is important is (1) knowing the nuclei involved, (2) knowing that many of the connections travel via the medial longitudinal fasciculus, and (3) knowing that the different semicircular canals control particular sets of extrinsic eye muscles (see later).

Different Semicircular Canals Control Particular Sets of Extrinsic Eye Muscles

The input from the afferent limb signals the direction of head rotation. The three semicircular canals are oriented in approximately the same planes as are the axes defined by the pulling direction of the different sets of extraocular muscles. Thus, (1) the horizontal canals sense movement in the horizontal plane in which the medial and lateral rectus muscles move the eyes, (2) the left anterior

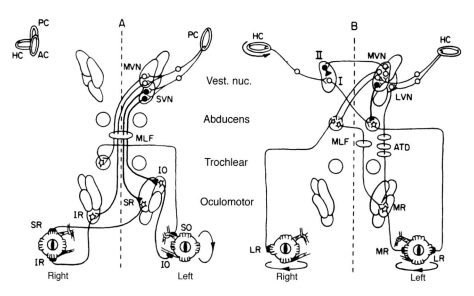

Figure 25.10. The circuits that mediate the VOR. The afferent limb of the VOR is provided by the first-order projections to the vestibular nuclei of the brainstem. The second-order neurons in these nuclei project (primarily via the medial longitudinal fasciculus) to the cranial nerve nuclei that control the extrinsic eye muscles. The efferent limb of the reflex is the projection of the motoneurons to the extrinsic eye muscles. The drawings illustrate the key circuitry involved in the vertical (left-hand side) and horizontal (right-hand side) VOR. Open symbols indicate excitatory projections; closed symbols indicate inhibitory projections. HC, PC, AC indicate horizontal, posterior, and anterior semicircular canals, respectively; IO, inferior oblique; SR, superior rectus; IR, inferior rectus; MLF, medial longitudinal fasciculus; ATD, ascending tract from Deiters' nucleus (the lateral vestibular nucleus). From Fuchs AF. The vestibular system. In: Patton HD, Fuchs AF, Hille B, Scher AM, Steiner R, eds. *Textbook of Physiology, 1: Excitable Cells and Neurophysiology.* Philadelphia, Pa: WB Saunders; 1989.

and right posterior canals sense movement in the plane in which the left superior and inferior rectus muscles and right oblique muscles move the eye, (3) the left posterior and right anterior canals sense movement in the plane in which the right vertical rectus and left oblique muscle move the eyes.

The different complementary pairs of semicircular canals control different extrinsic ocular muscle groups so that the eyes move exactly opposite to the direction of rotation. The physiological relationships between the different canals and the extrinsic muscles of the eye are summarized in Table 25.1.

The Nature of the Eye Movements That Occur During the VOR

The VOR triggers the two different types of coordinate movements of the eyes that are described in Chapter 21. To briefly review, as the eyes track a moving stimulus, there is a slow phase of eye movement to keep the gaze focused on the object *(smooth pursuit or tracking movements)*. When the eyes move to focus on a different point, the eyes snap to the new position in a rapid movement termed a *saccade.*

The same types of eye movements occur during rotation of the head. As the head turns, there is a slow phase of eye movement to keep the gaze focused on a point in space. The movement of the eyes that occurs during head rotation is timed exactly to coordinate with the movement of the head. If rotation continues beyond the point that a particular object can be seen, the eyes snap to a new position in a saccade. If rotation continues (for example, in a rotating chair), there are intervals of smooth pursuit followed by a saccade. The alternating slow

Table 25.1. Control of extrinsic eye muscles by the semicircular canals*

Canal	Excites	Inhibits
Horizontal	Ipsilateral medial rectus Contralateral lateral rectus	Ipsilateral lateral rectus Contralateral medial rectus
Posterior	Ipsilateral superior oblique Contralateral inferior rectus	Ipsilateral inferior oblique Contralateral superior rectus
Anterior	Ipsilateral superior rectus Contralateral inferior oblique	Ipsilateral inferior rectus Contralateral superior oblique

*From Kandel, ER, Schwartz, JH, and Jessel, TM, *Principles of Neuroscience*, 3rd ed. NY: Elsevier. 1991, p. 669.

tracking followed by a saccade is termed *nystagmus*. By convention, one refers to left or right nystagmus on the basis of the fast phase of eye movement. Hence, rotation to the left produces *left lateral nystagmus.*

Continuous rotation in one direction produces *rotational nystagmus* in the direction of the rotation. Because of the postrotational activation of the ampulae of the semicircular canal (described earlier), the cessation of rotation produces *postrotational nystagmus* in the direction opposite the rotation. One clinical test of vestibular function measures postrational nystagmus. The patient is subjected to rotation using a rotating chair; then the rotation is stopped. Postrotational nystagmus is then measured by recording eye movements. This is termed a *nystagmogram*.

Another clinical test of vestibular function involves *caloric stimulation* of the horizontal semicircular canal, which produces *caloric nystagmus*. Caloric stimulation is accomplished by irrigating the ear with warm or cold water while the head is tilted back about 60° so that the horizontal canal is near vertical. The outer edge of the horizontal canal is very close to the meatus, so that heat and cold are transferred. It is thought that the temperature differential causes convectional movement of the endolymph and, hence, deflects the cupula. Warm water causes nystagmus toward the side of the irrigation; cold water causes nystagmus away from the side of the irrigation. The mnemonic here is *COWS*: cold is opposite; warm is same. Each ear can be tested separately. Individuals with vestibular disorders exhibit abnormal nystagmus. This test is, by the way, a very unpleasant experience for the patient.

The VOR is also important for diagnosing injury to the brainstem in unconscious patients. If a patient without a brainstem injury is positioned on his or her back and the head is rotated to the left and the right, the VOR will cause the eyes to move coordinately opposite to the direction of rotation. This is termed the *doll's eye maneuver,* named after the dolls with weighted eyeballs that seem to remain in position when the doll is tilted. If the eyes do not move in the normal way, damage to the circuitry that mediates the VOR is indicated.

DISORDERS AFFECTING THE VESTIBULAR SYSTEM

The symptoms resulting from injury or disease that affects the vestibular receptors or the eighth nerve depend on the speed of onset of the pathology, and whether the dysfunction is acute or chronic. There is a wide variety of causes of vestibular dysfunction. Here we mention only a few, principal ones.

Acute unilateral injuries lead to feelings of vertigo (dizziness), nystagmus, and disequilibrium (falling toward the affected side), as well as vomiting and sweating because of abnormal activity in the circuitry involving the hypothalamus. All of this is extremely unpleasant for the patient. Indeed, it is usually impossible to stand or walk.

Chronic injuries or *slow-onset disease processes* produce much less severe symptoms than acute injuries or rapid-onset diseases. The reason is that there appears to be a mechanism through which the vestibular system compensates for imbalanced input from the two sides.

This is termed *vestibular compensation,* the mechanisms of which are not well understood. Typically, following an acute insult, young patients recover the ability to walk and may be able to suppress spontaneous nystagmus within about one week, and fairly complete compensation occurs within about one month. The speed and final extent of compensation are less in aged individuals.

Once the acute symptoms of asymmetric activation subside, the brain makes use of information from other sensory modalities (especially vision) for balance. However, in fully compensated individuals, problems with balance may reappear when visual cues are not available (for example, with eyes closed or in the dark). Symptoms may also reappear when patients have a cold.

Menier's disease has already been discussed in the context of the auditory system. The disease is due to a buildup of endolymphatic fluid, which leads to receptor activation.

Motion sickness (kinetosis) occurs when susceptible individuals experience a discrepancy between vestibular and visual inputs. The symptoms are similar to those produced by any of the diseases that affect the vestibular apparatus: dizziness, vomiting, and sweating. These usually subside over time. Some individuals are highly susceptible to motion sickness, whereas others are virtually immune. The reasons for the individual differences in susceptibility are not known. However, the condition can be treated by anticholinergic agents (dramamine or scopolamine).

Acute alcohol intoxication also produces vestibular symptoms, specifically nystagmus and vertigo. These are usually manifest when lying down (the spinning bed phenomenon). For reasons that are not clear, at least to the author, these symptoms can be reduced by hanging one leg over the edge of the bed and placing the foot on the floor. This is one of those pieces of folk knowledge that seems to have a father-to-son mode of communication across generations.

Aminoglycoside antibiotics are also extremely toxic to vestibular hair cells. For this reason, these antibiotics should be used with extreme care, especially in patients with kidney disorders. It is especially important to avoid chronic treatment because the antibiotics accumulate in the endolymph.

Vestibular hair cells also degenerate with advancing age. This is one of the reasons why elderly individuals experience problems with balance, especially in the dark. It is estimated that many of the falls that elderly individuals experience are due to vestibular dysfunction.

8

The Chemical Senses

The systems that underlie the chemical sensations of taste and smell are the *olfactory system* and the *gustatory system*. These are considered "general" rather than "special" senses.

The "senses" of taste and smell are closely intertwined. Indeed, much of what we perceive as taste is in fact due to the aroma of foods. You can demonstrate this to yourself by holding your nose while eating a jellybean. With your nose closed, there is little taste. When you release your nose, however, there is an immediate wash of the "fruity" taste that makes the jellybean appealing. This is why foods have little taste when you have a cold, why older people have trouble maintaining their food intake (there is an age-related loss of olfactory function), and why individuals with injuries to their olfactory system often have life-threatening loss of appetite. Hence, it is important for physicians to understand the consequences of the loss of chemical sensation so they can properly manage their patients' care.

In the following chapter, we consider the organization and functional features of the two chemical senses, beginning with gustation.

The components of the gustatory system that are important for the conscious perception of taste include (1) receptors in the tongue and oropharyngeal mucosa, (2) primary afferent fibers that innervate the receptors, (3) ganglion cells that give rise to the primary afferent fibers, (4) the central projections of the ganglion cells, which project to the brain via cranial nerves VII, IX, and X, (5) Second-order neurons in the solitary nuclear complex in the medulla, (6) Projections from the second-order neurons to the ventral posteromedial nucleus of the thalamus, and (7) the taste areas of the cortex, which are located in the postcentral gyrus and the insular lobe (see Figure 26.2)

The following are key concepts for physicians regarding gustation:

1. There are four "pure" taste sensations (qualities): salt, sour, sweet, and bitter. The complexity of our taste experience depends on an interaction between gustatory and olfactory stimuli.

2. The receptors for the different sensations are located in different parts of the tongue.

3. Information regarding taste is conveyed to the brain by three cranial nerves (VII, IX, and X).

4. The pathways that carry gustatory information project ipsilaterally.

The components of the olfactory system that are important for the conscious perception of smell include (1) the *olfactory receptors* in the sensory epithelium of the nose, (2) the projection from the receptors to the second-order neurons located in the *olfactory bulb* (termed the *olfactory nerve*), (3) the *olfactory tract* (cranial nerve I), which carries information from the olfactory bulb to the brain, and (4) the areas in the CNS that receive direct input from the olfactory bulb, including the *anterior olfactory nucleus, olfactory tubercle,* and *prepyriform cortex* (see Figure 26.7).

The following are key concepts for physicians regarding olfaction:

1. There are a large number of different olfactory sensations, which are mediated by specific *oderant receptors.*

2. Individual receptor cells express particular odorant receptors; the cells that express one type of receptor project to particular locations in the olfactory bulb. In this sense, there is an *aromotopic map* in the olfactory bulb.

3. The pathways carrying olfactory information are largely ipsilateral.

4. Information from the olfactory bulb reaches the olfactory portions of the cortex directly, without being relayed through the thalamus.

5. The areas that receive olfactory information are closely connected with areas in the *hypothalamus* and the *limbic system.* For this reason, olfactory stimuli have a strong *affective component* (that is, they impact on our motivational and emotional state).

26

The Chemical Senses

The systems that underlie the chemical sensations of taste and smell are present in the most primitive of organisms. Even single-cell organisms possess receptors that allow the organisms to respond to chemical signals. Indeed, in most organisms, the chemical senses play a pivotal role in locating food, discriminating acceptable foods from those that are toxic, motivating food intake (that is, regulating appetite), as well as regulating the aspects of social behavior that are critical for reproduction. Hence, the chemical senses underlie the activities that are most basic to the survival of the organism.

It is important to be aware that what we recognize as the sensation of taste in fact is a combination of gustatory and olfactory information. Hence, when patients refer to symptoms involving taste, these often reflect dysfunction of the olfactory system. Thus, it is important to keep in mind the distinction between the gustatory and olfactory systems and what we call the senses of taste and smell. In what follows, we begin our consideration with the gustatory system, and then consider the olfactory system.

THE GUSTATORY SYSTEM

The gustatory sense is mediated by chemoreceptors located primarily in the tongue; however, there are also some receptors in the palate, pharynx, epiglottis, and upper esophagus. The receptor cells are innervated by first-order neurons in *sensory ganglia,* which then send their axons via cranial nerves VII, IX, and X to second-order neurons in the *solitary nuclear complex* in the medulla. The second-order neurons then project to the *somatosensory thalamus,* which in turn projects to the cerebral cortex.

Gustatory chemoreceptors are specialized epithelial cells that detect soluble chemical substances that are present in the fluid environment in the mouth. Humans are able to distinguish four or perhaps five basic taste sensations. The primary four are bitter, salty, sour (acid), and sweet. A possible fifth is the sensation produced by monosodium glutamate, but there is not general agreement that this represents a separate category.

Gustatory receptors in the tongue are collected in *taste buds,* which in turn are localized in much larger structures called *papillae.* In regions other than the tongue, the receptors are collected in taste buds but are not part of papillae.

Taste buds are composed of collections of receptors and two other cell types: basal cells and supporting cells. These cells are arrayed in the manner illustrated in Figure 26.1. As the name implies, basal cells are located at the base of the taste bud. It is thought that they are precursor cells for the taste receptors (because the receptors are continuously turning over, with a lifetime of about 10 days). Supporting cells surround the collections of receptors and are thought to provide structural or trophic support.

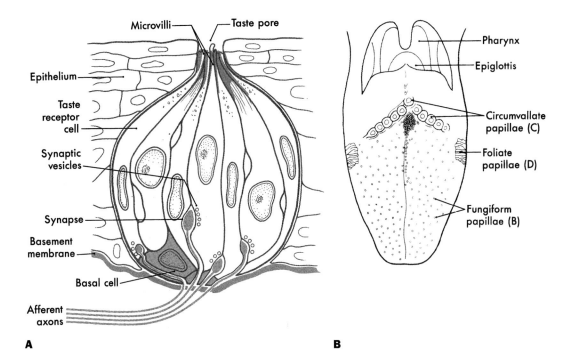

A   **B**

Figure 26.1. Organization of gustatory chemoreceptors into taste buds. A, Taste buds are composed of collections of receptors and two other cell types: basal cells and supporting cells. The outer taste pore opens into the mouth cavity. Afferent fibers synapse on the base of the taste cells. The receptor cells form chemical synapses with the afferent endings. **B,** Distribution of papillae and taste sensitivity across the tongue. Taste buds are located in three different types of papillae, which are differentially distributed across the tongue. Fungiform papillae are located on the anterior two-thirds of the tongue; foliate papillae are located on the posterior edge of the tongue; and cirvumvallate papillae are localized in the posterior one-third of the tongue. Different parts of the tongue exhibit different sensitivities for the principal taste sensations. Reproduced with permission from Haines DE. *Fundamental Neuroscience.* New York, NY: Churchill-Livingstone, New York; 1997.

The taste buds are embedded in the epithelium of the tongue. Openings into the mouth cavity, termed *outer taste pores,* provide a route for dissolved chemical substances to contact the receptive surface of the taste receptor. Microvillae extend from the apical surface of the taste receptors into the taste pore.

The base of the receptors are contacted by the afferent fibers. Collections of vesicles are present in the portion of the cytoplasm of the receptor cell that is apposed to the afferent fiber, forming the substrate for chemical neurotransmission between the receptor and the afferent ending. Each afferent fiber innervates a large number of receptors; hence, the activity induced in the fibers is the result of the summed inputs from a large number of individual receptors.

Taste buds are located in three different types of papillae, which are differentially distributed across the tongue (Figure 26.1B). Fungiform papillae are located on the anterior two-thirds of the tongue; foliate papillae are located on the posterior edge of the tongue; and cirvumvallate papillae are localized in the posterior one-third of the tongue. As we see in more detail later, different parts of the tongue contain receptors that are specialized for different substances (Figure 26.1B and Table 26.1).

The primary afferent fibers that innervate receptors in the tongue and oropharyngeal mucosa arrive via three nerves (Figure 26.2):

Table 26.1. Distribution of taste receptors on the tongue

Parts of the tongue	Type of papillae	Inner-vation	Sensitivity
Anterior tip	Fungiform	VII	Sweet
Anterior lateral	Fungiform	VII	Salty
Lateral margin	Fungiform	VII	Sour
Posterior margin	Foliate	IX	Bitter
Posterior middle	Circumvallate	IX	Bitter
Palate, pharynx, esophagus	none	X	

(1) the *chorda tympani,* which is a branch of the facial nerve. The greater superficial petrossal branch of the facial nerve innervates the taste buds on the palate; (2) the lingual branch of the glossopharyngial nerve (cranial nerve IX); and (3) the superior laryngeal branch of the vagus nerve (cranial nerve X).

The ganglion cells that give rise to the primary afferent fibers are located in the geniculate ganglion (cranial nerve VII), the petrosal ganglion (cranial nerve IX), and the nodose ganglion (cranial nerve X). The ganglion cells give rise to centrally projecting axons that travel via the respective cranial nerves and terminate in the solitary nuclear complex, which is located in the dorsal medulla (Figure 26.3).

The second-order neurons in the solitary nuclear complex are located in a cluster in the rostral and lateral part of the complex termed the *gustatory nucleus.* The neurons in other parts of the nucleus receive inputs from the gut, lungs, and cardiovascular system.

The second-order neurons in the solitary nuclear complex project to the *ventral posteromedial nucleus* of the thalamus via the *central tegmental tract.* Within the ventral posteromedial nucleus, the neurons that subserve taste

are located in a cluster that is separate from the neurons that represent somatic sensation for the tongue. The central projections of the gustatory system are exclusively ipsilateral.

Neurons in the gustatory nucleus also project to the *parabrachial nuclei,* so named because they lie medially adjacent to the *brachium conjunctivum* (also called the *superior cerebellar peduncle*). The parabrachial nuclei in turn project to nuclei involved in central autonomic control. This pathway is thought to mediate autonomic responses to taste stimuli.

Neurons in the ventral posteromedial nucleus that convey gustatory information project to two regions of the cerebral cortex (Figure 26.2): (1) the gustatory region of the postcentral gyrus, which lies just ventral and rostral to the region that receives somatosensory information from the tongue; and the frontal cortex and insula.

Sensory Transduction and Neural Processing in the Gustatory Pathway

Sensitivity to the four different taste sensations is differentially distributed across the tongue. The anterior tip of the tongue is most sensitive to sweet and salt; the lateral part of the tongue is most sensitive to sour; and the back of the tongue is most sensitive to bitter. By most sensitive, we mean that the threshold for detecting these tastes is lowest in the particular region. However, each part of the tongue can detect any of the four primary tastes if the chemical substance is present at high enough concentrations.

Different chemotransduction mechanisms exist for detecting the different chemical substances that underlie taste (Figure 26.4 and Table 26.2). Salt is detected as a result of the direct passage of Na^+ ions into the receptor cells through Na^+ channels in the membrane, leading to depolarization of the receptor cell. These channels can be blocked by amiloride: hence, the channels are termed *amiloride-blockable Na^+ channels.*

Sour is detected as a result of a direct action of acids on K^+ channels in the membrane of the receptor cells. Specifically, acids

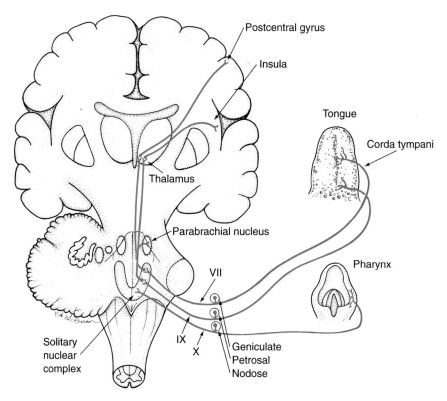

Figure 26.2. The organization of gustatory circuitry. Schematic illustration of the overall organization of the gustatory pathways originating from the right side of the tongue and pharynx. The primary afferent fibers that innervate receptors in the tongue and oropharyngeal mucosa arrive via three nerves: (1) the *chorda tympani*, which is a branch of the facial nerve; (2) the lingual branch of the glossopharyngial nerve (cranial nerve IX); and (3) the superior laryngeal branch of the vagus nerve (cranial nerve X). The ganglion cells that give rise to the primary afferent fibers are located in the geniculate ganglion (cranial nerve VII), the petrosal ganglion (cranial nerve IX), and the nodose ganglion (cranial nerve X). Centrally projecting axons terminate in the solitary nuclear complex in the medulla. The central projections of the gustatory system are ipsilateral. Modified from Kandel ER, Schwartz JH, Jessell TM. *Principles of Neural Science.* 3rd edition. New York, NY: Elsevier; 1991.

cause K^+ channels to close, causing depolarization of the receptor cell.

Sweet is detected through the activation of specific *metabotropic receptors* in the apical membrane, which activate adenylate cyclase. The activation of adenylate cyclase causes an increase in cyclic AMP, which in turn activates a kinase that phosphorylates voltage-dependent K^+ channels on the basolateral membrane. Phosphorylation then causes the channels to close, reducing K^+ currents, which again leads to receptor depolarization.

The transduction mechanisms for bitter have not yet been firmly established, but seem to involve a release of Ca^{2+} from intracellular stores. This release may be triggered by inositol trisphosphate or cyclic AMP.

All of these different transduction mechanisms operate through one final common path: depolarization of the receptor cell. The depolarization increases the rate of neurotransmitter release onto the afferent ending. The neurotransmitter released by the receptors is excitatory, so the increased release depolarizes the afferent ending. Action potentials are then triggered in the myelinated portion of the afferent nerve fiber, proportional to the level of depolarization of the afferent ending.

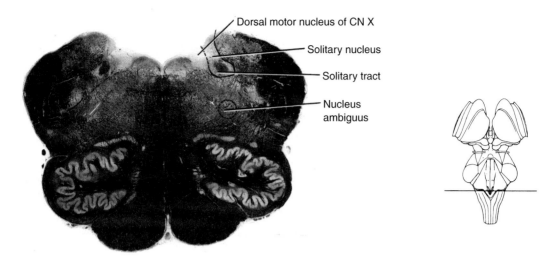

Dorsal motor nucleus of CN X

Solitary nucleus

Solitary tract

Nucleus
ambiguus

Figure 26.3. Location of the solitary nuclear complex in the medulla. The location of the gustatory component of the solitary nucleus in a myel-stained transverse section through the medulla. The inset shows the plane of the section. Reproduced with permission from Martin JH. *Neuroanatomy Text and Atlas.* New York, NY: Elsevier; 1989.

Integration and Decoding of Taste Information

It is not yet clear whether the different chemotransduction mechanisms are all present in a single taste receptor, or whether different receptors are specialized to respond to the different signals. The fact that different parts of the tongue are differentially responsive to the principal tastes suggests that receptors are specialized to at least some degree.

The representation of the four basic categories of gustatory stimuli is segregated at each synaptic station along the gustatory pathway. That is, the spatial segregation that is present in the tongue is "mapped" centrally by neurons that remain segregated from one another at the level of the gustatory nucleus, the thalamus, and the cortex. At the same time, however, each separate channel of information is broadly tuned. For example, individual fibers of the chorda tympani often respond to a broad range of stimuli.

It is not yet clear how the brain "reads" the information that is conveyed over the afferent fibers. It is clear, however, that neural coding in the gustatory system takes advantage of the same mechanisms of parallel and hierarchical processing that occur in other sensory systems.

Parallel processing in the gustatory system is reflected in the first place by the fact that different parts of the tongue are especially sensitive to particular substances. Then, information from the different parts of the tongue is conveyed to the brain by different "labeled lines" that travel through the different cranial nerves.

Hierarchical processing is reflected by the fact that the brain is able to decode complex stimulus patterns. For example, we perceive tastes that are made up of a composite of the different primary taste sensations. That is, we do not perceive tastes as "sweet, sour, and acid"; instead, we perceive the taste of grapefruit juice. The brain accomplishes using what is termed *across-fiber pattern coding,* which integrates information from different receptor types.

As we noted earlier, complaints regarding the loss of taste sensation are usually due to an inability to detect the aroma of foods (that is, due to problems with olfaction). To understand these, we move on to a consideration of the olfactory system.

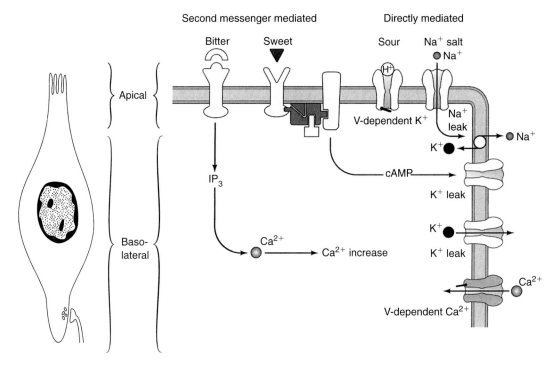

Figure 26.4. Different chemotransduction mechanisms exist for detecting the different chemical substances that underlie taste. Salt is detected as a result of the direct passage of Na^+ ions into the receptor cells through amiloride-blockable Na^+ channels in the membrane, leading to depolarization of the receptor cell. Sour is detected as a result of a direct action of acids on K^+ channels; acids cause K^+ channels to close, resulting in depolarization of the receptor cell. Sweet is detected through the activation of metabotropic receptors that activate adenylate cyclase, causing an increase in cyclic AMP. Cyclic AMP activates a kinase that phosphorylates voltage-dependent K^+ channels, causing the channels to close. The reduced flux of K^+ causes receptor depolarization. Bitter may activate receptors that trigger a release of Ca^{2+} from intracellular stores. From Kandel ER, Schwartz JH, Jessell TM. *Principles of Neural Science*. 3rd edition. New York, NY: Elsevier; 1991. Modified from Kinnamon JC. Taste transduction: a diversity of mechanisms. *Trends Neurosci*. 1988;11: 491–496.

Table 26.2. Transduction mechanisms in taste receptors

Taste sensation	Transduction mechanism
Salt	Direct passage of Na^+ ions through amiloride-blockable Na^+ channels
Sour	Acid modulation of K^+ channels
Sweet	Specific metabotropic receptors that activate adenylate cyclase
Bitter	Uncertain; probably involves the release of Ca^{2+} from intracellular stores, may be mediated by IP3 or cyclic AMP

THE OLFACTORY SYSTEM

The olfactory sense is mediated by chemoreceptors located in the olfactory epithelium in the nose. The receptor neurons then project to first-order neurons in the olfactory bulb. Neurons in the olfactory bulb then project to the CNS via the olfactory tract.

Olfactory chemoreceptors are specialized neurons that detect airborne chemical substances. In humans, the receptors are localized in the roof of the nasal cavity. They are true neurons in that they give rise to a typical axon that projects to, and forms synapses with, neurons in the olfactory bulb.

The olfactory neuroepithelium is made up of receptor neurons, supporting cells, and

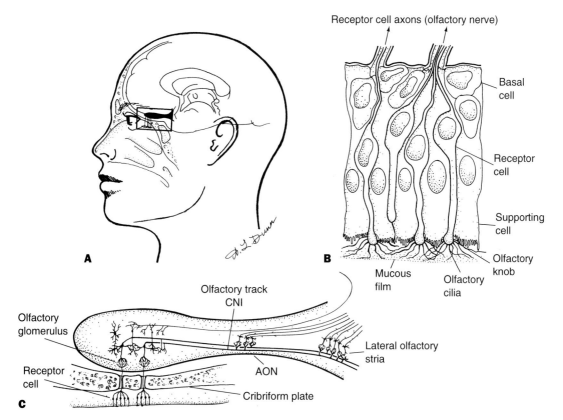

Figure 26.5. The cellular organization of the olfactory neuroepithelium in vertebrates. The olfactory neuroepithelium is located in the dorsal nasal cavity (**A**). The neuroepithelium is made up of receptor neurons, supporting cells, and basal cells (**B**). The receptor cells have a short peripheral process that extends toward the nasal cavity, and a longer central process that is an unmyelinated axon, which projects across the cribriform plate to the olfactory bulb. **C,** the projection from the receptor neuron to the bulb is termed the *olfactory nerve*. The peripheral process of the olfactory receptor cell ends in an olfactory knob, which is studded with cilia. The cilia extend into the mucous film of the nasal cavity. The odorant receptors are localized in the cilia. **B** is modified from Andres KH. Der Feinbau de Regio olfactoria von Makrosmatikern. *Z Zellforsch.* 1966;69:140–154.

basal cells (Figure 26.5). The receptor cells have a short peripheral process that extends toward the nasal cavity, and a longer central process that is an unmyelinated axon, which projects across the *cribriform plate* to the olfactory bulb. The projection from the receptor neuron to the bulb is termed the *olfactory nerve.*

The peripheral process of the olfactory receptor cells ends in an *olfactory knob,* which is studded with cilia. The cilia extend into the mucous film of the nasal cavity. The odorant receptors are localized in the cilia.

Olfactory receptors, like taste receptors, turn over continuously; they have a lifetime of about 60 days. The basal cells are thought to be the progenitors of the new receptor cells. This continuous turnover is remarkable in that the newly generated receptor cells have to regenerate their axon across the cribriform plate into the olfactory bulb. Hence, olfactory receptor neurons are extremely unusual in that they can be replaced by cellular proliferation and are capable of axonal regeneration in adult animals.

The centrally directed axons of the olfactory receptor neurons terminate in *synaptic glomeruli* in the olfactory bulb (Figure 26.5). The synaptic interactions in the glomeruli are complicated, but reasonably well under-

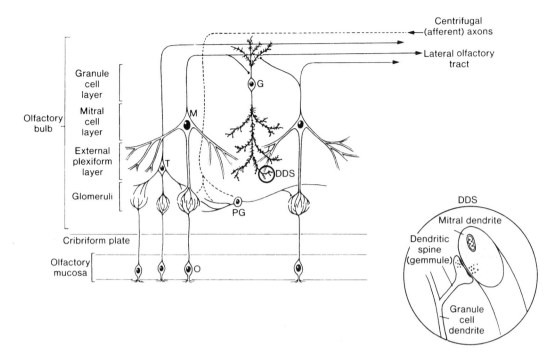

Figure 26.6. Synaptic organization in the olfactory bulb. Primary axons from the receptor neurons terminate in the glomeruli, where they synapse on the dendrites of mitral cells (M), tufted cells (T), and periglomerular cells (PG). Tufted cells and periglomerular cells form dendrodendritic synapses with the mitral cells (DDS). These mediate bidirectional chemical neurotransmission (inset). The deeper layers of the bulb contain the dendrites and cell bodies of the mitral cells and granule cells, which also form dendrodendritic synapses with mitral cells. The synaptic connections formed by periglomerular cells and granule cells are inhibitory; other synapses are excitatory. Mitral and tufted cells give rise to axons that project to the olfactory cortex via the olfactory tract (cranial nerve I). Reproduced with permission from Carpenter MB. *Core Text of Neuroanatomy.* Baltimore, Md: Williams & Wilkins; 1991. Based on diagrams in Shepherd GM. Synaptic organization of the mammalian olfactory bulb. *Physiol Rev.* 1972;52:864–917.

stood (Figure 26.6). The primary axons terminate in the glomeruli, where they synapse on the dendrites of large mitral cells and small tufted cells, both of which give rise to projections from the olfactory bulb to the brain. The primary axons also synapse on the dendrites of interneurons, termed *periglomerular cells,* which then form dendrodendritic synapses with the mitral cells. These dendrodendritic synapses mediate bidirectional chemical neurotransmission, and the olfactory bulb is one of the few places where this occurs in the CNS.

The deeper layers of the bulb contain the dendrites and cell bodies of the mitral cells and another cell type termed *granule cells.* The mitral and granule cells also form dendrodendritic synapses with one another. The synaptic connections formed by periglomerular cells and granule cells are inhibitory; other synapses are excitatory.

The mitral and tufted cells give rise to axons that project to five areas in the olfactory cortex via the *olfactory tract* (cranial nerve I). The basic circuitry is summarized in Figure 26.7. The axons of mitral and tufted cells project to (1) the anterior olfactory nucleus, (2) the olfactory tubercle, (3) the pyriform cortex, (4) the cortical nucleus of the amygdala, and (5) the entorhinal cortex.

The anterior olfactory nucleus is located at the base of the olfactory bulb. It gives rise to projections that cross the midline in the anterior commissure, enter the olfactory tract on the contralateral side, and then terminate in the contralateral olfactory bulb.

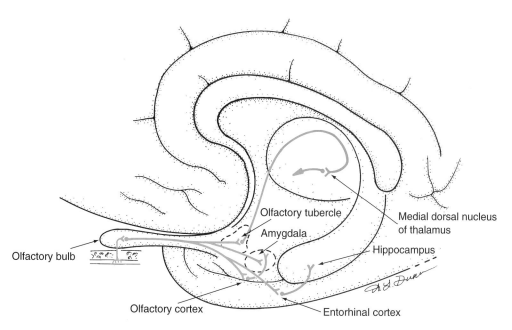

Olfactory tubercle

Amygdala

Medial dorsal nucleus of thalamus

Hippocampus

Olfactory bulb

Olfactory cortex

Entorhinal cortex

Figure 26.7. Central pathways conveying olfactory information. The main projections from the olfactory bulb terminate in (1) the anterior olfactory nucleus (not shown), which projects to the contralateral olfactory bulb (see Figure 26.5); (2) the olfactory tubercle; (3) the olfactory cortex; (4) the cortical nucleus of the amygdaloid complex; and (5) the entorhinal cortex. The olfactory tubercle then projects to the medial dorsal nucleus of the thalamus. The connections of the amygdaloid complex and entorhinal cortex are described in Chapter 29. As with the gustatory system, the central projections are largely ipsilateral. Modified from Noback CR and Demarest RJ. The Human Nervous System. New York, NY: McGraw-Hill; 1967.

The olfactory tubercle is located dorsal to the olfactory tract. It is the area that is termed the *anterior perforated space* because of the numerous small blood vessels that penetrate the brain in this region (see Chapter 3). Neurons in the olfactory tubercle give rise to projections to the medial dorsal nucleus of the thalamus that travel via the stria medullaris. The mediodorsal nucleus of the thalamus in turn projects to the orbitofrontal cortex.

The pyriform cortex is located just lateral to the olfactory tubercle. This is a phylogenetically old part of the cortex termed the *paleocortex*: it has three layers rather than the six layers that are characteristic of the neocortex. The pyriform cortex is considered to be the primary olfactory discrimination area. Neurons in the pyriform cortex project to the mediodorsal nucleus of the thalamus, the amygdala, and the entorhinal cortex.

The cortical nucleus of the amygdala is located in the amygdaloid complex in the ros-tral pole of the temporal lobe. The neurons in the cortical nucleus of the amygdala give rise to projections to the hypothalamus and midbrain tegmentum. These pathways are described in more detail in Chapter 29. The projections to the hypothalamus are thought to be the substrate for the affective component of olfactory stimuli.

The entorhinal cortex lies in the ventromedial part of the temporal lobe. Neurons in the entorhinal cortex receive input from the olfactory cortex and from other sensory association areas. The entorhinal cortical neurons then give rise to a major projection to the hippocampus, which is described in more detail in Chapter 29.

The central projections conveyed by the olfactory tract, along with most of the subsequent circuitry, are ipsilateral. However, there are extensive interhemispheric connections between the pyriform cortices on the two sides, between the amygdala, and between the two entorhinal cortices.

Sensory Transduction and Neural Processing in the Olfactory System

Oderants (the airborne chemicals that activate olfactory receptor neurons) are absorbed in the mucous layer overlying the receptor cilia. The odorants then either diffuse to the cilia or are carried there by an oderant binding protein called *olfactory binding protein.* It is thought that the olfactory binding protein binds lipophilic oderants and promotes their diffusion through the aqueous medium of the mucous layer. The oderant molecules then interact with specific oderant receptors that are present on the cilia of the receptor neurons.

Oderant receptors appear to be of the metabotropic type, and at least some operate by activating adenylate cyclase, leading to an increase in cyclic AMP concentrations in the receptor neurons. There is evidence, however, that the receptors for certain oderants may operate through other signal transduction pathways, especially inositol trisphosphate. Physiological studies reveal that the receptor potential generated by oderant binding is a depolarizing response due to the opening of Na^+ channels.

A large number of putative oderant receptor genes have been cloned; these encode proteins with seven transmembrane-spanning regions (the characteristic configuration for metabotropic receptors that are coupled to G proteins). The different receptor types are expressed by different populations of receptor neurons that are distributed across the olfactory neuroepithelium.

Individual receptor neurons respond to a number of different oderant molecules; it is not known, however, whether individual neurons express a number of different oderant receptors, or whether individual oderant receptors recognize a number of different oderant molecules.

Experimental studies reveal some degree of *aromotopic mapping* within the olfactory epithelium. For example, the anterior portions of the mucosa are more sensitive to certain molecules, and the posterior portions of the mucosa are more sensitive to others. However, these are relative sensitivities.

When the concentration of the oderant is high enough, responses to particular oderants are widespread.

The olfactory receptor neurons behave much like neurons whose dendrites are invested with excitatory synapses. Activation by oderants leads to depolarization of the peripheral process (functionally, the dendrite); this leads to generation of action potentials over the centrally projecting axon. The axons of olfactory receptor neurons then form excitatory synaptic connections on neurons in the olfactory bulb.

Integration and Decoding of Olfactory Information

As with the gustatory system, it is not yet clear how the brain "reads" the information that is generated within the olfactory bulb and olfactory cortical regions. But parallel and hierarchical processing surely play a role.

Parallel processing is reflected by the fact that olfactory neurons that express particular oderant receptors project to discrete regions in the olfactory bulb, perhaps to individual glomerluli. Hence there is an *aromotopic pattern* of innervation of neurons in the olfactory bulb. In this way, particular oderants activate the class of receptor neuron that expresses the appropriate oderant receptor, and this activation then causes the activation of particular collections of neurons in the bulb.

It is likely that a significant amount of neural processing goes on in the local circuitry of the olfactory bulb. Hence, the olfactory bulb functions in a way that is similar to the retina, by preprocessing information before it is conveyed to the brain. Neurons in the bulb, like ganglion cells in the retina, then project onto the various olfactory cortical areas, which eventually convey the information to the neocortex through polysynaptic pathways.

Hierarchical processing is again reflected by *across-fiber pattern coding,* which integrates information conveyed from different receptor types. In some ways, this type of process is especially clear in the case of the olfactory system. We do not perceive odors as "a mixture of organics that include ethanol, ketones . . ."; we perceive a singularity—the "bouquet" of a cabernet. And when we drink that cabernet, the sensation is a combination of gustatory and olfactory stimuli that combine to produce the sensations described so eloquently by wine tasters. There must be some site in the nervous system, presumably a population of neurons, that responds selectively to this particular constellation of stimuli.

The Affective Component of Olfactory Stimuli

Particular aromas evoke strong emotions, sometimes pleasurable, sometimes not. Some of these are species-specific responses that regulate social behaviors including reproductive behaviors. The molecules that operate in this way are termed *pheromones.*

The social behaviors of humans are not as sensitive to such olfactory influences as are the behaviors of other animals, but humans are certainly not insensitive. The fragrance industry is based on this fact. The particular salience of olfactory stimuli for triggering emotions is thought to be due to the interconnections between the olfactory cortex, the amygdala, and the hypothalamus.

Particular aromas also trigger specific memories. These are highly individual, very salient, and often unpredictable. All of us have our favorite examples. For the author, the smell of fish rotting in the sun (which most would consider unpleasant) triggers very pleasant childhood memories of going fishing with my father (as an aside, fish rotting in the sun smells very different from spoiled fish in your refrigerator). It is thought that these associations between particular aromas and specific memories reflect a close interaction between olfactory areas and the areas in the limbic system that are involved in laying down long-term memory traces.

Taste Preferences

To reiterate the point with which the consideration of the chemical senses began, olfaction plays a key role in the sense of taste. In fact, what we perceive as the distinct tastes of particular food substances is due to the aromas of the foods.

We perceive particular tastes as pleasurable and, hence, seek out the foods that produce those sensations. Moreover, taste preferences differ across individuals, and at different times in an individual's life. The most dramatic reflections of taste preferences are the "cravings" for particular foods.

Some taste preferences are in response to deficiencies in particular substances. These preferences are termed *specific hungers.* For example, individuals with salt deficiencies will seek out salt. This specific hunger, it turns out, is actually mediated by the salt receptors of the gustatory system. However, there are also preferences that cannot be directly attributed to specific dietary deficiencies.

It is obviously of adaptive value to avoid foods that produce illness. In this regard, there is a special form of learning that occurs when a particular food is "paired" with nausea. That is, if one eats a particular food and then becomes ill, even the thought of that food becomes unpleasant, and that food will be avoided. This phenomenon is called *taste aversion learning.*

There are several interesting and important features about taste aversion learning. Unlike most other forms of learning, it is fully established in one trial. If one becomes ill after eating snails, one develops an aversion to snails that can last a lifetime. At the same time, however, the pairing of a particular food and nausea does not inevitably lead to an aversion to that food. Why the learning is so strong in some cases and nonexistent in others is not entirely clear, but may reflect previous experience with the food substance. That is, if you've eaten snails in the past and enjoyed them, then a single bad experience doesn't necessarily cause aversion.

Another important feature about taste aversion learning is that aversions develop whether or not the food actually caused the illness. Many have had the experience of becoming nauseated as a result of a virus after eating a particular food, and then developing an aversion to that food. Intellectually, we know that the food did not cause the illness; however, this does not prevent the development of the aversion.

The other interesting feature of taste aversion learning is the very long time that separates the cue (the food substance) with the aversive experience (nausea). Taste aversion learning occurs even when the cue and the aversive experience are separated by many hours. How a specific association is made in this case is not clear. Even less clear is why the association is made to a particular food. For example, one might have eaten peas along with the snails, and nevertheless formed a selective aversion to the snails.

Disorders Affecting the Gustatory and Olfactory Systems

Olfactory function can be disrupted in a number of ways. The most common cause is a temporary disruption of olfactory receptor function; examples include the common cold and other upper respiratory infections.

Fractures involving the cribriform plate can injure olfactory nerve fibers, causing a loss of olfactory function. Because the olfactory receptors turn over continuously throughout life, injured olfactory axons can sometimes regenerate. If the injuries lead to the forma-tion of extensive scar tissue, however, regeneration may be blocked, leading to a permanent loss of olfactory function.

Tumors, for example meningiomas located in the olfactory groove, can compress the olfactory tract, disrupting the transmission of olfactory information to the brain. The probability of return of function after surgical removal of the tumor depends on the location of the tumor with respect to the olfactory bulb and tract.

Lesions in the pyriform cortex or temporal lobe (especially the area around the uncus and the amygdala) can lead to olfactory hallucinations. Lesions or other disorders affecting the temporal lobe also can lead to temporal lobe epilepsy. Some patients experience olfactory hallucinations at the onset of a seizure involving the temporal lobe. These are termed *uncinate fits*.

One of the most important consequences of a loss of olfactory function is a loss of appetite. We have all experienced the loss of appetite with colds. When olfaction is permanently disrupted, the resulting loss of appetite can be life-threatening.

Olfactory sensitivity decreases with age. It is thought that the deterioration of olfactory function accounts for the loss of appetite that often occurs in aged individuals. The age-related loss in olfactory function cannot at this time be treated. However, recognizing the cause of eating disorders in aged individuals can lead to appropriate intervention in the form of advice and counsel regarding diet.

9

The Autonomic Nervous System

The autonomic nervous system is made up of the efferent pathways through which the nervous system controls the viscera. For this reason, it is also termed the *visceromotor system*. The autonomic nervous system has three principal subdivisions: (1) the sympathetic (thoracolumbar), (2) the parasympathetic (craniosacral), and (3) the enteric. The sympathetic and parasympathetic divisions have opposing functions and exert reciprocal control on the viscera. The enteric division is a local neural network within the gut that controls local reflex activities such as peristalsis.

The components of the autonomic nervous system include (1) *postganglionic effector neurons* whose axons innervate viscera. The cell bodies of these neuron are located in autonomic ganglia; (2) peripheral nerves that contain the *postganglionic axons* that project from autonomic ganglia to the viscera, and *preganglionic axons* that project from the brain and spinal cord to the autonomic ganglia;

(3) preganglionic neurons located in the intermediolateral column of the spinal cord and in the parasympathetic cranial nerve nuclei; and (4) central pathways that control the output of the visceromotor neurons.

Also included in the autonomic nervous system is the adrenal medulla, which is innervated by preganglionic axons. The cells in the adrenal medulla have the same embryonic origin as the neurons in the autonomic ganglia (they migrate from the neural crest). Hence, the adrenal medulla can be thought of as one of the autonomic ganglia that differentiates in a somewhat different way.

The key concepts for physicians include the following:

1. The autonomic nervous system is the motor system for the viscera (that is, cardiac muscle, smooth muscle, and glands). The neural input from visceromotor neurons regulates the ongoing activity of the viscera.

2. The autonomic nervous system operates, for the most part, without conscious control (that is, it is involuntary). However, certain of its activities can be controlled voluntarily (micturition and defecation).

3. The sympathetic division of the autonomic nervous system prepares the body for *fight or flight* in emergency situations; the parasympathetic nervous system regulates and maintains visceral activities for *rest and digest.*

4. Preganglionic neurons are cholinergic. Postganglionic neurons of the sympathetic nervous system are noradrenergic except for the fibers that innervate the sweat glands, which are cholinergic; postganglionic neurons of the parasympathetic nervous system are cholinergic.

27

The Autonomic Nervous System

The autonomic nervous system is the motor system for the viscera (that is, cardiac muscle, smooth muscle, and glands). However, from the outset, it is important to recognize that the neural input does not directly drive the activity of the viscera; instead, the viscera have their own intrinsic activity, and this activity is regulated in both a positive and negative fashion by the two divisions of the autonomic nervous system. In this way, the two components of the autonomic system coordinate visceral activities so as to maintain homeostasis or to enable the body to respond to emergency situations.

Although the autonomic nervous system is by definition motor, its activities are regulated by sensory–motor integration. Examples range from simple reflexes mediated by local circuitry in the gut, to complicated reflex activities that involve coordination between multiple pathways. The sensory limb of these reflexes is represented by *visceral sensory fibers* that travel from the gut along the same peripheral nerves that carry the visceromotor efferents. Hence, our consideration of autonomic nervous system function will focus on both sensory and motor components.

In what follows, we begin by describing the components and overall organization of the autonomic nervous system. Then we outline some of the functions that it mediates, concentrating on examples that are of special importance for physicians.

COMPONENTS OF THE AUTONOMIC NERVOUS SYSTEM

The autonomic nervous system has three subdivisions: (1) the sympathetic, (2) the parasympathetic, and (3) the enteric. The sympathetic and parasympathetic subdivisions are the *final common pathways* through which the central nervous system controls the viscera. The enteric subdivision is a network of neurons within the wall of the gut (a *local circuit*), which mediates local reflex functions, but which is regulated by input from preganglionic parasympathetic inputs.

The sympathetic and parasympathetic divisions exert opposing actions on their targets. The sympathetic division is activated in emergency situations and prepares the body for fight or flight; the parasympathetic nervous system regulates and maintains visceral activities for rest and digest. Both of these output pathways operate without conscious effort, that is, they are involuntary.

The organization of the peripheral autonomic nervous system is the subject of gross anatomy. Hence, our consideration focuses on the preganglionic neurons in the brain and spinal cord, and the central circuitry that controls the output of these neurons. We begin by considering the general features of the autonomic nervous system and then consider the specific features and functions of the different subdivisions.

The Final Common Pathways of the Autonomic Nervous System

The final common pathways of the autonomic nervous system are made up of *preganglionic effector neurons* whose cell bodies lie within the CNS, and *postganglionic neurons* in autonomic ganglia that directly innervate the viscera (Figures 27.1 and 27.2); hence, the final common pathway is disynaptic. This is one of the ways that the visceromotor system differs from the somatomotor system.

The cell bodies of preganglionic neurons are located within the CNS. They are similar to the lower motoneurons of the somatic motor system in that their cell bodies are located in the spinal cord and cranial nerve nuclei; they send their axons into the ventral roots and respective cranial nerves; they use acetylcholine as their neurotransmitter; and they are controlled by descending pathways.

Preganglionic visceromotoneurons differ from somatic motoneurons in two ways: (1) the cell bodies of preganglionic neurons are located in *visceromotor columns* in the spinal cord and brainstem, and (2) the axons terminate on postganglionic neurons rather than directly on target organs.

The cell bodies of postganglionic neurons are located outside of the CNS in various ganglia. In the case of the sympathetic division, the ganglia lie some distance from the end organ. In contrast, the ganglia of the parasympathetic nervous system lie very near and sometimes even embedded within the respective viscera.

The Sympathetic Division

Preganglionic neurons of the sympathetic division of the autonomic nervous system are located in the intermediolateral column in the thoracic and lumbar segments of the spinal cord (Figure 27.3). The axons from these cells project out of the spinal cord in the ventral roots, along with the axons of the somatic motoneurons.

The preganglionic axons leave the ventral roots and enter the *paravertebral chain* in the *white communicating rami.* Some terminate within the paravertebral ganglia, whereas others continue through the ganglia to the *prevertebral ganglia* (for example, the celiac ganglion and the superior and inferior mesenteric ganglia, see Figure 27.2). One component of preganglionic axons also projects directly to the adrenal medulla. The axons of postganglionic neurons leave the paravertebral ganglia via the *gray communicating rami,* enter peripheral nerves, and then project to their target organs.

The direct preganglionic projection to the adrenal medulla provides the sympathetic division with a neuroendocrine component along with the neural component. The neural component directly contacts the individual viscera, whereas the endocrine component acts globally by releasing hormones into the circulation and, hence, amplifies the direct action of the neural component.

The Parasympathetic Division

Preganglionic neurons of the parasympathetic division are located in *nucleus X* in the sacral spinal cord and in various cranial nerve nuclei (Figure 27.3 and Table 27.1). The preganglionic axons leave the CNS via the respective cranial nerves (Figure 27.3 and Table 27.1) or via the ventral roots (in the case of the output from the sacral spinal cord) and project to the postganglionic neurons located in ganglia near, or embedded within, the respective viscera.

In contrast to the global responses elicited by the sympathetic division, the parasympathetic division exerts a much more precise control over the individual viscera. In keeping with this role, the innervation of the viscera by the parasympathetic division is more specific, so that each organ can be independently modulated.

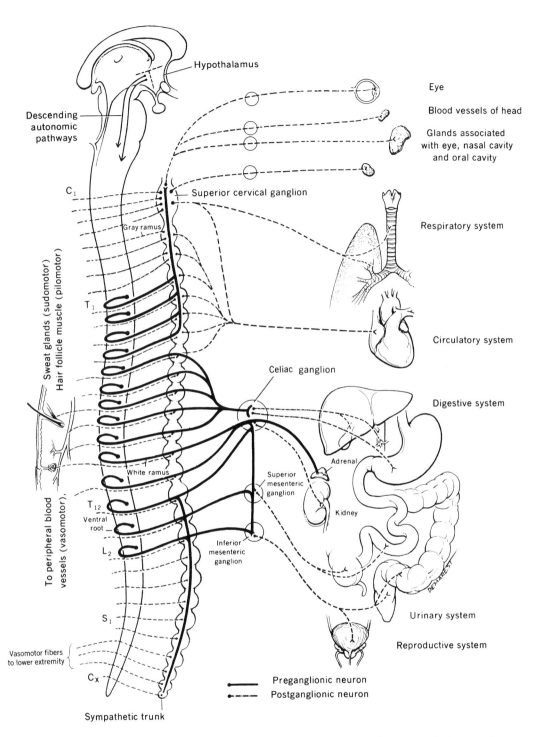

Hypothalamus

Descending autonomic pathways

Eye

Blood vessels of head

Glands associated with eye, nasal cavity and oral cavity

C_1

Superior cervical ganglion

Gray ramus

Respiratory system

T_1

Sweat glands (sudomotor)
Hair follicle muscle (pilomotor)

Circulatory system

Celiac ganglion

Digestive system

Adrenal

Superior mesenteric ganglion

Kidney

White ramus

T_{12}
Ventral root

L_2

Inferior mesenteric ganglion

To peripheral blood vessels (vasomotor)

S_1

Urinary system

Reproductive system

Vasomotor fibers to lower extremity

C_x

—— Preganglionic neuron
–·– Postganglionic neuron

Sympathetic trunk

Figure 27.1. Overview of autonomic nervous system organization: sympathetic division.
Preganglionic neurons of the sympathetic division of the autonomic nervous system are located in the intermediolateral column, which extends from the first thoracic to lower lumbar segments. Some terminate within the paravertebral ganglia, whereas others continue through the ganglia to the prevertebral ganglia. One component of preganglionic axons also projects directly to the adrenal medulla. Reproduced with permission from Nobak CR, Demarest RJ. *The Human Nervous System*. New York, NY: McGraw-Hill; 1967.

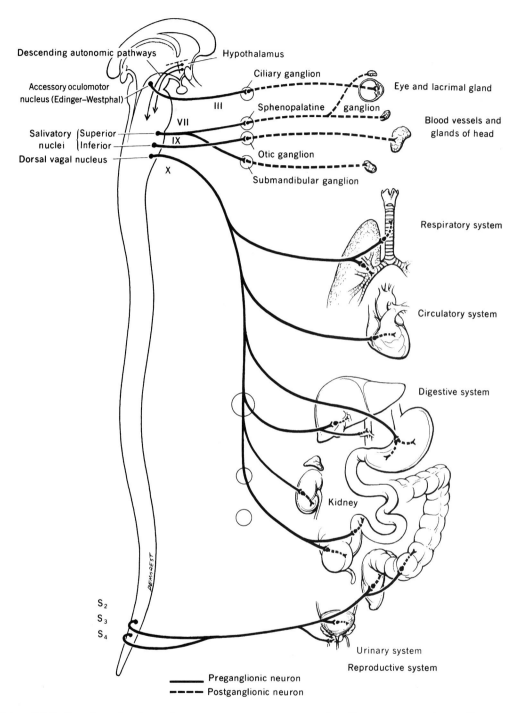

Figure 27.2. Overview of autonomic nervous system organization: parasympathetic division. Preganglionic neurons of the parasympathetic division are located in cranial nerve nuclei in the brainstem and in segments S2 through S4 of the spinal cord. The ganglia of the parasympathetic nervous system lie very near and sometimes even embedded within the respective viscera. Reproduced with permission from Nobak CR, Demarest RJ. *The Human Nervous System.* New York, NY: McGraw-Hill; 1967.

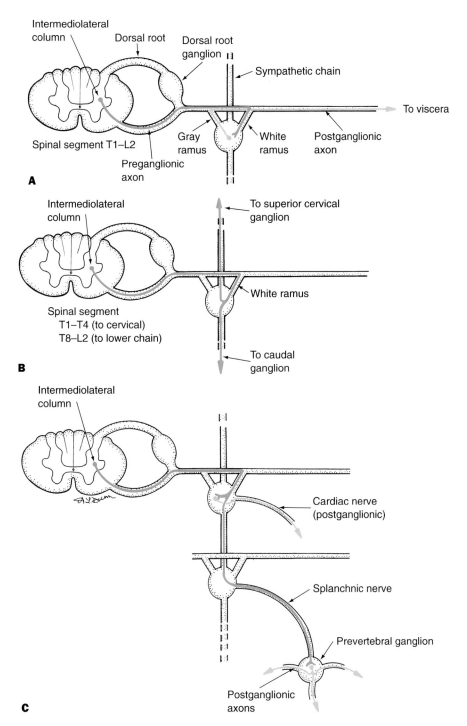

Figure 27.3. Organization of preganglionic sympathetic axons. Preganglionic neurons of the sympathetic division are located in the intermediolateral column of the thoracic and lumbar segments of the spinal cord. The axons from these cells project out of the spinal cord in the ventral roots, and then leave the ventral roots and enter the paravertebral chain in the white communicating rami. Preganglionic axons may (**A**) terminate in the ganglion at the level at which the axon enters the chain, (**B**) ascend or descend in the chain before terminating, or (**C**) traverse the ganglion, enter the splanchnic nerve, and terminate in the prevertebral ganglion. The axons of postganglionic neurons leave the paravertebral ganglia via the gray communicating rami, enter peripheral nerves, and project to their target organs. After Haines D. *Fundamental Neuroscience.* New York, NY: Churchill-Livingstone; 1997.

Table 27.1. Similarities and differences between visceromotor and somatomotor systems

Feature	Visceromotor	Somatomotor
Innervation	Disynaptic	Monosynaptic
Location in spinal cord	Intermediolateral column, nucleus X	Ventral horn
Cranial nerve nuclei	Edinger–Westphal (III n.) Salivatory (VII n.) Dorsal nuc. of vagus Nuc. ambiguus	Oculomotor, trochlear, trigeminal, abducens, facial, glossopharangeal, accessory, hypoglossal
Inhibition	Autonomic motoneurons can exert either excitatory or inhibitory influences on their target	Motoneurons are exclusively excitatory; inhibition is mediated by inputs to motoneurons
Type of control	Involuntary	Voluntary

The Enteric Division

The enteric division of the autonomic nervous system is a *local neural circuit* embedded within the gut (Figure 27.5). The neurons and interconnections are located between the layers of smooth muscle and endothelium that make up the visceral wall. The enteric nervous system has both sensory and motor components that mediate local visceral reflexes. This local circuit regulates the coordinated contraction–relaxation cycles that underlie peristalsis and also regulates gastric secretion and gastrointestinal blood flow.

Although the local neural network of the enteric nervous system can operate autonomously, it receives preganglionic inputs that modulate its activity. This input can override the intrinsic activity of the neural network in times of emergency or stress.

Central Circuitry of the Autonomic Nervous System

In considering central autonomic circuitry, it is useful to keep in mind that the visceromotor system mediates functions ranging from simple visceral reflexes to complex integrative behaviors involving multiple systems. Like other reflexes, visceral reflexes have an afferent limb, interpolated circuitry of varying complexity, and an efferent limb, the

final component of which is the visceromotoneuron. Hence, it is useful to begin our consideration of central autonomic circuitry with a consideration of visceral sensation.

Sensory information from the viscera is conveyed to the CNS in two ways: (1) First-order afferents carrying visceral sensory information enter the spinal cord via dorsal roots and terminate on second-order neurons in the dorsal horn. (2) Visceral sensory afferents also project to the brainstem via the same cranial nerves that carry visceromotor axons.

The Nucleus of the Solitary Tract Is a Key Relay Nucleus for Visceral Sensory Information

The visceral sensory afferents that enter the brain via the cranial nerves terminate in the *nucleus of the solitary tract,* which is located in the dorsal medulla (Figure 27.6 and see Chapter 26). The caudal end of the nucleus lies ventral and medial to the dorsal column nuclei. The nucleus then extends rostrally and laterally along the floor of the fourth ventricle.

The visceral sensory afferent from different organs terminate in different subnuclei of the nucleus of the solitary tract. In addition to this "viscerotopic" projection pattern to individual subnuclei, one subnucleus (the

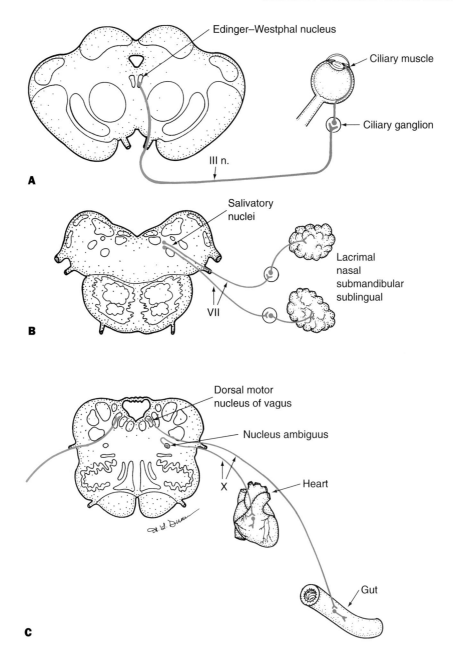

Figure 27.4. Preganglionic neurons of the parasympathetic division. Preganglionic parasympathetic neurons are located in several cranial nerve nuclei. The location of the Edinger–Westphal nucleus in the midbrain (**A**), the salivatory nuclei in the pontine tegmentum (**B**), and dorsal motor nucleus of the vagus and nucleus ambiguus in the medulla (**C**). The axons from these nuclei leave the brainstem via the oculomotor (III), facial (VII), glossopharyngeal (IX), and vagus (X) nerves. Neurons in the Edinger–Westphal nucleus innervate the ciliary ganglion, which in turn innervates the ciliary muscle. Neurons in the superior salivatory nucleus innervate the lacrimal gland, nasal glands, submandibular gland, and sublingual gland. Neurons in the inferior salivatory nucleus innervate the parotid gland via the glossopharyngeal nerve (not shown). Neurons in the dorsal motor nucleus of the vagus supply the gut. Neurons in the nucleus ambiguus supply the heart.

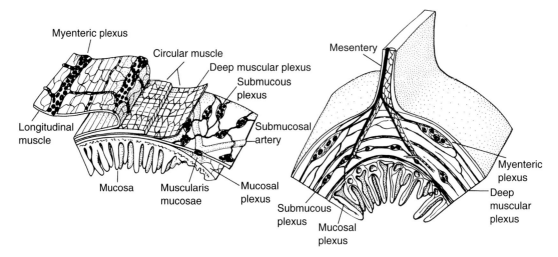

Figure 27.5. The enteric nervous system. Local neural circuits that control contraction–relaxation cycles during peristalsis. The enteric division of the autonomic nervous system is a local neural circuit embedded within the gut. The neurons and interconnections are located between the layers of smooth muscle and endothelium that make up the visceral wall. This local circuit regulates the coordinated contraction–relaxation cycles that underlie peristalsis. From Kandel ER, Schwartz JH, Jessell TM. *Principles of Neural Science.* 3rd ed. New York, NY: Elsevier; 1991. Adapted from Furness JB, Costa M. Types of nerves in the enteric nervous system. *Neuroscience.* 1980;5:1–20.

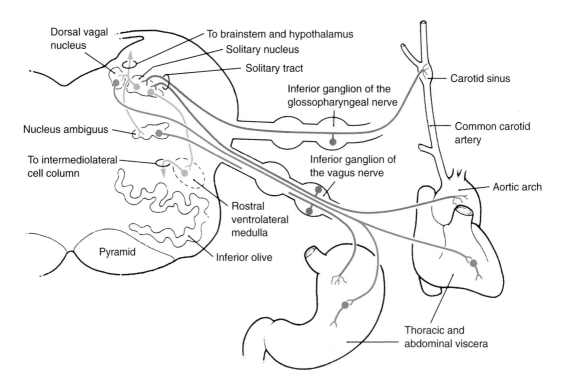

Figure 27.6. Visceral sensory inputs to the nucleus of the solitary tract. Visceral sensory afferents that enter the brain via the cranial nerves terminate in the nucleus of the solitary tract. From Haines D. *Fundamental Neuroscience.* New York, NY: Churchill-Livingstone; 1997.

commissural nucleus, the rostral continuation of which is the medial nucleus) receives overlapping projections from all of the major viscera except those of the pelvic region. Hence, the collection of subnuclei is organized so as to convey both organ-specific and generalized visceral sensory information.

In addition to receiving direct input from visceral sensory afferents that arrive via the cranial nerves, the nucleus of the solitary tract receives ascending viscerosensory information from the spinal cord. The nucleus also receives an important projection from the *area postrema,* which is a circumventricular organ that serves as a sensor for the chemical environment of the plasma and cerebrospinal fluid. In this way, the nucleus of the solitary tract serves as a point of convergence of information pertaining to visceral function.

As noted in Chapter 26, the nucleus of the solitary tract is also the major relay nucleus for afferents conveying taste information from the tongue. Conscious perception of taste is mediated by projections to the ventral posterolateral nucleus of the thalamus.

A different set of projections from the nucleus of the solitary tract is important for regulating autonomic function. Of particular importance for autonomic reflexes are the projections to cell groups in the pons and mesencephalon. Although these brainstem circuits are modulated by descending influences, they can also operate autonomously. For example, transection of the brainstem above the level of the pons in experimental animals leaves cardiovascular and respiratory functions largely intact.

The visceral sensory information that is important for regulating higher-level autonomic functions involving integrated activity in multiple systems is conveyed to the hypothalamus and various forebrain areas, especially the amygdala. Exactly how information is processed and integrated at higher levels of the autonomic nervous system is not fully understood. It is clear that various regions can influence autonomic output, including the amygdala, the limbic system, the cerebral cortex, and the basal ganglia. More-

over, all of these areas seem to mediate their effects on autonomic activity via the hypothalamus.

The Hypothalamus Plays a Key Role in Coordinating Autonomic Functions Involving Integrated Activity in Multiple Systems

The hypothalamus is sometimes called the *head ganglion* of the autonomic nervous system because of the key role that it plays in coordinating autonomic function. Neurons in the hypothalamus receive viscerosensory input from the nucleus of the solitary tract as well as input from a variety of forebrain structures. Nuclei in the hypothalamus then project back to nuclei in the brainstem that directly mediate autonomic reflex functions and also to brainstem regions that give rise to descending projections to preganglionic sympathetic and parasympathetic neurons in the spinal cord. The hypothalamus also controls the release of various hormones that compliment the action of the autonomic nervous system (see Chapter 28).

The descending projections from the hypothalamus to preganglionic neurons in the medulla and spinal cord travel primarily in the dorsal longitudinal fasciculus. The dorsal longitudinal fasciculus passes through the periaqueductal gray, forming connections with neurons in the periaqueductal gray along the way. The descending pathway then continues into the medulla. The periaqueductal gray itself serves as an important integrating center for various autonomic relays. Lesions involving the medulla can disrupt these descending pathways, leading to a loss of central regulation of autonomic function.

We consider the structure and functions of the hypothalamus in more detail in Chapter 28. In terms of its role in autonomic function, rather little is known about exactly how information is processed and integrated within the hypothalamus so as to coordinate autonomic output. Apparently, different nuclear groups within the hypothalamus control integrated autonomic responses. For example nuclei in the anterior hypothalamus are sensitive to body temperature. Thermal stimulation of the anterior hypothalamus produces an integrated autonomic response to reduce body temperature (sweating, cutaneous vasodilation, and panting). Lesions of the anterior hypothalamus in experimental animals disrupt these responses, causing animals to become hyperthermic.

Areas in the posterior hypothalamus seem to play a key role in coordinating the responses to cold. Reducing temperature in the posterior hypothalamus triggers a complex of responses that conserves heat (shivering, inhibition of sweating). Lesions of the posterior hypothalamus in experimental animals disrupt these heat-conserving responses.

In terms of sympathetic activation during fight-or-flight situations, the recognition of an emergency situation depends on sensory information and on the interpretation of that information. Seeing an automobile moving toward you on the opposite side of the road triggers no more than passing notice. Seeing the very same image coming toward you on what appears to be a collision course will likely trigger massive sympathetic activation and hopefully appropriate evasive responses. The visual stimulus in each case is virtually identical; what differs is the interpretation of the visual image and the resulting affective response (fear). Hence, the activation of the sympathetic nervous system in this case involves (1) the visual system proper; (2) visual association areas that interpret the significance of the image; (3) other cortical areas (especially the amygdala), which play a key role in the affective response (fear); and finally (4) the hypothalamus, which receives some combination of input from these higher structures.

It is important to note that massive sympathetic responses can be elicited by situations in which there is no direct physical threat. In modern society, we must often deal with situations that produce stress and trigger sympathetic activation; in many of these situations, either fight or flight may be completely inappropriate (for example, the final exam in your neuroscience course). Nevertheless, these situations can trigger full-blown sympathetic responses with all the at-

tendant signs. This is an example of a situation where sympathetic activation is triggered as a result of an interpretation of cues by higher cortical structures.

AUTONOMIC REFLEXES

As was the case with the somatic motor system, the visceromotor system mediates functions ranging from relatively simple reflexes to complicated and highly integrated functions involving multiple systems (see Tables 27.2 and 27.3). We consider several examples of autonomic reflexes that are of particular importance to physicians, beginning with a relatively simple reflex (regulation of the pupil) and then moving to more complicated functions (cardiovascular regulation and control of the urinary bladder).

In each example, it is helpful to keep in mind that the sympathetic and parasympathetic divisions generally have opposing (that is, reciprocal) effects. Hence, overall reflex control is mediated by integration of inputs from the two opposing divisions.

Regulation of the Pupil

Pupillary diameter is regulated by parasympathetic innervation that derives from the Edinger–Westphal nucleus (the visceromotor division of the oculomotor nucleus) and sympathetic innervation from the superior cervical ganglion. Parasympathetic fibers terminate on postganglionic neurons in the ciliary ganglion. Ciliary ganglion neurons in turn innervate the pupillary sphincter muscles. Activation of the parasympathetic fibers causes pupillary constriction. The circuitry responsible for the pupillary light reflex (the constriction of the pupil in response to light) is described in Chapter 21.

Sympathetic fibers innervate the pupillary dilator muscle, causing dilation of the pupil. This dilation occurs in situations in which there is global sympathetic activation. The pupil still constricts in response to light, even

Table 27.2. Autonomic efferent pathways

Preganglionic neurons	Efferent nerve	Ganglion	Target
Edinger–Westphal	III	Ciliary ganglion	Pupilloconstrictor muscle, ciliary muscle
Superior salivatory	VII	Pterygopalatine	Cerebral vasculature, lacrimal, nasal, palatine glands
		Submandibular	Submandibular, sublingual gland
Inferior salivatory	IX	Otic	Parotid gland
Dorsal nuc. of vagus	X	Ganglia embedded in viscera, cardiac	Gut
Nuc. ambiguus	X		Heart
Nucleus X of sacral cord	Ventral root	Pelvic ganglionic plexus	Colon, bladder, genitalia

Table 27.3. A summary of autonomic reflexes

Function	Parasympathetic	Sympathetic
Respiration	Bronchial constriction, increased secretion	Bronchial dilation, decreased secretion
Thermoregulation	Piloerection	Increases sweat production, cutaneous vasoconstriction, vasodilation in muscles
Sexual function	Erection	Glandular secretion
Digestion	Increases secretion of saliva and secretions of the gut, increases smooth muscle motility	Decreases secretion, decreases smooth muscle motility

under conditions of sympathetic activation; however, the constriction occurs from a relatively dilated baseline.

Lesions that interrupt the autonomic innervation of the pupil produce distinctive symptoms. Lesions that interrupt the parasympathetic inputs that travel via the third nerve lead to a dilated pupil ipsilateral to the lesion that is unresponsive to light and also lead to paralysis of the muscles innervated by the third nerve (see Chapter 21).

Lesions that interrupt the sympathetic supply cause pupillary constriction *(miosis)*, drooping of the eyelid *(ptosis)*, and retraction of the eye into the socket *(enopthalmos)*. This collection of signs is termed *Horner's syn-*

drome. Miosis is caused by loss of sympathetic innervation of the pupil, so the parasympathetic input is left unapposed. Ptosis is caused by a loss of sympathetic innervation of the tarsal muscle; enopthalmosis is a result of a loss of sympathetic innervation of the orbital muscle.

Lesions involving the brainstem or spinal cord can also disrupt the visceromotor pathways that regulate pupillary size. Lesions involving the midbrain can involve the oculomotor nucleus or tract. Lesions of the medulla can interrupt the descending pathways that project to the preganglionic sympathetic neurons in the spinal cord, causing Horner's syndrome.

Regulation of Cardiovascular Function

Heart rate is controlled by the balance of activity in sympathetic fibers, which increase heart rate, and parasympathetic fibers, which decrease heart rate. The circuitry that is responsible for regulating the activity of the two divisions is summarized in Figure 27.7.

The afferent input for cardiovascular reflexes derives from receptors in the carotid sinus and aorta that sense blood pressure (baroreceptors). These terminate in particular subnuclei of the nucleus of the solitary tract. Neurons in these subnuclei in turn give rise to projections to brainstem nuclei that control parasympathetic output, and to hypothalamic nuclei that give rise to descending projections to preganglionic sympathetic neurons in the spinal cord.

Activation of the sympathetic nervous system increases heart rate and cardiac contractility (that is, the force of contraction). This is mediated via the release of norepinephrine by postganglionic sympathetic fibers, which acts on β-adrenergic receptors. Activation of β-adrenergic receptors modulates several ionic currents within cardiac muscle (especially Ca^{2+} currents). Sympathetic activation also leads to a concomitant release of epinephrine from the adrenal medulla; in this way the noradrenergic activation by postganglionic sympathetic fibers is amplified by circulating epinephrine. In the fight-or-flight situations in which the sympathetic nervous system is activated, the activity over parasympathetic pathways is inhibited.

The catecholamines released as a result of sympathetic discharge also cause an increase in peripheral vascular resistance, increasing blood pressure. The increased pressure is sensed by the baroreceptors, which activate central circuitry that in turn decreases sympathetic vasomotor tone. This is termed the *carotid sinus reflex,* the purpose of which is to keep blood pressure from rising to dangerous levels.

Activation of the parasympathetic inputs to the heart decreases heart rate and cardiac contractility. This is mediated through the release of acetylcholine by the postganglionic fibers, which acts on muscarinic receptors on the cardiocytes of the sinoatrial and atrioventricular nodes. The muscarinic receptors increase a resting K^+ current, causing hyperpolarization, which slows conduction through the atrioventricular node.

Regulation of the Bladder

Bladder function is controlled by both autonomic and somatomotor pathways. Bladder emptying during urination involves coordinate regulation of both autonomic and somatomotor pathways by descending pathways.

Bladder function involves two separate processes: storage of urine and periodic elimination. During storage, it is important to maintain low intrabladder pressure to allow urine to flow from the kidney through the ureter. Bladder emptying then involves contraction of the smooth muscles of the bladder wall and relaxation of the internal and external sphincters so that urine can be expelled.

The transition between storage and expulsion modes can occur either voluntarily (as a result of descending influences) or involuntarily (when segmental reflex circuitry operates in the absence of descending control, for example, in infants, anesthetized individuals, or following spinal cord injury).

The autonomic control of bladder function is mediated by parasympathetic fibers deriving from preganglionic neurons in the sacral portion of the spinal cord, and sympathetic fibers deriving from preganglionic sympathetic neurons in the lower thoracic and upper lumbar spinal cord.

During bladder filling, activity over sympathetic fibers causes the release of noradrenaline. Noradrenaline acting via α-adrenergic receptors relaxes the muscles of the bladder wall and simultaneously activates the internal sphincter muscles, causing closure of the internal sphincter. Sympathetic activation also inhibits activation of the parasympathetic reflex pathways that cause bladder constriction. This combination of ac-

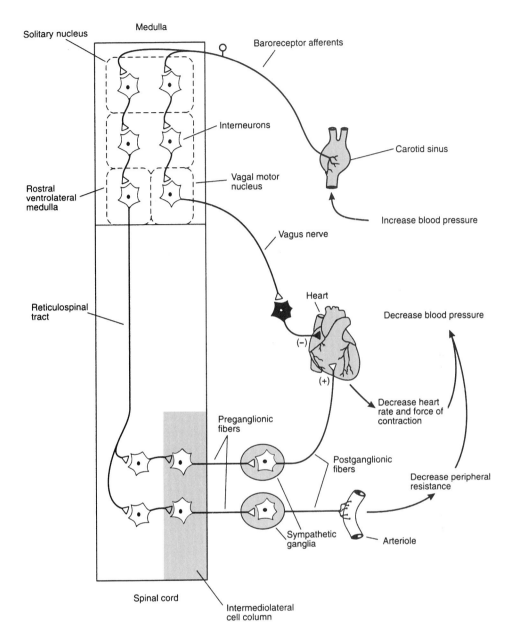

Figure 27.7. The circuitry that regulates cardiovascular function. Heart rate is controlled by the balance of activity in sympathetic fibers, which increase heart rate, and parasympathetic fibers, which decrease heart rate. From Kandel ER, Schwartz JH, Jessell TM. *Principles of Neural Science.* 3rd ed. New York, NY: Elsevier; 1991. From Patton HD. The autonomic nervous system. In: Patton HD, Fuchs AG, Hille B, Scher AM, Steiner R, eds. *Textbook of Physiology: Excitable Cells and Neuropyhysiology.* Vol 1. Philadelphia, Pa: WB Saunders; 1989:737–758.

tion creates an optimal situation for bladder filling.

The basic autonomic reflex pathways that operate when descending influences are absent involve the following: When the internal pressure of the accumulating urine causes distension of the bladder to a particular threshold level, tension receptors in the bladder wall are activated. These are innervated by viscerosensory fibers that project via the pelvic nerve to the sacral spinal cord. There appear to be two separate reflex circuits that control bladder emptying. One reflex pathway involves areas in the brainstem (termed the *micturition center*). Neurons in this region receive ascending sensory infor-

mation regarding bladder distension and then project back to preganglionic neurons in the cord (Figure 27.8). The other reflex circuit involves connections at the segmental level that can operate autonomously, at least in situations in which descending pathways are damaged (see later).

These micturition reflex circuits operate like on–off switches; a critical level of bladder distension triggers a burst of activity in the reflex circuit, which activates preganglionic parasympathetic neurons that project to the pelvic ganglionic plexus. The activation of parasympathetic postganglionic neurons causes the release of acetylcholine. The acetylcholine activates the muscles of

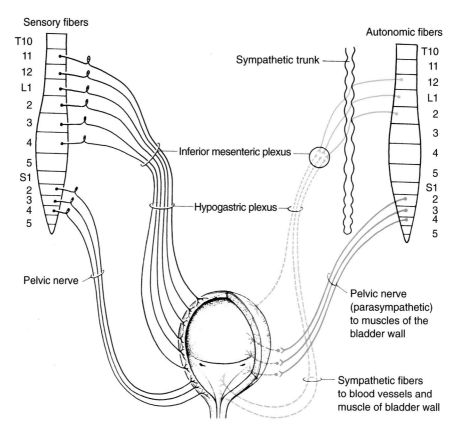

Figure 27.8. The circuitry that controls micturition. Autonomic control of bladder function is mediated by parasympathetic fibers deriving from preganglionic neurons in the sacral portion of the spinal cord, and sympathetic fibers deriving from preganglionic sympathetic neurons in the lower thoracic and upper lumbar spinal cord. Sensory information from stretch receptors enters upper lumbar and lower thoracic segments via the hypogastric plexus. Sensory information reaches sacral segments via the pelvic nerve. Reproduced with permission from Carpenter MB, Sutin J. *Human Neuroanatomy.* Baltimore, Md: Williams & Wilkins; 1983.

the bladder wall via muscarinic receptors, causing constriction.

Voluntary control of voiding involves two events: (1) descending pathways inhibit somatomotoneurons that innervate the external sphincter, causing relaxation of the sphincter muscles; and (2) simultaneously, the descending pathways inhibit the activity of preganglionic sympathetic neurons, removing their tonic inhibition of parasympathetic activity. This disinhibition releases the parasympathetic reflex, causing relaxation of the internal sphincter and bladder contraction.

Some of the other autonomic functions that are of importance to the physician are summarized in Table 27.3, focusing on the differential role of sympathetic versus parasympathetic divisions. An important feature of the autonomic reflexes is that they are modulated by descending influences from higher brain centers. Most of this descending modulation is relayed via the hypothalamus, which is also the area responsible for integrating the various reflexes into an adaptive behavioral response. The hypothalamus in turn receives descending inputs from forebrain structures including the cerebral cortex, which are relayed via the limbic system. Hence, in the next chapter we turn our attention to this central circuitry, considering first the hypothalamus and then the circuitry of the limbic system.

AUTONOMIC FUNCTION FOLLOWING SPINAL CORD INJURY

Spinal cord lesions above the level of the thoracic cause a loss of central regulation of other autonomic reflexes due to a loss of descending inputs to preganglionic neurons—a symptom complex termed *autonomic dysreflexia*. The viscera are not "paralyzed" because they are capable of considerable automatic function, but they are no longer regulated in the proper way.

Immediately after an injury, there is a period of spinal shock during which spinal autonomic reflexes are depressed, paralleling the depression of somatic reflexes:

1. Blood pressure drops precipitously as a result of a decrease in the sympathetic discharge to the vascular bed. Regulation of heart rate via the vagus nerve is spared.

2. Temperature regulation is lacking, sweating is absent, and body temperature drifts toward that of the environment.

3. Bladder function and bowel function are depressed. Part of this suppression presumably occurs because of the loss of the component of the micturition reflex circuitry that relays through the micturition center in the brainstem.

After a period of time ranging from days to weeks, spinal shock wanes and segmental autonomic reflexes reappear, although the reflexes are no longer modulated by descending influences.

1. Blood pressure rises from the low levels typical of spinal shock and fluctuates in response to noxious stimulation of the skin. However, gravitational modulation is lacking. The local vasoconstriction that normally prevents gravitational hypotension when the body is shifted from a horizontal to a vertical position is absent.

2. Sweating returns and noxious stimulation of the skin may elicit profuse sweating. However, the integrated heat loss and heat conservation reflexes, which involve sweating, piloerection, and the regulation of cutaneous vasculature, are no longer integrated by the hypothalamus. As a result, exposure to high or low ambient temperatures can cause body temperature to drift beyond acceptable limits.

3. Reflex bowel and bladder function returns, although voluntary control is still absent. Although segmental autonomic pathways do provide for reflex emptying of the bladder, the emptying is often not as complete as is the case with voluntary voiding. This incomplete voiding of the bladder increases the chances for urinary tract infections in patients with spinal cord injuries. In fact, urinary tract infections were among the most common causes of death for patients with spinal cord injuries before the development of modern antibiotics.

10

The Hypothalamus and Limbic System: Substrates of Motivation, Emotion, and Memory

The hypothalamus and limbic system are components of a complicated neural network that controls our emotions and motivations. Components of this network also seem to play a key role in establishing long-term memories, based on the fact that memory function is severely disrupted by lesions that affect certain structures.

The hypothalamus is a collection of nuclei in the basal diencephalon. Different nuclei within the hypothalamus play a key role in regulating the behaviors that contribute to the survival of the individual (that is, regulate visceral activity and trigger the ingestive behaviors of feeding and drinking) and the survival of the species (reproductive behaviors). The hypothalamus is sometimes called the *head ganglion of the autonomic nervous system* because of its role in regulating the output of lower autonomic centers so as to provide moment-by-moment control of visceral function. But to maintain internal homeostasis over the long term, it is also important to control and coordinate ingestive behaviors, for example, feeding and drinking. These behaviors are regulated by "feelings" of hunger and thirst, which motivate us to seek out food or drink. Specific neuron groups within the hypothalamus sense chemical imbalances (osmolarity of the blood, low blood sugar). These neurons then project to the autonomic centers that provide short-term regulation, and to other populations of neurons that somehow trigger the specific motivations that cause the appropriate ingestive behaviors.

The *limbic system* is a set of interconnected structures that were first defined as a "system" simply by virtue of the interconnections between structures. The components of the limbic system include the cingulate cortex, entorhinal cortex, hippocampus, amygdala, and certain nuclei in the thalamus and hypothalamus. How these different structures actually operate in a physiological sense is not yet clear. Moreover, clinical observations in patients with lesions in different structures, together with studies of experimental animals, have yielded complicated and sometimes contradictory hypotheses about the functions that this circuit mediates. Nevertheless, lesions of different structures produce striking and highly predictable symptoms. Damage to certain structures causes striking and predictable changes in affect; lesions in other structures cause profound disruption of memory function. Hence, although we cannot fully explain the operation of the system or exactly define the functions that it mediates, we can define the symptoms resulting from injury or disease to the different components of the system.

There are several key concepts for physicians:

1. The hypothalamus plays a central role in regulating autonomic and endocrine reflex functions and also plays a role in emotional and motivational states.

2. The hypothalamus receives ascending information from autonomic centers in the brainstem (primarily the nucleus of the solitary tract). Resident neurons within the hypothalamus also are sensitive to the chemical composition of the blood (osmolarity, glucose concentration) and to circulating hormones that are released by the viscera and that indicate visceral activities.

3. Information from cortical structures is relayed to the hypothalamus via the limbic system.

28

The Hypothalamus

Coordination of Visceral Operation and the Behaviors That Maintain Bodily Homeostasis

The hypothalamus is a collection of nuclei in the basal diencephalon (Figure 28.1). At the midline, the anterior boundary of the hypothalamus is marked by the lamina terminalis; laterally, the hypothalamus merges into the basal forebrain. Caudally, the hypothalamus merges into the midbrain tegmentum.

The hypothalamus is divided into three medio-lateral subdivisions, the periventricular, medial, and lateral subdivisions, and a basal region that forms the floor of the third ventricle, called the *tuberal region*. The internal circuitry of the hypothalamus has been difficult to define because the nuclei do not have the sort of highly organized structure that is conducive to defining interconnections. Moreover, most of the circuitry involves short axons, so it has been difficult to apply traditional tract-tracing techniques. Nevertheless, some of the key connections are described later. Figure 28.2 illustrates the location of the different hypothalamic nuclei in a series of frontal sections.

The *periventricular* region is a thin rim of gray matter that forms the walls of the lower part of the third ventricle. Neurons in the region give rise to the descending projections to autonomic centers

in the brainstem, which descend via the *dorsal longitudinal fasciculus*.

The *medial region* lies just lateral to the periventricular region and contains most of the well-defined nuclei of the hypothalamus. The anterior portion of the medial region contains the preoptic and suprachiasmatic nuclei, which lie above the optic chiasm. The middle portion of the medial region contains the supraoptic, anterior, dorsomedial, ventromedial, and paraventricular nuclei. The posterior portion of the medial region contains the posterior nuclei and mammillary nuclei. The *mammillary bodies* are the mounds on the base of the brain that represent the surface features that mark the location of the *mammillary nuclei* (the relationship here is similar to that of the olivary nuclei and the olive).

The *lateral region* is made up of a complicated collection of neurons that is not divided into discrete nuclei. Some of the neurons project locally so as to form multisynaptic ascending and descending relays; others give rise to ascending and descending projections to the forebrain and spinal cord. Some of the descending projections play a role in regulating sympathetic preganglionic neurons in the spinal cord.

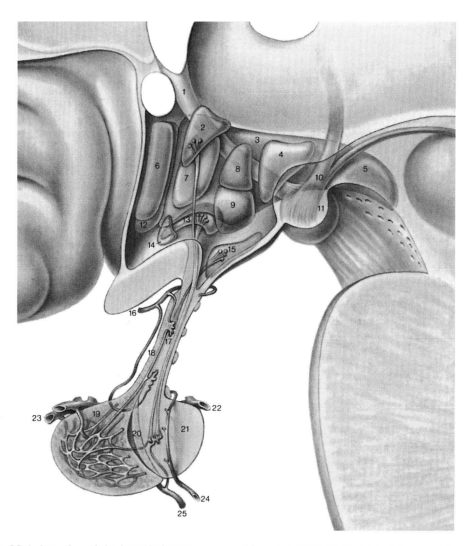

Figure 28.1. Location of the hypothalamus. Drawing of the location of the hypothalamus in midsagittal section. The anterior boundary of the hypothalamus is marked by the lamina terminalis. Laterally, the hypothalamus merges into the basal forebrain. Caudally, the hypothalamus merges into the midbrain tegmentum. 1, fornix; 2, paraventricular nucleus; 3, lateral hypothalamus; 4, posterior nucleus; 5, ventral tegmental area; 6, preoptic nucleus; 7, anterior nucleus; 8, dorsomedial nucleus; 9, ventromedial nucleus; 10, mammillothalamic tract; 11, mammillary body; 12, latera preoptic nucleus; 13, supraoptic nucleus; 14, suprachiasmatic nucleus; 15, arcuate nucleus, or infundibular nucleus; 16, superior hypophyseal artery; 17–20, anterior lobe of pituitary; 21, posterior lobe of pituitary; 22–23, venous sinuses; 24 and 25, inferior hypophyseal arteries. From Nieuwenhuys R, Voogd J, van Huijzen C. *The Human Central Nervous System.* Berlin, Germany: Springer-Verlag; 1988.

The *tuberal region* contains the *tuberal nuclei* and the *arcuate nucleus,* which is the rim of gray matter surrounding the base of the third ventricle. Neurons in the tuberal nuclei give rise to axons that terminate on the hypothalmohypophyseal portal system within the *median eminence* (Figure 28.1).

INPUTS AND OUTPUTS OF THE HYPOTHALAMUS

Most of the tracts carrying information to and from the hypothalamus are bidirectional (Table 28.1).

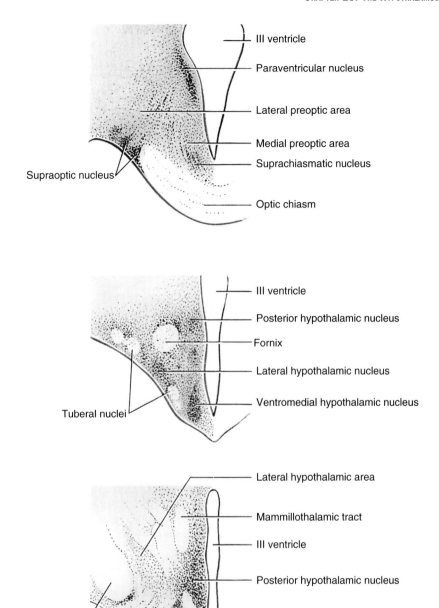

Figures 28.2. Hypothalamic nuclei in frontal sections. Drawings of a series of frontal sections through the hypothalamus. Reproduced with permission from Carpenter MB, Sutin J. *Human Neuroanatomy.* Baltimore, Md: Williams & Wilkins; 1983.

Table 28.1. Fiber tracts conveying information to and from the hypothalamus

Fiber Tract	Directionality	Source/target
Medial forebrain bundle	Bidirectional	To and from brainstem, septal nuclei, and basal forebrain
Fornix	Bidirectional	Primarily from subicular region of hippocampus; to hippocampus
Stria terminalis	Bidirectional	To and from amygdala
Ventral amygdalofugal pathway	Bidirectional	To and from amygdala
Mammillothalamic tract	Unidirectional	To thalamus
Mammillotegmental tract	Bidirectional	To and from brainstem tegmentum
Retinohypothalamic tract	Unidirectional	From retina
Hypothalamohypophyseal tract	Unidirectional	To neural lobe of pituitary

Inputs

The principal input pathways to the hypothalamus are illustrated in Figure 28.3. Ascending viscerosensory information is relayed to the hypothalamus from brainstem regions (especially the nucleus of the solitary tract). These ascending projections are conveyed via the *medial forebrain bundle* (MFB), which projects through the lateral hypothalamus. This pathway also conveys information from dopaminergic neurons in the ventral

Figure 28.3. Principal inputs to the hypothalamus. The major tracts conveying input to the hypothalamus are illustrated. MB indicates mammillary bodies; MFB, medial forebrain bundle. Modified from Noback CR and Demarest RS. *The Human Nervous System.* New York, NY: McGraw-Hill, 1975.

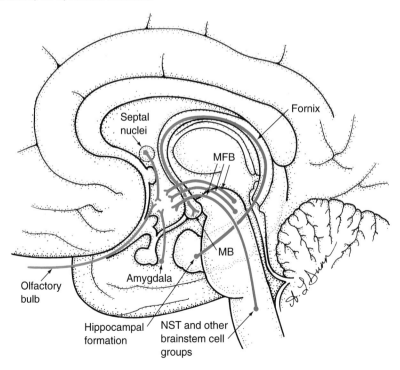

tegmental area, and noradrenergic neurons in the locus ceruleus (see Chapter 30). The pathway also contains ascending and descending fibers that interconnect nuclear groups in the hypothalamus with areas in the forebrain, midbrain, and brainstem.

Descending information from forebrain structures is conveyed via three major fiber tracts. The most prominent tract is the *fornix,* which contains fibers that project from the subicular region of the hippocampal formation to the mammillary nuclei of the hypothalamus. The fornix also carries ascending projections from the hypothalamus to the hippocampal formation. Other important inputs derive from the amygdala; these are conveyed via the *stria terminalis* and the *ventral amygdalofugal pathway.* We consider this circuitry again when considering the outputs from the limbic system. Finally, there are

also projections from the septal nuclei that are conveyed via the MFB.

One other important input to the hypothalamus is the *retinohypothalamic tract,* which terminates in the suprachiasmatic nucleus. This pathway relays information that is important for regulating circadian (24-hour) rhythms.

Finally, certain neuron groups within the hypothalamus are sensitive to the chemical composition of the blood (especially osmolarity and glucose concentrations). Certain neurons possess receptors for circulating hormones that reveal visceral function (for example, angiotensin II).

Outputs

The principal output pathways from the hypothalamus are illustrated in Figure 28.4. As

Figure 28.4. Principal outputs from the hypothalamus. The principal output pathways of the hypothalamus are illustrated. Thal indicates thalamus; DLF, dorsal longitudinal fasciculus.

Modified from Noback CR and Demarest RS. *The Human Nervous System.* New York, NY: McGraw-Hill, 1975.

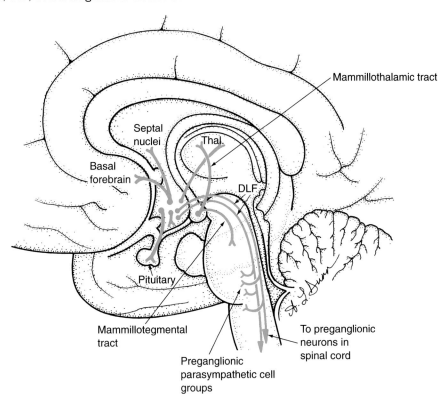

already noted, the MFB contains axons that project into and out of the hypothalamus. Included in this fiber tract are important descending projections to autonomic (sympathetic) cell groups in the spinal cord. The MFB also contains axons that project from nuclear groups in the hypothalamus to basal forebrain structures.

The mammillary nuclei give rise to two prominent projection systems: the *mammillothalamic tract,* which projects from the mammillary nuclei to the anterior nucleus of the thalamus, and the *mammillotegmental tract,* which contains axons that project to and from the midbrain tegmentum. The mammillothalamic tract is one of the important relays in the so-called Papez circuit, which is discussed in more detail in the section on the limbic sys-

tem. Other areas of the hypothalamus give rise to projections to the dorsomedial nucleus of the thalamus, thus providing a route to convey information to the frontal cortex.

The dorsal longitudinal fasciculus is a diffuse fiber system that carries descending projections from neurons in the periventricular nucleus to autonomic nuclei in the brainstem. It travels through the periventricular region itself.

The *hypothalamo-hypophyseal tract* is a collection of axons that project from nuclei in the hypothalamus to the *neural lobe* of the pituitary gland (Figure 28.5). These axons travel through the *stalk of the pituitary* (the *infundibular stalk*) and terminate on capillaries within the neural lobe. These axon terminals release hormones directly into the circula-

Figure 28.5. Neural inputs to the pituitary gland. The hypothalamo-hypophyseal tract carries axons from nuclei in the hypothalamus to the neural lobe of the pituitary gland. These axons travel through the stalk of the pituitary (the infundibular stalk) and terminate on capillaries within the neural lobe. These axon terminals release hormones directly into the circulation. Other

populations of neurons give rise to projections that terminate on the capillary bed within the median eminence (where the pituitary stalk connects with the base of the hypothalamus). Reproduced with permission from Carpenter MB, Sutin J. *Human Neuroanatomy.* Baltimore, Md: Williams & Wilkins; 1983.

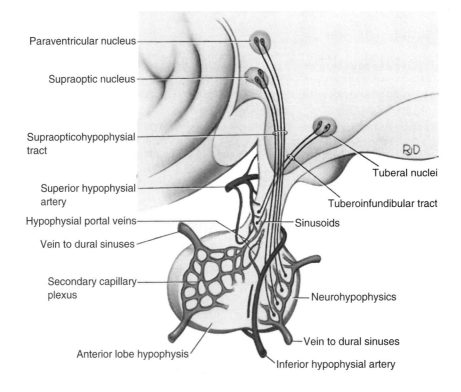

Paraventricular nucleus

Supraoptic nucleus

Supraopticohypophysial tract

Superior hypophysial artery

Hypophysial portal veins

Vein to dural sinuses

Secondary capillary plexus

Anterior lobe hypophysis

Tuberal nuclei

Tuberoinfundibular tract

Sinusoids

Neurohypophysics

Vein to dural sinuses

Inferior hypophysial artery

tion (see later). Other populations of neurons give rise to projections that terminate on the capillary bed within the median eminence (where the pituitary stalk connects with the base of the hypothalamus).

NEURAL CONTROL OF THE PITUITARY GLAND

The pituitary gland is connected with the base of the hypothalamus by the stalk of the pituitary (the infundibular stalk). The stalk carries the *hypothalamo-hypophyseal tract,* and also contains the vasculature comprising the *hypothalamo-hypophyseal portal system.*

Hormone release from the pituitary is regulated differently in the anterior and posterior lobes. In the posterior pituitary (the neural lobe, or neurohypophysis), peptide hormones are released directly into the capillary bed from the axon terminals that project to the neurohypophysis from peptidergic neurons in the paraventricular and supraoptic nuclei (Figure 28.4). The hormones that are released in this way include vasopressin and oxytocin (more on this later).

In the *anterior pituitary (adenohypophysis),* hormone release is regulated indirectly. Neurons within the hypothalamus release peptide factors into the capillary bed of the hypothalamo-hypophyseal portal system in the median eminence (Figure 28.4). These factors are then conveyed to the anterior pituitary via the portal vessels. The factors that are conveyed in this way can either stimulate or inhibit release of particular hormones and are, hence, called *releasing factors* or *release-inhibiting factors,* respectively.

THE FUNCTIONS MEDIATED BY THE HYPOTHALAMUS

When considering the functions that are mediated by the hypothalamus, it is important to recall that the hypothalamus operates at several levels. First, the hypothalamus plays a role in coordinating autonomic reflex responses. Second, the hypothalamus serves as the relay between forebrain structures and lower autonomic centers. Third, the hypothalamus controls the release of hormones from the pituitary gland. By controlling endocrine function, it regulates a wide variety of systems throughout the body. Fourth, the hypothalamus plays an as yet poorly understood role in conjunction with forebrain structures to create the "emotions" (for example, anger and fear) and the "motivations," or "drives" (for example, hunger and thirst) that cause us to engage in particular behaviors, including reproductive behaviors.

We will consider several examples that illustrate the roles that different nuclei within the hypothalamus play in particular functions (see Table 28.2 for a summary). We begin by considering the areas of the hypothalamus that are most directly involved in the control of lower autonomic centers.

Hypothalamic Control of Autonomic Responses

As already noted, neurons distributed in a number of different nuclei give rise to descending projections to autonomic centers in the brainstem and preganglionic autonomic neurons in the spinal cord. Sites of termination include the nucleus of the solitary tract, parabrachial nucleus, dorsal motor nucleus of the vagus, and the medullary centers that control respiration and cardiovascular function. Rather little is known about how activity over these descending pathways is actually controlled in physiological settings. The information that is available is based on studies in experimental animals in which particular parts of the hypothalamus were either stimulated electrically via indwelling electrodes or destroyed by producing focal lesions.

As noted in Chapter 27, stimulation of the hypothalamus produces integrated autonomic responses that are similar to those seen in natural settings—what we interpret as anger or fear. For example, electrical stimulation of the lateral hypothalamus in cats produces piloerection, arching of the back, pupillary constriction, raising of the tail, and snarling—exactly the responses that humans

Table 28.2. A summary of some of the principal connections and functions of hypothalamic nuclei.*

Nucleus	Principal inputs	Principal outputs	Functions
Periventricular	Multiple	DLF to midbrain	Descending input to autonomic cell groups in the brainstem
Preoptic	SFO, baroreceptors, amygdala		Regulate fluid consumption
Suprachiasmatic	Retina		Sensory input for circadian behaviors
Paraventricular and supraoptic (vasopressin)	Osmotic (blood Na$^+$) baroreceptors	Neural lobe of pituitary	Secrete vasopressin into circulation to modulate kidney function
Paraventricular and supraoptic (oxytocin)	Somatosensory	Neural lobe of pituitary	Secrete oxytocin into circulation to trigger milk ejection and uterine contraction
Dorsomedial ventromedial	Amygdala		Glucoreceptors; role in regulating food intake
Anterior	Amygdala		
Posterior		Dorsal longitudinal fasciculus hypothalamus–brainstem	Input to autonomic nuclei of brainstem and spinal cord
Mammillary	Limbic system (via fornix)	Tegmentum, DM thalamus	An important relay for limbic input
Lateral	Amygdala	Amygdala	Bidirectional pathway interconnecting hypothalamus and limbic system
Tuberal (mainly arcuate)	Amygdala	Hypophyseal portal system	Secrete releasing factors that control hormone release from anterior pituitary

*Where there are no notations, the nuclei are not known to be involved in the principal afferent or efferent circuitry.

(and other cats) interpret as anger or fear. Following lesions of the lateral hypothalamus, animals seem more placid and are difficult to arouse to anger. In contrast, following lesions of the medial hypothalamus, animals appear more aggressive and/or fearful.

These findings do not tell us how the circuits within the hypothalamus actually operate or what other circuitry may be involved in eliciting the responses; they do, however, strengthen the hypothesis that the hypothalamus plays a key role in integrating the autonomic responses that we associate with emotional states.

Regulation of Fluid Balance

It is important to maintain osmolarity of the blood and total blood volume within defined limits. This is accomplished in two ways. Over the short term, osmolarity is controlled by regulating water recovery by the convoluted tubules of the kidney. Over the long term, however, maintenance of water balance requires fluid intake.

Neurons within the paraventricular and supraoptic nuclei of the hypothalamus play a central role in short-term regulation of fluid balance by synthesizing and releasing *vasopressin* (also called *antidiuretic hormone*) into the capillary bed of the posterior pituitary. The vasopressin is then conveyed to the kidney where it regulates kidney tubule function.

The neurons that release vasopressin function as *osmoreceptors;* their firing rate is directly modulated as a function of the Na^+ concentration of the blood. Increases in blood osmolarity (Na^+ concentration) lead to increases in firing rate, causing an increased release of vasopressin. Firing rate is also modulated by blood volume receptors in the blood vessels, so decreases in blood volume also trigger an increased release of vasopressin.

Fluid intake (drinking) appears to be initiated by two mechanisms, one involving osmoreceptors and the other involving blood volume receptors.

The osmoreceptive neurons within the paraventricular and supraoptic nuclei not only function as neuroendocrine cells; they also function as traditional neurons that receive and transmit information to other neurons within the hypothalamus. These projections represent one of the ways that information regarding the need to ingest fluids may be communicated to the neurons that trigger the feeling that we recognize as thirst, which motivates us to drink. In this way, the osmoreceptive neurons within the paraventricular and supraoptic nuclei can mediate both the short-term autonomic responses and the long-term behavioral responses that restore fluid balance.

Blood volume changes trigger drinking via an indirect mechanism involving the kidney (Table 28.3). Low blood volume triggers the secretion of renin from the kidney. Renin is a proteolytic enzyme that cleaves plasma

Table 28.3. Steps through which drinking is triggered by decreases in blood volume

1. Decreases in blood volume
2. Release of renin from kidney
3. Renin cleaves circulating angiotensinogen to angiotensin I
4. Angiotensin I is hydrolyzed to angiotensin II
5. Angiotensin II activates receptors on neurons in subfornical organ
6. Neurons in SFO project to preoptic nucleus

angiotensinogen into angiotensin I, which is then hydrolyzed to the octapeptide angiotensin II. Circulating angiotensin operates at one of the specialized regions of the brain that lacks a blood–brain barrier—a region bordering the third ventricle termed the subfornical organ (SFO). Neurons within the SFO are activated by antiotensin II; these neurons then project to the preoptic nucleus of the hypothalamus. The preoptic nucleus also receives information regarding blood volume from baroreceptors. These two types of information are integrated by the preoptic neurons and conveyed to other brain regions, which induces drinking.

The central circuitry that is actually involved in triggering the perception of thirst and the behaviors that lead to drinking are not known. Certainly, important components of the circuitry are located in the hypothalamus. However, thirst is consciously perceived, and so the cortex must also become involved.

Regulation of Milk Ejection and Uterine Contraction

There are two separate populations of neurons within the paraventricular and supraoptic nuclei that project to the neural lobe of the pituitary gland: the vasopressin neurons already described, and a different population of neurons that synthesize and release the hormone oxytocin. This hormone plays a key role in regulating milk ejection from the mammary gland and also regulates uterine contraction.

The *milk ejection reflex* in lactating females is triggered by suckling. Information from sensory receptors in the nipple is conveyed via ascending sensory pathways to the hypothalamus. This sensory stimulation triggers periodic bursts of action potentials in oxytocin-containing neurons in the paraventricular and supraoptic nuclei. The bursts of action potentials trigger the "pulsatile" release of oxytocin into the circulation. Within a few seconds after oxytocin release into the circulation, there is an increase in intramammary pressure, causing milk ejection.

Milk ejection can be triggered by a variety of stimuli other than suckling, including the sight or sound of an infant. This indicates that the secretion of oxytocin can also be regulated by higher cortical structures.

Oxytocin also plays an important role during childbirth. Specifically, the amplitude of contraction of the uterine muscles is increased by oxytocin when the muscles are appropriately primed by estrogen. Exogenous oxytocin is sometimes given during labor if uterine contractions are not sufficiently vigorous.

Regulation of Food Intake

The hypothalamus also plays an important role in regulating food intake. The regulation of feeding behavior is very complex, however, and the neural mechanisms that determine what and how much we eat are not well understood. At the most basic level, maintenance of bodily homeostasis requires the ingestion of the appropriate amounts of fats, proteins, and total calories. But there are a large number of other factors that influence food intake that have nothing to do with need. Here we consider only a few of the principals.

Each individual appears to regulate food intake so that body weight is maintained around a set point. Individuals will regulate their eating habits over long periods of time so as to maintain body weight at a particular level plus or minus a few pounds. This level is termed the set point. These set points change over long periods of time, however, as a function of emotional state, level of exercise, and other factors. The most common change is a slow, upward drift.

Different regions of the hypothalamus are thought to be involved in initiating and suppressing feeding. Experimental studies have revealed that animals with lesions involving the medial hypothalamus overeat and become obese; animals with lesions involving the lateral hypothalamus ignore food. Indeed, such animals die unless force-fed. Humans with lesions involving the hypothalamus may exhibit similar symptoms. Electrical stimulation of the respective regions has the opposite effects: stimulation of the medial hypothalamus suppresses feeding; stimulation of the lateral hypothalamus elicits feeding.

Originally, these findings were interpreted as indicating the existence of a discrete "feeding center" and "satiety center" within the hypothalamus. It is now recognized, however, that the brain does not function in such a simplistic way. In particular, no brain region operates in isolation; functions are mediated by systems made up of interconnected neuronal groups, and different regions of the hypothalamus are in fact part of circuits that operate in an integrated fashion in the control of behavior. A more accurate way to think of the areas that play a role in initiating feeding is that they contain neurons that possess receptors that detect particular metabolic deficiencies and, hence, signal the need to ingest food substances. Other regions contain neurons that possess receptors for substances that signal that particular food substances have been recently consumed (see later).

It is noteworthy that chemical stimulation of the hypothalamus with neurotransmitters alters feeding in a neurotransmitter-specific fashion. For example, stimulation with norepinephrine causes animals to eat more carbohydrates; stimulation with opiates causes animals to eat more protein; and stimulation with a peptide termed *galanin* causes animals to eat more fat. These findings suggest the existence of neurotransmitter-specific neuron types that play a key role in triggering each of these behaviors.

Neurons within the hypothalamus possess receptors for circulating glucose and various hormones that are released by the gut. Neurons that possess receptors that are sensitive to circulating glucose (glucoreceptors) may play a role in triggering eating when blood glucose falls below critical levels. Also, certain hypothalamic neurons have receptors for hormones that are released by the gut during digestion. One example is cholecystokinin, which is released from the intestine when amino acids and fatty acids are present. Injection of cholecystokinin, as well as certain other peptides, into the hypothalamus inhibits feeding. Hence, the release of gut hormones into the circulation may serve as a feedback pathway that informs the hypothalamus that food is being digested (a mechanism for signaling satiety).

Self-stimulation and the Concept of "Pleasure Centers"

In the mid-1950s, a series of striking experiments revealed that experimental animals with electrodes implanted into certain regions in the hypothalamus (the MFB) would aggressively press a level to deliver electrical stimulation to their brains. Moreover, stimulation of particular sites could be substituted for other "rewards" in learning tasks. Psychologists interpreted these findings as indicating that the stimulation was pleasurable, and postulated the existence of "pleasure centers" within the hypothalamus. As with the concept of feeding and satiety centers, the concept of pleasure centers has evolved such that we think of these areas as part of a larger circuit. Still, these early experiments clearly implicated the hypothalamus as an important nodal point in circuitry that mediates motivational states.

29

The Limbic System

Circuits That Underlie Affect and Memory

The term *limbic system* implies a functional system like the sensory and motor systems. In fact, however, the term refers to a collection of structures with extensive interconnections that seem to be involved in a number of complex functions. Before beginning our consideration of the limbic system, it is important to briefly review how the concept of a limbic system arose.

The term *limbic system* derives from the concept of the *limbic lobe*, enunciated by the French neuroanatomist P. Broca. Broca recognized that there was a set of structures that formed the *limbus* (border, or fringe) of the cerebral cortex, which included the cingulate gyrus, parahippocampal gyrus, and olfactory cortex. These structures are cortical, but have fewer layers than the neocortex.

Broca proposed that the limbic lobe was made up of two concentric rings of structures. The inner ring included structures such as the hippocampal formation, which was representative of a primitive type of cortex termed the *allocortex*. The outer ring included structures such as the cingulate cortex, which exhibited a lamination pattern that was transitional between allocortex and neocortex. These were termed *transitional cortex,* or *juxtallocortex* (next to the cortex). Broca also emphasized the

close relationship between the structures of the limbic lobe and the olfactory system.

The limbic system concept emerged in the 1930s. Implicit in the use of the term *system* was the idea that there was a group of interconnected structures that were functionally related. Two papers published in 1937 were especially influential in shaping the concept. In one paper, Kluver and Bucy reported that injuries involving structures in the temporal lobe (especially the amygdala and surrounding regions) produced alterations in emotional responses, abnormal social behaviors, and hypersexuality—complex and unexpected symptoms that were difficult to explain. In the second paper, Papez defined a "circuit" that involved interconnections between the hippocampus, amygdala, septum, mammilary bodies, and olfactory bulbs. This came to be known as the *Papez circuit* (see later). Papez further proposed that this circuit played an important role in controlling emotional behaviors. This idea was consistent with the observations of Kluver and Bucy. In this way, a concept was launched that has shaped our thinking even up to the present.

We now know that components of the limbic system have extensive interconnections with one another and also with the olfactory system, the hypothalamus, and the cerebral cortex. Indeed, the components of the limbic system serve as key relays between the olfactory system, cerebral cortex, and the autonomic nervous system. We also know a good deal about the symptoms that result from selective lesions at various sites. Nevertheless, we still do not have a clear understanding of how the system actually operates. Even though we cannot explain the operation of the system, it is important for physicians to be aware of the dramatic and unusual symptoms that result from injury to the system's various components.

In what follows, we (1) describe the structure and cellular organization of each component of the limbic system, (2) summarize the principal interconnections between these structures, (3) consider some of the functional characteristics of the circuits, and (4) review some of the symptoms that result from lesions involving the different structures.

COMPONENTS OF THE LIMBIC SYSTEM

We consider the following structures as components of the limbic system. Some texts include additional structures, but mosts texts will agree on these core components:

1. The *hippocampal formation*
2. The *entorhinal cortex* and related areas in the *parahippocampal gyrus*
3. The *amygdaloid complex*
4. The *septal nuclei*
5. The *cingulate cortex*

In addition, certain nuclei in the thalamus and hypothalamus represent key relays in the circuits of the limbic system:

1. The *mediodorsal* and *anterior nuclei* of the thalamus
2. The *mammillary* and *ventromedial nuclei* of the hypothalamus

STRUCTURE AND CELLULAR ORGANIZATION OF THE COMPONENTS OF THE LIMBIC SYSTEM

The Hippocampal Formation

The hippocampal formation is one of the most striking structures in the brain, owing to the prominent, arching fiber tract that interconnects the hippocampal formation and the mammillary body (the *fornix*). The hippocampal formation is located along the medial edge of the temporal lobe (Figure 29.1). The hippocampal formation merges ventrally with the entorhinal cortex; rostrally, it merges with the amygdalar complex. The fornix extends from its dorsal end and then arches around to eventually end in the mammillary body.

The hippocampal formation has three subdivisions: (1) the *dentate gyrus,* (2) the *hippocampus proper,* and (3) the *subiculum* (Figure 29.1). The hippocampus proper is further subdivided into subregions: CA1, CA2, CA3, and CA4. The principal neurons in the dentate gyrus are small *granule neurons;* the principal neurons in the hippocampus proper and subiculum are *pyramidal neurons.*

The hippocampal formation is a strikingly laminated structure. In each subdivision, the cell bodies of the principal neurons are collected in discrete layers. The principal neurons radiate into neuropil layers in which the incoming fibers terminate. This simple lamination pattern is the hallmark of *archicortex.*

The internal circuitry of the hippocampal formation is complicated, but in essence involves a *four-synapse relay.* The principal input to the hippocampal formation arises from the entorhinal cortex in the parahippocampal gyrus. There are actually two separate input pathways. The major pathway terminates on the granule cells of the dentate gyrus (synapse 1 in the four-synapse relay). This pathway is called the *perforant path.* There is also a smaller projection to the distal dendrites of the pyramidal neurons in the hippocampus proper. The granule cells in turn send a powerful excitatory projection to the pyramidal neurons in the CA3 and CA4 regions of the hippocampus proper (synapse 2). This is termed the *mossy fiber projection* because of the appearance of the fibers themselves. The pyramidal neurons in the CA3 region then project to pyramidal neurons in CA1 (synapse 3). Pyramidal neurons in CA1 in turn project to neurons in the subiculum (synapse 4).

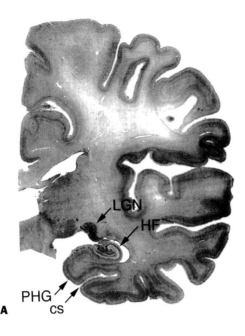

A

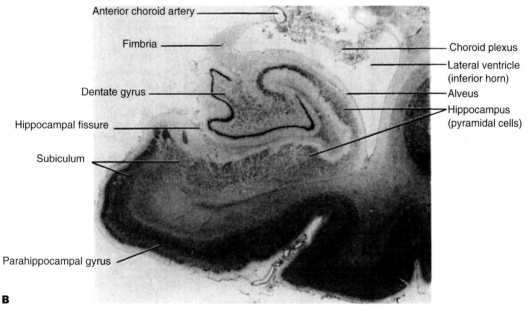

B

Figure 29.1. The hippocampus and entorhinal cortex. A illustrates the location of the hippocampal formation and entorhinal cortex in the temporal lobe. PHG indicates parahippocampal gyrus; LGN, lateral geniculate nucleus; HF, hippocampal formation; CS, collateral sulcus. **B** illustrates a higher-magnification photomicrograph illustrating the cellular organization of the hippocampal formation. The principal neurons in the dentate gyrus are small "granule neurons"; the principal neurons in the hippocampus proper and subiculum are pyramidal neurons. **A** is from Corkin S, Amaral DG, Bonzales G, Johnson KA, Hyman BT. H. M.'s medial temporal lobe lesion: findings from magnetic resonance imaging. *J Neurosci.* 1997;17:3964–3979. **B** is from Waxman SG. *Correlative Neuroanatomy.* Norwalk, Conn: Appleton & Lange; 1996.

In addition to the major input from the entorhinal cortex, the hippocampus also receives projections from the septal nuclei and from nuclear groups in the brainstem. These arrive via the fornix. The projections from the septal nuclei arise from two populations of neurons, one cholinergic and one GABAergic. The projections from the brainstem arise from noradrenergic and serotonergic cell groups.

Neurons in the subiculum give rise to the efferent projections out of the hippocampus. One component of axons projects via the fornix to terminate in the mammillary nuclei of the hypothalamus. Another component of axons project back to the entorhinal cortex and adjacent cortical areas. Indeed, despite the fact that the fornix is so visually striking, most of the output from the hippocampal formation is actually to cortical areas in the temporal lobe.

The Entorhinal Cortex

The entorhinal cortex is the region of cortex that makes up the parahippocampal gyrus (Figure 29.1). The entorhinal cortex has a much more complicated lamination pattern than the hippocampus proper. Indeed, the overall lamination pattern resembles that of the neocortex. Nevertheless, the lamination pattern is subtly different from that in the neocortex and, hence, is termed *transitional*.

The principal neurons of the entorhinal cortex include large stellate cells of layer II, medium-sized pyramidal cells in layer III, and polymorph cells in layer V. The large stellate neurons of layer II are the cells of origin of the projections to the dentate gyrus. The pyramidal neurons of layer III give rise to the projections to the hippocampus proper. Medially adjacent to the entorhinal cortex are two regions that contain cells that project to the entorhinal cortex: the *presubiculum* and *parasubiculum*.

The entorhinal cortex receives direct projections from the olfactory cortex and indirect projections from a large number of *cortical association areas* that are relayed by areas in the temporal lobe. These cortical association areas in turn receive input from each primary sensory cortex. In this way, the entorhinal cortex is positioned so as to relay highly processed sensory information from the association areas of the cortex to the hippocampal formation. It is noteworthy that the entorhinal cortex does not receive a major projection from the thalamus.

In terms of limbic circuitry, the most important efferent projection is to the hippocampal formation.

The Amygdaloid Complex

The amygdaloid complex is a collection of nuclei within the rostral pole of the temporal lobe (the area that lies beneath the surface feature of the temporal lobe called the *uncus*). The complex is noteworthy because much of it is not laminated, despite the fact that it is embedded in a cortical structure. The shape and cellular organization of the amygdala distinguish it from surrounding cortical regions. The general shape of the complex resembles an almond, hence the name *amygdala,* which is the Greek word for almond.

There are two major divisions of the amygdaloid complex: the *corticomedial* and the *basolateral.* The corticomedial division is the site of termination of most of the projections from the olfactory system (described in Chapter 26). The basolateral subdivision receives inputs from and gives rise to projections to cortical structures. The basolateral subdivision is the most pronounced nuclear group of the amygdaloid complex in the human. In terms of internal organization, 22 separate nuclear groups have been described within the amygdaloid complex. The internal circuitry of the complex is not well understood.

Inputs to the amygdaloid complex include the projections from the olfactory bulb and olfactory cortex (described in Chapter 26), projections from various cortical regions in the temporal lobe and from the cingulate cortex, projections from the hypothalamus, projections from the thalamus, and projec-

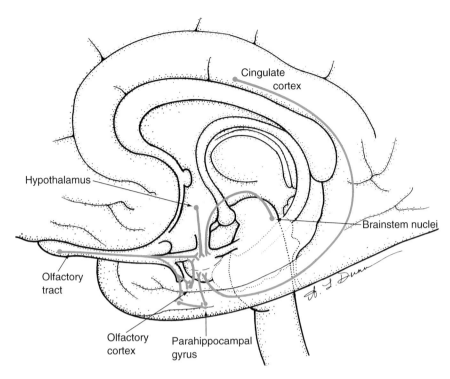

Figure 29.2. Inputs to the amygdaloid complex. Inputs to the amygdaloid complex include the projections from the olfactory bulb and olfactory cortex (described in Chapter 26), projections from various cortical regions in the temporal lobe and from the cingulate cortex, projections from the hypothalamus, projections from the thalamus, and projections from various brainstem nuclei. After Carpenter MB, Sutin J. *Human Neuroanatomy*. Baltimore, Md: Williams & Wilkins; 1983.

tions from various brainstem nuclei. These are summarized in Figure 29.2.

There are two principal output pathways from the amygdaloid complex (Figure 29.3): the *stria terminalis* and the *ventral amygdalofugal pathway.*

The stria terminalis is the prominent fiber pathway that follows the tail of the caudate nucleus as it curves around the lateral ventricle and then curves ventrally around the rostral border of the thalamus to enter the hypothalamus. The stria terminalis carries axons that originate primarily from neurons in the corticomedial division. An especially prominent component of fibers terminates in and around the ventromedial nucleus of the hypothalamus. Other fibers terminate in other regions of the hypothalamus, especially the preoptic area, in the bed nucleus of the stria terminalis, and in basal forebrain structures including the septal nuclei.

The ventral amygdalofugal pathway leaves the amygdala ventromedially. One component of fibers projects to the dorsomedial nucleus of the thalamus. Other components project rostrally to the prefrontal cortex and caudally to areas in the brainstem including the periaqueductal gray, nucleus of the solitary tract, and dorsal motor nucleus of the vagus.

It is noteworthy that many of the efferent pathways from the amygdaloid complex project to structures that are part of the autonomic nervous system or to the hypothalamus, which is considered to be the head ganglion of the autonomic nervous system. It is thought that these projections are the substrate through which forebrain structures influence autonomic functions.

The basolateral division also gives rise to important projections to areas in the parahippocampal gyrus of the temporal lobe. These

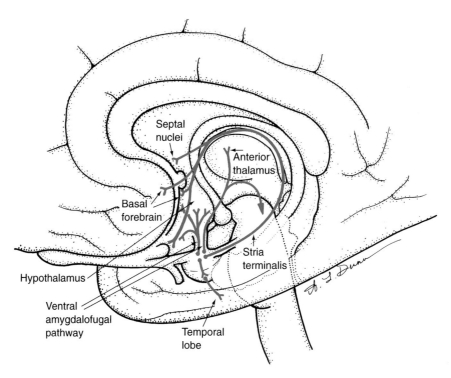

Figure 29.3. Outputs from the amygdaloid complex. The two principal output pathways from the amygdaloid complex are the stria terminalis and the ventral amygdalofugal pathway. Other outputs include the projections to the parahippocampal gyrus of the temporal lobe and to the anterior nuclear group of the thalamus. Modified from Noback CR, Demarest RJ. *The Human Nervous System.* New York, NY: McGraw-Hill; 1975.

are likely to relay information to the hippocampus via the projections from the entorhinal cortex.

and cingulate cortex. There are also important inputs from the brainstem and hypothalamus that arrive via the medial forebrain bundle.

The Septal Nuclei

The septal nuclei are located within the base of the *septum pellucidum,* the thin sheet of tissue that separates the anterior horns of the two lateral ventricles. Cholinergic and GABAergic neurons in the septal nuclei give rise to a sparse but functionally important input to the hippocampus. These projections are thought to play a key role in regulating the overall level of activity of hippocampal neurons.

Inputs to the septal nuclei derive from the hippocampus and from a number of other limbic structures, including the amygdala

The Cingulate Cortex

The cingulate cortex is the area defined by the cingulate gyrus. The lamination pattern in the cingulate cortex is transitional, as in the entorhinal cortex, but the cellular organization is somewhat different from that of the entorhinal cortex. Two cytoarchitectonic subdivisions are recognized, one in the anterior cingulate gyrus and one in the posterior. There are extensive interconnections between the two subdivisions, but each has somewhat different inputs and outputs.

The cingulate cortex is similar to the neocortex in that it receives projections from specific nuclei within the thalamus (in particular, the anterior nucleus). Other important inputs to the cingulate gyrus derive from the subiculum and from various association cortical areas. The projection from the anterior nucleus is of particular note because the anterior nucleus receives a prominent projection from the mammillary bodies (the *mammillothalamic tract*). This tract and the projections from the anterior nucleus to the cingulate cortex are important relays in the Papez circuit (see later).

The principal outputs of the cingulate cortex are to areas in the temporal lobe that in turn project to the entorhinal cortex (specifically the presubiculum), to the amygdala, and to the dorsomedial nucleus of the thalamus. The dorsomedial nucleus in turn projects to the prefrontal cortex. In this way, the cingulate cortex is positioned so as to relay information between limbic structures and other areas of the neocortex.

THE PRINCIPAL CIRCUITS OF THE LIMBIC SYSTEM

We have considered each component of the limbic system and the principal inputs and outputs of each structure, and this information is summarized in Table 29.1 It is now useful to consider how these connections can be organized into circuits. In this regard, it is useful to distinguish four principal circuits.

1. The *amygdala circuit*
2. The *midline circuit*
3. The *olfactory circuit*
4. The *Papez circuit*

The amygdala circuit is schematized in Figures 29.2 and 29.3. The circuit includes the stria terminalis, the ventral amygdalofugal pathway, and other projections that travel via unnamed tracts.

The midline circuit is schematized in Figure 29.4. This circuit is the one that interconnects the limbic system with the hypothalamus and autonomic nervous system. It comprises several projection systems that interconnect the telencephalic and diencephalic portions of the

limbic systems with the "limbic midbrain." The major relays include the *medial forebrain bundle*, the *mammillotegmental tract*, the interconnections between the septum and the habenula that travel via the *stria medullaris thalami*, and the habenular projection to the interpeduncular nucleus (the *habenulointerpeduncular* projection system).

The principal connections of the olfactory circuit are schematized in Figure 29.2. Important components of olfactory tract fibers terminate in the amygdaloid complex (especially the corticomedial division) and in the lateral part of the entorhinal cortex. The entorhinal cortex also receives projections from the primary olfactory cortex.

The Papez circuit is summarized in Figure 29.5. This is probably the most visually prominent circuit of the limbic system. The relays of the circuit include (1) the hippocampal projection to the mammillary nuclei via the fornix; (2) the projection from the mammillary bodies to the anterior nucleus of the thalamus (the *mammillothalamic tract*); (3) the projection from the anterior nucleus of the thalamus to the cingulate cortex; and (4) the projection from the cingulate cortex back to the presubiculum, which then projects to the entorhinal cortex. Also schematized in the figure are the projections from the septal nuclei to the hippocampal formation.

Functional Characteristics of Limbic Circuitry

Essentially all of the projections described above are excitatory. The only exception is the projection from the septal nuclei to the hippocampus, which has an inhibitory component (noted earlier).

Limbic Pathways Exhibit Activity-Dependent Synaptic Plasticity

It is noteworthy that certain pathways of the limbic system exhibit *activity-dependent synaptic plasticity,* which includes *long-term potentiation* (LTP) and *long-term depression* (LTD).

Table 29.1. Connections of the Limbic System. The principal interconnections between the different components of the limbic system are summarized

Structure	Afferents	Efferents
Hippocampal formation	1. From entorhinal cortex via perforant path 2. From septal nuclei via fornix (ACh and GABA) 3. From brainstem nuclei via fornix	1. To entorhinal cortex (from subiculum) 2. To mammillary body via fornix
Entorhinal cortex	1. From areas in the temporal lobe; these relay information from cortical association areas 2. From olfactory cortical areas	1. To hippocampus via perforant path
Amygdaloid complex	1. From olfactory bulb (to corticomedial division) 2. From cortical areas (to basolateral division)	1. To preoptic and ventromedial nuclei of hypothalamus septal nuclei and basal forebrain via stria terminalis 2. To dorsomedial nucleus of the thalamus and brainstem autonomic centers via the ventral amygdalofugal pathway
Septal nuclei	1. From amygdala 2. From cingulate cortex 3. From brainstem and hypothalamus via medial forebrain bundle	1. To hippocampal formation 2. To hypothalamus and brainstem via medial forebrain bundle
Cingulate cortex	1. From anterior nucleus of thalamus 2. From subiculum (of hippocampal formation) 3. From association cortex	1. To presubiculum 2. To amygdaloid complex 3. To anterior nucleus of thalamus

These forms of plasticity are considered to represent the kinds of processes that may play a role in long-term information storage.

LTP and LTD were initially discovered through studies of two pathways in the hippocampal formation: (1) the pathway from the entorhinal cortex to the granule cells of the dentate gyrus (the perforant path), and (2) the projection from the CA3 region of the hippocampus proper to the CA1 region.

LTP is a very long-lasting increase in synaptic efficacy that can be induced by stimulating the pathways with brief high-frequency trains. Figure 29.6 illustrates an example of the induction of LTP in the perforant path. LTP can be induced with only a few trains, yet can persist for hours and even days.

Importantly, the enduring phase of LTP depends on RNA and protein synthesis. When either transcription or translation is blocked by means of drugs, the changes in synaptic efficacy occur only transiently. These results suggest that the laying down of this cellular memory trace depends on the expression of particular genes.

It is noteworthy that when animals are given drugs that block protein synthesis after receiving training in simple memory tasks, memory storage is disrupted. This is one of the key pieces of information that suggests that the synaptic changes induced during

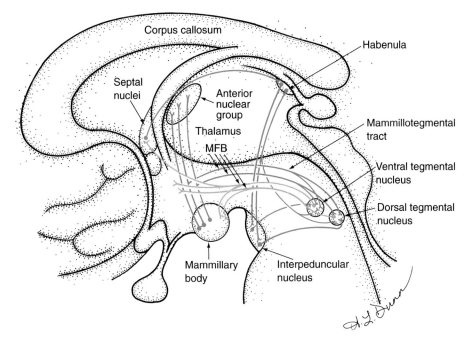

Figure 29.4. The midline circuit of the limbic system. The major relays of the midline circuit include the medial forebrain bundle, the mammillotegmental tract, the interconnections between the septum and the habenula that travel via the stria medullaris thalami, and the habenular projection to the interpeduncular nucleus (the habenulointerpeduncular projection system). Modified from Carpenter MB, Sutin J. *Human Neuroanatomy.* Baltimore, Md: Williams & Wilkins; 1983.

LTP have characteristics that are similar to the changes that are responsible for information storage.

LTD is a long-lasting decrease in synaptic efficacy. One form of LTD occurs in the dentate gyrus in populations of synapses that remain inactive at the same time that LTP is induced in other synapses. This is termed *heterosynaptic LTD* (heterosynaptic because activation of one population of synapses affects synaptic efficacy in another population). Another form of LTD has been reported following prolonged low-frequency stimulation of the CA3–CA1 projection. This is termed *homosynaptic LTD* because the change occurs in the same population of synapses that was activated by the low-frequency stimulation.

Although hippocampal pathways have been a favorite site for experimental studies of LTP and LTD, other pathways exhibit similar forms of synaptic plasticity. LTP has been described in the pathways emanating from the amygdala and in several synaptic relays of the olfactory cortex. Similar long-lasting changes in synaptic efficacy have been seen in the synaptic relays of the neocortex. Nevertheless, the characteristic features of the LTP and LTD differ in important ways at different types of synapses.

Activity-dependent synaptic plasticity can be studied in brain slices maintained in vitro. Much of our knowledge about the synaptic mechanisms of LTP and LTD has come from neurophysiological studies of hippocampal slices studied in vitro. For these experiments, the hippocampus is quickly removed from the brain and cut into slices that are about 400 μm thick. The slices are then transferred to an oxygenated physiological medium where they can be maintained for hours. The in vitro approach facilitates the neurophysiological experiments and also offers the advantage that drugs can be delivered directly.

In vitro approaches have recently been adapted for studies of human tissue that has

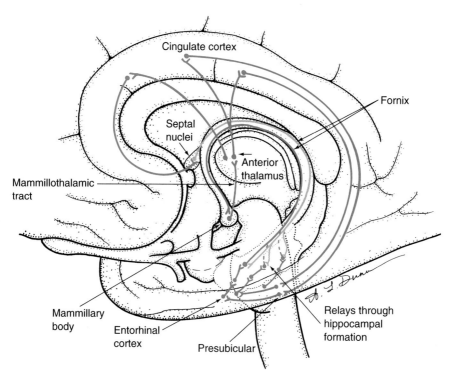

Figure 29.5. The Papez circuit. The relays of the Papez circuit include (1) the hippocampal projection to the mammillary nuclei via the fornix; (2) the projection from the mammillary bodies to the anterior nucleus of the thalamus (the *mammillothalamic tract*); (3) the projection from the anterior nucleus of the thalamus to the cingulate cortex; and (4) the projection from the cingulate cortex back to the presubiculum, which then projects to the entorhinal cortex. Modified from Noback CR, Demarest RJ. *The Human Nervous System.* New York, NY: McGraw-Hill; 1975.

been removed during neurosurgical procedures. These types of studies are providing important new insights in the cellular mechanisms that lead to the uncontrolled neuronal firing that leads to epilepsy.

Limbic structures in the temporal lobe (the hippocampus, entorhinal cortex, and amygdaloid complex) are especially prone to abnormal electrical activity. One common seizure disorder is *temporal lobe epilepsy.* In this disorder, an injury or disease process affects structures in the temporal lobe so that they begin to exhibit increased excitability. The exact mechanism through which this initial event causes an increased excitability is not entirely known. These hyperexcitable sites can be periodically triggered into an uncontrolled pattern of discharge termed an *electrographic seizure.*

The electrographic seizures are initially restricted to the site of the initial abnormality. Neurologists refer to these as *focal electrographic seizures;* the site in which focal seizures originate is termed a *focus.* If focal seizures occur repetitively, other areas are often recruited into the abnormal activity pattern. Eventually, a bout of abnormal activity in the temporal lobe can trigger seizure activity that spreads to the cerebral cortex. In this situation, a behavioral seizure occurs.

An important feature of temporal lobe epilepsy is its progressive nature. If untreated, both the frequency and severity of seizures of temporal lobe origin increase. Experimental studies have revealed that this is not a progression of the disease process that caused the focal abnormality in the first place. For example, when focal electrical

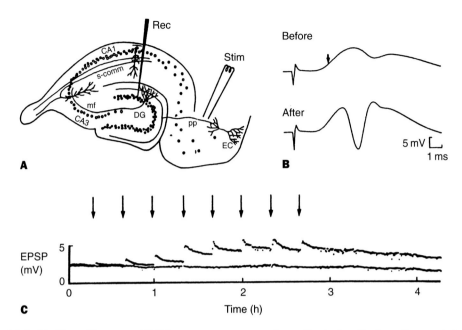

Figure 29.6. Limbic pathways exhibit long-lasting forms of synaptic plasticity (long-term potentiation). An example of long-term potentiation in the perforant path—the projection from the entorhinal cortex to the dentate gyrus. **A,** Schematic diagram of the hippocampal formation showing the major excitatory pathways. The perforant path (pp) originates in the entorhinal cortex (EC) and terminates on granule cells in the dentate gyrus (DG). **B** illustrates evoked potentials generated in the dentate gyrus by stimulating the perforant path. These are extracellular recordings in which the excitatory postsynaptic currents generated in granule cells produce large extracellular potentials. The rising phase of the response is an indication of the summed excitatory postsynaptic potentials (EPSPs) from a large number of granule neurons. The downward-going deflection is a "population spike" generated by the near-simultaneous discharge of large numbers of granule cells. The upper trace illustrates the baseline response before delivering brief high-frequency trains. The lower trace illustrates the response after delivering high-frequency stimulation. **C** illustrates an experiment in which recordings were made on two sides of the brain, only one of which received high-frequency stimulation. The dots indicate measurements of the amplitudes of the EPSPs over time. On the control side, there is no change in response amplitude. On the experimental side, high-frequency trains were delivered at each downward arrow. Note the progressive increase in response amplitude, and the fact that the "potentiation" is maintained for up to 2 hours after the last train. From Bliss TVP, Lomo T. Long-term potentiation of synaptic transmission in the hippocampus: properties and mechanisms. In: Lanfield PW, Deadwyler SA, eds. *Long-term Potentiation: From Biophysics to Behavior.* New York, NY: Alan R Liss; 1988:3–72.

stimulation is delivered repeatedly via electrodes that are chronically implanted into limbic structures in experimental animals, there is a similar progressive development of seizures. Initially, stimulus trains cause only focal electrographic discharges. If, however, the stimulation is delivered once per day for several days, the same stimulus trains come to elicit full-blown behavioral seizures. The process through which there is a progressive development of seizure severity is called *kin-dling.* Essentially the same process is thought to be responsible for the development of temporal lobe epilepsy.

Activity-dependent synaptic plasticity could contribute to the progressive increase in seizure severity. Specifically, the seizure activity at the initial focus could lead to LTP-like enhancement of synaptic efficacy at other sites. In this way, the consequences of the abnormal activity at the initial focus would be progressively amplified.

BEHAVIORAL FUNCTIONS OF THE LIMBIC SYSTEM

As we have seen in previous sections, three techniques have been used to elucidate the functional/behavioral role of particular structures or circuits: (1) one can damage particular regions and evaluate how behavior is altered (the experimental lesion technique), (2) one can stimulate particular regions and observe the behaviors that are induced, and (3) one can observe the signs and symptoms that result from disease processes that affect the structures or circuits in question.

There is an enormous literature pertaining to the limbic system in which each approach has been used alone or in combination. This body of work suggests certain general conclusions:

1. The limbic system, and especially the amygdaloid complex, is an important relay through which conscious activities influence autonomic functions. Stimulation of the amygdala triggers integrated autonomic responses that resemble the responses that occur naturally in situations that cause anger or fear.

2. Components of the limbic system (especially the cingulate cortex and amygdala) are involved in regulating affect (emotion). Circuits involving these structures play an important role in behaviors that are essential to the self-preservation of the organism (feeding, fighting, procreating) and behaviors that contribute to the preservation of the species (emotions that trigger appropriate care of offspring, social behaviors, and so on). Electrical stimulation of these areas produces complicated integrated responses that appear identical to the responses that are seen in situations that cause anger or fear. Experimental lesions of the structures cause the opposite response—a placid and emotionally unresponsive animal.

3. Components of the limbic system, especially the hippocampal formation and entorhinal cortex, play an important role in memory. In particular, as we see in more detail later, damage to the hippocampal formation and/or entorhinal cortex produces characteristic *amnesic syndromes.*

Nevertheless, despite intense effort, there is still no clear understanding of exactly how the different structures and circuits actually operate. For this reason, the most important concerns for the physician are the symptoms that result from lesions or disease processes that affect the limbic system.

Symptoms Resulting From Damage to Components of the Limbic System

Damage to components of the limbic system may cause changes in affect, changes in social behavior (patients may develop sociopathic or psychopathic personalities), or changes in memory function. In most cases, symptoms either do not appear at all or are very minor after unilateral lesions and are problematic only when lesions occur bilaterally.

Lesions involving the hippocampal formation and entorhinal cortex cause amnesic syndromes. Bilateral injury to the hippocampal formation causes a remarkable memory disorder in which patients are unable to form new memories. Such bilateral injuries can occur as a result of episodes of transient ischemia (for example, as a result of cardiac insufficiency or near-drowning). Neurons in the hippocampus are especially vulnerable to such ischemic episodes. After prolonged, transient ischemic episodes, there is extensive neuronal loss in "vulnerable" sectors of the hippocampus, producing an extensive but selective lesion. Patients who experience such bilateral hippocampal lesions retain their long-term memories of people, places, and events that occurred before the injury, but have difficulty forming new memories. This is termed *anterograde amnesia.*

A remarkable illustration of this type of memory deficit is provided by a very famous patient. H. M. began to experience seizures about one year after being hit by a bicycle at age 9. As is typical of temporal lobe epilepsy, the frequency and severity of the seizures increased progressively. By the time H. M. was in his early 20s, his seizures had become frequent and severe, and the available anticonvulsant medications did not provide adequate seizure control. At that time (the early

1950s), surgical strategies were being considered to treat intractable temporal lobe epilepsy. The idea was that if seizures were triggered by certain structures, the removal of those structures would eliminate the seizures. Hence, H. M. was offered a surgical procedure that had previously been used only in psychotic patients: bilateral resection of the temporal lobe.

H. M. underwent this experimental surgical procedure at the age of 27. The surgeon (W. B. Scoville) removed about 8 to 10 cm of brain tissue, which included the hippocampal formation, the amygdala, and much of the parahippocampal gyrus. The procedure had the desired effect of reducing the frequency of H. M.'s seizures. Unfortunately, however, the lesion also caused a severe amnesic syndrome characterized by profound anterograde amnesia. Although H. M.'s IQ remained in the normal or above-normal range after the surgery, H. M. was essentially completely unable to form new memories about people, places, and events. He was and is literally trapped in time.

MRI studies have only recently been undertaken with H. M. because of concerns about metal clips left in the dura at the time of the surgery. However after ensuring that the clips would not be displaced by the magnetic fields generated, a full MRI was completed and the complete extent of the lesion is now known. These studies confirmed extensive destruction of the hippocampal formation, entorhinal cortex, and other areas in the parahippocampal gyrus.

H. M. is undoubtedly one of the most thoroughly studied patients in history. A battery of psychological tests over the 40-year period since his surgery have provided considerable insights into the nature of human memory. His remarkable symptoms also triggered an avalanche of studies in experimental animals that attempted to duplicate the memory disorder. These experimental studies implicated the hippocampal formation and the entorhinal cortex and surrounding structures as being especially important.

The concept that has emerged from the combined evidence from human and experimental animal studies is that there is a *temporal lobe memory circuit* that involves the entorhinal cortex and related structures, and the hippocampal formation. This circuit is thought to play a key role in what is now called *declarative memory* (memories of people, places, and events).

It is important to emphasize that damage to limbic structures in the temporal lobe causes deficits in memory acquistition; the lesions usually do not disrupt established memories. Hence, the structures are not the site at which memories are stored. It is thought that memories are stored in a distributed fashion in the cerebral cortex, and that the temporal lobe memory circuit is important for laying down the memory trace.

Damage to other components of the limbic system also causes memory disorders. One example is the memory disruption seen in *Korsakoff's syndrome*. This syndrome is due to prolonged thiamine deficiency usually brought about by chronic alcohol consumption. Hence the syndrome is sometimes called *alcoholic dementia*. The thiamine deficiency causes the degeneration of neurons in the mammillary bodies, dorsomedial nucleus of the thalamus, and the hippocampus.

Patients with Korsakoff's syndrome have deficits in both short-term and long-term memory. They also exhibit an interesting tendency to *confabulate,* that is, fill in the gaps in their memory with inaccurate details. They also often appear demented and may exhibit ataxia and gaze disorders due to damage to the cerebellum.

Bilateral lesions involving the cingulate cortex cause changes in *emotional tone* (affect). Bilateral lesions involving the cingulate cortex diminish normal emotional responses. Patients sometimes become akinetic to the point of remaining immobile for long periods. They may also be mute and unresponsive, although they are not comatose. This condition is called *akinetic mutism.*

Bilateral lesions involving the amygdala produce a complicated set of symptoms (the *Kluver–Bucy syndrome*). The syndrome was initially described in experimental animals but is also seen in patients with bilateral damage to the amygdala and surrounding structures. The symptoms include (1) *hypersexuality,* which is expressed by inappropriately suggestive behavior and inappropriate contact, (2) a tendency to place objects in the mouth *(hyperorality),* (3) a tendency to explore the immediate environment in a compulsive fashion, and (4) an unusually placid demeanor similar to the decreased emotional tone that occurs with lesions of the cingulate cortex. Patients may also be unable to recognize objects by sight *(visual agnosia)* or decode sounds *(auditory agnosia).*

It is noteworthy that many of the symptoms that result from lesions of limbic structures resemble the symptoms of common psychiatric disorders. This similarity has led to the concept that some psychiatric disorders may be due to pathological processes involving the components of the limbic system. The idea that psychiatric diseases have an organic basis is still controversial, however.

11

Arousal, Attention, Consciousness

30

Modulatory Systems of the Brainstem

As noted in previous chapters, the brainstem contains components of a number of different sensory and motor systems, including (1) second-, and third-order neurons of the somatosensory, auditory, and vestibular systems; (2) lower motor neurons in the cranial nerve nuclei; (3) preganglionic neurons of the parasympathetic nervous system; (4) circuitry that mediates autonomic reflexes (for example, those that control breathing and cardiovascular function); (5) neurons that give rise to the descending brainstem pathways of the motor system (reticulospinal and vestibulospinal tracts); (6) neuronal groups that are part of cerebellar circuitry (olivary complex and the pontine nuclei); and (7) various sensory and motor tracts that relay information to and from the spinal cord.

The brainstem also contains the cells of origin of systems that play an important role in modulating neuronal activity throughout the brain. These systems play a role in mediating the overall level of arousal (alertness), in regulating sleep and wakefulness, and in attentional processes. In this chapter we briefly consider these modulatory systems.

THE BRAINSTEM RETICULAR FORMATION

The cells of origin of the modulatory brainstem systems are located in an area that is called the *brainstem reticular formation,* or just *reticular formation*. The reticular formation is defined by exclusion; it is the gray matter in the core of the medulla, pontine tegmentum, and midbrain that is not organized into discrete sensory and motor nuclei or named ascending and descending sensory, motor, and autonomic fiber tracts. The name derives from the fact that the area appears to be a diffuse collection of neurons intermixed with a network of axons. The axonal network and intercalated neuronal cell bodies have a "reticulated" appearance, hence, the name *reticular formation* was used to refer to the area in which this mixed network is found. Figure 30.1A illustrates the area termed the *reticular formation* in sections through the brainstem.

The reticular formation is phylogenetically the oldest part of the brain; it has a cellular organization that is similar to the spinal

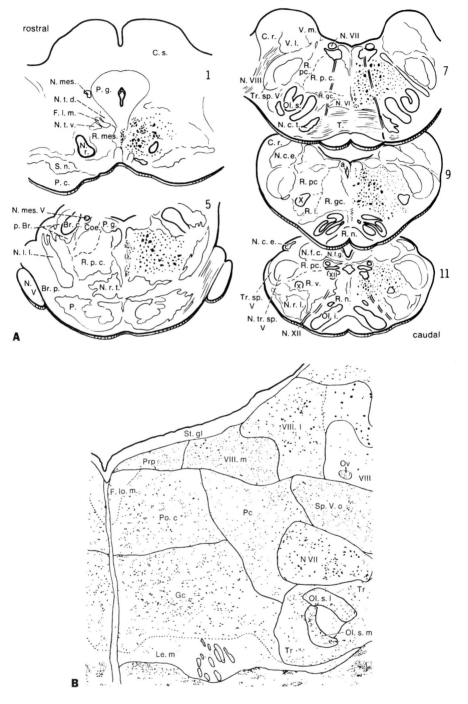

Figure 30.1. Location of the brainstem reticular formation. A illustrates the location of the cells comprising the reticular formation in a series of sections through the brainstem of the cat. Nuclei of the reticular formation include R.gc, gigantocellular nucleus; R.l., lateral nucleus; R.mes., mesencephalic reticular formation; R.n, raphe nucleus; R.p.c., caudal pontine nucleus; R.pc., parvicellular nucleus; R.v., ventral nucleus; N.r.t., reticular nucleus of the pontine tegmentum. Other abbreviations: Coe, nucleus subceruleus; F.l.m., medial longitudinal fasciculus; N.c.e., external cuneate nucleus; N.f.c., cuneate nucleus; N.f.g., gracile nucleus; N.r., red nucleus;

cord. Neurons in the reticular formation are characterized by having straight, poorly ramified dendrites; these dendrites usually extend more or less perpendicular to the long axis of the brainstem. The axons of these neurons often project over wide areas and have numerous collaterals. In fact, individual neurons often give rise to both ascending and descending projections (see Figure 30.2).

Although conventional histological methods do not reveal a clear organization within the reticular formation, organization is evident when the region is studied with more sophisticated staining techniques. Histochemical and immunocytochemical techniques that identify particular transmitter

systems were especially useful for defining several discrete nuclear groups that use particular neurotransmitters. Hence, we now recognize a number of nuclear groups, each of which includes several individual nuclei:

1. The *precerebellar* reticular nuclei
2. The *central* reticular nuclei
3. The *lateral* reticular nuclei
4. The *serotonergic cell groups* in the raphe nuclei
5. The *noradrenergic cell groups* (primarily the locus ceruleus)
6. The *dopaminergic cell groups* in the midbrain

Certain of these cell groups are involved in motor function. For example, neurons in the precerebellar reticular nuclei receive input from the spinal cord, vestibular nuclei,

Figure 30.2. The axonal projections of a single neuron in the reticular formation. The drawing is a reconstruction of a Golgi-stained neuron in the magnocellular nucleus of the reticular formation. This individual neuron gives rise to an axon that branches into ascending and descending collaterals. N.gc, the gigantocellular nucleus of the reticular formation; N.g, nucleus gracilis; P.g., periaqueductal gray; Pf, Pc, and Re, various thalamic nuclei; H, hypothalamus. From Scheibel ME, Scheibel AB. Structural substrates for integrative patterns in the brainstem reticular core. In: Jasper HH, Proctor LD, et al, eds. *Henry Ford Hospital Symposium: Reticular Formation of the Brain.* Boston, Mass: Little & Brown; 1958:31–55.

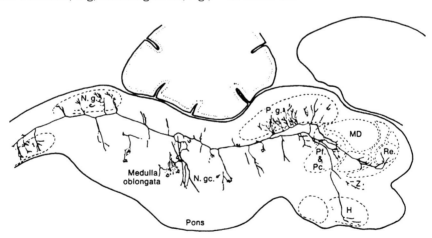

Figure 30.1. *(Continued)*
N.tr.sp.V, spinal trigeminal nucleus; P., pontine nuclei; P.g, periaqueductal gray. **B** illustrates a schematic representation of a Nissl-stained section through the pontine tegmentum of a human. The components of the reticular formation at this level include Po.c, caudal pontine nucleus; Gc, Gigantocellular nucleus; Pc, Parvicellular nucleus. Other abbreviations: N.VII, facial nucleus; Sp.V.o, spinal trigeminal nucleus; VIII.l and VIII.m, vestibular nuclei; Prp, nucleus prepositus; F.lo.m, medial longitudinal fasciculus; Le.m, medial lemniscus. **A** is reproduced with permission from Brodal A. *Neurological Anatomy.* 3rd ed. New York, NY: Oxford University Press; 1981. **B** is from Olszewski J, Baxter D. *Cytoarchitecture of the Human Brain Stem.* New York, NY: S Karger; 1954.

and cerebral cortex and project to the cerebellum. Neurons in the central reticular nuclei (including the gigantocellular reticular nucleus) are the cells of origin of the reticulospinal tract.

The other cell groups are not immediately assignable to sensory or motor systems; instead, these neurons seem to be part of modulatory systems that play a key role in regulating the overall level of activity in other brain regions:

1. The cholinergic system
2. The serotonergic system
3. The noradrenergic system
4. The dopaminergic system

We consider each of these systems in turn.

The Cholinergic System

The cholinergic neurons of the reticular formation are situated primarily in the dorsomedial tegmentum (Figure 30.3). The identification of the neurons as cholinergic is based on histochemical and immunocytochemical studies. These neurons give rise to projections that project to the thalamus. Many of these cholinergic fibers terminate in the intralaminar, midline, and reticular nuclei of the thalamus, although there are also projections to other thalamic relay nuclei. The intralaminar nuclei in turn contain the neurons that give rise to *nonspecific thalamocortical projections.*

To recall the distinction between *specific* and *nonspecific* thalamocortical projections: Specific thalamocortical projections arise from *thalamic relay nuclei* and project to specific areas of the cortex. Nonspecific thalamocortical projections arise from the intralaminar, midline, and reticular nuclei of the thalamus. Neurons in these nuclei project to widespread cortical areas. Specific projections terminate in discrete tufts in layer IV; nonspecific projections terminate diffusely in layer I.

The Concept of a Reticular Activating System

The nonspecific thalamocortical projections modulate the overall level of activity through-out the cerebral cortex. This activation plays a key role in regulating our level of consciousness. Indeed, either naturally occurring or experimental lesions of the reticular formation result in a permanent state of behavioral coma. Hence, the cholinergic cell groups of the reticular formation, together with the intralaminar nuclei of the thalamus, are thought to be components of a system that activates the cerebral cortex.

The concept of an activating system based in the reticular formation first arose as a result of studies by G. Moruzzi and H. Magoun in the late 1940s that revealed that stimulation of the core of the brainstem in anesthetized animals caused changes in the electroencephalogram (EEG) that resembled the transition from deep sleep to an alert, aroused state. These results, together with the fact that lesions of the brainstem cause a permanent state of coma, led to the concept of an "activating system" that began in the reticular formation and relayed through the thalamus. The term *reticular activating system* was used to refer to the part of this system within the reticular formation. It is important to emphasize that this is a functional concept and does not "map" exactly onto the anatomical entity called the reticular formation.

Neurons in the reticular formation receive input from sensory systems. Activation of the cortex occurs in response to events in the environment. A painful stimulus, a loud noise, a flash of light that surprises or frightens us can trigger an enhanced state of "alertness" and also trigger autonomic responses of the fight-or-flight response (increased heart rate, pupillary dilation). For this reason, considerable experimental effort was directed toward identifying how sensory information might reach the reticular formation. A host of physiological and anatomical studies carried out in the 1950s and 1960s revealed that neurons in the reticular system receive inputs from all sensory systems. Somatosensory information is conveyed by spinoreticular projections that travel with ascending spinothalamic pathways; visual information is conveyed via pathways that relay through the superior colliculus; auditory and vestibular information is conveyed

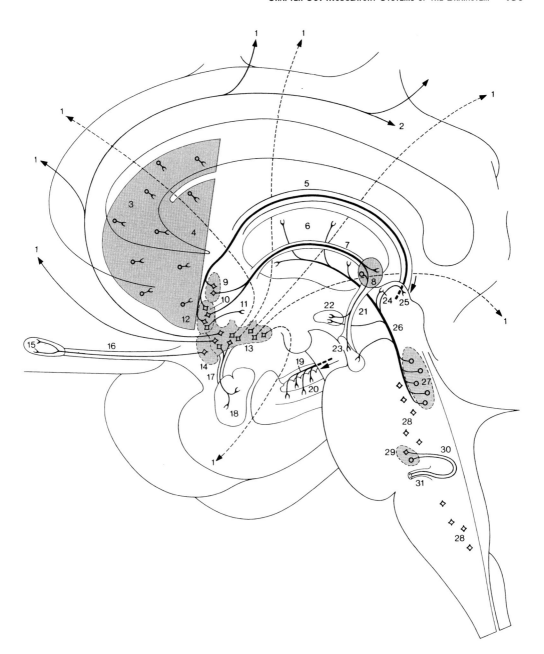

Figure 30.3. Cholinergic cell groups in the reticular formation give rise to ascending projections to the thalamus. Schematic illustration of the distribution of cholinergic neurons and their axonal projections in the human brain. Note the collection of cholinergic neurons in the dorsal tegmental region (27). Cholinergic cell groups in other brain regions are also shown. 1, neocortex; 2, cingulate cortex; 3, caudate nucleus; 4, putamen; 5, cholinergic axons in the fornix; 6, thalamus; 7, stria medularis; 8, habenular nucleus; 9, septal nuclei; 10, diagonal band; 11, lateral hypothalamus; 12, nucleus accumbens; 13 and 14, basal forebrain cholinergic system; 15–26, various cholinergic pathways; 29, 30, and 31, cell bodies and axonal projections of periolivary nucleus. From Nieuwenhuys R. *Chemoarchitecture of the Brain.* Berlin, Germany: Springer-Verlag; 1985.

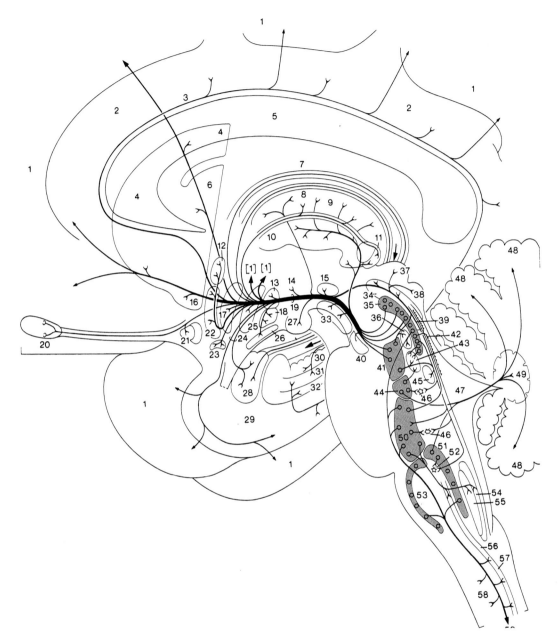

Figure 30.4. The serotonergic system. 35, 41, 44, 50, 51, and 53, serotonergic cell groups in the brainstem raphe. Other abbreviations: 1, neocortex; 2, cingulate cortex; 3, serotonin axons in cingulate cortex; 4, caudate nucleus; 5, corpus callosum; 6, putamen; 7, fornix; 8, stria terminalis; 9, thalamus; 10, stria medularis; 11, hebenula; 12, septal nuclei; 13, dorsomedial nucleus of hypothalamus; 14, lateral hypothalamus; 15, ventral tegmental area; 16, nucleus accumbens; 17, preoptic nucleus; 18, ventromedial nucleus of hypothalamus; 19, medial forebrain bundle; 20, olfactory bulb; 21, anterior olfactory nucleus; 22, basal forebrain; 23, suprachiasmatic nucleus; 24, ventral amygdalofugal pathway; 25, anterior hypothalamus; 26, arcuate nucleus (also called the *infundibular nucleus*); 27, mammillary body; 28, amygdaloid complex; 29, parahippocampal gyrus; 30, dentate gyrus; 31, hippocampus proper; 32, subiculum; 33, substantia nigra; 34, periaqueductal gray; 35, dorsal raphe nucleus; 36, dorsal tegmental nucleus; 37, superior colliculus; 38, inferior colliculus; 39, dorsal

via projections from the respective brainstem nuclei. Hence, individual neurons in the reticular formation are said to receive *multimodal inputs.*

To further discuss the ways that the activating system modulates cortical activity, it is necessary to understand how cortical electrical activity is actually measured. Hence, we postpone further consideration of how the activating system modulates cortical activity until after we consider the cortex itself in Chapter 31.

Before leaving the topic of the activating system, it is worth emphasizing that the role of the system is more than the term *activating* implies. By regulating the overall level of activity in the cerebral cortex, the activating system plays a key role in regulating our level of consciousness. This is amply demonstrated by the fact that injury to the brainstem, and especially the mesencephalic reticular formation, can cause permanent loss of consciousness (coma).

The Serotonergic System

The serotonergic neurons of the brainstem are collected in a group of nuclei at the midline termed the *raphe nuclei.* These neurons were initially identified through histofluorescent techniques that revealed the presence of the neurotransmitter serotonin (5-hydroxytryptamine, or 5-HT). Serotonergic neurons give rise to widespread projections that extend throughout most of the brain and spinal cord (Figure 30.4).

The descending projections to the spinal cord play an important role in modulating transmission over pain pathways (see Chapter 12). The exact functional role of the serotonergic projections to other brain regions is less clear. Neurophysiological studies indicate that serotonin can induce either slow depolarization or slow hypolarization as a result of the modulation of K^+ channels. These changes in membrane potential cause sustained increases or decreases in neuronal firing rates.

It is thought that the serotonergic system plays a role in regulating the phase of sleep characterized by slow waves in the EEG (*slow-wave sleep*). Lesions of the serotonergic cell groups in experimental animals cause insomnia. Also noteworthy is the fact that the drug LSD operates on serotonin receptors.

The Noradrenergic System

The noradrenergic neurons of the brainstem are collected in two nuclei: (1) the *locus ceruleus* (the *blue locus,* so called because the nucleus appears blue in fresh brain sections), and (2) the lateral tegmental nucleus. The locus ceruleus is located in the dorsolateral tegmentum just rostral to the principal trigeminal nucleus. The lateral tegmental nucleus is located in the lateral pontine tegmentum. There are also scattered noradrenergic neurons throughout the ventral tegmentum. It is noteworthy that virtually all the cell bodies of noradrenergic neurons are located in the brainstem.

The noradrenergic neurons, especially those of the locus ceruleus, give rise to extensive projections to widespread regions in the brain and spinal cord (Figures 30.5 and 30.6). In many cases, the collaterals of a single neuron project to very different structures (the cerebral cortex and the cerebellum, for example).

Neurophysiological studies indicate that noradrenalin, like serotonin, can induce

---

Figure 30.4. (Continued)
longitudinal fasciculus; 40, interpeduncular nucleus; 41, choroid plexus of the fourth ventricle; 43, locus ceruleus; 45, parabrachial nucleus; 46, pontine reticular formation; 47, fourth ventricle; 48, cerebellar cortex; 49, deep nuclei of the cerebellum; 52, medullary reticular formation; 54, nucleus of the solitary tract; 55, dorsal motor nucleus of the vagus; 56, spinal trigeminal nucleus; 57, dorsal horn of spinal cord; 58, ventral horn of spinal cord; 59, intermediolateral nucleus. From Nieuwenhuys R. *Chemoarchitecture of the Brain.* Berlin, Germany: Springer-Verlag; 1985.

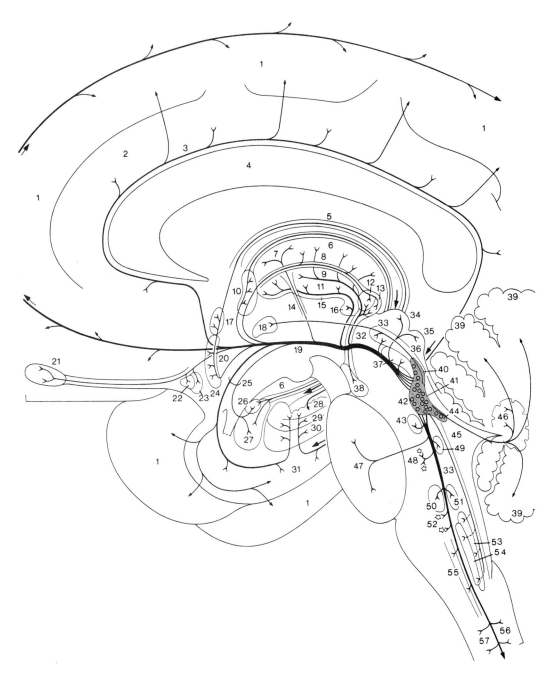

Figure 30.5. Neurons in the locus ceruleus give rise to projections that extend throughout most of the brain and spinal cord. 40–42, and 44, the locus ceruleus complex. Other abbreviations: 1, neocortex; 2 and 3, cingulate gyrus; 4, corpus callosum; 5, fornix; 6, stria terminalis; 7, anterior nucleus of the thalamus; 8, stria medularis; 9, thalamus; 10, bed nucleus of the stria terminalis; 11, internal medullary lamina of thalamus; 12 and 13, habenula; 14, mammillothalamic tract; 15 and 16, axonal projections to various thalamic nuclei; 17, septal nuclei; 18, paraventricular nucleus of hypothalamus; 19, medial forebrain bundle; 20, nucleus of the diagonal band; 21, olfactory bulb; 22, anterior olfactory nucleus; 23, olfactory tubercle; 24–30, various noradrenergic pathways; 31, parahippocampal gyrus; 32, projection to habenula; 33, dorsal longitudinal fasciculus; 34, superior colliculus; 35, inferior colliculus; 36, periaqueductal gray; 37, dorsal

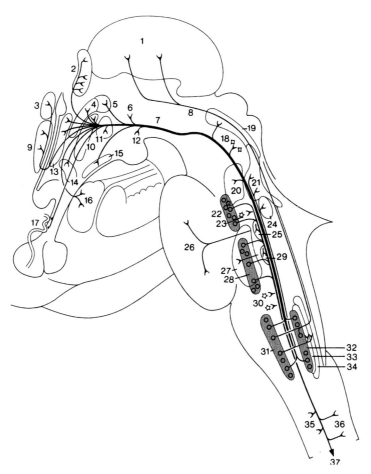

Figure 30.6. Other noradrenergic cell groups of the brainstem. 22, 28, 31, 32, noradrenaline cell groups of the brainstem. Other abbreviations: 1, thalamus; 2, bed nucleus of the stria terminalis; 3, septal nuclei; 4 and 5, paraventricular nucleus of hypothalamus; 6, lateral hypothalamus; 7, medial forebrain bundle; 8, dorsal longitudinal fasciculus; 9, basal forebrain; 10, anterior hypothalamus; 11, dorsomedial nucleus of the hypothalamus; 12, caudal hypothalamus; 13, preoptic nucleus; 14, supraoptic nucleus; 15, arcuate nucleus (also called the *infundibular nucleus*); 16, amygdaloid complex; 17, median eminence; 18, mesencephalic reticular formation; 19, periaqueductal gray; 20 and 21, locus ceruleus and subceruleus; 23, pontine reticular formation; 24, parabrachial nuclei; 25, motor nucleus of V; 26, pontine nuclei; 27, raphe magnus nucleus; 29, facial nucleus; 30, medullary reticular formation; 33, dorsal motor nucleus of the vagus; 34, nucleus of the solitary tract; 35, central gray; 36, dorsal horn of the spinal cord; 37, intermediolateral column. From Nieuwenhuys R. *Chemoarchitecture of the Brain.* Berlin, Germany: Springer-Verlag; 1985.

Figure 30.5. *(Continued)*
raphe nucleus; 38, interpeduncular nucleus; 39, cerebellar cortex; 43, nucleus of the lateral lemniscus; 45, superior cerebellar peduncle; 46, deep cerebellar nuclei; 47, pontine nuclei; 48, mesencephalic reticular formation; 49, principal nucleus of V; 50, ventral cochlear nucleus; 51, dorsal cochlear nucleus; 52, medullary reticular formation; 53, nucleus of the solitary tract; 54, dorsal motor nucleus of the vagus; 55, spinal trigeminal nucleus; 56, dorsal horn of the spinal cord; 57, ventral horn of the spinal cord. From Nieuwenhuys R. *Chemoarchitecture of the Brain.* Berlin, Germany: Springer-Verlag; 1985.

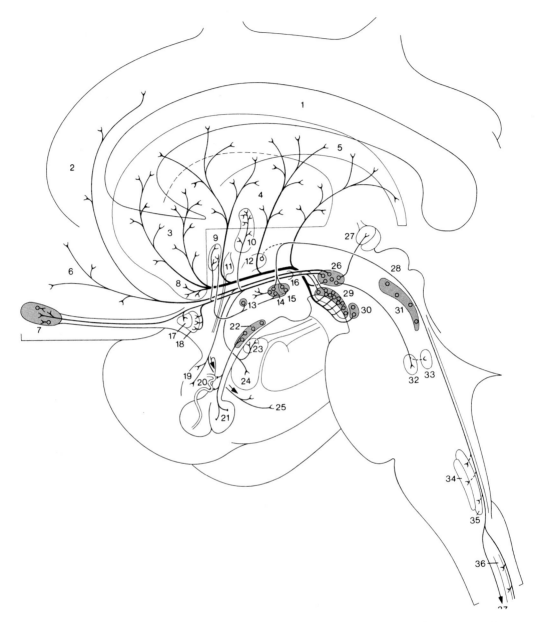

Figure 30.7. The dopaminergic system. 26, 29, and 30, dopaminergic cell groups of the brainstem. Note also that there are a few dopaminergic neurons in other locations (the diencephalon, 13–15, and the olfactory bulb, 7). Other abbreviations: 1, corpus callosum; 2, cingulate gyrus; 3, head of caudate nucleus; 4, putamen; 5, tail of caudate nucleus; 6, frontal cortex; 7, olfactory bulb; 8, nucleus accumbens; 9, septal nuclei; 10, bed nucleus of the stria terminalis; 11, anterior commissure; 12, paraventricular nucleus; 13–15, dopaminergic cell groups in the diencephalon; 16, medial forebrain bundle; 17, anterior olfactory nucleus; 18, olfactory tubercle; 19, prepiriform cortex; 20, median eminence; 21, posterior lobe of the pituitary; 22, arcuate nucleus (also called the *infundibular nucleus*); 23 and 24, amygdaloid complex; 25, entorhinal cortex; 27, habenula; 28, dorsal longitudinal fasciculus; 29, substantia nigra; 31, dorsal raphe nucleus; 32, locus ceruleus; 33, parabrachial nucleus; 34, dorsal motor nucleus of the vagus; 35, nucleus of the solitary tract; 36, dorsal horn of the spinal cord; 37, intermediolateral column. From Nieuwenhuys R. *Chemoarchitecture of the Brain.* Berlin, Germany: Springer-Verlag; 1985.

either slow depolarization or slow hyperpolarization, leading to sustained increases or decreases in neuronal firing rates.

The Dopaminergic System

The dopaminergic system originates from dopamine-containing cell groups in the midbrain (Figure 30.7). Three projection systems have been identified: (1) the *mesostriatal system*, (2) the *mesolimbic system*, and (3) the *mesocortical system*. The prefix *meso* reflects the location of the cell bodies in the mesencephalon (midbrain).

The mesostriatal system refers to the projections from the dopaminergic neurons in the substantia nigra and midbrain tegmentum to various parts of the striatum. The main component that originates in the substantia nigra has been described in the basal ganglia chapter (Chapter 16). The dopaminergic neurons in the midbrain tegmentum are probably a component of this system whose cell bodies happen to be located out of the main nucleus of the substantia nigra.

Like serotonin and norepinephrine, dopamine can produce either depolarization or hyperpolarization, causing sustained increases or decreases in neuronal firing, respectively. As further discussed in Chapter 16, the nature of the physiological responses that are produced depend on the receptors that are present on the postsynaptic cell (D1 or D2).

MODULATORY SYSTEMS AND THE REGULATION OF CONSCIOUSNESS

The original concept of the reticular activating system has undergone significant modification since it was first proposed. Initially, the idea was that the system activated the cortex; the implication of this simple formulation was that the cortex is quiescent in the absence of the activating system. It is now clear that the different modulatory systems actually operate in a cooperative way to upregulate or downregulate cortical activity. Hence, both activation and inactivation are active processes; the modulatory systems already described play a key role in this balanced regulation. Exactly what role the different systems play, or for that matter, exactly what "consciousness" really means, remains to be established.

31

Higher Cortical Functions, Localization of Function, and Regulation of Consciousness

The cerebral cortex reaches its highest level of specialization in humans. It is intimately involved in all aspects of behavior, and particularly those that are definingly human: language, personality, learning and memory, and consciousness. Indeed, C. Judson Herrick, who is considered to be the founding father of American neuroscience, termed the cerebral cortex "the organ of civilization."

The cerebral cortex is of immense clinical importance not only because of its prominent role in brain function, but also because it is vulnerable. The cortical mantle lies adjacent to the skull and, hence, is vulnerable to contusion injuries resulting from sudden deceleration or acceleration, or penetrating injuries. It is also often affected by stroke, and common functional disorders also involve the cerebral cortex (for example, epilepsy). Hence, an understanding of the functional organization of the cortex is important for the physician.

This chapter provides an overview of the cerebral cortex (neocortex), with emphasis on its functional properties. We begin with a brief review of cortical organization. We then consider how the so-called higher cortical functions are represented (that is, which cortical areas are especially important for particular functions). Finally, we consider how the electrical activity of the neocortex is controlled.

ORGANIZATION OF THE NEOCORTEX

Following are the defining structural characteristics of the cerebral cortex:

1. The cells and fibers systems in the cortex are laminated; indeed, the cortex is defined as such on the basis of its lamination.

2. The cortex is made up of functional modules in which neurons are interconnected across laminae.

3. The cerebral cortex is parcellated; different cortical areas have specialized structural features and functional properties.

Lamination

As noted in Chapter 2, the cerebral cortex (neocortex) is defined as such on the basis of its lamination. The nature of the layering provides a tool for parcellating the cortex into different areas. Figure 31.1 schematically illustrates the cellular organization that is typical of neocortical areas.

Figure 31.1. Cellular organization of the neocortex. The left panel illustrates how neurons appear in a Golgi preparation. The center panel illustrates the appearance in a Nissl preparation. The right panel illustrates the appearance in a section stained for myelin. The six layers are indicated in roman numerals on the left. The arabic numerals on the right indicate sublayers. Reproduced with permission from Brodal A. *Neurological Anatomy.* 3rd ed. New York, NY: Oxford University Press; 1981.

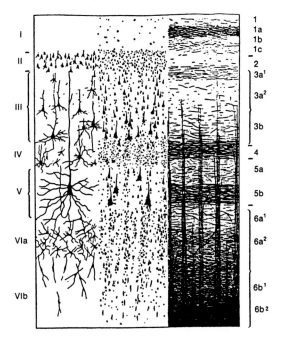

Cellular Organization

There are essentially three types of neurons in the cerebral cortex: (1) *pyramidal cells,* found principally in layers III and V, but some of the cells in layer VI are also pyramidal cells; (2) *stellate cells,* or *granule cells,* found principally in layers II and IV; (3) other cell types (polymorphic in form), which are found principally in layers V and VI and, to a lesser extent, in layer I.

Each cell type is named on the basis of the shape of its cell body; this is the case because cell types were categorized by early neuroanatomists on the basis of their physical appearance in Nissl-stained material, where only the cell body can be seen. Thus, pyramidal cells are so named because they have a pyramid-shaped cell body; stellate cells have a starlike shape (also, stellate cells are small, and in the areas where they are highly concentrated, the cortex has a granular appearance, hence, the name *granule cells*). Fusiform cells have a football shape, and so forth.

Pyramidal cells are oriented strictly perpendicular to the cortical surface. They have two types of dendrites: (1) a long, apical dendrite that may extend across many layers, sometimes reaching to the cortical surface; and (2) several basal dendrites, which arborize for the most part within the same layer as the cell body. This structural organization makes them well suited to integrate inputs of various sorts that arrive in different laminae.

Pyramidal cells give rise to the principal outputs of the cortex. Pyramidal neurons in layers V and VI give rise to the projections to the striatum, brainstem, and spinal cord. The largest neurons of layer V in the motor cortex, which give rise to corticospinal projections, are termed *Betz cells.* Pyramidal neurons in layers II and III are the primary source of association and commissural projections, although some of the pyramidal neurons of layer V may also contribute.

Stellate cells (also called granule cells) have dendrites that radiate in all directions from the cell body. These dendrites ramify primarily within the layers in which the cell body is found. In sensory areas, the stellate, or granule, cells receive thalamic inputs (see later).

Stellate cells are interneurons of the cortex. Most other axonal projections are local (hence the term *interneuron*). Many of these cells are inhibitory; one frequent type of stellate cell is the basket cell, which gives rise to axons that project horizontally through the cortex. These terminate in pericellular baskets around the cell body of other cells. Many of these cells use GABA as their neurotransmitter, and the cells are thought to contribute to sharpening foci of activity (for example, between adjacent functional columns). Other stellate cells are presumably excitatory and contribute to processing within the functional columns of the cerebral cortex.

Polymorphic cells are found in all cortical layers, but particularly in layers I and VI. Many have "fusiform" cell bodies and bipolar dendritic arbors. In layer I, the dendritic arbors are oriented horizontally to the cortical surface. In other layers, their orientation is vertical to the cortical surface.

Fiber Layers

Layer I of the cortex is primarily a "neuropil" layer; it contains mainly axons and dendrites of the cortical neurons, with few cell bodies. Some of the axons originate within the cortex (from interneurons), whereas some come from elsewhere (for example, the non-specific nuclei of the thalamus).

Layer IV is the site of termination of the thalamic afferents, particularly the sensory relay nuclei. The fiber plexus that can be seen in this layer is termed the *external band of Baillarger*. The layer can be seen with the naked eye in the visual cortex, where it is termed the *stripe of Gennari*.

Layer V contains a band of intracortical association fibers termed the *inner band of Baillarger*. In contrast to the association/commissural fibers, these do not leave the cortical area in which they originate. Many of

these are from the stellate/granule cells, and their site of termination is on the basal dendrites of layer V pyramidal cells, as well as on the dendrites of other stellate/granule cells.

Association/commissural projections terminate throughout layers II through V, but mainly in layers II and III. There is no identifiable fiber plexus associated with their site of termination.

The white matter of the cortex is found just deep to layer VI and is not considered one of the layers.

Functional Modules

The fact that neurons in the cortex are organized in functional modules has been discussed with respect to neural processing in sensory systems (see especially the discussion of the visual system). The functional modules derive from the *columnar organization* of the interconnections between resident neurons. It is thought that this same basic organizational plan is present in all cortical areas.

Parcellation

Brodmann divided the neocortex into areas on the basis of "cytoarchitectonics"—differences in the relative size of the different layers (see Chapter 2). Originally, the different areas were defined on the basis of anatomical differences alone. It is noteworthy that modern studies of the functional organization have largely validated the parcellation scheme. A good example of this validation can be seen in the primary somatosensory cortex. Areas 1, 2, 3A, and 3B, were initially distinguished from one another on the basis of subtle anatomical differences, and physiological studies have revealed that each area represents a particular type of cutaneous receptor (see Chapter 11).

There is a good correspondence between cytoarchitectonic areas and functional areas, but the cytoarchitectonic boundaries are not tightly correlated with the gyral and sulcal pattern. Thus, except in exceptional circumstances (that is, the central sulcus), the gyral and sulcal divisions do not indicate functional boundaries in the cortex.

With regard to the parcellation of the cortex, we distinguish between areas in which the six layers are easily recognizable (termed *homotypic cortex*), and areas in which certain layers are highly represented and others are virtually absent *(heterotypic cortex)*. Heterotypic cortex includes the following:

1. *Agranular cortex* (prototype "motor" or "projection") are areas in which the small (granular) neurons that are the targets of thalamic afferents are sparsely represented, and projection neurons are heavily represented.

2. *Granular cortex* (prototype "sensory") are areas in which the granule neurons are

heavily represented, and projection neurons are sparse.

Between the two extremes lie varieties of cortex that are homotypic but tend toward one end of the spectrum or the other (Figure 31.2). These are the "association" areas of the cortex. The association areas can be further divided based on whether they are more related to sensory processing (sensory association) or motor integration (motor association).

It is again noteworthy that the functional distinctions are predicted by the cytoarchitectonics. That is, areas that are more toward the granular end of the spectrum are in general *sensory association areas*; areas that are more toward the agranular end of the spectrum are in general *motor association areas*.

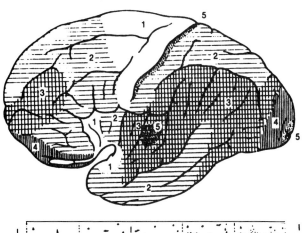

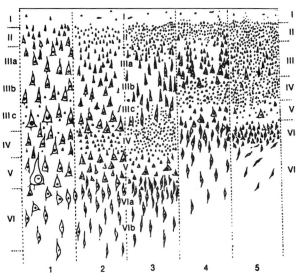

Figure 31.2. Homotypic versus heterotypic neocortex. Homotypic cortex refers to areas in which the six layers are easily recognizable; heterotypic cortex refers to areas in which certain layers are highly represented and others are virtually absent. One type of heterotypic cortex is the agranular cortex (prototype "motor" or "projection"). These are areas in which the small (granular) neurons that are the targets of thalamic afferents are sparsely represented, and projection neurons are heavily represented (1 in drawing). Another type of heterotypic cortex is the granular cortex (prototype "sensory"). These are areas in which the granule neurons are heavily represented, and projection neurons are sparse (5 in drawing). Between the two extremes lie "association" areas of cortex that are homotypic, but tend toward one end of the spectrum or the other (2–4). The association areas can be further divided based on whether they are more related to sensory processing (sensory association) or motor integration (motor association). Reproduced with permission from Brodal A. *Neurological Anatomy*. 3rd ed. New York, NY: Oxford University Press; 1981.

Sensory Association Areas

Sensory association areas have been mentioned in the chapters describing the particular sensory systems. These are the areas that are "appurtenant to" (pertain to) the primary sensory cortices. For example, area 17 is the primary visual cortex; the immediately adjacent area 18 is a visual association area that receives input from area 17. Brodmann's area 41 is the primary auditory cortex; a region in the temporal lobe surrounding area 41 is important for higher-order processing of auditory information.

The sensory association areas then project to areas that are even more eclectic in their functions (that is, further removed from the processing of a particular sensory modality). For example, the parietotemporo-occipital association cortex (basically the area that lies around the three-way boundary between these areas) plays a role in processing higher-order somatosensory, visual, and auditory information; that is, the parietotemporal-occipital cortex is a *multimodal association cortex*.

Motor Association Areas

Motor association areas are the areas anterior to the primary motor cortex that play a role in the initiation of a movement (the supplementary and premotor areas). Rostral to these lies the prefrontal association cortex. The prefrontal cortex is thought to play an important role in cognitive behaviors and in planning, that is, being able to anticipate the consequences of particular actions and plan one's behavior accordingly (more on this later).

FUNCTIONAL FEATURES THAT ARE COMMON TO ALL AREAS OF THE NEOCORTEX

There are several important functional features that apply to all areas of cortex.:

1. Localization of function
2. Hemispheric specialization (lateralization)
3. Topographic mapping
4. Columnar organization

Localization of Function

Our thinking about how the brain carries out its functional role has undergone several evolutionary changes. An important development in the 1800s was the emergence of the concept of localization of function within the brain, that is, that particular brain areas are specialized to carry out particular functions.

Gall and Spurtheim were the proponents of one line of thinking that held that functions were parceled into different brain regions, and that different segments of the cortex were specialized for different "attributes." The basic concept was sound, but they extrapolated on it to say that the size of various cortical regions is associated with the prominence of an attribute (musical aptitude, for example), and that this is reflected by enlargements of the skull. Thus, *phrenology* was the study of the architecture of the skull, with the hope of determining the personality, capabilities, and so forth of the individual.

The basic concept of localization of function evolved from this concept, and even late in the current century, much thinking was directed toward attributing functional capabilities to particular brain regions (good examples include the hypothalamic centers). In this notion, psychological processes are carried out by anatomically defined brain regions. Thus, there are centers for hunger, thirst, pleasure, sex, memory, and so on.

The alternate conceptual strategy was that of a functional "system." Here, the critical assumption is that areas that are connected with one another operate as an integrated system, with the different areas serving as components of the system. Thus, for example, there is a visual system that includes the eye and all of its central circuitry, which is involved in all visual perception and visuomotor integration. Of course the brain is not divisible entirely into discrete systems, because integration occurs at all levels of a system.

For example, while it is useful to think of separate sensory and motor systems, as we have seen, there is sensory motor integration at all levels. The rise of "functional neuroanatomy" in the early 1900s further solidified the idea that systems can be identified on the basis of interconnections between component structures.

Our current concept of localization of function is that different parts of the cortex are devoted to different functional systems. Thus, part of the cortex is devoted to somatic sensation, part to motor control, part to visual representation, and so forth.

But there is also hemispheric specialization of the cerebral cortex, in that the two sides are not mirror images of each other. Certain cortical areas are actually more developed on one side than the other. Not only is there structural specialization, but the two sides of the cerebral cortex actually mediate different functions. All of this together is what is meant by the term *lateralization of function,* but again, it is important to recall that the different cortical areas do not operate in isolation; they are part of functional systems.

Hemispheric Specialization and Lateralization of Function

Lateralization of function is reflected in a number of ways. In terms of motor and sensory abilities, humans exhibit decided *handedness*. Almost everyone has a preferred hand that is used for writing, throwing a ball, or handling a fork, a preferred side on which to swing a bat or a golf club, a dominant eye, and so forth. In some cases, the handedness is extreme. Many of us are virtually incapable of effectively throwing a ball with our nonpreferred hand. A small number of individuals are *ambidextrous* and can work with either hand or on either side (for example, a switch hitter in baseball). These are all basic sensory–motor skills and are lateralized despite the fact that sensory and motor cortices are represented bilaterally.

Lateralization is even more prominent with "higher cortical functions" such as expressive language (the ability to speak) and receptive speech (the ability to understand the spoken word). Paul Broca was the first to demonstrate that damage to a small portion of the left hemisphere immediately anterior to the margin of the Sylvian fissure leads to severe dysfunction of expressive language. This area is now called *Broca's area.*

Carl Wernicke then made extensive studies of patients with small lesions involving another area in the left hemisphere just behind the Sylvian fissure in the superior temporal gyrus (now called *Wernicke's area*). These patients had difficulty comprehending spoken language (termed *word deafness*) as well as difficulties in reading (called *alexia*) and writing (called *agraphia*).

Additional insights into the control of speech were gleaned from experiments in which areas of the cortex were stimulated in human volunteers during neurosurgical procedures. This approach, used extensively by the neurosurgeon W. Penfield and his colleagues, provided extensive information on the organization of the cerebral cortex. For example, Figure 31.3 illustrates the areas that, when stimulated, interfere with speech.

Other higher cortical functions such as perceptual and performance skills, spatial abilities, and so forth are also lateralized. The details of their localization are not as well known as those of the sensory and motor systems. Nevertheless, clinical data present a clear and consistent picture of the functional specialization of some areas.

Hemispheric Specialization in Patients with "Split Brains"

Much of what we know about localization of function and hemispheric specialization comes from the work of Roger Sperry, for which he shared the Nobel Prize in Physiology and Medicine with David Hubel and Torsten Wiesel. Sperry worked with patients whose corpus callosum had been completely sectioned for treatment of intractable epilepsy. These so-called split-brain patients exhibited few overt symptoms following surgery; nevertheless, by devising special

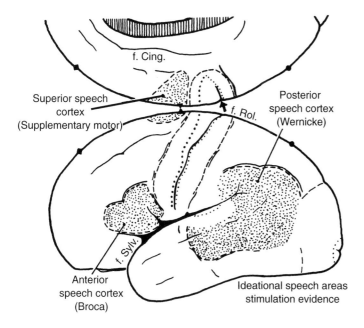

Figure 31.3. Areas of the cortex that are specialized for speech. The drawing illustrates areas of the cortex in the left (dominant) hemisphere that, when stimulated electrically, interfere with speech. Reproduced with permission from Penfield W, Roberts L. *Speech and Brain Mechanisms*. Princeton, NJ: Princeton University Press; 1959.

testing procedures that took advantage of the organization of sensory systems, Sperry and his colleagues were able to demonstrate a dramatic division of labor between the two cerebral hemispheres.

The testing device that provides separate access to the two hemispheres is illustrated in Figure 31.4. Patients are asked to focus on a spot in the center of two screens. Words or pictures are then flashed on the screen; the rapidity of the flash and the focus on the center spot means that the stimulus activates only one half of the visual field. Because of the distribution of projections from the two eyes, such stimuli gain access to only one side of the cortex. (To understand this, you'll need to recall the organization of the central visual pathways; see Chapter 19.)

The perceptions of the patient can be tested in several ways. A picture of an object can be presented, and the patient can be asked to name the object or retrieve the named object with one hand or the other (the motor system is crossed; thus, the left hand is controlled exclusively by the right cortex). This involves tactile object recognition based on shape (stereognosis). Alternatively, the patient can be asked to palpate and then identify the object verbally. Sperry found that patients could name what was seen in the right visual field (which therefore reached the left cortex), but could not easily name objects in the left visual field. On the other hand, the right cortex (as demonstrated by the skill of the left hand) was much better at stereognosis. Furthermore, the right hemisphere was better at visual recognition (discriminating one face from another), whereas the left hemisphere was better at visual naming.

Interestingly, when each hemisphere is tested separately, the right hemisphere can actually perform better on spatial tasks than the whole patient using free vision and unrestricted hand use. How the two independent hemispheres in these individuals interact in gaining control of "self" is a fascinating and largely unresolved question.

Functional Consequences of Injuries to Different Cortical Areas

We have considered the functional consequences of injuries to sensory and motor cortices in previous chapters. Here we briefly summarize the consequences of injury to association cortices. Before beginning, it is important to note that the symptoms resulting from injuries to the association cortex are very complicated. Indeed, neuropsychologists have developed a complicated vocabulary just to describe the symptoms (see box).

This is only a sampling. Obviously, it is far beyond the scope of this chapter to fully review even the language referring to disorders that result from cortical injury. Hence, in what follows, only some key points are considered.

Injuries to the Frontal Lobe

The consequences of injury to the frontal lobe are usually detailed using a remarkable anecdote—the story of Phineas Gage. Phineas Gage was a railroad foreman. In 1848, he was tamping dynamite into a hole in a rock using a metal rod (1.25 in wide and 3.5 ft long). The dynamite exploded, and the metal rod was driven through his orbit, up through

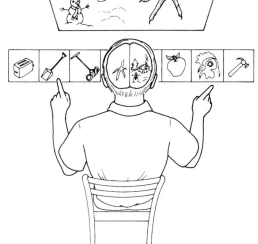

Figure 31.4. Assessing lateralization of function in split-brain patients. The figure illustrates the testing paradigm used to assess information processing by the two hemispheres in individuals in whom the corpus callosum has been transected. Because of the representation of the visual fields (see Chapter 18), a stimulus presented in the left visual field is represented exclusively in the right cerebral cortex. Hence, the subject is asked to focus on a spot, and stimuli are presented in one visual field. He or she is then asked to match the object by pointing with one or the other hand or to name the object seen. Reproduced with permission from Springer SP, Deutsch G. *Left Brain, Right Brain*, New York, NY: WH Freeman; 1981.

An Abbreviated Glossary

Agnosia	Literally, without knowledge; the inability to perceive or attach meaning to sensory stimuli.
Alexia	A reading disorder that is acquired (compare with dyslexia).
Agraphia	Inability to write.
Anomia	Inability to name objects.
Aphasia	A disorder of language; there are seven different types of aphasia: (1) expressive (Broca's), (2) receptive (Wernicke's), (3) conduction, (4) global, (5) anomic, (6) transcortical motor, (7) sensory.
Aprosodia	Disorders in the affective components of language.
Asomatognosia	Absence of body awareness, unawareness of sensory stimulation of a particular part of the body.
Astereognosia	Inability to recognize the form of an object by touch.
Ataxia	Disorders of motor function (clumsiness).
Dyslexia	A reading disorder that is congenital.
Dysgraphia	Writing disorder.
Dyscalculia	Inability to carry out calculations.
Paraphasia	Using the wrong word or combination of words.
Prosopagnosia	Inability to recognize familiar faces.

his frontal lobes, and through the superior surface of his skull, completely destroying most of his frontal lobes. Surprisingly, Mr. Gage survived and recovered.

Gage was followed for a number of years by the physician who treated him initially (J. M. Harlow). Although Gage recovered neurologically, his personality was fundamentally altered by the accident. Before the accident, he had been considered competent, energetic, and well balanced. After the accident, he was virtually sociopathic. His physi-

cian described him as being "fitful, irreverent, indulging at times in the grossest profanity . . . , manifesting but little deference for his fellows, impatient of restraint or advice when it conflicts with his desires"(see Harlow JM. Recovery from the passage of an iron bar through the head. *Mass Med Soc Publ.* 1868;2:327–346). Also evident were deficiencies in planning abilities. Again, to quote, Gage was described as "devising many plans of future operation, which are no sooner arranged than they are abandoned in turn for others."

The constellation of symptoms is now called the *prefrontal lobe syndrome.* The key features of the syndrome are personality changes. Patients often exhibit a lack of social sensitivity; they seem to either be unaware or uncaring about the consequences of their actions. They seem to have a general lack of ability to comprehend the consequences of actions, and they exhibit deficits in tasks that require planning.

Patients with frontal lobe lesions may perseverate (engage in repetitive actions) and have difficulty shifting their attention or their strategies. For example, if they have succeeded in solving a problem using a particular strategy, and the problem is changed so as to require a different strategy, the patients have great difficulty modifying their approach.

Patients with frontal lobe injuries may also become socially withdrawn. Some patients may lie motionless and mute for weeks. It is of considerable interest that this and other symptoms of frontal lobe lesions are reminiscent of the symptoms of psychiatric illnesses, especially schizophrenia. This similarity has led to the idea that schizophrenia may result from a disease process that affects the frontal lobe. This idea has been supported by studies that indicate that the frontal lobe is often smaller in schizophrenics than in normal individuals.

The experimental evidence regarding personality changes after frontal lobe lesions motivated the first *psychosurgical procedures* (surgical procedures designed to alter behavior). In the 1930s, a series of experimental studies by J. Fulton revealed that lesions of the frontal lobe in chimpanzees seemed to have a "calming" effect. These data were presented at a meeting in 1935, where a Portuguese neuropsychiatrist named Egas Moniz was in attendance. Moniz was so struck by the finding that he returned to Portugal and arranged the first prefrontal lobotomies in human patients with psychiatric disorders.

Initial findings suggested that lobotomies did reduce anxiety, and so surgeons throughout the world began to follow up on the initial studies by evaluating various surgical procedures that interrupted the white matter tracts projecting to and from the frontal lobe. Frontal lobotomies continued until the 1950s, but the procedure was generally abandoned because of the development of effective antipsychotic drugs, and also because a number of studies indicated that the lobotomy procedure had numerous complications and produced inconsistent benefits.

The symptoms described earlier are not generally seen unless the injury to the frontal lobe is bilateral. There is, however, one important symptom of injuries that affect the portion of the left frontal lobe that contains Broca's area. These injuries lead to *expressive aphasia*—an inability to speak. Patients can understand spoken and written language but cannot speak; they also sometimes have difficulty writing. If the lesion does not extend into the primary motor cortex, motor functions other than speech and writing are unimpaired.

In virtually all right-handed individuals, speech is represented on the left side. Speech is also represented on the left side in the majority of left-handed individuals; however, in a small percentage of individuals, speech is represented on the right side. These conclusions derive from studies in normal human volunteers in which speech function was tested when one or the other side of the cortex was anesthetized using a very short-acting anesthetic (amytal).

Injuries to the Parietal Association Cortex

The parietal lobe contains the primary somatosensory cortex, somatosensory association cortex, and a higher-order association

area that receives multimodal input (higher order in the sense that this part of the cortex receives highly processed information from somatosensory, visual, and auditory cortices). The symptoms resulting from lesions in the primary somatosensory and somatosensory association cortices have been discussed in Chapter 11. Here we briefly consider the symptoms resulting from lesions involving the higher-order association cortex of the parietal lobe.

One general conclusion is that lesions in either parietal lobe produce deficits in spatial perceptions. For example, these patients have difficulties perceiving spatial relationships between objects in the environment. Patients may also exhibit *sensory neglect*—a failure to notice stimuli presented on the side contralateral to the lesion. Patients may completely ignore the side of their body that is contralateral to the lesion, failing to wash or dress that side. They may even deny ownership of their own arm or leg when it is passively brought into their field of view. If the somatosensory cortex itself is not involved, these types of sensory neglect may be present even if somatosensory perception is intact. These kinds of symptoms strengthen the general conclusion that the cortex is important for *conscious awareness.*

The symptoms resulting from injuries to the parietal lobe also depend on the hemisphere that is involved. Lesions on the left side can affect language; lesions on the right side spare language but may cause more severe deficits in spatial perception than similar lesions on the left side.

The reason that a parietal lobe lesion can affect language is that the fibers that interconnect Wernike's area in the temporal lobe (the area that is important for receptive language) with Broca's area in the frontal lobe pass through the parietal lobe. These fibers can be interrupted by lesions that extend into the white matter of the parietal lobe.

Injuries to the Occipital Association Cortex

The association cortical areas that are present in the occipital lobe are the visual association areas that receive highly processed information from the primary and secondary visual cortices (areas 17 and 18, respectively). One such area lies in the inferior occipital lobe at the junction between the occipital and temporal cortices.

Injuries to the occipital association cortex cause deficits in higher-order visual perception. For example, patients may be unable to recognize faces *(prosopagnosia)* or objects *(object agnosia)*.

Injuries to the Temporal Association Cortex

The temporal lobe contains auditory association areas (including Wernicke's area) and higher-order association areas that receive multimodal sensory information.

The memory deficits produced by lesions of the temporal lobe have been discussed in the chapter pertaining to the limbic system. The only additional point is that although memory deficits are seen following injuries that affect the hippocampal formation, the deficits are much more severe when the temporal neocortex is involved.

Lesions in the left temporal lobe that involve Wernicke's area lead to *receptive aphasia*—the inability to understand speech. There are also deficits in expressive speech, however. Patients can speak fluently in that the pace and rhythm of speech seems normal. However, patients may make up words (neologisms) or add syllables to words. Moreover, the speech of a patient with an injury involving Wernicke's area often conveys little meaning and is, hence, termed *empty speech.*

This very brief summary conveys something of the complexity of the symptoms that result from cortical injuries. Obviously, this is an incomplete summary; complete volumes have been written on the symptoms resulting from cortical injury. For the general physician, the key point is that the symptoms are usually complex yet selective. Patients with small lesions may seem fairly normal in most regards, yet have very specific and severe deficits that affect particular abilities. Understanding the nature of these

deficits and learning to adjust to them is a challenge for the patient.

LEVELS OF CONSCIOUSNESS ARE RELATED TO NEOCORTICAL ELECTRICAL ACTIVITY

The term *consciousness* is used in two ways. One refers to *conscious awareness.* The other usage is with reference to *levels of consciousness* (that is, whether one is awake or asleep). In this context, our level of consciousness is closely related to the pattern of activity in the cerebral cortex. This has been demonstrated by studies that correlate levels of consciousness with cortical electrical activity.

Cortical electrical activity is typically measured using the electroencephalogram (EEG). The EEG measures the electrical activity generated by conglomerates of neurons. In humans, the EEG is recorded by placing an array of scalp electrodes at different points overlying various brain regions. By comparing the electrical activity at these leads with an "indifferent" site (the earlobe, for example) it is possible to sample from specific brain regions(Figure 31.5).

The EEG has been a very important clinical tool because it is entirely noninvasive. Hence, it can be used both as a diagnostic tool and a research tool. Moreover, the EEG can detect functional abnormalities involving the cortex (for example, the abnormal electrical discharges that are associated with epilepsy; see Figure 31.5).

The Sleep-Waking Continuum

It is intuitively apparent that there are more than two levels of consciousness (asleep and awake). When we are awake, there are different levels of alertness ranging from relaxed/drowsy to highly alert/vigilant. Also, the EEG has revealed a number of different stages of sleep. Different levels of consciousness are the result of the combined activity of a number of brain systems, especially the

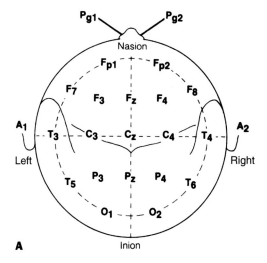

A

Calibration: 50 μV (vertical) and 1 s (horizontal)

LF–LAT

RF–RAT

LAT–LT

RAT–RT

LT–LO

RT–RO

LT–LPc

RT–RPc

B **Normal adult**

Figure 31.5. The EEG provides a measure of cortical electrical activity. A, Illustration of the standard placements of scalp electrodes for recording the EEG. A indicates ear; C, central; Cz, central at midline; F, frontal; Fp, frontal pole; Fx, frontal at midline; O, occipital; P, parietal; Pg, nasopharyngeal; Pz, parietal at midline; T, temporal. The records in **B** illustrate the EEG of a normal adult; the records in **C** illustrate the record of a 6-year-old boy during a *petit mal* seizure. During a petit mal seizure, patients typically are unaware of their surroundings but do not lose consciousness or exhibit overt motor seizures *(grand mal)*. From Wasman SG, DeGroot J. *Correlative Neuroanatomy.* Norwalk, Conn: Appleton & Lange; 1995. The drawing in **A** is courtesy of Grass Medical Instruments Co., Quincy Mass. (manufacturers of the polygraphs used for EEG recording).

LF–LAT

RF–RAT

LAT–LT

RAT–RT

LT–LO

RT–RO

LO–LPc

RO–RPc

C

Petit mal epilepsy. Record of a 6-year-old boy during one of his "blank spells," in which he was transiently unaware of surroundings and blinked his eyelids during the recording.

modulatory systems of the brainstem described in Chapter 30.

Some Key Points

1. Sleep is not just the absence of consciousness. Indeed, there are several different stages of sleep, each of which appears to be actively regulated.

2. The electrical activity of the cortex (as measured by the EEG) is closely correlated with the level of consciousness. Five stages of sleep can be recognized on the basis of the characteristic patterns of activity as recorded by the EEG. In general, arousal is characterized by low-voltage fast activity, and sleep (or relaxation) by higher-voltage slow activity. The higher the voltage of cortical activity and the slower the waves, the deeper the stage of sleep.

3. Low-voltage fast activity reflects cortical desynchrony; high-voltage activity reflects synchrony. Synchrony is a measure of the number of cortical neurons that are firing together. Thus, desynchrony means that cells are being activated more independently than during a synchronized EEG.

4. The EEG is thought to represent summed synaptic potentials, not the summed discharge (action potentials).

5. In addition to the stages of sleep that are characterized by high-voltage slow activity, there is one stage of sleep that is

characterized by a low-voltage fast activity that is almost indistinguishable from the waking state. This is call *paradoxical sleep* (so called because sleep exists when the EEG suggests wakefulness), or *rapid eye movement* (REM) *sleep*. REM sleep is the time during which most dreaming occurs (see later).

In normal individuals, sleep occurs in stages with fairly stereotyped properties. Furthermore, sleep has certain stereotyped structural characteristics. During a typical sleeping session, a human will progress through a series of EEG changes that are characterized by increasing amplitude of the EEG waves, and a slowing of activity. The stages are quantitatively characterized by the frequency spectrum of the EEG (the percentage of the waves that are high frequency versus low frequency). After reaching stage 4, a sleeper will typically exhibit a reverse sequence of changes that ends with EEG activity that is similar to that in the waking state. The individual does not awaken at this stage but instead progresses into a stage of sleep characterized by rapid eye movements (REM sleep). These cycles may be repeated several times during a sleep session, and typically toward the end of a session, there is a much greater preponderance of REM sleep.

How Is the Level of Consciousness Regulated?

Until the late 1940s, the prevalent notion was that sleep is essentially a passive process, somewhere near the low end of a continuum of vigilance. Research carried out during the 1950s and 1960s changed this view by revealing that sleep is actively regulated by several different circuits, especially the modulatory circuits of the brainstem discussed in the previous chapter.

The foundation of our understanding of sleep regulation came from experiments by F. Bremer. Bremer found that when the brainstem of a cat was transected at the midcollicular level, the disconnected forebrain displayed an EEG pattern typical of deep sleep (high-voltage slow activity). If a transection was made at the spinomedullary

junction (and the animal was maintained on a respirator) the animals continued to exhibit EEG changes typical of normal cycles of sleeping and waking. These experiments suggested that an area between the two transections was responsible for regulating sleep and waking. Subsequent studies implicated the reticular formation.

The Activating System

As discussed in Chapter 30, experiments by G. Moruzzi and H. W. Magoun demonstrated that lesions involving areas in the reticular core of the midbrain resulted in coma. Measurements of EEG in the comatose animals revealed cortical activity typical of slow-wave sleep. As we have seen, Moruzzi and Magoun proposed the concept of an "activating system" that controlled cortical activity and thus the level of consciousness. As noted in Chapter 30, the components of the system include neuron groups in the mesencephalic reticular formation that operate in conjunction with the nonspecific nuclei of the thalamus to modulate neuronal activity throughout the cortex.

Sleep-Inducing Systems

In the late 1950s, work in Moruzzi's lab demonstrated other brain regions that participate in the regulation of vigilance. If the brainstem was transected just a few millimeters caudal to the midbrain transections that caused coma, animals remained permanently awake. Hence, the concept emerged that a collection of neurons in the rostral reticular formation is necessary for wakefulness, and a different population of neurons in the caudal brainstem (midpontine level) is necessary for sleep.

Subsequent experiments, carried out largely by Jouvet and Dement, suggested that at least one sleep-including area is the group of serotonergic neurons in the raphe nuclei. This conclusion was based on several observations: (1) lesions involving the raphe nuclei produced insomnia; (2) injections of

5-hydroxytryptophan (a serotonin precursor) into the systemic circulation induced sleep; and (3) injections of agents that disrupt the production of serotonin produced insomnia.

It is of interest that animals that are deprived of REM sleep often exhibit behavioral activities that suggest hallucinations (aggressive behavior directed toward nonexistent stimuli). Humans deprived of sleep also hallucinate. In this regard, it is noteworthy that the hallucinogenic drug LSD acts on serotonin receptors.

Sleep Disorders

All animals require sleep. If humans are deprived of sleep for long periods, cognitive and performance skills deteriorate, individuals become more and more disoriented, and hallucinations may appear. For this reason, sleep disorders can be extremely debilitating. There are several common sleep disorders.

Insomnia is an inability to fall asleep and/or remain asleep. Insomnia at night is often accompanied by *hypersomnia* (falling asleep) during the day. Insomnia is a very common disorder and is often due to psychological stress.

Sleep apnea refers to conditions that lead to oxygen deprivation. The most common type is *obstructive sleep apnea,* which is caused by excessive relaxation of the pharyngeal muscles, which causes upper airway constriction. The resulting oxygen deprivation disrupts the sleep cycle. *Nonobstructive sleep apnea* is thought to be due to decreased respiratory drive; the decreased respiration leads to hypoxemia and hypercapnia, which causes the patient to awaken and thus disrupts sleep patterns.

Narcolepsy is a disorder that is characterized by sudden and uncontrollable onset of sleep bouts, sleep paralysis (a condition in which patients are unable to move as they are falling asleep), and hallucinations while falling asleep. Narcolepsy has a genetic basis and is inherited as an autosomal dominant trait.

A common cause of transient sleep disorders is travel across time zones (jet lag). The reason is that sleep–waking cycles are circadian rhythms that are entrained by the light–dark cycle. When there is a shift in the relationship between the behavioral rhythm and the light cycle, the behavioral rhythm must be reentrained. This usually occurs within a few days of the shift (just in time to return home and start the process again).

Disruptions of Consciousness Following Brain Injury

A common consequence of brain injury is a loss or disruption of consciousness. These range from the brief losses of consciousness that result from *concussions* to the permanent unconsciousness characteristic of coma. Disruptions of consciousness are categorized according to duration and severity:

1. *Confusion.* A "clouding" of consciousness, disorientation regarding time and place.
2. *Stupor.* Minimal consciousness. Patients can be aroused, but only by strong stimuli.
3. *Coma.* A persistent unconscious state. The comatose individual cannot be aroused and is unresponsive to painful stimuli.

Disruption of consciousness can result from sudden acceleration or deceleration, blows to the head, stroke, or metabolic disorders. Small lesions in critical structures can also cause a disruption of consciousness or coma. One example is a lesion that affects the midbrain reticular formation.

Index